JOURNAL OF REPRODUCTION AND FERTILITY

SUPPLEMENT 52

CONTROL OF PIG REPRODUCTION V

Proceedings of the Fifth International Conference on Pig Reproduction

held at
Congress Centre Rolduc
Kerkrade, The Netherlands
June 1997

Edited by G. R. Foxcroft, R. D. Geisert and C. Doberska

COVER ILLUSTRATION

Laser scanning confocal micrograph of cytochalasin-b treated and cryopreserved pig embryo at the hatched blastocyst stage. Courtesy of J. R. Dobrinsky, Germplasm & Gamete Physiology Laboratory, Agricultural Research Service, US Department of Agriculture, Beltsville, MD 20705, USA

Journal of **REPRODUCTION** *and fertility*
1997

First published 1997

ISSN 0449 3087
ISBN 0 906545 32 3

JOURNALS OF REPRODUCTION AND FERTILITY LTD

The Journals of Reproduction and Fertility Ltd produce the *Journal of Reproduction and Fertility* and *Reviews of Reproduction.*

The *Journal of Reproduction and Fertility* publishes novel and significant papers on the molecular biology, biochemistry and physiology of reproduction, including lactation, and early embryogenesis in man and other animals. The *Journal* is the official organ of the Society for the Study of Fertility.

The *Journal* also publishes Supplements which are distinct from the regular issues, are not associated with any particular volume and are known by their serial number and date.

Reviews of Reproduction publishes topical, concise, well-illustrated reviews, contributed by leading research workers, on basic mechanisms in reproductive biology and comments on recent developments in the field, new hypotheses and highlights of scientific meetings. Articles are commissioned by the Editorial Board in consultation with the International Advisory Panel, and are peer-reviewed by experts in the field.

Published by **The Journals of Reproduction and Fertility Ltd.**

Agents for distribution: **Portland Press, PO Box 32, Commerce Way, Whitehall Industrial Estate, Colchester, CO2 8HP, Essex, UK; Tel: 44(0)1206 796351; Fax: 44(0)1206 799331; e-mail: sales@portlandpress.co.uk**

Printed by **Cambridge University Press**

CONTENTS

Acknowledgements v

Preface vi

List of delegates vii

REGULATION OF REPRODUCTIVE FUNCTION

Role of neuropeptides and amino acids in controlling secretion of hormones from the anterior pituitary gland in pigs
M. J. Estienne, J. M. Harter-Dennis and C. R. Barb 3–17

Immune–endocrine interactions affecting luteal function in pigs
W. Wuttke, L. Pitzel, I. Knoke, K. Theiling and H. Jarry 19–29

Control of follicular development and ovulation rate in pigs
N. M. Cox 31–46

Mechanisms mediating nutritional effects on embryonic survival in pigs
G. R. Foxcroft 47–61

Poster titles 63–64

FACTORS AFFECTING BREEDING EFFICIENCY IN THE PIG

Management of boars for efficient semen production
W. L. Flowers 67–78

Consequences of variation in interval from insemination to ovulation on fertilization in pigs
B. Kemp and N. M. Soede 79–89

Expression of oestrus and timing of ovulation in pigs
N. M. Soede and B. Kemp 91–103

Effects of semen components on ovulation and fertilization
D. Waberski 105–109

Poster titles 111–112

EMBRYONIC AND FETAL DEVELOPMENT IN THE PIG

Role of uterine immune cells in early pregnancy in pigs
H. Engelhardt, B. A. Croy and G. J. King 115–131

Regulation of conceptus development and attachment in pigs
R. D. Geisert and J. V. Yelich 133–149

Extracellular matrix and the implantation cascade in pigs
R. C. Burghardt, J. A. Bowen, G. R. Newton and F. W. Bazer 151–164

Embryonic and fetal development in different genotypes in pigs
S. P. Ford 165–176

Poster titles 177–178

BOAR FERTILITY AND SEMEN PRESERVATION

Physiology of the Meishan boar
D. D. Lunstra, J. J. Ford, J. Klindt and T. H. Wise 181–193

Sperm plasma membrane characteristics and boar semen fertility
R. A. P. Harrison 195–211

Assessment of boar sperm function in relation to freezing and storage
W. V. Holt and A. Medrano 213–222

Poster titles 223–224

ADVANCES IN BIOTECHNOLOGY IN PIG REPRODUCTION

Molecular approaches to improved pig fertility
M. F. Rothschild, L. A. Messer and A. Vincent 227–236

Developments in transgenic techniques in pigs
M. B. Nottle, H. Nagashima, P. J. Verma, Z. T. Du, C. G. Grupen, R. J. Ashman and S. MacIlfatrick 237–244

Advances in the generation of transgenic pigs via embryo-derived and primordial germ cell-derived cells
J. A. Piedrahita, K. Moore, C. Lee, B. Oetama, R. Weaks, J. Ramsoondar, J. Thomson and J. Vasquez 245–254

Advances in gender preselection in swine
L. A. Johnson 255–266

Poster titles 267

ASPECTS OF OOCYTE AND EMBRYONIC DEVELOPMENT IN THE PIG

Advances in *in vitro* production of pig embryos
H. Funahashi and B. N. Day 271–283

Oviductal regulation of fertilization and early embryonic development
W. C. Buhi, I. M. Alvarez and A. J. Kouba 285–300

Cryopreservation of pig embryos
J. R. Dobrinsky 301–312

Poster titles 313

Author Index 315

Subject Index 317

ACKNOWLEDGEMENTS

We are grateful to those who made the Vth International Conference on Pig Reproduction a success: the invited speakers for the quality of their presentations at the meeting and timely submission of the manuscripts contained in this Supplement, those who presented posters for sharing their research with delegates at the meeting, and those who contributed to discussion of both spoken papers and posters. We would like to thank Drs R. R. Kraeling, T. van der Lende, K. F. Weitze, F. W. Bazer, L. A. Johnson, H. Niemann and B. N. Day for agreeing to act as Chairmen of the sessions and Dr H. Niemann in addition for his insightful summary of the meeting.

The success of the meeting was in great part due to the efforts of the Local Organizing Committee in The Netherlands. Their willingness to work with the International Organizing Committee to provide an excellent environment for the Vth International Conference was a new and very successful venture and we applaud the excellent support of colleagues from their home institutions.

The Conference was organized under the auspices of the Netherlands Foundation for Animal Reproduction and we are appreciative that this arrangement was possible. Financially, the meeting would not have been possible without the support of our three major sponsors, IMV (France), Intervet International BV (The Netherlands) and PIC Group (UK).

Finally, the contributions of Drs D. J. A. Cole and S. K. Webel as retiring, founder members of the International Organizing Committee, to the ongoing success of the International Conferences on Pig Reproduction, has been much appreciated.

D. J. A. Cole, B. N. Day,
G. R. Foxcroft, R. D. Geisert,
T. van de Lende, H. Niemann,
S. K. Webel
International Organizing Committee

T. van de Lende, B. Kemp,
S. J. Dieleman, N. M. Soede,
P. C. Vesseur
Local Organizing Committee

PREFACE

The Fifth International Conference on Pig Reproduction was held at the Congress Centre Rolduc in the monastery complex of the former 12th century Augustinian abbey, in Kerkrade, The Netherlands, on 2–4 June, 1997. The Conference represented the continuation of the International Organizing Committee's plans to convene meetings at four year intervals with the primary objective to provide a forum for interaction among scientists with an interest in pig reproduction and to promote research in this area.

Organization of the Conference was initiated at the Fourth International Conference in 1993 in Missouri, when the International Organizing Committee invited the Dutch community involved in pig reproduction research to prepare a proposal to act as a Local Organizing Committee for the Fifth International Conference in 1997. This represented the first time that the venue for the Conference had not alternated between the Universities of Nottingham and Missouri-Columbia but still ensured that the Conference was organized by a group with a very active involvement in pig reproduction research. The IOC was also keen to ensure that the location and facilities would continue to provide for intensive interaction among the delegates to the Conference from both a scientific and social perspective. The outstanding success of the Conference in meeting these objectives testifies to the success of the LOC in perpetuating the success and traditions of the International Conferences on Pig Reproduction.

The IOC retained the responsibility for developing the scientific programme under the leadership of Heiner Niemann and the programme again evolved from a comprehensive survey of those actively involved in research in pig reproduction. The main scientific programme comprised six sessions in which 22 invited speakers from seven different countries presented state-of-the-art lectures on a wide range of fundamental research and technological developments in pig reproduction. The full edited text of these presentations forms the basis of the present Supplement. The 183 delegates from 19 countries who attended the Conference supported the main programme with the presentation of the 73 submitted posters listed by title in this Supplement. An additional feature of the Fifth International Conference was the inclusion of an informal and well supported evening workshop which focused on the application of existing knowledge to the improvement of breeding herd efficiency in practice.

The IOC has again cooperated with the *Journal of Reproduction and Fertility* in the publication of the papers presented at the Fifth International Conference on Pig Reproduction as Supplement 52 of the *Journal*. We believe that the timely publication of this Supplement will continue to make the proceedings of the International Conferences on Pig Reproduction an up-to-date and authoritative source of information on this topic.

B. N. Day
Chairman, International Organizing Committee

LIST OF DELEGATES

Aas, A. H. G.	Norsvin, N-2301 Hamar, Norway
Almond, G.	North Carolina State University, Raleigh, NC 27606, USA
Arbov, T.	Pig Improvement (Canada) Ltd., Box 266, ACME, Alberta, Canada
Armstrong, T.	North Carolina State University, Dept Animal Science, Raleigh, NC 27695–7621, USA
Ashworth, C. J.	Rowett Research Institute, Bucksburn, Aberdeen AB21 9SB, UK
Bacci, M. L.	Dimorfipa, 40064 Ozzano Emilia, Bologna, Italy
Barb, C. R.	USDA–ARS Animal Physiology, Athens, GA 30604–5677, USA
Barre, M.	IMV, 61302 L'Aigle, France
Bazer, F.	Texas A&M University, College Station, TX77843–2471, USA
Beckers, J-F.	University of Liege, Faculty of Veterinary Medicine, B4000 Sart-Tilman, Belgium
Bloor, A. P.	Cotswold Pig Development Co Ltd, Market Rasen LN1 6BJ, UK
Bobb, J. D.	Pipestone Veterinary Clinic, Pipestone MN 56164, USA
Boerjan, M.	Institute for Animal Science and Health, 8200 AB Lelystad, The Netherlands
Borchardt, G.	Institute for Reproductive Medicine, 30173 Hannover, Germany
Bouwman, E.	Wageningen Agricultural University, 6700 AH Wageningen, The Netherlands
Bridoux, N.	Hoechst Roussel Vet, 93235 Romainville Cedex, France
Britt, J. H.	North Carolina State University, Raleigh, NC 27606, USA
Brüssow, K-P.	Research Institute Biology of Farm Animals, 18196 Dummerstorf, Germany
Buhi, W. C.	University of Florida, Dept of Obstetrics & Gynaecology, Gainesville 32610, USA
Burghardt, R.	Texas A&M University, College Station, TX 77843–4458, USA
Burke, P.	PIC-USA, Franklin KY 42135–0348, USA
Bussière, J. F.	INRA-SEIA, 86480 Rouille, France
Cameron, J.	University of Nottingham, Loughborough, LE12 5RD, UK
Cassou, B.	IMV, 10, Rue Clemenceau, BP 81, 61302 L'Aigle Cedex, France
Cerovsky, J.	Research Institute of Animal Production, 517 41 Kostelec nad Orlici, Czech Republik
Christenson, R. K.	USDA-ARS, Clay Center, Nebraska 68933–0166, USA
Cliff, A.	JSR Healthbred Ltd, Selby YQ8 9NN, UK
Colenbrander, B.	Utrecht University, 3508 TD Utrecht, The Netherlands
Cox, N. M.	Mississippi State University, Dept Animal & Dairy Sciences, MS 39762, USA
Dalin, A-M.	Swedish Agricultural University, Dept Obstetrics and Gynaecology, S-75007 Uppsala, Sweden
Dawson, M.	Royal Veterinary College, Potters Bar, Herts EN6 1NB, UK
Day, B.	University of Missouri, 159 Animal Sciences Center, Columbia MO 65211, USA

Delhomme, G.	IMV, 61302 L'Aigle Cedex, France
Deschamps, J. C.	Universidade Federal de Pelotas, Faculdade de Veterinaria, 96010–900 Pelotas RS, Brasil
Didion, B.	Dekalb, IL 60115, USA
Dieleman, S. J.	Utrecht University, Dept Herd Health and Reproduction, 3508 TD Utrecht, The Netherlands
Doberska, C. A.	Journal of Reproduction and Fertility, Cambridge CB5 8DT, UK
Dobrinsky, J. R.	USDA–ARS–GGPL, Beltsville MD 20705, USA
Drapeau, R.	Centre d'Insémination porcine Inc, Saint-Lambert, Québec, GOS 2WO, Canada
Druart, X.	UNCEIA, 94703 Maisons-Alfort, France
Dumont, P.	UNCEIA, 94703 Maisons-Alfort, France
Dyck, G. W.	Spar Research Ltd, Brandon, Manitoba R7B 1B1, Canada
Dziuk, P.	University of Illinois, Urbana, IL 61801, USA
Einarsson, S.	Swedish Agricultural University, S-750 07 Uppsala, Sweden
Eliasson-Selling, L.	Swedish Animal Health Service, S-75105 Uppsala, Sweden
Elsaesser, F.	Institute Animal Husbandry and Animal Behaviour, Mariensee, D-31535 Neustadt, Germany
Engelhardt, H.	University of Guelph, Guelph Ontario N1G 2W1, Canada
Englund, P.	Swedish Agricultural University, S-75007 Uppsala, Sweden
Enright, W.	Intervet International B.V., 5830 AA Boxmeer, The Netherlands
Eriksson, B.	Swedish Agricultural University, S-75007 Uppsala, Sweden
Estienne, M. J.	University of Maryland Eastern Shore, Salisbury MD 21801, USA
Evans, G.	University of Sydney, Sydney NSW 2006, Australia
Feitsma, H.	Varkens KI Noordbrabant bv, 5268 ZH Helvoirt, The Netherlands
Ferreira, F. M.	Institute for Reproductive Medicine, D-30559 Hannover, Germany
Fioretti, D.	University La Sapienza, 00161 Roma, Italy
Flechon, J. E.	INRA, Biologie Cellulaire, 78352 Jouy en Josas, France
Flowers, W. L.	North Carolina State University, Raleigh NC 27695–7621, USA
Ford, J.	USDA–ARS, Clay Center, NE 68933–0166, USA
Ford, S.	Iowa State University, Ames, IA50010, USA
Forni, M.	Dimorfipa, 40064 Ozzano Emilia (BO), Italy
Foxcroft, G.	University of Alberta, Edmonton, Alberta T8G 2P5, Canada
Funahashi, H.	Okayama University, 1-1-1 Tsushima-naka, Okayama 700, Japan
Galeati, G.	Universita di Bologna, 50 Ozzano (BO), Bologna, Italy
Garcia Casado, P.	INIA Area de Reproduccion Animal, 28040 Madrid, Spain
Geisert, R. D.	Oklahoma State University, Stillwater, OK 74074, USA
Gommert, A.	Deutsche Pig Improvement, D-24837 Schleswig, Germany
Grooten, H.	Dutch Pig AI Association, 6093 AD Heijthuysen, The Netherlands
Guillouet, P.	INRA-SEIA, 86480 Rouille, France
Harrison, R. A. P.	Babraham Institute, Babraham, Cambridge CB2 4AT, UK
Hazeleger, W.	Wageningen Agricultural University, 6700 AH Wageningen, The Netherlands
Hiatt, M.	PIC-USA, Franklin KY 42135–0348, USA

Hochereau-de Reviers, M-T.	INRA-Tours, 37380 Nouzilly, France
Hofmo, P. O.	Norsvin, N-2301 Hamar, Norway
Holt, W. V.	Institute of Zoology, London NW1 4RY, UK
Hoste, S.	PIC-UK, Abingdon, Oxon OX13 5NA, UK
Hultén, F.	Swedish Agricultural University, S-75007 Uppsala, Sweden
Hunter, M. G.	University of Nottingham, Sutton Bonington Campus, LE12 5RD, UK
Jansen, G.	Wageningen Agricultural University, 6700 AH Wageningen, The Netherlands
Johansson, M.	Swedish Agricultural University, S-75007 Uppsala, Sweden
Johnson, L. A.	ARS – USDA, Beltsville, MD 20705, USA
Jukola, E.	00241 Helsinki, Finland
Karlberg, K.	College of Veterinary Medicine, N-0033 Oslo, Norway
Kemp, B.	Wageningen Agricultural University, 6700 AH Wageningen, The Netherlands
Kerkaert, B.	Pipestone Veterinary Clinic, Pipestone MN 56164, USA
Kikuchi, K.	National Institute of Agrobiological Resources, Kannondai 2-1-2, Tsukuba, Ibaraki 305, Japan
Kim, J-H.	University of Missouri, 162 Animal Sciences Research Center, Columbia, MO 65211, USA
King, G.	University of Guelph, Animal Science, Guelph, ON N1G 2W1, Canada
Kraeling, R.	USDA–ARS, Richard Russell Agricultural Research Center, Athens GA 30604, USA
Kunavongkrit, A.	Chulalongkorn University, Dept Obstetrics and Gynecology, Bangkok 10330, Thailand
Laforest, J-P.	Université Laval, Dept Sciences Animales, Ste-Foy, Quebec G1K 7P4, Canada
Lamberson, W. R.	University of Missouri, Animal Sciences Department, Columbia, MO 65211, USA
Lancaster, R. T.	Imutran Ltd, Cambridge CB2 2AH, UK
Lavitrano, M. L.	University La Sapienza, Dept Experimental Medicine, 00161 Roma, Italy
Leenhouwers, J. J.	Wageningen Agricultural University, Dept Animal Breeding, 6700 AH Wageningen, The Netherlands
Lidický, J.	Biopharm, Pohori Chotoun, 25449 Jlove near Prague, Czech Republic
Lorton, S. P.	Minitube of America, Inc., Verona, WI 53593, USA
Lunstra, D.	USDA–ARS, Meat Animal Research Center, Clay Center, Nebraska 68933, USA
Machaty, Z.	University of Missouri, 162 Animal Science Research Center, Columbia MO 65211, USA
Madsen, K. S.	Danske Slagterier, DK 6600 Vejen, Denmark
Martens, M.	Intervet International bv, 5830 AA Boxmeer, The Netherlands
Martinat-Botté, F.	INRA, PRMD, 37350 Nouzilly, France
Mathias, K.	Dekalb Swine Breeders Inc., Plains Kansas 67869, USA
Mburu, J. N.	Swedish Agricultural University, Dept Obstetrics and Gynaecology, S-75007 Uppsala, Sweden
Meinecke, B.	Institute for Reproductive Medicine, D-30559 Hannover, Germany
Melli, F.	Medichimica, 42010 Gavassa, Reggio Emilia, Italy
Mercat, M-J	Cobiporc, 35590 St Gilles, France

Merks, J. W. M.	IPG -Institute for Pig Genetics, 6640 AA Beuningen, The Netherlands
Messias de Braganca, M.	INRA-Recherches Porcines, 35590 Saint-Gilles, France
Mileham, A.	University of Cambridge, Dept of Pathology, Cambridge CB2 1QP, UK
Miller, H.	University of Leeds, Dept of Animal Physiology and Nutrition, Leeds LS2 9JT, UK
Moeller-Holtkamp, P.	Hoechst Roussel vet gmbh, D-65203 Wiesbaden, Germany
Murphy, P.	Hermitage AI and Pedigree Pigs, Killkenny, Ireland
Mwanza, A. M.	Swedish Agricultural University, Dept Obstetrics and Gynaecology, S-75007 Uppsala, Sweden
Nakanishi, Y.	Kagoshima University, Laboratory Animal Reproduction, Kagoshima 890, Japan
Niemann, H.	Institut fuer Tierzucht und Tierverhalten, Dept Biotechnology, 31535 Neustadt, Germany
Nolibois, P.	Hoechst Roussel Vet, 93235 Romainville, Cedex, France
Nottle, M. B.	BresaGen Ltd, Adelaide SA 5000, Australia
Nyby Pedersen, P.	Danske Slagterier, DK-1609 Kopenhagen V., Denmark
Oskam, J. P. H.	Cofok bv, 6679 AD Oosterhout, The Netherlands
Overstrom, E. W.	Tufts University, School Veterinary Medicine, North Grafton MA 015316, USA
Oving, L.	Cofok bv, 6679 AD Oosterhout, The Netherlands
Palomo, M-J.	Babraham Institute, Babraham, Cambridge CB2 4AT, UK
Parmeggiani, A.	Dipto di Morfofisiol. Vet./Produz. Anim., 50 Ozzano Emilia (BO), Italy
Pere, M-C.	INRA-Recherches Porcine, 35590 Saint-Gilles, France
Perez Llano, B.	Gestion Veterinaria Porcina, 28039 Madrid, Spain
Persson, E. M.	Swedish Agricultural University, Dept Anatomy and Histology, S-75007 Uppsala, Sweden
Peters, A.	Royal Veterinary College, Potters Bar, Herts EN6 1NB, UK
Petrounkina, A.	Institute for Reproductive Medicine, 30559 Hannover, Germany
Piedrahita, J. A.	Texas A&M University, Veterinary Anatomy, College Station, TX 77843, USA
Polge, C.	Landbeach, Cambridge, CB4 4EA, UK
Poolperm, P.	North Carolina State University, College Veterinary Medicine, Raleigh, NC 27606, USA
Prather, R. S.	University of Missouri, Columbia MO 65211, USA
Prunier, A.	INRA-Recherches Porcines, 35590 St Gilles, France
Pucci, L.	University La Sapienza, Dept Experimental Medicine, 00161 Roma, Italy
Quesnel, H.	INRA-Recherches Porcines, 35590 St Gilles, France
Rampacek, G.	University of Georgia, Animal and Dairy Science Dept, Athens, GA 30602, USA
Rath, D.	Institut fur Tierzucht und Tierverhalten, 31535 Neustadt, Germany
Ratky, J.	A'TK, 2053 Herceghalom, Hungary
Richardson, J. S.	Intervet UK Ltd, Cambridge, CB4 4FP, UK
Rodriguez-Martinez, H.	Swedish Agricultural University, Dept Obstetrics/Gynecology, 75007 Uppsala, Sweden
Rothschild, M.	Iowa State University, Dept Animal Science, Ames, IA 50011, USA
Safranski, T.	University of Missouri, S-133 Animal Sciences Center, Columbia, MO 65211, USA

Sedlak, L.	Biopharm, Pohori Chotoun, 25449 Jlove near Prague, Czech Republic
Shimada, A.	National Institute of Agrobiological Resources, Dept Genetic Resources II, Tsukuba Ibaraki, Japan
Simmet, C.	Minitub, D-84184 Tiefenbach b. Landshut, Germany
Smith, P.	Bunge Meat Industries, Corowa NSW 2646, Australia
Soede, N. M.	Wageningen Agricultural University, Dept Animal Husbandry, 6700 AH Wageningen, The Netherlands
Sterle, J.	University of Missouri, 159 ASRC, Columbia MO 65211, USA
Steverink, D. W. B.	Wageningen Agricultural University, Dept Animal Husbandry, 6700 AH Wageningen, The Netherlands
Stumpf, T. T.	Continental Grain Company, Fairmont MN 56031, USA
Swan, W. P.	JSR Healthbred Ltd, Northern SDS Centre Thoipe Willoghby, Selby YQ8 9NN, UK
Tantasuparuk, W.	Swedish Agricultural University, Dept Obstetrics and Gynaecology, S-75007 Uppsala, Sweden
Taverne, M.	Utrecht University, Dept Herd Health and Reproduction, 3584 CL Utrecht, The Netherlands
Terlouw, S.	Genetic Improvement, Princeton MO 104673, USA
Terqui, M.	INRA PRMD, 37380 Nouzilly, France
Thacker, P.	University of Saskatchewan, Dept Animal Science, Saskatchewan, Canada
Thilmant, P.	CIAP, B-4601 Argenteau, Belgium
Thorup, F.	Danske Slagterier, DK-1609 Copenhagen V., Denmark
Trout, B.	University of Missouri, 164 Animal Sciences Research Center, Columbia, MO 65211, USA
Vallet, J. L.	USDA–ARS, US Meat Animal Research Center, Clay Center, NE 68933, USA
van der Weijden, B. C.	Utrecht University, Dept Herd Health and Reproduction, 3508 TD Utrecht, The Netherlands
van Schie, F. W.	Animal Health Service Center, 7400 AA Deventer, The Netherlands
van den Brand, H.	Wageningen Agricultural University, Dept Animal Husbandry, 6700 AH Wageningen, The Netherlands
van der Lende, T.	Wageningen Agricultural University, Dept Animal Breeding, 6700 AH Wageningen, The Netherlands
van Asten, P.	Euribrid, Boxmeer, The Netherlands
van Nes, A.	Utrecht University, 3584 CL Utrecht, The Netherlands
van der Steen, H.	PIC Group, Abingdon, Oxfordshire, OX13 5NA, UK
van Rens, B. T. T. M.	Wageningen Agricultural University, Dept Animal Breeding, 6700 AH Wageningen, The Netherlands
Varley, M. A.	University of Leeds, Dept Animal Physiology and Nutrition, Leeds LS2 9ST, UK
Vazquez, J. M.	Universidad de Murcia, Reproduccion y Obstetricia, 30071 Murcia, Spain
Vermeulen, F.	Wageningen Agricultural University, 6700 AH Wageningen, The Netherlands
Vesseur, P.	Research Institute for Pig Husbandry, 5240 AB Rosmalen, The Netherlands
Waberski, D.	Institute for Reproductive Medicine, Veterinary School of Hannover, D-30559 Hannover, Germany
Wähner, M.	Fachhochschule Anhalt, Fachbereich LOEL, 06406 Bernburg, Germany

Weitze, K-F.	Institute of Reproductive Medicine, 30559 Hannover, Germany
Wilson, M. E.	United Feeds, Gridley, IL 61744, USA
Woelders, H.	Institute for Animal Science and Health, 8200 AB Lelystad, The Netherlands
Wuttke, W.	Medical University Goettingen, Dept of Obstetrics and Gynaecology, D-37075 Goettingen, Germany
Yang, L-G.	Rowett Research Institute, Aberdeen AB2 9SB, UK

REGULATION OF REPRODUCTIVE FUNCTION

Chair
R. Kraeling

Journal of Reproduction and Fertility Supplement **52**, 3–17

Role of neuropeptides and amino acids in controlling secretion of hormones from the anterior pituitary gland in pigs

M. J. Estienne[1], J. M. Harter-Dennis[1] and C. R. Barb[2]

[1]*Department of Agriculture, University of Maryland Eastern Shore, Princess Anne, MD 21853; and* [2]*Animal Physiology Unit, Richard B. Russell Agriculture Research Center, United States Department of Agriculture, Agricultural Research Service, Athens, GA 30613, USA*

All reproductive processes involve one or more of the protein hormones secreted from the anterior pituitary gland: LH, FSH, prolactin, growth hormone, ACTH and thyroid-stimulating hormone (TSH). Primary hormones of reproduction, such as LH and FSH, directly regulate a reproductive activity. For example, LH and FSH stimulate follicular growth and the associated secretion of oestradiol in sows. In contrast, secondary hormones of reproduction such as TSH are permissive and regulate other physiological systems that indirectly, but profoundly, influence reproduction. Reproduction in pigs can be enhanced by developing strategies to alter and control secretion of hormones from the anterior pituitary gland. However, the successful manipulation of adenohypophysial hormone secretion will require a sound understanding of the mechanisms controlling the function of the hypothalamic–pituitary axis. Hypothalamic hormones including GnRH, dopamine, growth hormone-releasing hormone (GHRH), somatostatin, corticotrophin-releasing hormone (CRH) and thyrotrophin-releasing hormone (TRH) are synthesized in perikarya that possess axons that terminate at the median eminence. These hormones are released into the hypothalamo–hypophysial portal vasculature, travel to the anterior pituitary gland and stimulate or inhibit secretion of adenohypophysial hormones. Secretion of hypothalamic hormones is ultimately controlled by a variety of neurotransmitters and neuropeptides, the most studied in swine being the endogenous opioid peptides (EOP) and more recently, the excitatory amino acids (ExAA). In general, EOP inhibit GnRH and hence LH secretion, and this effect involves the central catecholaminergic system. A definitive role for EOP in the modulation of FSH release remains to be determined. EOP stimulate secretion of GHRH and thus growth hormone release, and depending on the animal model studied, EOP exert either stimulatory or inhibitory influences on prolactin secretion. ExAA, working via *N*-methyl-D-aspartate (NMDA) receptors at the central nervous system, stimulate secretion of LH, FSH, growth hormone and prolactin in appropriate animal models. However, in certain situations, an inhibitory effect of ExAA on LH secretion has been demonstrated. The modulation of growth hormone and prolactin secretion by ExAA involves EOP. Research investigating the function of ExAA and EOP in the physiological control of swine reproduction warrants further scrutiny.

Neuroendocrine Control of Adenohypophysial Hormone Secretion

Hypothalamic hormones, including GnRH, dopamine, GHRH, somatostatin, CRH and TRH, are synthesized in perikarya that possess axons that terminate at the median eminence. These hormones are released into the hypothalamo–hypophysial portal blood vessels, travel to the anterior pituitary gland, and either stimulate or inhibit secretion of adenohypophysial hormones. Anterior pituitary gland hormones stimulate target tissues to release hormones and metabolites that in turn modify adenohypophysial hormone secretion via negative and positive feedback mechanisms. Because of space limitations, these intricate feedback loops will not be discussed in this review.

Gonadotrophins

Administration of GnRH stimulates LH secretion, follicular growth and ovulation in prepubertal gilts (Lutz *et al.*, 1985) and hypophysial stalk-transected gilts injected with pregnant mares' serum gonadotrophin (Kraeling *et al.*, 1990). In pigs, as in other mammals, there are two distinct modes of LH secretion that presumably reflect two patterns of GnRH secretion from the central nervous system: (1) episodes of LH release (i.e. pulsatile secretion), the frequency and amplitude of which vary with the reproductive status of the animal, circulating steroidal milieu or with both factors, and (2) preovulatory surge secretion. With regard to gonadotrophins, this review focuses primarily on the neuroendocrine control of pulsatile secretion of LH. The neuroendocrine control of the preovulatory surge of gonadotrophins in pigs has been thoroughly covered in the review by Kraeling *et al.* (1992).

Kineman *et al.* (1988) described GnRH-immunostained cell bodies in the medial preoptic area, diagonal band of broca, lateral hypothalamic area, paraventricular nucleus, periventricular zone, suprachiasmatic nucleus and medial basal hypothalamus of ovariectomized and ovary-intact gilts. These perikarya had axons that projected to the median eminence.

Using push–pull cannulae, Leshin *et al.* (1992) demonstrated that pulsatile release of LH is associated with pulsatile secretion of GnRH, as detected in the vasculature of the anterior pituitary gland of ovariectomized gilts. Disrupting the GnRH signal to the anterior pituitary gland by transection of the hypophysial stalk (Kraeling *et al.*, 1990) or by active or passive immunization against GnRH (Esbenshade *et al.*, 1990) abolished pulsatile release of LH in gilts. Secretion of FSH was also suppressed after immunization against GnRH (Esbenshade *et al.*, 1990).

Prolactin

Prolactin secretion is tonically inhibited by signals from the central nervous system in pigs. Kraeling *et al.* (1994) reported high circulating concentrations of prolactin in ovariectomized gilts that were subjected to transection of the hypophysial stalk.

Several studies suggest that dopamine inhibits prolactin secretion in swine. For example, bromocryptine, a dopamine agonist, suppressed prolactin secretion in lactating sows, ovariectomized gilts and ovariectomized gilts that were subjected to transection of the hypophysial stalk (Kraeling *et al.*, 1982, 1994).

In pigs, TRH is a putative prolactin-releasing hormone (Dubreuil *et al.*, 1990; Kraeling *et al.*, 1994). The prolactin response to a TRH challenge was attenuated by bromocryptine treatment in ovariectomized, hypophysial stalk-transected gilts, suggesting that the suppression of dopaminergic pathways is a necessary antecedent to the mechanism by which TRH stimulates prolactin secretion (Kraeling *et al.*, 1994).

Growth hormone

GHRH stimulates, and somatostatin inhibits, growth hormone secretion in swine (Dubreuil *et al.*, 1990). Immunocytochemical studies have revealed GHRH neurones in the arcuate and ventromedial nuclei, and somatostatin neurones in the periventricular region of gilts (Leshin *et al.*, 1994). GHRH and somatostatin nerve fibres project ventrally into the median eminence.

Active immunization against GHRH abolished episodic secretion of growth hormone in the lactating sow (Armstrong *et al.*, 1990). In contrast, active immunization against somatostatin increased growth hormone secretion in prepubertal boars and gilts (Dubreuil *et al.*, 1989). Disrupting the GHRH signal to the anterior pituitary gland by transection of the hypophysial stalk resulted in a loss of pulsatile growth hormone secretion in ovariectomized gilts. Nevertheless, basal growth hormone concentrations were increased, perhaps due to the decreased amount of somatostatin reaching the adenohypophysis (Klindt *et al.*, 1983).

Other anterior pituitary gland hormones

Relatively little is known about the neuroendocrine control of ACTH and TSH in swine. However, CRH stimulated secretion of cortisol in ovariectomized gilts and in gilts that were ovariectomized and subjected to transection of the hypophysial stalk, presumably by enhancing ACTH secretion (Estienne *et al.*, 1988). Serum concentrations of cortisol before CRH administration were similar for hypophysial stalk-transected and control gilts, indicating that basal secretion of cortisol was maintained in the absence of central inputs (Estienne *et al.*, 1988). Finally, treatment of gilts or lactating sows with TRH enhanced thyroxine secretion, probably as a consequence of increased secretion of TSH (Dubreuil *et al.*, 1990; Barb *et al.*, 1991a).

Catecholamines

The catecholamines noradrenaline, adrenaline and dopamine are synthesized by neurones within the central nervous system. The conversion of tyrosine to dihydroxyphenylalanine (DOPA) is catalysed by tyrosine hydroxylase and is the rate-limiting step in catecholamine biosynthesis. DOPA is converted to dopamine via the action of aromatic decarboxylase. Dopamine is converted to noradrenaline and then adrenaline via the action of dopamine β-hydroxylase and phenylethanolamine-*n*-methyltransferase, respectively. Leshin *et al.* (1996) located tyrosine hydroxylase and dopamine-β-hydroxylase immunopositive neurones in the hypothalamus of the gilt. It has been hypothesized that in pigs, nerve cells that release catecholamines act as interneurones and participate in the control of secretion of hypothalamic hormones and hence anterior pituitary gland function by EOP and ExAA. Thus, a brief description of the effects of catecholamines on secretion of adenohypophysial hormones is warranted.

Central administration of noradrenaline to adult, gonadectomized miniature pigs had dichotomous effects on LH release (Parvizi and Ellendorff, 1982). Whether noradrenaline stimulated or inhibited secretion of LH depended on the dose administered and specific site of hypothalamic injection.

Reports concerning the effects of dopamine on LH secretion in swine are equivocal. Kraeling *et al.* (1982) reported that the dopamine agonist bromocryptine decreased serum concentrations of LH in lactating sows. However, Bevers *et al.* (1983) reported that LH secretion increased in bromocryptine-treated lactating sows. Treatment with bromocryptine had no effect on LH secretion in ovariectomized gilts (Kraeling *et al.*, 1982).

AIMAX (synonym = methallibure) is a derivative of dithiocarbamoylhydrazine and has actions similar to diethyldithiocarbamate, a potent inhibitor of dopamine β-hydroxylase. Treatment of rats with AIMAX decreased LH secretion, and suppressed noradrenaline synthesis and increased dopamine content in the hypothalamus (Chang *et al.*, 1995). When fed to ovariectomized gilts with (Chang *et al.*, 1993a) or without (Kesner *et al.*, 1987) progesterone replacement therapy, AIMAX suppressed pulsatile gonadotrophin secretion. Moreover, the oestradiol-induced LH surge was abolished and the FSH surge attenuated, in AIMAX-fed, ovariectomized gilts (Kesner *et al.*, 1987). Effects of AIMAX were manifested at the hypothalamus since pituitary responsiveness to exogenously administered GnRH was uncompromised by treatment (Kesner *et al.*, 1987). These findings suggest that catecholamines stimulate both pulsatile and surge secretions of gonadotrophins in swine.

Ingestion of AIMAX had no effect on serum concentrations of growth hormone and prolactin in ovariectomized or ovariectomized, steroid-treated gilts (Kesner *et al.*, 1987; Chang *et al.*, 1993a), and failed to alter pulsatile secretion of growth hormone in boars (D. S. Broughton, M. J. Estienne, J. M. Harter-Dennis, C. R. Barb and T. G. Hartsock, unpublished). AIMAX also had no effect on circulating concentrations of the thyroid hormones and cortisol, suggesting that secretion of TSH and ACTH was unaffected in ovariectomized gilts (Kesner *et al.*, 1987).

Endogenous Opioid Peptides

Endorphins, enkephalins and dynorphins are collectively called EOP and are natural ligands for receptors that also bind opiates. Proopiomelanocortin (POMC) is the precursor for β-endorphin. Using immunocytochemical techniques, Kineman *et al.* (1989) located POMC perikarya within the arcuate area in gilts. These cell bodies had fibres that projected to the medial basal hypothalamus, periventricular zone, preoptic area and median eminence.

Effect of EOP and EOP agonists on secretion of anterior pituitary gland hormones

The role of EOP in controlling LH secretion in pigs has been scrutinized in several thorough reviews (Barb *et al.*, 1991b; Kraeling *et al.*, 1992). In general, treatment of pigs with EOP or their agonists decreased secretion of LH. For example, morphine, an agonist of EOP, suppressed LH secretion when administered to ovariectomized gilts via the lateral cerebral ventricle (Estienne *et al.*, 1990). However, intravenous injections of morphine failed to alter LH secretion in immature boars (Trudeau *et al.*, 1988).

Responses to EOP agonists with regard to FSH secretion have been less consistent. For example, intracerebroventricular injection of morphine suppressed FSH release in ovariectomized gilts in some experiments, but not others (Estienne *et al.*, 1990; Barb *et al.*, 1992a). Morphine inhibited FSH secretion when intravenously administered to immature boars (Trudeau *et al.*, 1988).

Results of several studies suggest that EOP stimulate secretion of prolactin and growth hormone. Intracerebroventricular administration of morphine increased circulating concentrations of prolactin and growth hormone in ovariectomized gilts (Estienne *et al.*, 1990; Barb *et al.*, 1992a).

Effect of EOP antagonists on secretion of anterior pituitary gland hormones

The classic method for studying the physiological role of an endogenous neuromodulator is to use a selective antagonist to block its effects. Several studies have been conducted in which male and female pigs in various reproductive states were treated with the EOP receptor antagonist naloxone.

Stimulation of LH secretion by treatment with naloxone supports the concept that EOP have a physiological role in inhibiting gonadotrophin secretion. Intravenous treatment with naloxone increased LH secretion in luteal phase, but not follicular phase, gilts (Barb *et al.*, 1986a). Injections of naloxone also increased serum concentrations of LH in lactating sows (Barb *et al.*, 1986b; Mattioli *et al.*, 1986; De Rensis *et al.*, 1993).

Treatment with naloxone stimulated LH release in ovariectomized gilts only if the animals were treated with exogenous progesterone (Barb *et al.*, 1986a, 1988). In contrast, naloxone treatment failed to alter LH release in ovary-intact and ovariectomized, progesterone-treated prepubertal gilts (Barb *et al.*, 1988). However, naloxone increased LH secretion when gilts were ovariectomized before puberty and were treated with progesterone and the EOP antagonist at an age when contemporary gilts had begun to display oestrous cycles (Barb *et al.*, 1988). This study suggests that, in gilts, the development of EOP modulation of LH secretion is a brain maturational process that is independent of gonadal influences. In contrast, naloxone had no effect on LH release in immature boars or barrows, but increased LH secretion in barrows treated with testosterone so as to establish adult concentrations of androgen (Patchev *et al.*, 1987; Trudeau *et al.*, 1989).

Similar to experiments in which equivocal responses to EOP agonists were found, studies examining the effects of naloxone on FSH secretion have yielded conflicting results. For example, naloxone increased FSH secretion in immature boars (Trudeau *et al.*, 1989) and lactating sows (Barb *et al.*, 1987), but had no effect on FSH release in ovariectomized gilts, or ovariectomized, prepubertal gilts with or without progesterone therapy (Barb *et al.*, 1992a). De Rensis *et al.* (1993) reported that naloxone had no effect on FSH secretion in postpartum sows. Interestingly, Barb *et al.* (1992a) also reported that naloxone treatment suppressed FSH secretion in prepubertal gilts and in mature, ovariectomized gilts treated with progesterone.

Changes in LH release without concomitant alterations in FSH secretion and vice versa support the notion that there are separate releasing hormones for the two gonadotrophins in pigs. Alternatively, divergent patterns of LH and FSH secretion following treatment with EOP agonists and antagonists could be related to circulating concentrations of steroids and gonadal peptides or the existing pattern of GnRH release. Jayes *et al.* (1997) suggested that high GnRH pulse frequency was more effective in acutely releasing LH than FSH, and inhibits FSH synthesis and secretion. In contrast, low GnRH pulse frequency supports FSH synthesis and release but is not effective in increasing LH concentrations.

Naloxone treatment suppressed secretion of growth hormone (Armstrong *et al.*, 1990; Barb *et al.*, 1991a) and prolactin (Mattioli *et al.*, 1986; Barb *et al.*, 1987) in lactating sows. De Rensis *et al.* (1993) reported that naloxone treatment suppressed prolactin secretion when administered at day 10 of lactation but not when given during the first 78 h postpartum.

Naloxone increased prolactin secretion in luteal phase gilts (Barb *et al.*, 1986a) and in mature, ovariectomized, progesterone-treated gilts (Barb *et al.*, 1992a), but had no effect on prolactin release in mature, ovariectomized gilts or ovariectomized, prepubertal gilts with or without progesterone treatment (Barb *et al.*, 1992a). Thus, it appears that EOP exert both stimulatory and inhibitory effects on prolactin secretion in swine.

The effects of naloxone on ACTH and TSH secretion in swine have not been reported. However, naloxone treatment stimulated cortisol secretion in cyclic and ovariectomized gilts (Barb *et al.*, 1986a; Estienne *et al.*, 1988) and inhibited cortisol secretion in lactating sows (Barb *et al.*, 1991a).

Site of action for effects of EOP on adenohypophysial hormone secretion

The notion that the EOP restraint on gonadotrophin secretion is manifested primarily at the hypothalamus is supported by studies *in vitro* in which naloxone stimulated GnRH secretion from the hypothalamic–preoptic area collected from gilts (Barb *et al.*, 1994). Moreover, naloxone failed to increase LH secretion in gilts treated with antisera to GnRH (Chang *et al.*, 1993a). Naloxone had no effect on the LH and FSH response to GnRH when the EOP receptor antagonist and secretogogue were administered concomitantly in immature boars (Trudeau *et al.*, 1989).

Treatment with naloxone in sows injected with control serum decreased growth hormone concentrations and the frequency of growth hormone pulses to values reported for saline-treated, GHRH-immunized sows (Armstrong *et al.*, 1990). This finding suggests that high circulating concentrations of growth hormone in the lactating sow are primarily due to an opioid-modulated increase in GHRH secretion.

Naloxone-stimulated secretion of cortisol in ovariectomized gilts was abolished by transection of the hypophysial stalk (Estienne *et al.*, 1988). Since serum cortisol concentrations increased in these animals following CRH and ACTH administration, Estienne *et al.* (1988) concluded that naloxone increased cortisol secretion in swine by acting at the hypothalamus to alter release of CRH and subsequent ACTH secretion.

Despite the abundance of evidence suggesting that modulation of anterior pituitary gland hormone secretion by EOP in swine is primarily a consequence of central actions, there may be subtle effects of EOP directly on the adenohypophysis. β-Endorphin suppressed basal and GnRH-induced LH secretion by pig pituitary cells *in vitro* (Barb *et al.*, 1990). In contrast, chronic exposure to naloxone increased basal secretion of LH and enhanced pituitary responsiveness to GnRH (Barb *et al.*, 1990).

Mechanism of action for effects of EOP on adenohypophysial hormone secretion

Catecholamines stimulate, and EOP inhibit, secretion of GnRH and hence LH in pigs. Several lines of evidence suggest that EOP and catecholamines interact in regulating GnRH secretion (Fig. 1). Anatomical studies have revealed that POMC, tyrosine hydroxylase and dopamine

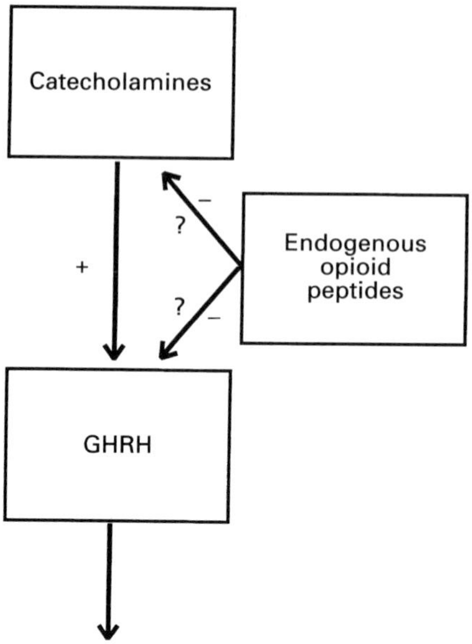

Fig. 1. Possible neuronal pathways by which endogenous opioid peptides inhibit secretion of GnRH, and hence LH, release in swine.

β–hydroxylase immunopositive neurones are located in the vicinity of GnRH neurones in the hypothalamus (Kineman *et al.*, 1988, 1989; Leshin *et al.*, 1996).

Treatment with AIMAX abolished naloxone stimulation of LH secretion in ovariectomized, progesterone-treated gilts (Chang *et al.*, 1993a). The prolactin response to naloxone was not attenuated in ovariectomized, progesterone-treated gilts fed AIMAX, suggesting EOP modulation of prolactin secretion is independent of the noradrenaline system (Chang *et al.*, 1993a).

Excitatory Amino Acids

A group of amino acids that includes glutamate and aspartate has been implicated as the major family of neurotransmitters in the mammalian central nervous system. The ExAA satisfy the main criteria for classification as neurotransmitters, namely: (1) presynaptic localization in specific nerve terminals, (2) specific release by physiological stimuli in concentrations great enough to evoke a postsynaptic response, (3) identity of action (i.e., postsynaptic actions of the naturally occurring transmitter are mimicked by the candidate substance, including response to antagonists), and (4) existence of neuronal mechanisms, such as re-uptake and metabolism, that terminate neurotransmitter action.

Studies in other species have demonstrated that glutamate and aspartate, and receptors for these ExAA, appear in many regions of the brain including the hypothalamus and median eminence. There are several types of receptor that are stimulated by glutamate and these are named according to their selective agonists. Receptor types include *N*-methyl-D-aspartate (NMDA), kainate, and D,L-amino-3-hydroxy-5-methyl-4-isoxazole propionic acid (AMPA) (Petralia and Wenthold, 1996).

Secretion of hormones from the anterior pituitary gland following administration of ExAA

The ExAA agonist, *N*-methyl-D,L-aspartate (NMA) is a potent activator of the NMDA receptor. The ability of intravenously administered NMA to alter secretion of hormones from the anterior pituitary gland has been confirmed in several experiments in pigs (Table 1). In contrast to the

Table 1. Effects of intravenous injections of *N*-methyl-D,L-aspartate (NMA) on circulating concentrations of hormones secreted from the anterior pituitary gland in swine

Hormone and model	Dose of NMA (mg kg^{-1} body weight)	Result	References
LH			
Prepubertal gilts	1.25, 2.5 or 5.0	No effect	Estienne *et al.*, 1995
	10.0	Increased	Estienne *et al.*, 1995
Cyclic gilts – luteal phase	10.0	Increased	Hurlock *et al.*, 1995
– follicular phase	10.0	No effect	Hurlock *et al.*, 1995
Lactating sows	1.5 or 3.0	No effect	Sesti and Britt, 1992
	5.0	Increased	Sesti and Britt, 1992
	10.0	Increased	Sesti and Britt, 1993, 1994
Ovariectomized gilts	10.0	Decreased	Barb *et al.*, 1992b
		Decreased	Chang *et al.*, 1993b
		Decreased	Hurlock *et al.*, 1995
		Decreased	Popwell *et al.*, 1996
Ovariectomized, oestradiol-treated gilts	10.0	Increased	Sesti and Britt, 1992
	10.0	No effect	Barb *et al.*, 1992b
Ovariectomized, progesterone-treated gilts	10.0	Decreased	Barb *et al.*, 1992b; Chang *et al.*, 1993b
Boars	1.25, 2.5, 5.0 or 10.0	No effect	Broughton *et al.*, 1996
Barrows	2.5	No effect	Popwell *et al.*, 1996
FSH			
Lactating sows	10.0	Increased	Sesti and Britt, 1993, 1994
Growth Hormone			
Prepubertal gilts	1.25 or 5.0	No effect	Estienne *et al.*, 1995
	2.5 or 10.0	Increased	
Ovariectomized gilts	10.0	Increased	Barb *et al.*, 1992b; Chang *et al.*, 1993b
Ovariectomized, oestradiol-treated gilts	10.0	Increased	Barb *et al.*, 1992b
Ovariectomized, progesterone-treated gilts	10.0	Increased	Barb *et al.*, 1992b; Chang *et al.*, 1993b
Boars	1.25, 2.5, 5.0 or 10.0	Increased	Broughton *et al.*, 1996
	2.5	Increased	Estienne *et al.*, 1996c
Barrows	1.25	No effect	Estienne *et al.*, 1996b
	2.5 or 5.0	Increased	Estienne *et al.*, 1996b
Prolactin			
Ovariectomized gilts	10	Increased	Barb *et al.*, 1992b; Chang *et al.*, 1993b
Ovariectomized, oestradiol-treated gilts	10	Increased	Barb *et al.*, 1992b
Ovariectomized, progesterone-treated gilts	10	No effect	Barb *et al.*, 1992b
	10	Increased	Chang *et al.*, 1993b

extensive studies on NMDA receptor regulation of adenohypophysial hormone secretion, the role of other ExAA receptors in controlling secretion of anterior pituitary gland hormones has received scant attention.

Effects of ExAA on gonadotrophin secretion. Typically, secretion of LH and FSH is increased by NMA in animal models that are characterized as having suppressed gonadotrophin secretion. For

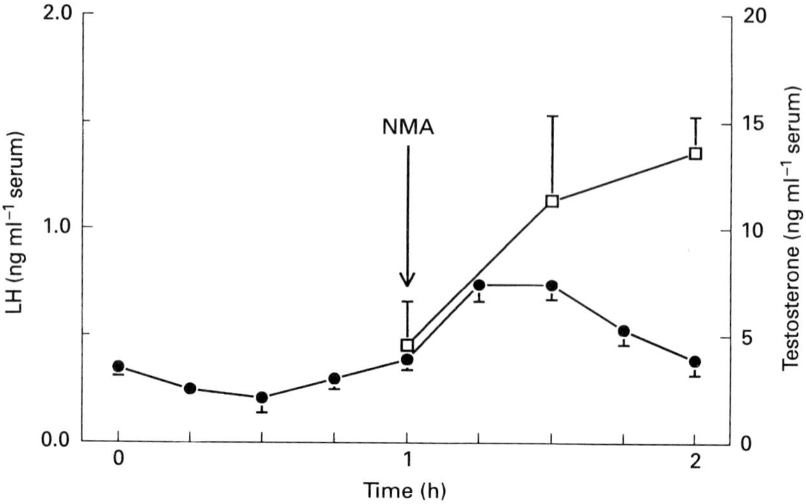

Fig. 2. Serum concentrations of LH (●) and testosterone (□) in mature boars receiving an intravenous injection of *N*-methyl-D,L-aspartate (NMA; 10 mg kg^{-1} body weight). Blood samples were collected at 15 min intervals for 2 h and NMA was administered at 1 h (arrow). All of the samples were analysed for concentrations of LH, and testosterone concentrations were determined in samples collected at 1.0, 1.5 and 2.0 h. Values are means±standard errors (n = 5). NMA administration resulted in an increase in LH ($P < 0.01$) and testosterone ($P < 0.05$) secretion.

example, NMA stimulated LH release in ovariectomized, oestradiol-treated gilts (Sesti and Britt, 1992), and enhanced secretion of both LH and FSH in lactating sows (Sesti and Britt, 1992, 1993, 1994). NMA stimulated LH release throughout lactation in both multiparous and primiparous sows, while FSH secretion increased when the secretogogue was administered between day 7 and day 21 postpartum in multiparous sows and between day 14 and day 21 postpartum in primiparous sows (Sesti and Britt, 1993, 1994).

Work in our laboratory demonstrated that administration of NMA increased LH secretion in prepubertal gilts (Estienne *et al.*, 1995) and in gilts treated during the luteal, but not follicular, phase of the oestrous cycle (Hurlock *et al.*, 1995). Failure of follicular phase gilts to respond to NMA in that study was perhaps due to oestradiol-induced alterations in pituitary sensitivity to GnRH. Indeed, the response to a GnRH challenge was compromised in follicular phase compared with luteal phase or ovariectomized gilts (Hurlock *et al.*, 1995).

Broughton *et al.* (1996) reported that NMA at doses of 1.25–10 mg kg^{-1} body weight failed to alter LH release in immature boars that were approximately 6 months of age. However, we recently demonstrated that NMA, at a dose of 10 mg kg^{-1} body weight, increased secretion of LH and testosterone in mature boars that were approximately 13 months of age (Fig. 2). These findings are consistent with the concept that, in boars, there is an age-related change in sensitivity of the gonadotrophin response to NMA.

In contrast to the studies showing stimulation of LH secretion by NMA, Barb *et al.* (1992b) and Chang *et al.* (1993b) reported that NMA had no effect on LH release in ovariectomized, oestradiol-treated gilts and decreased LH secretion in ovariectomized, progesterone-treated gilts. Treatment with NMA also decreased circulating concentrations of LH in ovariectomized gilts (Barb *et al.*, 1992b; Chang *et al.*, 1993b; Hurlock *et al.*, 1995; Popwell *et al.*, 1996). These studies suggest that, in addition to well-documented stimulatory effects, under appropriate conditions, ExAA may also suppress gonadotrophin secretion.

Effect of ExAA on growth hormone secretion. The effects of ExAA on growth hormone secretion, growth performance and carcass quality in domestic animals were summarized by Estienne *et al.* (1996a). Barb *et al.* (1996) reported that intravenous administration of aspartate (50–150 mg kg^{-1} body weight) or glutamate (100–150 mg kg^{-1} body weight) increased serum concentrations of growth hormone in prepubertal gilts. Moreover, infusion of NMA evoked growth hormone release in ovariectomized gilts receiving no treatment or steroid replacement therapy (Barb *et al.*, 1992b; Chang *et al.*, 1993b), prepubertal gilts (Estienne *et al.*, 1995), barrows (Estienne *et al.*, 1996b) and boars (Broughton *et al.*, 1996; Estienne *et al.*, 1996c).

The D,L racemic mixture of NMA has been used in neuroendocrine experiments but no attempt was made to determine which specific isomer was responsible for increasing blood concentrations of adenohypophysial hormones. However, using barrows, we demonstrated that the pure D isomer of NMA, at a dose of 1.25 mg kg^{-1} body weight, increased serum concentrations of growth hormone in a manner comparable to that which occurs after treatment with the D,L racemic mixture of NMA at a dose of 2.5 mg kg^{-1} body weight (Estienne *et al.*, 1996b). In contrast, injection of the pure L isomer of NMA (1.25 mg kg^{-1} body weight) did not increase secretion of growth hormone. These results do not preclude the possibility that higher doses of the pure L isomer of NMA may evoke hypersecretion of growth hormone.

Effect of ExAA on secretion of other anterior pituitary gland hormones. Barb *et al.* (1992b) reported that the injection of NMA increased serum concentrations of prolactin in ovariectomized and ovariectomized, oestradiol-treated gilts, but not in ovariectomized, progesterone-treated gilts. However, in a subsequent study, NMA increased prolactin secretion in both ovariectomized and ovariectomized, progesterone-treated gilts (Chang *et al.*, 1993b).

Circulating concentrations of ACTH and TSH following treatment with NMA have not been reported for swine. However, Barb *et al.* (1992b), Chang *et al.* (1993b) and Popwell *et al.* (1996) demonstrated that NMA increased serum concentrations of cortisol in ovariectomized and ovariectomized, steroid-treated gilts, and barrows, perhaps as a consequence of pituitary release of ACTH.

ExAA-induced secretion of hormones from the adenohypophysis: mechanism of action

Secretion of adenohypophysial hormones induced by exogenous ExAA is primarily a consequence of action at the central nervous system rather than a direct pituitary effect. Administration of NMA had no effect on serum concentrations of LH in ovariectomized gilts that were passively immunized against GnRH (Sesti and Britt, 1992). Active immunization against GHRH abolished the ability of NMA to increase circulating concentrations of growth hormone in ovariectomized gilts (Barb *et al.*, 1996). Similarly, NMA-induced increases in growth hormone secretion were abolished by treating barrows with antiserum to GHRH (Estienne *et al.*, 1996b).

Although most of the evidence suggests a central site for the effects of ExAA on release of pituitary hormones, subtle effects of the compounds directly on the anterior pituitary gland cannot be discounted. Studies *in vitro* demonstrated a positive effect of aspartate and glutamate on growth hormone secretion by pituitary cells collected from prepubertal gilts (Barb *et al.*, 1996). Barb *et al.* (1993) also reported that NMA stimulated growth hormone release from pituitary cells collected from ovariectomized gilts and gilts in the luteal phase of the oestrous cycle, but not from pituitary cells obtained from gilts in the follicular phase of the oestrous cycle. Treatment with NMA increased LH secretion from pituitary cells collected from ovariectomized, luteal phase and follicular phase gilts (Barb *et al.*, 1993).

Neural loci at which intravenously administered ExAA may act to stimulate release of hypothalamic hormones include the areas of the brain that lack a distinct blood–brain barrier. Price *et al.* (1981) reported preferential uptake of subcutaneously administered aspartate and glutamate by circumventricular organs of the rat brain. Therefore, exogenously administered ExAA may affect hypothalamic hormone secretion in pigs by influencing function of cell bodies in specific nuclei and(or) nerve terminals in the median eminence. However, these results do not preclude the existence of endogenous ExAA neurotransmission within the blood–brain barrier.

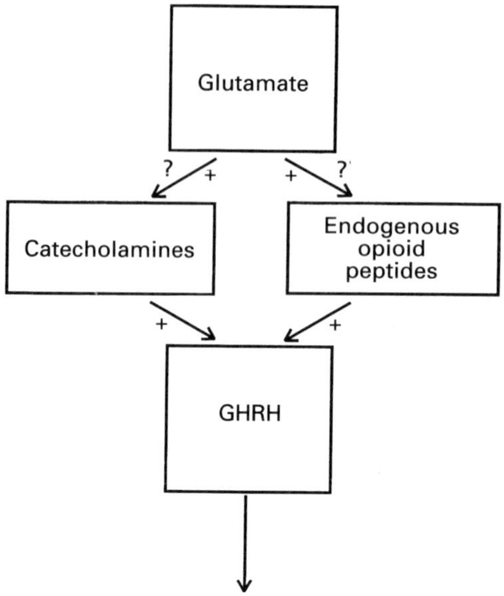

Fig. 3. Possible neuronal pathways by which excitatory amino acids such as glutamate stimulate release of growth hormone-releasing hormone (GHRH), and hence, growth hormone secretion in pigs.

Results from our laboratory are consistent with the concept that stimulation of adenohypophysial hormone secretion (at least growth hormone) by NMA is modulated by NMDA receptors. Treatment of barrows with ketamine hydrochloride, a noncompetitive NMDA receptor antagonist, attenuated the growth hormone response to NMA (Estienne *et al.*, 1996b).

It is unclear whether ExAA affect hypothalamic hormone secretion by a direct effect on neurones that secrete these hormones or indirectly through effects on interneurones (Fig. 3). As previously mentioned, drugs with opiate-like activity alter adenohypophysial hormone secretion in swine and it has been hypothesized that NMA-induced effects on secretion of LH, growth hormone and prolactin are mediated through EOP. In support of this hypothesis, administration of naloxone attenuated the growth hormone and prolactin responses to NMA in ovariectomized and ovariectomized, progesterone-treated gilts (Chang *et al.*, 1993b). In that study, however, responsiveness to NMA with regard to LH secretion was unaffected by naloxone.

We recently conducted an experiment to test the hypothesis that NMA-induced growth hormone secretion is modulated by catecholamine neurotransmission (D. S. Broughton, M. J. Estienne, J. M. Harter-Dennis, C. R. Barb and T. G. Hartsock, unpublished). Boars were fed a daily ration containing 0 or 125 mg AIMAX daily for 9 days ($n = 5$ per treatment). On day 8, blood samples were collected every 15 min for 8 h. Four and six hours after sampling was initiated, boars allowed each ration received intravenous injections of NMA (2.5 mg kg^{-1} body weight) or saline. The next day the experiment was repeated, but boars that previously received NMA were administered saline and vice versa. Before NMA or saline injections, mean serum growth hormone concentrations and the frequency and amplitude of growth hormone pulses were similar for AIMAX and control groups. Injections of NMA significantly increased mean serum growth hormone concentrations and growth hormone pulse amplitude in both groups in a similar fashion (Fig. 4). Treatment with NMA did not alter frequency of growth hormone pulses. These results suggest that in boars, catecholamines: (1) do not play a significant role in modulating pulsatile secretion of growth hormone, and (2) do not mediate the effects of NMA on growth hormone secretion.

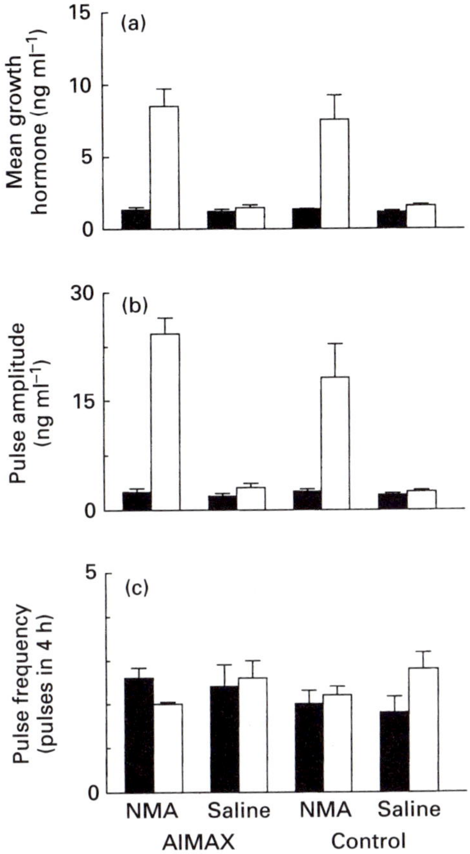

Fig. 4. Mean growth hormone concentrations (a) and the (b) amplitude and (c) frequency of growth hormone pulses in immature boars fed the central catecholamine synthesis inhibitor AIMAX (125 mg daily for 9 days) or a control diet, and injected intravenously with *N*-methyl-D,L-aspartate (NMA; 2.5 mg kg^{-1} body weight) or saline. Blood samples were collected every 15 min for 8 h and NMA was administered at 4 h and 6 h. (■): the 4 h period before NMA or saline injections; (□): the 4 h period after initiation of NMA or saline injections. Values are means ± standard errors ($n = 5$). Administration of NMA resulted in an increase ($P < 0.01$) in mean growth hormone concentrations and growth hormone pulse amplitude in both AIMAX-fed and control boars. The frequency of growth hormone pulses was unaffected ($P > 0.1$) by NMA injections.

ExAA as physiological modulators of adenohypophysial hormone secretion

Although exogenous administration of ExAA alters secretion of adenohypophysial hormones, a physiological role for aspartate and (or) glutamate in control of the hypothalamic–pituitary axis in pigs remains to be determined. If ExAA are involved in physiological control of a particular hormone, then treatment with an ExAA antagonist should have the opposite effect to the ExAA agonist.

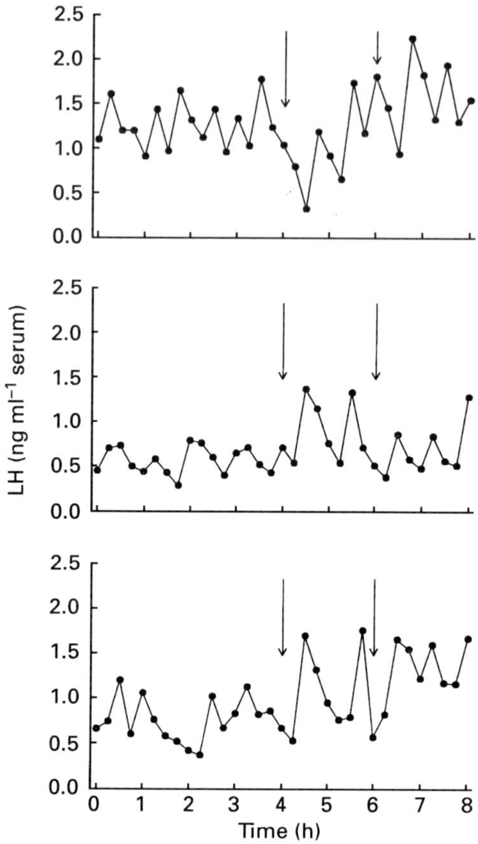

Fig. 5. Serum LH profiles in individual, ovariectomized gilts receiving intravenous injections of the competitive *N*-methyl-D-aspartate receptor antagonist D,L-2-amino-5-phosphonovaleric acid (AP5; 13 mg kg^{-1} body weight). Blood samples were collected at 15 min intervals for 8 h. AP5 injections were administered at 4 h and 6 h and are indicated by the arrows. Frequency of LH pulses was not altered ($P > 0\ .1$) by AP5 injections. There was, however, a tendency ($P < 0\ .09$) for LH pulses of greater amplitude after AP-5 treatment.

Experiments during which gonadotrophin secretion was assessed in pigs treated with ExAA antagonists yielded conflicting data. Equivocal results could arise because ExAA neurotransmission may have dichotomous effects on LH secretion. As mentioned earlier, exogenously administered ExAA can both stimulate and inhibit LH secretion. In addition, exogenously administered ExAA and ExAA antagonists may act at different areas of the brain owing to the ability or inability to cross the blood–brain barrier.

Popwell *et al.* (1996) reported that intramuscular administration of ketamine hydrochloride (20 mg kg^{-1} body weight) decreased concentrations of LH in serum of ovariectomized gilts, a finding consistent with the notion that ExAA are involved in modulating gonadotrophin secretion. In contrast, the frequency of LH pulses was unaltered in a recently conducted experiment during which ovariectomized gilts received intravenous injections of D,L-2-amino-5-phosphonovaleric acid (AP5), a competitive NMDA receptor antagonist. There was, however, a tendency for greater

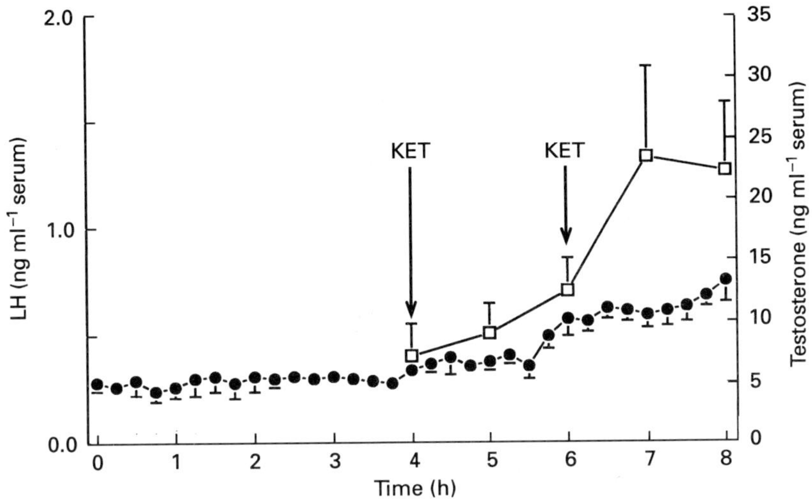

Fig. 6. Serum concentrations of LH (●) and testosterone (□) in immature boars receiving intramuscular injections of the noncompetitive *N*-methyl-D-aspartate receptor antagonist ketamine hydrochloride (KET; 19.9 mg kg^{-1} body weight). Blood samples were collected at 15 min intervals for 8 h and KET was administered at 4 h and 6 h. All samples were analysed for concentrations of LH, and testosterone concentrations were determined in samples collected each hour beginning at 4 h. Values are means±standard errors ($n = 4$). Administration of KET increased LH ($P < 0.01$) and testosterone ($P < 0.07$) secretion.

amplitude LH pulses after AP5 treatment (Fig. 5; M. J. Estienne, J. M. Harter-Dennis, and C. R. Barb, unpublished). The dose of AP5 used (13 mg kg^{-1} body weight) was equimolar to the dose of NMA used previously to stimulate secretion of LH in gilts (Estienne *et al.*, 1995; Hurlock *et al.*, 1995).

We conducted an experiment in which fasted boars, weighing approximately 125 kg each, were treated intramuscularly with ketamine hydrochloride (19.9 mg kg^{-1} body weight) or saline (Estienne *et al.*, 1996c). Fasting increased serum concentrations of growth hormone. However, circulating growth hormone concentrations were similar for boars treated with ketamine hydrochloride and saline, suggesting that endogenous ExAA acting via the NMDA receptor are not involved in the physiological control of growth hormone secretion. Interestingly, ketamine hydrochloride increased circulating concentrations of both LH and testosterone (Fig. 6), a finding consonant with the theory that ExAA inhibit LH release in pigs.

Conclusion

Evidence has been presented that supports the concept that EOP and ExAA affect secretion of hypothalamic hormones and hence function of the anterior pituitary gland in pigs. In general, EOP inhibit LH secretion and this effect involves the central catecholaminergic system. A definitive role for EOP in the modulation of FSH release remains to be determined. EOP stimulate secretion of growth hormone and, depending on the animal model studied, EOP exert either a stimulatory or an inhibitory influence on prolactin secretion. ExAA, acting via NMDA receptors, stimulate secretion of LH, FSH, growth hormone and prolactin in appropriate animal models. However, in certain situations an inhibitory effect of ExAA on LH secretion has been demonstrated. The modulation of growth hormone and prolactin secretion by ExAA involves EOP. Research investigating the role of non-NMDA ExAA receptors in the control of adenohypophysial hormone secretion and the function of ExAA in the physiological control of swine reproduction warrant scrutiny.

References

Armstrong JD, Esbenshade KL, Coffey MT, Heimer E, Campbell R, Mowles T and Felix A (1990) Opioid control of growth hormone in the suckled sow is primarily mediated through growth hormone releasing factor *Domestic Animal Endocrinology* **7** 191–198

Barb CR, Kraeling RR, Rampacek GB and Whisnant CS (1986a) Influence of stage of the estrous cycle on endogenous opioid modulation of luteinizing hormone, prolactin and cortisol secretion in the gilt *Biology of Reproduction* **35** 1162–1167

Barb CR, Kraeling RR, Rampacek GB and Whisnant CS (1986b) Opioid inhibition of luteinizing hormone secretion in the postpartum lactating sow *Biology of Reproduction* **35** 368–371

Barb CR, Kraeling RR, Rampacek GB and Leshin LS (1987) Opioid modulation of follicle stimulating hormone (FSH) and prolactin (PRL) secretion in the postpartum sow. In *Regulation of Ovarian and Testicular Function* pp 647–652 Ed. VB Mahesh, DS Dhindsa, E Anderson and SP Kalra. Plenum Publishing Corporation, New York

Barb CR, Rampacek GB, Kraeling RR, Estienne MJ, Taras E, Estienne CE and Whisnant CS (1988) Absence of brain opioid peptide modulation of luteinizing hormone secretion in the prepubertal gilt *Biology of Reproduction* **39** 603–609

Barb CR, Barrett JB, Wright JT, Kraeling RR and Rampacek GB (1990) Opioid modulation of LH secretion by pig pituitary cells *in vitro*. *Journal of Reproduction and Fertility* **90** 213–219

Barb CR, Estienne MJ, Kraeling RR, Marple DN, Rampacek GB, Rahe CH and Sartin JL (1991a) Endocrine changes in sows exposed to elevated ambient temperature during lactation *Domestic Animal Endocrinology* **8** 117–127

Barb CR, Kraeling RR and Rampacek GB (1991b) Opioid modulation of gonadotropin and prolactin secretion in domestic farm animals *Domestic Animal Endocrinology* **8** 15–27

Barb CR, Kraeling RR and Rampacek GB (1992a) Opioid modulation of FSH, growth hormone and prolactin secretion in the prepuberal gilt *Journal of Endocrinology* **133** 13–19

Barb CR, Derochers GM, Johnson B, Utley RV, Chang WJ, Rampacek GB and Kraeling RR (1992b) *N*-methyl-D,L-aspartate stimulates growth hormone and prolactin but inhibits luteinizing hormone secretion in the pig *Domestic Animal Endocrinology* **9** 225–232

Barb CR, Barrett JB, Rampacek GB and Kraeling RR (1993) *N*-methyl-D,L-aspartate modulation of luteinizing hormone and growth hormone secretion from pig pituitary cells in culture *Life Sciences* **53** 1157–1164

Barb CR, Chang WJ, Leshin LS, Rampacek GB and Kraeling RR (1994) Opioid modulation of gonadotropin releasing hormone release from the hypothalamic preoptic area in the pig *Domestic Animal Endocrinology* **11** 375–382

Barb CR, Campbell RM, Armstrong JD and Cox NM (1996) Aspartate and glutamate modulation of growth hormone secretion in the pig: possible site of action *Domestic Animal Endocrinology* **13** 81–90

Bevers MM, Willemse AH and Kruip AM (1983) The effect of bromocriptine on luteinizing hormone levels in the lactating sow: evidence for a suppressive action by prolactin and the suckling stimulus *Acta Endocrinologica* **104** 261–265

Broughton DS, Estienne MJ, Harter-Dennis JM and Barb CR (1996) Concentrations of growth hormone (GH) and LH in serum of boars administered *N*-methyl-D,L-aspartate (NMA) *Journal of Animal Science* **74** (Supplement 1) 91 (Abstract)

Chang WJ, Barb CR, Kraeling RR, Rampacek GB and Leshin LS (1993a) Involvement of the central noradrenergic system in opioid modulation of luteinizing hormone and prolactin secretion in the pig *Biology of Reproduction* **49** 176–180

Chang WJ, Barb CR, Kraeling RR, Rampacek GB and Asanovich KM (1993b) *N*-methyl-D,L-aspartate modulation of pituitary hormone secretion in the pig: role of opioid peptides *Domestic Animal Endocrinology* **10** 305–313

Chang WJ, Barb CR, Kraeling RR, Rampacek GB and Wright JT (1995) Biogenic amines in the hypothalamus of rats after diethyldithiocarbamate or AIMAX treatment, an alternative for norepinephrine depletion *Journal of Animal Science* **73** 1147–1151

De Rensis F, Cosgrove JR and Foxcroft GR (1993) Luteinizing hormone and prolactin responses to naloxone vary with stage of lactation in the sow *Biology of Reproduction* **48** 970–976

Dubreuil P, Pelletier G, Petitclerc D, Lapierre H, Gaudreau P and Brazeau P (1989) Effects of active immunization against somatostatin on serum growth hormone concentration in growing pigs: influence of fasting and repetitive somatocrinin injections *Endocrinology* **125** 1378–1384

Dubreuil P, Couture Y, Pelletier G, Petitclerc D, Delorme L, Lapierre H, Gaudreau P, Morisset J and Brazeau P (1990) Effect of long-term administration of porcine growth hormone-releasing factor and(or) thyrotropin-releasing factor on growth hormone, prolactin and thyroxine concentrations in growing pigs *Journal of Animal Science* **68** 95–107

Esbenshade KL, Ziecik AJ and Britt JH (1990) Regulation and action of gonadotrophins in pigs *Journal of Reproduction and Fertility Supplement* **40** 19–32

Estienne MJ, Kesner JS, Barb CR, Kraeling RR and Rampacek GB (1988) On the site of action of naloxone-stimulated cortisol secretion in gilts *Life Sciences* **43** 161–166

Estienne MJ, Kesner JS, Barb CR, Kraeling RR, Rampacek GB and Estienne CE (1990) Gonadotropin and prolactin secretion following intraventricular administration of morphine in gilts *Proceedings of the Society for Experimental Biology and Medicine* **193** 92–97

Estienne MJ, Harter-Dennis JM, Barb CR and Hartsock TG (1995) Luteinizing hormone and growth hormone concentrations in serum of prepubertal gilts treated with *N*-methyl-D,L-aspartate *Domestic Animal Endocrinology* **12** 207–213

Estienne MJ, Harter-Dennis JM and Barb CR (1996a) Endocrine responses, growth performance, and carcass quality in domestic animals treated with excitatory amino acids *Proceedings 1996 Maryland Nutrition Conference for Feed Manufacturers* pp 122–131 Maryland Nutrition Council, College Park, Maryland

Estienne MJ, Harter-Dennis JM, Barb CR, Hartsock TG, Campbell RM and Armstrong JD (1996b) *N*-methyl-D,L-aspartate-induced growth hormone secretion in barrows: possible mechanisms of action *Journal of Animal Science* **74** 597–602

Estienne MJ, Harter-Dennis JM and Barb CR (1996c) Modulation of growth hormone (GH) secretion by excitatory amino acids (ExAA) in boars *Journal of Animal Science* **74** (Supplement 1) 229 (Abstract)

Hurlock WF, Estienne MJ, Harter-Dennis JM and Barb CR (1995) LH concentrations in serum of gilts treated with *N*-methyl-D,L-aspartate (NMA) during the estrous cycle or following ovariectomy (OVX) *Journal of Animal Science* **73** (Supplement 1) 212 (Abstract)

Jayes FCL, Britt JH and Esbenshade KL (1997) Role of gonadotropin-releasing hormone pulse frequency in differential regulation of gonadotropins in the gilt *Biology of Reproduction* **56** 1012–1019

Kesner JS, Kraeling RR, Rampacek GB and Johnson B (1987) Absence of an estradiol-induced surge of luteinizing hormone in pigs receiving unvarying pulsatile gonadotropin-releasing hormone stimulation *Endocrinology* **121** 1862–1869

Kineman RD, Leshin LS, Crim JW, Rampacek GB and Kraeling RR (1988) Localization of luteinizing hormone-releasing hormone in the forebrain of the pig *Biology of Reproduction* **39** 665–672

Kineman RD, Kraeling RR, Crim JW, Leshin LS, Barb CR and Rampacek GB (1989) Localization of proopiomelanocortin (POMC) immunoreactive neurons in the forebrain of the pig *Biology of Reproduction* **40** 1119–1126

Klindt J, Ford JJ, Berardinelli JG and Anderson LL (1983) Growth hormone secretion after hypophysial stalk transection in pigs *Proceedings of the Society for Experimental Biology and Medicine* **172** 508–513

Kraeling RR, Rampacek GB, Cox NM and Kiser TE (1982) Prolactin and luteinizing hormone secretion after bromocryptine (CB-154) treatment in lactating sows and ovariectomized gilts *Journal of Animal Science* **54** 1212–1220

Kraeling RR, Kesner JS, Estienne, MJ, Estienne CE, Barb CR and Rampacek GB (1990) Follicle growth in hypophysial stalk-transected pigs given pulsatile GnRH and pregnant mares' serum gonadotropin *Domestic Animal Endocrinology* **7** 395–402

Kraeling RR, Barb CR, Leshin LS and Rampacek GB (1992) Central nervous system peptide and amino acid modulation of luteinizing hormone and prolactin secretion in the pig *Journal of Physiology and Pharmacology* **43** (Supplement 1) 79–103

Kraeling RR, Estienne MJ, Barb CR and Rampacek GB (1994) Prolactin secretion after bromocryptine in the hypophysial stalk-transected gilt *Life Sciences* **54** 1681–1685

Leshin LS, Kraeling RR, Barb CR and Rampacek GB (1992) Associated luteinizing hormone-releasing hormone and luteinizing hormone secretion in ovariectomized gilts *Domestic Animal Endocrinology* **9** 77–88

Leshin LS, Barb CR, Kiser TE, Rampacek GB and Kraeling RR (1994) Growth hormone-releasing hormone and somatostatin neurons within the porcine and bovine hypothalamus *Neuroendocrinology* **59** 251–264

Leshin LS, Kraeling RR, Kineman RD, Barb CR and Rampacek GB (1996) Immunocytochemical distribution of catecholamine-synthesizing neurons in the hypothalamus and pituitary gland of pigs: tyrosine hydroxylase and dopamine-β-hydroxylase *Journal of Comparative Neurology* **364** 151–168

Lutz JB, Rampacek GB and Kraeling RR (1985) Induction of ovulation in the prepuberal gilt by pulsatile administration of gonadotropin releasing hormone *Domestic Animal Endocrinology* **2** 61–65

Mattioli M, Conte F, Galeati G and Seren E (1986) Effect of naloxone on plasma concentrations of prolactin and LH in lactating sows *Journal of Reproduction and Fertility* **76** 167–173

Parvizi N and Ellendorff F (1982) Further evidence on dual effects of norepinephrine on LH secretion *Neuroendocrinology* **35** 48–55

Patchev V, Parvizi N and Ellendorff F (1987) Naloxone effects on luteinizing hormone secretion in immature male pigs: an androgen independent process *Neuroscience* **22** (Supplement 1) 5156 (Abstract)

Petralia RS and Wenthold RJ (1996) Types of excitatory amino acid receptors and their localization in the nervous system and hypothalamus. In *Excitatory Amino Acids: Their Role in Neuroendocrine Function* pp 55–101 Ed. DW Brann and VB Mahesh. CRC Press, Boca Raton, Florida

Popwell JM, Estienne MJ, Kraeling RR, Barb CR, Whitley NC, Utley RV and Rampacek GB (1996) The role of excitatory amino acids in pulsatile secretion of luteinizing hormone in gilts and barrows *Journal of Animal Science* **74** 1067–1073

Price MT, Olney JW, Lowry OH and Buchsbaum S (1981) Uptake of exogenous glutamate and aspartate by circumventricular organs but not other regions of brain *Journal of Neurochemistry* **36** 1774–1780

Sesti LAC and Britt JH (1992) Elicitation of release of luteinizing hormone by *N*-methyl-D,L-aspartic acid during three paradigms of suppressed secretion of luteinizing hormone in the female pig *Domestic Animal Endocrinology* **9** 105–114

Sesti LAC and Britt JH (1993) Agonist-induced release of gonadotropin-releasing hormone, luteinizing hormone, and follicle-stimulating hormone and their associations with basal secretion of luteinizing hormone and follicle-stimulating hormone throughout lactation in sows *Biology of Reproduction* **49** 332–339

Sesti LAC and Britt JH (1994) Secretion of gonadotropins and estimated releasable pools of gonadotropin-releasing hormone and gonadotropins during establishment of suckling-induced inhibition of gonadotropin secretion in the sow *Biology of Reproduction* **50** 1078–1086

Trudeau VL, Meijer JC, van de Wiel DFM, Bevers MM and Erkens JHF (1988) Effects of morphine and naloxone on plasma levels of LH, FSH, prolactin and growth hormone in the immature male pig *Journal of Endocrinology* **119** 501–508

Trudeau V, Pharazyn A, Beltranena E and Aherne FX (1989) Naloxone elevates plasma follicle stimulating hormone but not luteinizing hormone levels in the immature male pig *Canadian Journal of Animal Science* **69** 1095–1098

Journal of Reproduction and Fertility Supplement **52**, 19–29

Immune–endocrine interactions affecting luteal function in pigs

W. Wuttke, L. Pitzel, I. Knoke, K. Theiling and H. Jarry

Division of Clinical and Experimental Endocrinology, Department of Obstetrics and Gynecology, University of Göttingen, Germany

The formation, normal function and destruction of corpora lutea are essential features of normal reproduction. Although the formation of corpora lutea from follicles is largely dependent on pituitary gonadotrophins, the process of luteolysis is locally regulated and poorly understood. The corpus luteum consists of several steroidogenic and non-steroidogenic cell types that interact with each other in a paracrine manner. Under cell culture conditions, large luteal cells that stem from follicular granulosa cells can be identified easily under the microscope and collected individually for single cell RT-PCR. As each of the 120 large luteal cells express the gene encoding 3β-hydroxysteroid dehydrogenase, it appears that all large luteal cells are steroidogenic. Large luteal cells also express the oestrogen receptor gene and as they are known to produce oestradiol, it can be concluded that the steroid acts in an auto- or intracrine manner in large luteal cells. Since we showed previously that oestradiol stimulates progesterone release under *in vitro* and *in vivo* conditions, it can be concluded that the steroid is an important intraluteally acting luteotrophic signal. At the time of luteal regression, macrophages invade the corpora lutea and their cytokine products, particularly tumour necrosis factor α (TNFα), appear to be involved in reduced steroid secretion. Indeed, TNFα inhibits production of progesterone and oestradiol from cultivated luteal cells. In sows, oestradiol is a strong luteotrophic factor and the production of oestradiol and of its receptor is downregulated by TNFα. Thereby, TNFα not only exerts direct luteolytic effects but also prevents the luteotrophic effects of oestradiol. Hence, it has an anti-luteotrophic action. In most species, functional luteolysis is accompanied by morphological regression of the corpus luteum. This structural luteolysis also appears to involve TNFα, as we have shown in pigs that expression of TNFα gene is high during luteolysis. Furthermore, TNFα stimulates programmed cell death (apoptosis) in luteal cells kept under culture conditions.

Introduction

It is well known that the normal function of the corpus luteum is largely dependent on pulsatile LH secretion. However, there is evidence that paracrine regulatory processes are also essential for normal luteal function (Adashi *et al.*, 1990). For example, it has been demonstrated that oxytocin, a peptide produced in steroidogenic large luteal cells, is an important luteotrophic factor for the normal function of the corpus luteum (Jarry *et al.*, 1990); however, towards the end of its lifespan the corpus luteum becomes luteolytic (Wuttke *et al.*, 1995). Little is known about the process of luteolysis, which occurs spontaneously even though normal pituitary gonadotrophin support is maintained. In pigs, the presence of the uterine endometrium, which secretes the luteolytic substance prostaglandin $F_{2\alpha}$ ($PGF_{2\alpha}$), is essential for luteolysis. However, there is also evidence that $PGF_{2\alpha}$ is produced not only by the endometrium but also by the cells of the corpus luteum (Patek and Watson, 1976). When $PGF_{2\alpha}$ is administered directly into the corpus luteum, it increases the secretion of oestradiol and progesterone (Jarry *et al.*, 1990). However, under monolayer cell culture conditions, the ecosanoid has a direct inhibitory effect on progesterone secretion, whereas oestradiol is stimulated (Pitzel *et al.*, 1993a,b). In addition, it was shown that oestradiol stimulates progesterone

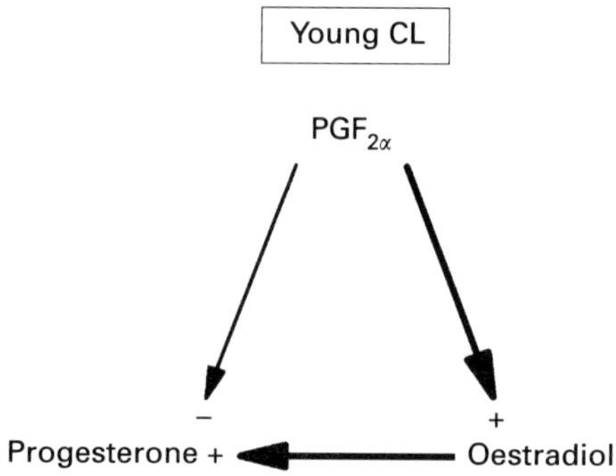

Fig. 1. Prostaglandin $F_{2\alpha}$ has a direct stimulatory effect on oestradiol production by luteal cells but indirectly inhibits progesterone secretion. In the fully functioning corpus luteum, the $PGF_{2\alpha}$-stimulated oestradiol overrides the direct inhibitory effects of the eicosanoid and this is indicted by the thickness of the arrows. Hence, $PGF_{2\alpha}$ stimulates progesterone secretion.

secretion. On the basis of these results, we hypothesized that in fully functioning corpora lutea, $PGF_{2\alpha}$ stimulates so much oestradiol that the direct inhibitory effect on progesterone secretion is overcome by oestradiol such that $PGF_{2\alpha}$ has a direct oestradiol-mediated stimulatory effect on progesterone secretion (see Fig. 1). Hence in non-regressing corpora lutea , $PGF_{2\alpha}$ has a luteotrophic function. The mechanisms by which $PGF_{2\alpha}$ becomes luteolytic are largely unknown and the evidence that cells stemming from the white blood cell line, particularly macrophages, are important factors for the induction of luteolysis will be reviewed.

The Luteotrophic Oestrogenic Mechanism

We have recently shown that oestradiol stimulates progesterone release from pig luteal cells under *in vitro* culture conditions (Pitzel *et al.*, 1993a,b). Conley and Ford (1989) and Pusateri *et al.* (1996) have demonstrated that oestradiol lengthens the luteal phase of gilts *in vivo*, which also points to a luteotrophic or anti-luteolytic effect of the steroid. Using a microdialysis system that can be implanted in corpora lutea (CL) and that functions like an artificial capillary with exteriorized inlets and outlets (Jarry *et al.*, 1985), we demonstrated that an intraluteal application of oestradiol stimulates progesterone release in freely moving sows (Jarry *et al.*, 1990). Furthermore, we demonstrate now that this luteotrophic effect of oestradiol is most prominent during the early and mid-luteal phases and declines at the time of luteal regression (Fig. 2); at day 8 of the luteal phase oestradiol stimulates progesterone secretion. At the time of luteolysis this effect is no longer seen. The mechanism of action and the luteal cell type in which oestradiol stimulates progesterone secretion remains unknown. Earlier attempts to demonstrate oestradiol receptors in luteal tissue by means of immunocytochemistry or with molecular tools failed (Hild-Petito *et al.*, 1988; Chandrasekher *et al.*, 1994). Therefore, we applied the reverse transcription-polymerase chain reaction (RT-PCR) using primers that identified expression of the oestradiol receptor gene in tissues known to contain such receptors and tried to quantify mRNA encoding the oestradiol receptor in pig luteal tissue at different times of the luteal phase and in corpora lutea of pregnancy. The intensity of the PCR product staining was high in luteal tissue obtained during the early and mid-luteal phase and low in corpora lutea undergoing regression (Fig. 3). Expression of the oestradiol receptor gene remained high in luteal tissue obtained from pregnant gilts. A quantitative evaluation of all PCR

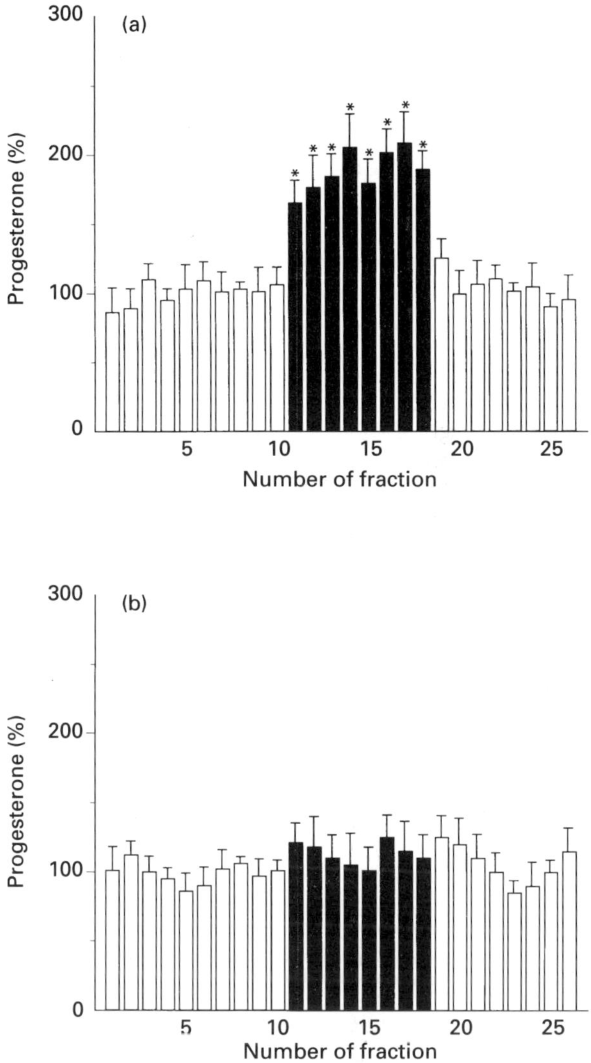

Fig. 2. Effects of an intraluteal application of oestradiol on progesterone secretion of corpora lutea at (a) day 8 and (b) day 12 of the luteal phase. The corpora lutea (n = 8) of three sows were implanted with a microdialysis system (Jarry *et al.*, 1985, 1990) at day 7 of the luteal phase and oestradiol (10^{-6} mmol l^{-1}) dialysed (■) into the corpora lutea at day 8 and 12. Effluent fractions were collected at intervals of 30 min. Note that oestradiol stimulates progesterone secretion at day 8 but not at day 12. Progesterone values in the ten fractions collected before application of oestradiol were set as 100%. Asterisks indicate significantly increased progesterone concentrations in the dialysates ($P < 0.05$).

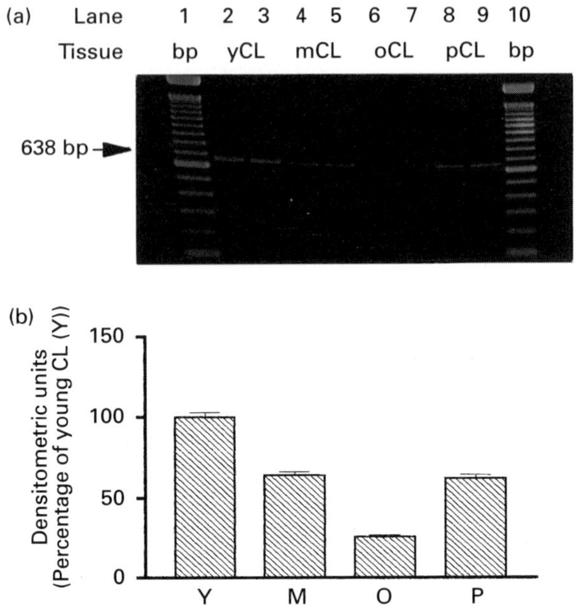

Fig. 3. Expression of the oestradiol receptor gene in pig corpora lutea at various stages of the luteal phase. (a) RT-PCR products using primers for mRNA encoding the oestradiol receptor from pig luteal tissues at various times of the luteal phase. Lanes 1 and 10 show the 100 bp ladder. Lanes 2 and 3 show products of young corpora lutea (yCL, collected at day 3 or 4 of the luteal phase); lanes 4 and 5 show products of middle-aged corpora lutea (mCL, collected at day 7 or 8) and lanes 6 and 7 show products of old, regressing corpora lutea (oCL, collected at day 12 or 13). Lanes 8 and 9 show products of corpora lutea of pregnancy (pCL, collected at day 27 of pregnancy). (b) Densitometric quantification of the PCR products shows progressively lower expression of the oestrogen receptor gene as the corpora lutea age; expression remains high in the corpora lutea of pregnancy. For statistical evaluation, the mean densitometric units obtained for PCR products from young corpora lutea were set as 100% and the other values adjusted accordingly. Asterisks indicate significantly different values ($P < 0.05$) compared with the values obtained in young corpora lutea.

products by densitometry ($n = 10$) confirms these observations (Fig. 3). Hence, at times of the luteal phase when oestradiol can stimulate progesterone secretion (see Fig. 2), expression of the oestradiol receptor gene is high and decreases at the time of luteal regression when oestradiol no longer stimulates progesterone secretion. In corpora lutea of pregnancy, oestradiol remained stimulatory to progesterone secretion (data not shown) and consequently expression of the oestradiol receptor gene is high in these corpora lutea (Fig. 3).

Which Luteal Cell Type Expresses the Oestradiol Receptor?

The corpora lutea of all species studied so far consist of several steroidogenic and non-steroidogenic cell types (Fitz *et al.*, 1982; Alila and Hansel, 1984; Brannian *et al.*, 1993; Schwall *et al.*, 1986; Pitzel *et*

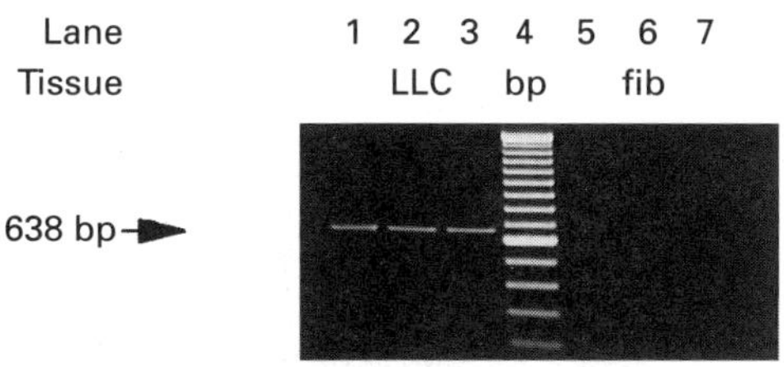

Fig. 4. Expression of oestrogen receptor gene in 3 × 30 large luteal cells (LLC) and in 3 × 30 fibroblasts (fib). Note that the LLC express the oestrogen receptor gene, which is not the case in fibroblasts.

al., 1990). The two steroidogenic cell types stem from the two follicular compartments, i.e. granulosa cells develop to so-called large luteal cells (LLC), whereas the follicular theca cells develop to small luteal cells (SLC). These two types of cells are not the most abundant cells in the corpus luteum; because of the rich capillary blood supply, the non-steroidogenic endothelial cell is the most common cell in luteal tissue (O'Shea *et al.*, 1986). Besides these non-steroidogenic cells, there are fibroblasts and cells stemming from the white blood cell line (Kirsch *et al.*, 1981; Hehnke *et al.*, 1994; Adashi *et al.*, 1990). Under the microscope, because of their size (> 25 μm), the LLC are easiest to distinguish from other cell types maintained under culture conditions. We therefore developed a tool that allowed us to collect individual LLC using a stereotaxic microdrive and a micropipette that enabled us to aspirate individual LLC into the pipette and to transfer them individually into vials for mRNA extraction (Theiling *et al.*, 1996). Using primers for the 3β-hydroxysteroid dehydrogenase (3β-HSD), we submitted each extract of an individual LLC to a RT-PCR procedure (60 cycles). We were able to demonstrate that each of 120 LLC submitted to this procedure expressed the 3β-HSD gene. Hence, we conclude that each large luteal cell is steroidogenic. For comparison, we submitted fibroblasts of luteal origin, which can also be distinguished from other cell types under the microscope, to the same procedure and demonstrated that none of the fibroblasts express the 3β-HSD gene. With this knowledge, we harvested greater numbers of LLC (30 per vial), extracted their RNA and studied whether less abundantly expressed genes, such as the oestradiol receptor gene, are also expressed by this cell type. Figure 4 shows clearly that oestradiol receptor gene expression can be demonstrated in LLC but not in fibroblasts. Hence, the steroidogenic LLC also express the oestradiol receptor gene and it can be concluded that oestradiol, which is produced by LLC, exerts an auto- or intracrine effect in LLC. Unfortunately, other cell types, such as the SLC, the endothelial cells and cells stemming from the white blood cell line, including macrophages, cannot be easily distinguished from each other in freshly prepared and dispersed luteal cells. Therefore, we were unable to perform single cell PCR in any of these cell types. Hence, the question of whether or not the steroidogenic SLC and other non-steroidogenic cells also express the oestradiol receptor gene remains unanswered.

What Causes Downregulation of the Expression of the Oestradiol Receptor Gene at the Time of Luteal Regression?

There is ample evidence that, at the time of luteal regression, macrophages invade the pig corpus luteum (Adashi *et al.*, 1990). The major products of macrophages are tumour necrosis factor α (TNFα) and a variety of interleukins (IL), including IL-1 (Kennedy and Jones, 1991). Both cytokines have been shown to inhibit progesterone and oestradiol release (Pitzel *et al.*, 1993b). An increased production of TNFα at the end of the luteal phase was demonstrated using an *in vivo* microdialysis system (Shaw and Britt, 1995). During inflammatory processes, macrophages can be activated to increase TNFα secretion by bacterial lipopolysaccharides (LPS) (Kennedy and Jones, 1991). Since

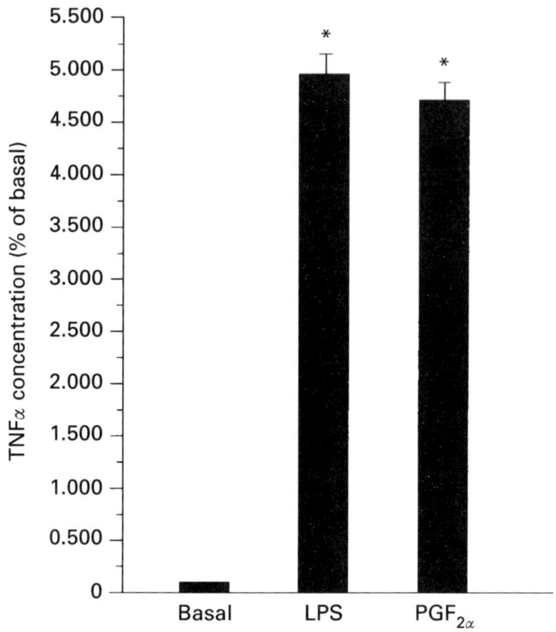

Fig. 5. The growth of a mouse fibrosarcoma cell line (WeHi cell) is inhibited by TNFα in a dose-dependent manner and this can be used as a basis for a highly sensitive *in vitro* bioassay. When macrophages are kept under culture conditions, their TNFα production is low compared with production under the influence of bacterial lipopolysaccharides (LPS 5 μg ml^{-1}) or $PGF_{2\alpha}$ (10^{-7} mol l^{-1}). Hence, $PGF_{2\alpha}$ is a potent stimulator of macrophage TNFα secretion. Basal TNFα production by macrophages is set as 100% and the LPS and $PGF_{2\alpha}$-stimulated values are corrected accordingly. Asterisks indicate significantly stimulated TNFα production ($P < 0.01$).

bacteria are not present during the process of luteolysis, we investigated other compounds that might stimulate macrophage action. One compound known to be involved in luteolysis is $PGF_{2\alpha}$ (Moeljono *et al.*, 1976) and this eicosanoid can stimulate macrophage invasion in the corpus luteum (Hehnke *et al.*, 1994) and also TNF production (Fig. 5). In pigs as well as in ruminants, $PGF_{2\alpha}$ is known to be of endometrial origin (Moeljono *et al.*, 1976) but it can also be produced by luteal cells (Patek and Watson, 1976). The eicosanoid is known to inhibit progesterone release from luteal cells under culture conditions (Pitzel *et al.*, 1993a) but it also causes luteal regression when administered systemically; this effect is exerted only during the late luteal phase (Moeljono *et al.*, 1976). Hence, it appears that the direct luteolytic effects of $PGF_{2\alpha}$ are augmented by its stimulatory effect on macrophage TNFα production. Indeed, increased expression of the gene encoding TNFα can be demonstrated in regressing corpora lutea (Fig. 6). The observation that the expression of the oestradiol receptor gene remains high (Fig. 3) and that of TNFα low (Fig. 6) in the corpus luteum of early pregnancy indicates that the early conceptus signals its presence by inhibiting expression of the gene encoding TNFα and therefore probably the action of TNFα is to reduce the luteotrophic oestrogenic machinery. The conceptuses may achieve this by inhibiting uterine $PGF_{2\alpha}$ secretion (Gross *et al.*, 1988) and thereby preventing the stimulation of macrophage TNFα production. In

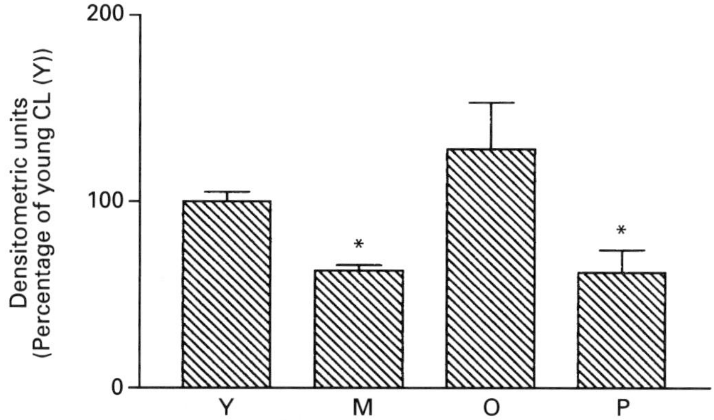

Fig. 6. Expression of the gene encoding TNFα in luteal tissue at various times of the luteal phase and in corpora lutea of early pregnancy. Note that expression of TNFα gene is highest at the time when the corpora lutea undergo luteolysis. The densitometric units indicating that TNFα gene expression obtained for PCR products of young corpora lutea (Y) were set as 100% and the other values were related accordingly. Asterisks indicate significantly ($P < 0.05$) lower expression of the TNFα gene in middle-aged (M) corpora lutea and in corpora lutea of pregnancy (P). O: old regressing.

addition, we have preliminary evidence that another cytokine, namely interferone γ which is of embryonic origin (La Bonnardière, 1993), has powerful luteotrophic effects that may override the potentially deleterious effects of remaining macrophage TNF production. To further our knowledge about whether TNFα shuts off the oestradiol-mediated luteotrophic effects, we studied the effects of TNFα on oestradiol production, aromatase activity and gene expression as well as on oestradiol receptor gene expression under cell culture conditions. TNFα has a profound inhibitory effect on luteal cell oestradiol secretion (Fig. 7); it inhibits aromatase activity as shown by the decreased oestradiol production by luteal cells when androstenedione (the precursor for oestradiol synthesis) is added to the culture media. Finally, Fig. 7 shows that expression of the gene encoding aromatase (the enzyme that converts androstenedione to oestradiol) as well as expression of the gene encoding the oestradiol receptor is profoundly inhibited by TNFα. These data strongly suggest that TNFα not only exerts its luteolytic effects by a direct inhibition of progesterone secretion but that it also has strong anti-luteotrophic effects in that it inhibits the whole oestrogen machinery that is required for normal luteal function.

In summarizing this part of the review, we refer back to the ideas that were shown in Fig. 1 about how $PGF_{2\alpha}$ may stimulate progesterone secretion. It was noted that $PGF_{2\alpha}$ is a strong stimulator of oestradiol secretion (it also stimulates aromatase activity; (Pitzel *et al.*, 1993) and this is essential for $PGF_{2\alpha}$ to be a luteotrophic signal. It has further been shown that TNFα inhibits not only oestradiol secretion but also aromatase activity and the expression of the oestrogen receptor gene. Hence all mechanisms by which $PGF_{2\alpha}$ may indirectly stimulate progesterone secretion are largely abolished and this results in direct inhibition of progesterone secretion. Hence macrophage-derived cytokines, particularly TNFα, are the signals that switch the action of $PGF_{2\alpha}$ from luteotrophic to luteolytic (Fig. 8).

Functional Versus Structural Luteolysis

Thus far only functional aspects of luteal regression have been discussed and it has been shown that TNFα is directly involved in reduced progesterone secretion by a direct inhibitory effect as well as by inhibiting the oestradiol-mediated luteotrophic effects. However, luteolysis involves not only

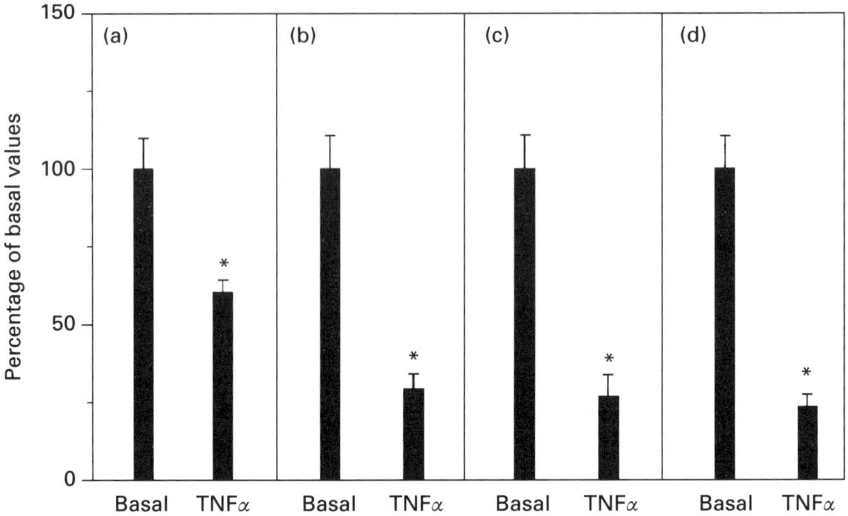

Fig. 7. Effects of TNFα on (a) oestradiol secretion (b) aromatase activity (c) aromatase gene expression and (d) oestrogen receptor gene expression of cultivated luteal cells. Note the strong inhibitory effect of TNFα on all of these parameters. In each case the mean basal values were set as 100% and the influence of TNFα is related to this. Asterisks indicate that $P < 0.01$ versus the 100% values.

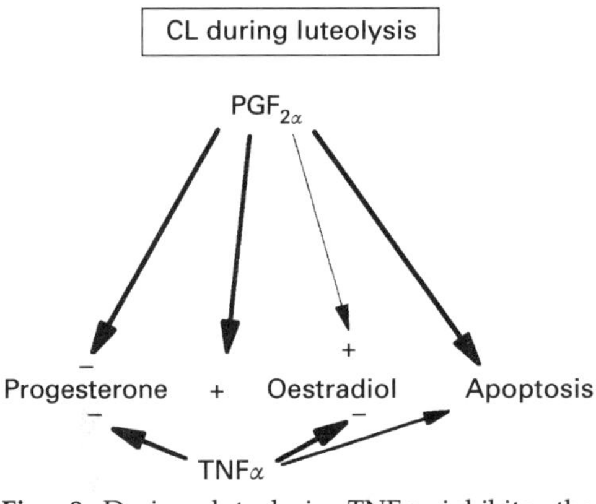

Fig. 8. During luteolysis TNFα inhibits the oestrogenic luteotrophic complex (aromatase activity, expression of the genes encoding aromatase and oestrogen receptor). Thereby the direct inhibitory action of $PGF_{2\alpha}$ and progesterone secretion becomes dominant. Therefore, the action of $PGF_{2\alpha}$ (which in young corpora lutea is luteotrophic, see Figs 1 and 2) becomes luteolytic. In addition, TNFα also inhibits progesterone secretion. Both the eicosanoid and the cytokine induce apoptosis (see Fig. 10) and thereby also cause structural luteolysis.

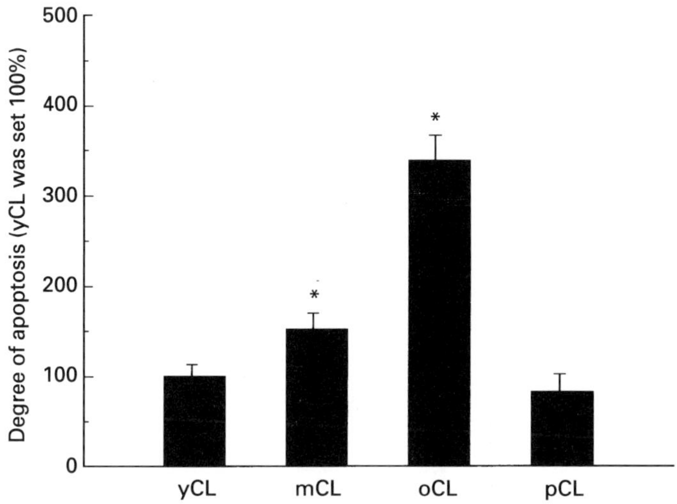

Fig. 9. Quantitative evaluation of apoptosis in corpora lutea obtained during the early (y), midluteal (m) phase and in regressing (o) corpora lutea. The densitometric units of DNA fragments obtained from yCL were set as 100% and the other values related accordingly. Note the much higher degree of apoptosis at the end of the luteal phase, which was not seen in the corpora lutea of pregnancy (pCL). Asterisks indicate that values are significantly different from yCL.

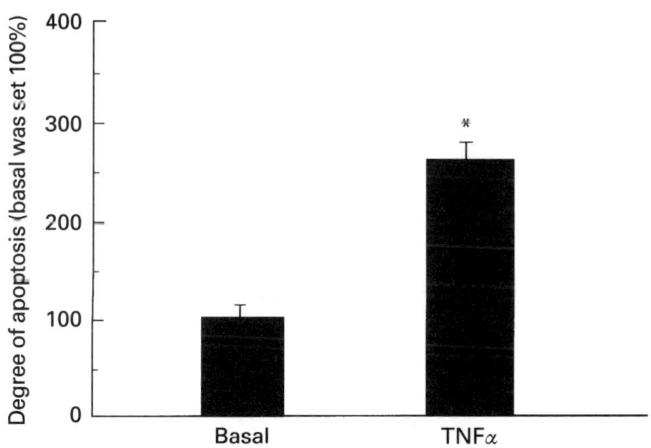

Fig. 10. Luteal cells were kept under culture conditions with or without TNFα (10^{-7} mol l^{-1} for 48 h). The cells were then harvested, the DNA subjected to electrophoresis and the DNA fragments measured densitometrically. Note that TNFα has a strong apoptotic effect compared with the basal degree of apoptosis. The basal level of apoptosis was set as 100% and the TNFα-induced values related accordingly. The asterisk indicates a significant difference ($P < 0.01$) compared with the basal value.

decreased steroid production but also morphological destruction of the corpus luteum, which is called structural luteolysis. As detailed above, TNFα is involved in functional luteolysis and it is possible that this cytokine may also induce structural luteolysis. Therefore, we studied the effects of TNFα on apoptotic processes. Apoptosis (programmed cell death) involves activation of an endonuclease (Juengel *et al.*, 1993) that causes DNA sequestration into 180 bp fragments. These DNA fragments can be electrophoretically separated and quantified by densitometry. Endonuclease activity is indeed higher in regressing corpora lutea than in young or middle aged corpora lutea (Fig. 9). Such an increased degree of apoptosis does not occur in corpora lutea of pregnancy (Fig. 9), again indicating that an embryonic signal prevents structural luteolysis. Under conditions *in vitro*, TNFα induces apoptosis (Fig. 10) which clearly demonstrates that this cytokine is involved not only in functional but also in structural luteolysis.

Conclusion

In summary, we have shown that the production of TNFα by macrophages can be stimulated by $PGF_{2\alpha}$ and that this cytokine has direct inhibitory effects on luteal function and on its structure. Furthermore, TNFα inhibits the expression and function of the oestrogenic luteotrophic machinery by inhibiting aromatase and oestradiol receptor gene expression as well as aromatase activity. Thereby, it inhibits $PGF_{2\alpha}$ stimulated oestradiol production, which allows the eicosanoid to exert its direct luteolytic function and, in addition, TNFα itself inhibits progesterone secretion. Both the macrophage-derived cytokine TNFα and $PGF_{2\alpha}$ also increase the degree of apoptosis, which is essential for morphological luteal regression.

References

Adashi EY, Resnick CE, Packman JN, Hurwitz A and Payne DW (1990) Cytokine-mediated regulation of ovarian function: tumor necrosis factor α inhibits gonadotropin-supported progesterone accumulation by differentiating and luteinized murine granulosa cells *American Journal of Obstetrics and Gynecology* **162** 889–899

Alila HW and Hansel W (1984) Origin of different cell types in the bovine corpus luteum as characterized by specific monoclonal antibodies *Biology of Reproduction* **31** 1015–1025

Brannian JD, Stouffer RL, Shiigi SM and Hoyer PB (1993) Isolation of ovine cell subpopulations by flow cytometry *Biology of Reproduction* **48** 495–502

Chandrasekher YA, Melner MH, Nagalla S and Stouffer RL (1994) Progesterone receptor, but not estradiol receptor, messenger ribonucleic acid is expressed in luteinizing granulosa cells and the corpus luteum in rhesus monkeys *Endocrinology* **135** 307–314

Conley AJ and Ford SP (1989) Direct luteotrophic effect of estradiol-17ß on pig corpora lutea *Journal of Reproduction and Fertility* **87** 125–131

Fitz TA, Mayan MH, Sawyer HR and Niswender GD (1982) Characterization of two steroidogenic cell types in the ovine corpus luteum *Biology of Reproduction* **27** 703–711

Gross TS, Lacroix MC, Bazer FW, Thatcher WW and Harney JP (1988) Prostaglandin secretion by perifused porcine endometrium: further evidence for an endocrine versus exocrine secretion of prostaglandins *Prostaglandins* **35** 327–341

Hehnke KE, Christenson LK, Ford SP and Taylor M (1994) Macrophage infiltration into the porcine corpus luteum during prostaglandin $F_{2\alpha}$-induced luteolysis *Biology of Reproduction* **50** 10–15

Hild-Petito S, Stouffer RL and Brenner RM (1988) Immunocytochemical localization of estradiol and progesterone receptors in the monkey ovary throughout the menstrual cycle *Endocrinology* **123** 2896–2905

Jarry H, Düker E-M and Wuttke W (1985) Adrenal release of catecholamines and metenkephalin before and after stress as measured by a novel *in vivo* dialysis method in the rat *Neuroscience Letters* **60** 273–278

Jarry H, Einspanier A, Kanngießer L, Dietrich M, Pitzel L, Holtz W and Wuttke W (1990) Release and effects of oxytocin on estradiol and progesterone secretion in porcine corpora lutea as measured by an *in vivo* microdialysis system *Endocrinology* **126** 2350–2358

Juengel JL, Garverick A, Johnson AL, Youngquist RS and Smith MF (1993) Apoptosis during luteal regression in cattle *Endocrinology* **132** 249–254

Kennedy RL and Jones TH (1991) Cytokines in endocrinology: their roles in health and in disease *Journal of Endocrinology* **129** 167–178

Kirsch TM, Friedman AC, Vogel RL and Flickinger GL (1981) Macrophages in corpora lutea of mice: characterization and effects on steroid secretion *Biology of Reproduction* **25** 629–638

La Bonnardière C (1993) Nature and possible functions of interferons secreted by the preimplantation pig blastocyst *Journal of Reproduction and Fertility Supplement* **48** 157–170

Moeljono PME, Bazer FW and Thatcher WW (1976) A study of prostaglandin $F_{2\alpha}$ as the luteolysin in swine: I. Effect of prostaglandin $F_{2\alpha}$ in hysterectomized gilts *Prostaglandins* **11** 737–743

O'Shea JD, Rodgers RJ and Wright PJ (1986) Cellular composition of the sheep corpus luteum in the mid- and late luteal phases of the oestrous cycle *Journal of Reproduction and Fertility* **76** 685–691

Patek CE and Watson J (1976) Prostaglandin F and progesterone secretion by porcine endometrium and corpus luteum *in vitro. Prostaglandins* **12** 97–111

Pitzel L, Jarry H and Wuttke W (1990) Effects of oxytocin on *in vitro* steroid release of midstage small and large porcine luteal cells *Endocrinology* **126** 2343–2349

Pitzel L, Jarry H and Wuttke W (1993a) Different steroidogenic response of young and aged porcine small and large luteal cells to prostaglandin $F_{2\alpha}$, oxytocin, and estradiol *Experimental and Clinical Endocrinology & Diabetes* **101** 255–261

Pitzel L, Jarry H and Wuttke W (1993b) Effects and interactions of $PGF_{2\alpha}$, oxytocin and cytokines on steroidogenesis of porcine luteal cells *Endocrinology* **132** 751–756

Pusateri AE, Smith JM, Smith II JW, Thomford PJ and Diekman MA (1996) Maternal recognition of pregnancy in swine. I. Minimal requirement for exogenous estradiol-17ß to induce either short or long pseudopregnancy in cycling gilts *Biology of Reproduction* **55** 582–589

Schwall RH, Sawyer HR and Niswender GD (1986) Differential regulation by LH and prostaglandins of steroidogenesis in small and large luteal cells of the ewe *Journal of Reproduction and Fertility* **76** 821–829

Shaw DW and Britt JH (1995) Concentrations of tumor necrosis factor alpha and progesterone within the bovine corpus luteum sampled by continuous-flow microdialysis during luteolysis *in vivo. Biology of Reproduction* **53** 847–854

Theiling K, Knoke I, Jarry H and Wuttke W (1996) Demonstration of estrogen and progesterone receptor gene expression in porcine luteal cells. In *Program & Abstracts* Vol. II, *10th International Congress of Endocrinology (ICE 96)*, P3–330, p. 837

Wuttke W, Jarry H, Knoke I, Pitzel L and Spiess S (1995) Luteotrophic and luteolytic effects of oxytocin in the porcine corpus luteum. In *Oxytocin, Cellular and Molecular Approaches in Medicine and Research* pp 495–506 Eds R Ivell and A Russell. Plenum Press, New York

Journal of Reproduction and Fertility Supplement **52**, 31–46

Control of follicular development and ovulation rate in pigs

Nancy M. Cox

Department of Animal and Dairy Sciences, Mississippi State University, Mississippi State, MS 39762, USA

There is considerable evidence that nutritional and metabolic control of follicular growth is mediated by metabolic hormones and growth factors, particularly with processes mediated by insulin-like growth factor I (IGF-I) and its binding proteins (IGFBPs). From knowledge that hormones and growth factors which can be affected by diet also positively affect ovarian function, the concept has emerged that metabolic modifiers of gonadotrophin action, rather than gonadotrophins themselves, could affect follicle development. While ovulation rate can be enhanced under certain conditions in cyclic gilts, assessing influences of metabolic modifiers on the post-lactational sow is confounded by variability in the return to oestrus after weaning. In a series of studies involving insulin administration between weaning and oestrus, successive experiments produced different results, but several measures of reproductive performance were enhanced. Administration of somatotrophin (ST) has also been shown to increase follicular development in both gilts and sows. Both insulin and ST increase IGF-I production by pig ovarian follicles, and insulin is more effective than IGF-I in reducing atresia and increasing progesterone in cultured pig follicles. Whether increases in litter size are achieved after an increase in ovulation rate involves many factors, including the quality of ova and whether the increase in ovulation rate exceeds the uterine capacity to maintain pregnancy. Given the variation in genetics and management practices, development of treatments to enhance follicle quality leading to maximal litter size is challenging.

Introduction

Although it has long been known that exogenous gonadotrophins increase follicular development and ovulation rate, the number of ovulations is variable and may have negative effects on embryo survival. Nutritional and hormonal manipulation may enhance follicular development, both by enhancing gonadotrophin secretion directly and by enhancing the ovarian response in the face of unvarying gonadotrophin concentrations. Given the knowledge that hormones and growth factors that can be affected by diet also positively affect ovarian function, the concept that metabolic modifiers, rather than gonadotrophins themselves, could be applied to affect follicle development has emerged. This review will concentrate mainly on metabolic interfaces between nutrition and reproduction rather than nutritional manipulations themselves. Emphasis will be on follicle development rather than embryo survival; effects of metabolic manipulations on embryo survival are discussed by Foxcroft (this volume).

When consideration is given to increasing ovulation rate, there are practical limitations. First, while ovulation rate and litter size are positively correlated up to approximately 18 ovulations, ovulations beyond that point result in little or no increases in litter size (Wu *et al.*, 1987). In mature sows and gilts on full feed, ovulation rates are usually maximal, and manipulation of follicular

Journal Series J-9182 of the Mississippi Agricultural and Forestry Experiment Station.

development to increase ovulation rate will not be useful. However, in animals with lower ovulation rates, such as gilts on the first or second oestrous cycle and primiparous sows, augmentation of ovulation rate may be useful.

Current Concepts of Follicle Growth, Atresia and Ovulation Rate

It is well established that the ovulatory population of follicles increases growth between about day 14 and day 16 of the oestrous cycle. Granulosa and theca cell proliferation decreases between day 15 and oestrus, which indicates differentiation (Fricke *et al.*, 1996). Between day 16 and oestrus, there is a loss of 40–50% of the medium-sized follicles due partly to growth of follicles to the next size category but mostly to atresia (Guthrie *et al.*, 1995a). Exposure to FSH during the last stages of granulosa cell mitosis appears to be required so that some follicles can escape atresia (Hsueh *et al.*, 1994). In pigs, FSH patterns were temporally correlated with atresia, decreasing at the time atresia increased (Guthrie *et al.*, 1993). Apoptotic cell death in the ovary has been characterized by several biochemical markers including decreased DNA turnover, decreased oestrogen production, increased progesterone production, a decrease in the number of gonadotrophin receptors and increased production of IGF-binding protein (Hsueh *et al.*, 1994). Thus follicular atresia is considered to be a result of a balance between survival and atretogenic factors. Follicle survival factors include, but are not limited to, epidermal growth factor, nerve growth factor, insulin-like growth factor I (IGF-I), gonadotrophins, activin and oestrogens, while atretogenic factors include testosterone, GnRH and interleukins (Hsueh *et al.*, 1994).

In the pig ovary as in the ovaries of other species, apoptotic cell death is the mechanism by which follicular atresia occurs (Guthrie *et al.*, 1995a). Guthrie *et al.* (1995a) demonstrated that apoptosis could be assessed in gilts by flow cytometric analysis and that these measures were correlated highly with oestradiol production and with morphological assessment of atresia. Atresia was highest in medium-sized (3–6 mm) follicles (87%) on day 5 after withdrawal of Altrenogest (Guthrie *et al.*, 1994), which corresponds to the late stages of preovulatory maturation before the LH surge. In contrast, the percentage of atretic granulosa cells was 17% in medium-sized follicles on days 1 and 3 after Altrenogest withdrawal, which corresponds to earlier stages of preovulatory follicular development (Guthrie *et al.*, 1994). These researchers also demonstrated that during the early luteal phase, there is a wave of follicular growth followed by atresia (Guthrie *et al.*, 1995a). In contrast, during follicular growth in the latter part of the luteal phase between day 7 and day 15, there was no change in the percentage of atretic follicles, indicating that follicular growth in pigs may not be characterized by repeated waves of follicular development (Guthrie and Cooper, 1996). The most severely affected enzyme was p450 aromatase, which was undetectable in atretic follicles (Garrett and Guthrie, 1996). In contrast, 3β-hydroxysteroid dehydrogenase and $P450_{c17}$ lyase were not affected by atresia, which is in agreement with other work which demonstrates that progesterone production continues in atretic follicles (Guthrie *et al.*, 1994). Figure 1 illustrates identification of Ki-67, a nuclear antigen associated with cell proliferation, aromatase and detection of apoptosis in atretic and nonatretic follicles. Since oestrogen is a follicle survival factor that reduces atresia (Hsueh *et al.*, 1994), aromatase expression may be critical for survival of follicles.

Techniques for enhancing follicular development include methods that may either influence negative feedback relationships with gonadotrophins or intrafollicular control mechanisms. Immunization against inhibin (King *et al.*, 1993) and androstenedione (McKinnie *et al.*, 1988) increase ovulation rate but their effects on litter size have not been evaluated. However, immunization against the α-subunit of inhibin prevented the compensatory increase in corpora lutea following unilateral ovariectomy, suggesting a local effect of inhibin (King *et al.*, 1995). There is increasing evidence for intraovarian paracrine regulation by inhibin and related peptides, and further refinement of these relationships may result in more predictable control of ovulatory processes (Findlay *et al.*, 1996). Manipulation of the steroid environment may also influence ovulation rate. When testosterone was administered between day 13 and oestrus, ovulation rate and the number of recovered blastocysts were increased (Cardenas and Pope, 1997). Part of the mechanism may be provision of a substrate for oestradiol production, but the mechanism is not known.

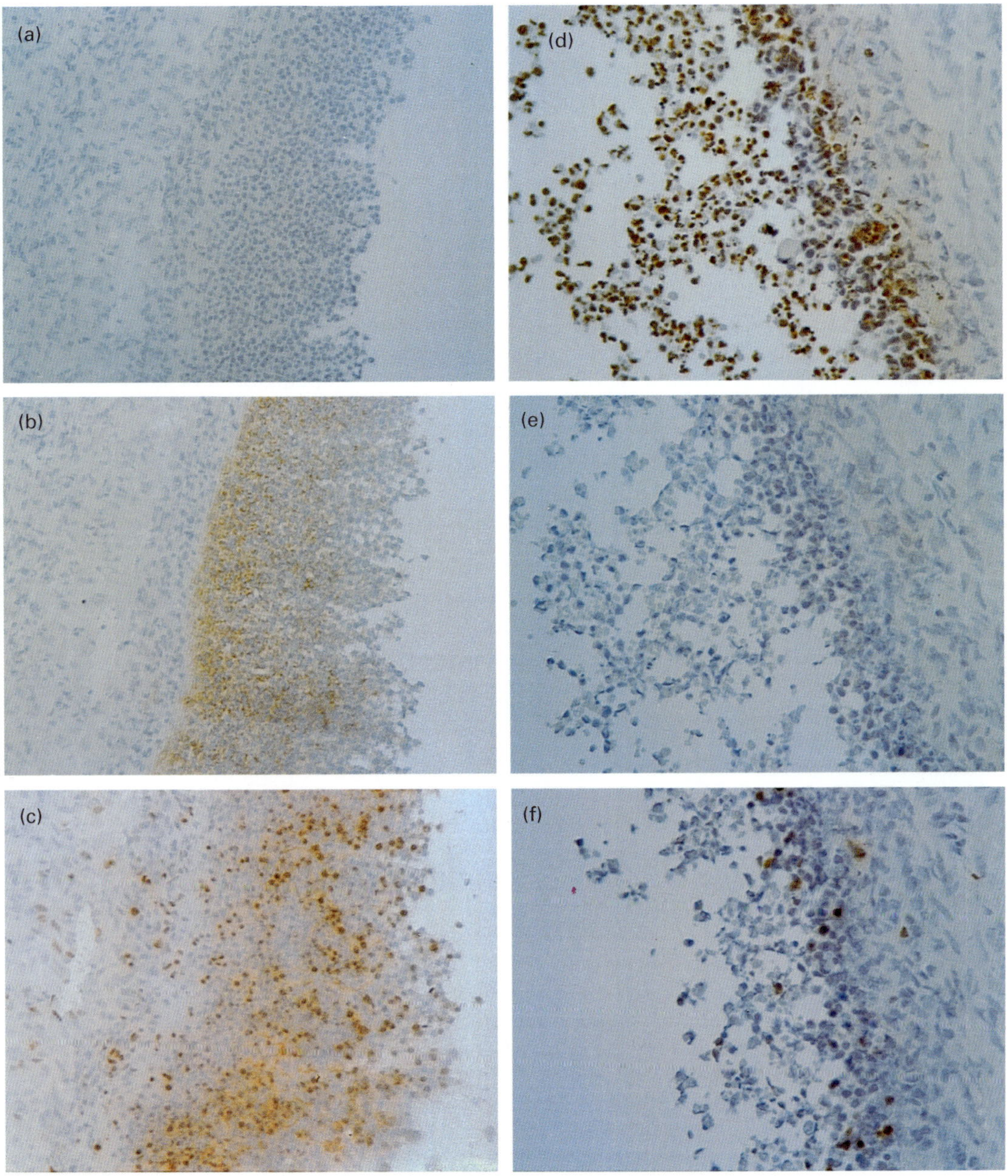

Fig. 1. Photomicrographs of sections from ovarian follicles from non-atretic (a–c) and atretic (d–f) follicles. Sections (a) and (d) are 3′ end-labelled to detect apoptosis; sections (b) and (e) are immunostained for aromatase, and sections (c) and (f) are immunostained for cell proliferation-associated antigen, Ki-67 (W. Garrett and D. Guthrie, unpublished).

Metabolic Influences on Components of the Hypothalamo–Hypophyseal–Ovarian Axis

Although positive associations between level of feed intake and ovulation rate or litter size are well established (reviewed by Einarsson and Rojkittikhun, 1993; Cosgrove and Foxcroft, 1996), the mechanisms by which increased nutritional status affects ovarian function have not been established. In gilts, the principal effect of feed restriction is to compromise the GnRH pulse generator, and realimentation or glucose infusion restores LH pulses (Booth, 1990), while insulin restores follicular growth (Britt *et al.*, 1988). Charlton *et al.* (1993) observed lowered ovarian IGF-I mRNA with restricted feeding, but refeeding did not increase IGF-I mRNA. In feed-restricted lactating sows, glucose infusion did not affect pulsatile LH secretion (Tokach *et al.*, 1992b). These results suggest that metabolic state can affect the LH response to a stimulus such as glucose.

When extra feed was given to cyclic gilts, increases in ovulation rate were accompanied by increases in LH pulse frequency and FSH, as well as insulin (Cox *et al.*, 1987; Flowers *et al.*, 1989). However, in the studies of Cox *et al.* (1987), increases in ovulation rate stimulated by dietary energy or exogenous insulin were not necessarily accompanied by increased LH. In ovariectomized gilts with insulin depleted by diabetes mellitus and without insulin therapy for 4 days, pulsatile LH secretion was not altered (Angell *et al.*, 1996). When ovary-intact diabetic gilts were depleted of insulin on day 12 of the oestrous cycle, LH pulse frequency increased between day 12 and day 18, and LH increased again at the time at which normal gilts had preovulatory LH surges (Cox *et al.*, 1994). In spite of the absence of adverse effects on gonadotrophins, follicle growth decreased and atresia increased (Meurer *et al.*, 1991; Cox *et al.*, 1994). Taken together, these results suggest that the effect of metabolic modifiers of reproduction on ovarian function does not require changes in gonadotrophins.

In considering the applicability of using metabolic modifiers to enhance ovulation rate in gilts, it must be considered that maximal feed intake will produce maximal ovulation rates (Beltranena *et al.*, 1991). In contrast, sows, in which weaning triggers preovulatory development, are more vulnerable to the metabolic influences of the previous lactation. Despite the increased LH pulsing that is stimulated by weaning, there is considerable variability in follicular development and steroid secretion (Foxcroft *et al.*, 1987). The primiparous sow is particularly susceptible to delayed oestrus after weaning or smaller litter sizes, suggesting that when metabolic factors interface with enhanced gonadotrophin secretion at weaning, follicle development can be affected. In support of this concept, Clowes *et al.* (1994) observed greater litter sizes (12.8 versus 10.4) and increased insulin concentrations when remating was delayed until the second post-weaning oestrus. Understanding of metabolic influences in sows is further complicated by a non-linear relationship between interval to oestrus and litter size and perhaps ovulation rate (Vesseur *et al.*, 1994). Thus metabolic manipulations that affect the interval from weaning to oestrus could alter litter size.

Research assessing influences of nutritional manipulations on ovarian function in sows has demonstrated that ovarian function after weaning can be affected by metabolic state during lactation. Tokach *et al.* (1992a) and Koketsu *et al.* (1996) demonstrated that concentrations of insulin and glucose and the number of LH pulses during lactation were greater in sows with normal intervals to oestrus (Fig. 2). Zak *et al.* (1997) observed that restricting feed during weeks 1–3 or week 4 of lactation decreased ovulation rate and increased the weaning to oestrus interval in primiparous sows. Circulating insulin and IGF-I were reduced during feed restriction, and embryo survival was decreased only in sows receiving less feed during the last week of lactation. Feeding starch as an energy source during lactation may also facilitate LH production and ovarian function after weaning (Kemp *et al.*, 1995). Thus dietary manipulation clearly can affect ovarian function after weaning, perhaps by affecting follicles during the lactation phase or before, as well as embryo survival.

Role of Metabolic Hormones on Ovarian Function

The insulin-like growth factor I system

A preponderance of evidence confirms positive associations between IGF-I and follicle function in swine, although most of these studies were performed *in vitro* (reviewed by Hammond *et al.*, 1993; Spicer and Echternkamp, 1995). Although, as indicated above, nutrition modulates systemic concentrations of IGF-I, there is local control of IGF-I as well as other intraovarian compounds, and this explains the marked heterogeneity of follicles in the face of a similar endocrine milieu (Tonetta

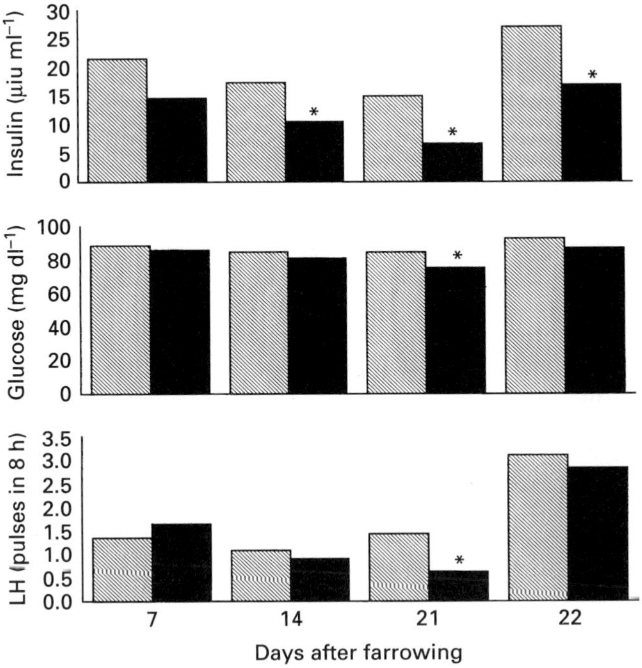

Fig. 2. Patterns of insulin, glucose and LH in sows categorized on the basis of days to oestrus (▨ in oestrus < 7 days after weaning $n = 23$; and ■, in oestrus > 7 days after weaning, $n = 11$). Day 22 represents the first day after weaning. In sows with delayed oestrus, serum insulin was lower on days 14, 21 and 22, and glucose and number of LH pulses were lower on day 21 ($P < 0.05$). SEMs ranged from 1.43 to 2.67, 1.35 to 2.88, and 0.16 to 0.35 for insulin, glucose and LH pulses, respectively. Redrawn from Koketsu *et al.* (1996).

and diZerega, 1990). Synthesis of IGF-I by granulosa cells from pigs has been demonstrated, and production of IGF-I is increased by gonadotrophins and growth hormone *in vitro* (Hammond *et al.*, 1993). Messenger RNA for both IGF-I and -II increased with increasing size of follicles (Yuan *et al.*, 1996). In contrast, Zhou *et al.* (1996) observed that IGF-I and IGF-I receptor mRNA were concentrated in healthy follicles, whereas mRNA encoding IGF-II was found in all follicles regardless of the degree of atresia. This finding suggests a functional paracrine role for IGF-I but does not allow conclusions for IGF-II, although IGF-II affects steroidogenesis (Spicer and Echternkamp, 1995). The IGF-binding proteins modulate effects of IGF-I systemically as well as in the ovary and are in general considered to inhibit follicular development (Hammond *et al.*, 1993).

Recent evidence indicates that IGFBPs, as well as IGF-I, are related to follicular function in swine. Expression of mRNA encoding IGF-I increased as follicles increased in size during the oestrous cycle, whereas expression of the IGFBP-2 gene was greater in smaller follicles, and IGFBP-3 was not detected in follicles using Northern blot analysis (Hammond *et al.*, 1993). Yuan *et al.* (1996) demonstrated that most IGFBP-2 gene expression was in the theca interna cells and was lower in large antral than in small antral follicles in that cell type, but decreased less markedly in granulosa cells. Concentrations of IGFBP-2 in follicular fluid were also inversely related to follicular diameter or oestradiol production (Howard and Ford, 1992) and were positively related to atresia in the early luteal phase as well as during preovulatory maturation (Guthrie *et al.*, 1995b). Despite the association with atresia, mRNA encoding IGFBP-2 was detected in healthy, preovulatory follicles (Zhou *et al.*,

1996). In contrast, IGFBP-3 was not associated with atresia (Guthrie *et al.*, 1995b; Edwards *et al.*, 1996), although addition of IGFBP-3 to cultured rat follicles prevented IGF-I from acting as a survival factor (Chun *et al.*, 1994). Messenger RNA for IGFBP-3 was not detected in the ovary at the follicular phase (Zhou *et al.*, 1996), but recent evidence from our laboratory using a sensitive ribonuclease protection assay in individual follicles suggests that mRNA for IGFBP-3 is present in the walls of preovulatory follicles (Cox and Qiu, 1995). Associations between atresia and IGFBP-4 and -5 have been reported, but these IGFBPs are present at low concentrations in the follicular fluid of pigs (Guthrie *et al.*, 1995b). However, Zhou *et al.* (1996) observed that mRNA for IGFBP-4 was associated with the presence of LH receptor gene expression in granulosa cells, indicating a role that is more complex than mediation of atresia. In addition to alterations in gene expression, IGFBPs in follicular fluid may be altered by post-translational processing, including proteolysis. Recent evidence in swine indicates that proteolytic activity for IGFBP -2, -4 and -5 increased during follicular growth and was decreased during atresia (Besnard *et al.*, 1997). Thus factors that influence activity of proteases for IGFBPs represent another level of control of follicle function.

Insulin-like growth factor I is a strong survival factor for rat follicles, attenuating apoptosis in a similar way to gonadotrophins (Chun *et al.*, 1994). Insulin also acted as a survival factor in that study, but was less potent than IGF-I. Growth hormone, which increased IGF-I, nevertheless did not reduce atresia in the rat early antral or preovulatory follicle (Chun *et al.*, 1994, 1996). In contrast to the rat model, we have obtained evidence that insulin is more potent in reducing apoptosis and increasing progesterone than IGF-I in pig follicles (Purvis *et al.*, 1997). The model we developed was aimed at studying the population of medium follicles (4 mm diameter) representative of days 16–18 of the oestrous cycle. It was considered that this population, although non-atretic when obtained, has a high potential for atresia. We have previously determined that follicles cultured in the presence of FSH (100 ng ml^{-1}) produce oestradiol linearly over time in culture (Edwards *et al.*, 1996). Addition of insulin to FSH reduced apoptosis, whereas IGF-I had no effect (Fig. 3). Insulin (50 ng ml^{-1}) also increased progesterone concentrations, but 5000 ng IGF-I ml^{-1} did not affect progesterone. The mechanism of insulin action does not appear to involve mediation by IGF-I, because intrafollicular IGF-I concentrations were higher in insulin-treated follicles but were similar to those produced by addition of IGF-I. Interestingly, insulin increased IGFBP-2 in medium after culture for 23 h, but not in follicular fluid, whereas IGF-I did not alter IGFBP-2 compared with controls. In cultures of pig preantral follicles that already contained 1% insulin in the medium, IGF-I enhanced growth although the increase was less than with FSH or epidermal growth factor (Flowers and Turner, 1996). There have been few attempts to administer IGF-I *in vivo* to affect follicular function in pigs.

Both insulin and ST increase IGF-I production systemically and in the ovarian follicle, and both affect IGFBP production, although each hormone responds to circulating glucose in opposite ways (Spicer and Echternkamp, 1995). In our laboratory the combined influences of pST and insulin on follicular development were examined by inducing diabetes mellitus in gilts that had been immunized against growth hormone-releasing hormone (provided by Jeffrey D. Armstrong of North Carolina State University) (Howell *et al.*, 1993). The two treatments were compared with control animals in a 2×2 factorial arrangement. Immunoneutralization of GHRH lowered follicular IGF-I concentrations from 245 to 113 ng ml^{-1} but did not affect oestradiol concentrations on day 18 of the oestrous cycle, after half the diabetic gilts were without insulin therapy from day 12 (Fig. 4). In contrast, diabetes mellitus significantly lowered both IGF-I (104 versus 254 ng ml^{-1}) and oestradiol (71 versus 301 ng ml^{-1}), as well as the number of follicles greater than 2 mm (31.5 versus 58.3). Another unique effect of diabetes mellitus, but not GHRH immunoneutralization, was that IGFBP-2 was increased in follicles to 186% of control values. However, follicular IGFBP-3 was reduced to 50% of control values as a result of GHRH immunoneutralization. Therefore, although the functional absence of either ST or insulin decreased IGF-I in follicular fluid, reduction in IGFBP-3 activity may have ameliorated negative effects of lowered IGF-I. In contrast, the combined influence of increased IGFBP-2 and decreased IGF-I in follicles of diabetic gilts may account for decreased follicular function. Interactions among systemic insulin, ST, IGFBPs and IGF-I as well as intrafollicular interactions beween the last two factors may function to control normal follicular responses to alterations in metabolism.

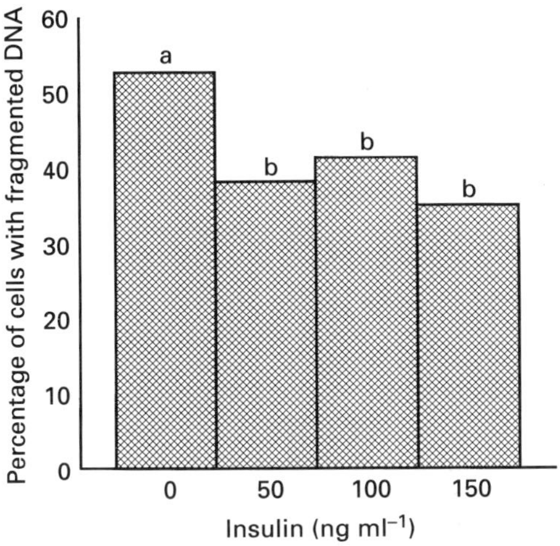

Fig. 3. Percentage of granulosa cells with fragmented DNA identified using cell flow cytometry in 4 mm follicles (six per treatment) from luteal phase ovaries cultured for 23 h with 100 ng ml^{-1} porcine FSH supplemented with insulin. All doses of insulin decreased apoptosis ($P < 0.05$). The SEM was 5.3%. Redrawn from Purvis *et al.* (1997).

Insulin

Follicle development in gilts. We have previously observed that administration of insulin to cyclic gilts during the first pubertal oestrous cycle increased ovulation rate (Cox *et al.*, 1987) and reduced follicle atresia (Matamoros *et al.*, 1990) in gilts at the second postpubertal oestrus. In prepuberal, eCG-treated gilts, we observed an increase of IGF-I in medium-sized follicles of gilts given insulin (Matamoros *et al.*, 1991). In contrast, when insulin was administered to gilts during days 13–17 of the fifth postpubertal oestrous cycle, IGF-I and oestradiol concentrations were decreased and atresia was not affected (Hughey *et al.*, 1993). This result suggests that effects of insulin may be related to age, maturity and (or) related metabolic changes.

We demonstrated that normal ovarian function could be maintained in streptozotocin-induced diabetic pigs given daily insulin injections from 35 days of age until initiation of oestrous cycles, but that withdrawal of insulin therapy reduced follicle diameter and oestradiol production and increased atresia (Meurer *et al.*, 1991; Cox *et al.*, 1994; Edwards *et al.*, 1996). Upon withdrawal of insulin replacement therapy, the incidence of follicular atresia increased within 2 days. Gene expression for IGF-I in follicle walls, as well as IGF-I in follicular fluid, decreased rapidly upon withdrawal of insulin, and IGFBP-2 in follicular fluid increased (Cox and Qiu, 1995; Fig. 5). Neither mRNA encoding IGFBP-3 nor IGFBP-3 in follicular fluid (not shown) was affected by diabetes. Both lowered IGF-I and increased IGFBP-2 probably contributed to the increased atresia and lowered oestradiol.

There is evidence that 4–6 mm follicles are particularly affected by insulin. First, it was the medium sized follicle population that was maintained from day 17 (39.8% of total follicles) to 19 (35.1%) of the cycle by insulin injections beginning on day 15, compared with a reduction from 41.7 to 16.6% in saline-treated animals. A corresponding increase in atretic follicles from 15.5 to 38.2% occurred in control animals, while a reduction in atretic follicles on both days 17 (6.3%) and 19 (10.7%) was observed in insulin-treated animals (Matamoros *et al.*, 1991). This evidence is supported by the *in vitro* data presented in Fig. 3. Further evidence that the 5 mm follicle population is

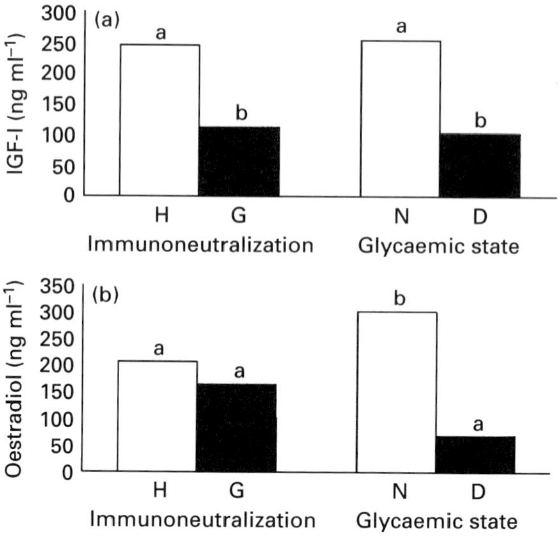

Fig. 4. Influence of immunoneutralization of GHRH on concentrations of (a) IGF-I and (b) oestradiol in the five largest follicles present on day 18 of the oestrous cycle in gilts ($n = 22$). The letters H and G designate gilts immunized against human serum albumin (HSA) and an HSA–GHRH conjugate, respectively, and the letters N and D represent normal and streptozotocin-diabetic gilts, respectively. Insulin replacement therapy was removed from diabetic gilts on day 12 of the oestrous cycle. There were no interactions, and both main effects significantly reduced IGF-I, but only diabetes mellitus reduced oestradiol ($P < 0.05$). SEMs ranged from 15.3 to 25.0 and from 13.3 to 19.1 for IGF-I and oestradiol, respectively. Redrawn from Howell *et al.* (1993).

vulnerable to altered metabolic support was obtained by Edwards *et al.* (1996). In diabetic gilts, the largest follicles present on the ovary were 5 mm in diameter, and oestradiol production *in vitro* was similar to that of follicles from control and diabetic insulin-treated gilts. However, IGF-I was severely reduced in those follicles, indicating that either insulin, IGF-I or a combination thereof is necessary for follicle growth beyond the 5 mm stage. While research has demonstrated that insulin is required for normal follicle function, manipulating follicle development in gilts with insulin is not of practical use, because, as noted above, maximal ovulation rates can be obtained with feeding. Table 1 illustrates the lack of an effect of insulin on ovulation rates and litter size in gilts fed *ad libitum*, for which ovulation rates were maximal, and is in agreement with the observations of Beltranena *et al.* (1991) cited above. Insulin was not given to limit-fed gilts because of the danger of hypoglycaemia.

Effects on reproductive performance in primiparous sows. In sows after weaning, there is potential for manipulating reproduction with metabolic modifiers such as insulin, because ovulation occurs in the face of the recent negative energy balance of lactation. On the basis of initial studies with gilts, in several studies in commercial settings we have administered insulin during the period between weaning and oestrus and have observed inconsistent, but positive effects on reproductive function (Table 2). These effects include increasing farrowing rate in a herd with typically low farrowing rates and increasing litter size in another herd with greater than 90% farrowing rate (Ramirez *et al.*, 1997). These results contrast with those of Kirkwood and Thacker (1991), who did not find a significant effect of exogenous insulin on litter size, using a higher daily dose of insulin (0.75 versus 0.40 iu kg^{-1} day^{-1})

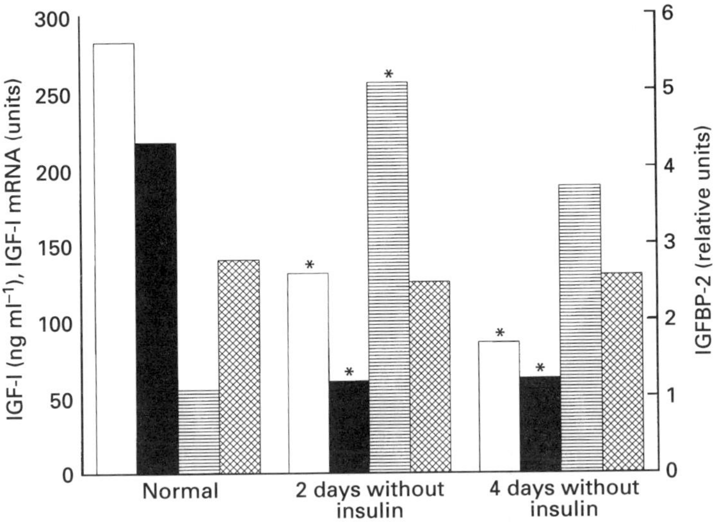

Fig. 5. Influence of removal of insulin therapy from diabetic gilts on (□) IGF-I and (▤) IGFBP-2 in follicular fluid and mRNA for (■) IGF-I and (⊠) IGFBP-2 in follicle walls. Insulin removal for 2 or 4 days decreased IGF-I and its mRNA and increased follicular fluid IGFBP-2 ($P < 0.05$) but did not affect its mRNA. SEMs ranged from 9.8 to 15.6 and 4.9 to 10.4 for IGF-I concentration and mRNA and from 0.3 to 0.5 and 0.9 to 1.5 for IGFBP-2 in follicular fluid and mRNA, respectively. Redrawn from Cox *et al.* (1995).

and mixed parity sows. Their results indicate that insulin did not influence ovulation rate, that uterine capacity was insufficient, or that mature sows already ovulated maximally. We also failed to increase litter size in multiparous sows treated with insulin (Cox *et al.*, 1995). When primiparous and multiparous sows were compared, total litter size was not affected by insulin but was higher, as expected, for multiparous sows compared with primiparous sows (11.8 and 9.8 pigs born in total, SEM = 0.3; $P < 0.001$). Farrowing rates (89.7 ± 3% overall) were not influenced by parity or insulin treatment.

In a study with primiparous sows lactating for 30 days, doses of insulin were higher and given for longer than in other studies (Table 2), and insulin treatment lowered ovulation rate (Rojkittikhun, 1992). This finding may be related to production of a negative metabolic signal by the increased duration of injection, the small sample sizes or to the fact that some of the treated sows gave birth to a smaller number of piglets in their previous litter, indicating a predetermined low ovulation rate.

Since increased feed intake alone between weaning and oestrus might be sufficient to increase insulin, we examined the influence of insulin injections daily for 4 days between weaning and postweaning oestrus in primiparous sows given extra feed between weaning and postweaning oestrus in a 2 × 2 arrangement (Cox *et al.*, 1995). Total litter size in primiparous sows (in oestrus within 10 days after weaning) was affected differently by insulin depending on the level of feed intake between weaning and oestrus (treatment by feed level interaction, $P < 0.02$) and was also positively related to feed intake during lactation ($P < 0.05$). When litter sizes, adjusted for lactation feed intake, were compared with the mean of insulin-treated sows given 3.63 kg feed (11.6 piglets in total), means for the insulin/2.72 kg and saline/3.63 kg treatments (9.6 piglets each) were lower ($P < 0.05$) and the mean for the saline/2.72 kg (10.5 piglets) treatment was similar ($P = 0.10$; overall SEM = 0.7). It should also be noted that despite the larger litter size in the insulin-treated sows fed 3.63 kg, farrowing rate tended ($P < 0.07$) to be lower in this treatment group (74 ± 8%), compared with 86 ± 5, 89 ± 5, and 93 ± 8% for saline/2.72 kg, insulin/2.72 kg and saline/3.63 kg treatment combinations, respectively. Although the greatest litter size was achieved with insulin plus the high level of feed, the tendency for lower farrowing rates suggests that there are negative effects of this combination of insulin and extra feed. The results show that primiparous sows may have a critical need to be in a

Table 1. Effects of exogenous insulin and feed intake on ovulation rate and litter size in gilts

Treatment	Number of gilts	Corpora lutea	Number of fetuses at 60 days
Limit-fed [a]	14	13.2[b]	Not mated
Ad libitum feed	36	17.5	12.0
Ad libitum+insulin once a day [c]	37	17.2	11.4
Ad libitum+insulin twice a day	31	17.9	12.3
SEM		0.9	0.5

[a]A diet based on corn and soybean meal (14% protein) was limit-fed to provide 5400 kcal metabolizable energy (1.8 kg day^{-1}) or fed *ad libitum*. Data from Ramirez, 1994.
[b]Significantly different from the other treatments ($P < 0.05$).
[c]Each insulin treatment provided a total of 0.80 IU kg^{-1} day^{-1} Lente insulin (Eli Lilly and Company, Indianapolis, IN).

certain metabolic condition for insulin to be effective. More appropriate feeding regimens may permit an augmented and repeatable reproductive response to insulin. More recently interval to oestrus was decreased but ovulation rate (18.3 average) and litter size (10.3) were unaffected by insulin (N. Whitley, unpublished observation, Table 2). Taken together, these results are consistent with the notion that insulin increased the anabolic state of sows, but the positive effects were manifested in different ways.

Manipulation of post-weaning follicle development in sows. In lactating sows, follicle growth advances as lactation advances, but follicles reach only about 5 mm in diameter by 3–4 weeks of lactation (Britt *et al.*, 1985). Follicular growth patterns during lactation and after weaning measured ultrasonically illustrate that within 2 or 3 days after weaning, the 4–5 mm follicles are replaced by larger follicles (M. Lucy, personal communication). We have conducted one study in which follicle populations were monitored after insulin administration to primiparous sows for 3 or 5 days after weaning (Whitley *et al.*, 1997a). Insulin increased atresia and the number of 6 mm follicles at 3 days after weaning and the overall number of follicles present at day 5, but at both 3 and 5 days after weaning follicular fluid IGF-I, IGFBP-2 and oestradiol were reduced. These results suggest that insulin may have increased numbers of follicles without increasing steroidogenesis by attenuating IGF-I. Perhaps the lowered IGFBP-2 allowed follicles to escape atresia. However, in another study, insulin increased oestradiol in medium- and large-sized follicles but did not affect IGF-I or IGFBP-2 (Whitley *et al.*, 1997b). It is assumed that in the latter study insulin may not have lowered peripheral glucose concentrations as much (not measured), because animals received 0.45 kg more feed per day between weaning and oestrus, and the duration of lactation was 21 days, compared with 30 days in the former study. In addition to increasing oestradiol, insulin administration in the second study also resulted in increased follicular testosterone (not shown) and progesterone in a pattern similar to that for oestradiol.

The effects of insulin reported to date, while generally positive, are not consistent. However, the results are consistent with an overall anabolic effect that could variously affect follicles, oocytes or embryos, reproductive tract environment or gonadotrophin actions. Administration of insulin may provide a physiological signal of increased anabolism before the time such a state would normally occur after weaning, but coincident with preovulatory follicle growth.

Pig somatotrophin

Results of studies investigating the control of follicular development and ovulation rate after administration of pig somatotrophin (pST) are contradictory. Positive effects of pST have been

Table 2. Summary of effects of exogenous insulin on reproductive function of sows during lactation or after weaning

Number in experiment	Insulin dose and duration	Significant effects of insulin	Authors
120 Mixed parity	0.40 iu kg^{-1} day^{-1} 2 days before to 4 days after weaning	No effects	Kirkwood and Thacker, 1991
8 Primiparous	0.75 iu kg^{-1} day^{-1} 1day pre-weaning through oestrus	Lowered ovulation rate	Rojkittikhun, 1992
26 Primiparous	0.75 iu kg^{-1} day^{-1} from weaning to oestrus (up to 30 days)	Higher incidence of anoestrus (10/13 versus 5/13)	Johnston *et al.*, 1994
28 Primiparous	Infusion 130 iu day^{-1}, day 20 to 26 post-farrowing	No effects	Tilton *et al.*, 1996
138 Primiparous	0.40 iu kg^{-1} day^{-1} 4 days after weaning	Farrowing rate increased (92.3 versus 76.7%)	Ramirez *et al.*, 1997
491 Primiparous	0.40 iu kg^{-1} day^{-1} 2 or 4 days after weaning	Insulin for 4 days increased total litter size (10.3 versus 9.3) and live born pigs (10.0 versus 9.0)	
231 Multiparous	0.40 iu kg^{-1} day^{-1} 4 days after weaning	None	Cox *et al.*, 1995
171 Primiparous	0.40 iu kg^{-1} day^{-1} 4 days after weaning	Increased return to oestrus in 10 days (86 versus 78%)	
143 Primiparous	0.40 iu kg^{-1} day^{-1} 4 days, increased feed	Interaction to increase litter size in insulin-treated sows on higher feed intake	
43 Primiparous	0.40 iu kg^{-1} day^{-1} 4 days after weaning	Decreased interval to oestrus (5.0 versus 6.9 days)	N. Whitley and N. Cox, unpublished

observed for IGF-I production *in vitro* by pig granulosa cells (Spicer and Echternkamp, 1995) as well as other species, and synergism of ST with gonadotrophins has been demonstrated in humans undergoing superovulation (Homburg *et al.*, 1988). In early antral rat follicles in culture, rbST increased IGF-I messenger RNA but did not reduce apoptosis (Chun *et al.*, 1996).

Many attempts *in vivo* to increase follicular development involved administration of pST during follicular development (Table 3). Administration during the prepubertal period has had few permanent effects on reproductive characteristics, with the exception of the study of Echternkamp *et al.* (1994), who used low doses of pST in sustained release implants. Treatment with 4 mg pST day^{-1} interacted with body type to increase the number of medium-sized follicles, with a greater initial number in control compared with lean and obese body types (12, 2 and 4, respectively) and 23, 9 and 11 follicles in control, lean and obese gilts treated with 4 mg pST day^{-1}. Importantly, pST decreased IGFBP-2 concentration in follicular fluid but not in serum.

In cyclic animals, pST can have potent effects on the occurrence of oestrus. Kirkwood *et al.* (1988) administered pituitary-derived pST from day 14 of the oestrous cycle to 24 h after oestrus. This dose increased insulin and glucose concentrations on day 20 of the oestrous cycle, and ovulation rate was increased by pST (12.4 for control and 14.3 for pST-treated). However, nine of 20 pST-treated gilts were anoestrous, whereas none of the control animals were anoestrous. The anoestrous gilts tended to have lower insulin and glucose values, which may have been beneath a threshold necessary for continued cycling. Despite *ad libitum* feeding, pigs did not gain weight during the experimental period, adding additional credence to the idea that there is a critical threshold of metabolism below which metabolic modifiers can have harmful peripheral effects. A critical period for inhibition of oestrus with pST was determined by Kirkwood *et al.* (1989). They observed that administration of pST beginning on day 14, but not day 17, suppressed the subsequent oestrus, illustrating that pST may affect follicles negatively during follicle selection. In a follow-up experiment, treatment was shortened to days 14–17 of the oestrous cycle, and that treatment resulted in an increase in percentage in oestrus (91%) compared with gilts treated from day 14 to day 22 (73%), but no increase in ovulation rate. It should be noted that the ovulation rate of the controls in this experiment was 14.0, compared to 12.3 in the study of Kirkwood *et al.* (1988). Thus ovulation rate must be below a physiological maximum in order to be increased by metabolic modifiers, which may explain the inconsistent results for the two studies.

Preovulatory development in sows may also be manipulated by pST, although farrowing rate and litter size in multiparous sows were not affected by treatment with 6 mg pST day^{-1} from 2 days before weaning to 4 days after weaning (Kirkwood *et al.*, 1993). Recently we treated primiparous sows with 40 μg pST kg^{-1} day^{-1} (dose of approximately 6 mg day^{-1}) beginning one day after weaning and observed increased IGF-I on day 5 (Whitley *et al.*, 1997b). The only IGFBPs increased by pST treatment in this model were the 30 and 22 kDa species, but the relevance of this increase is questionable since these IGFBPs are present in small quantities and proteolysis of these IGFBPs is related to follicle development (Besnard *et al.*, 1997). Follicular fluid oestradiol and testosterone were also increased by pST and insulin, indicating that ST may have positive effects on follicular development in sows if administered during the 5 days after weaning. In summarizing the potential applicability of manipulation with ST, increased follicular development is possible with appropriate dosages and administration, but the influence on pregnancy establishment has not been evaluated.

Conclusion

Parallel with information on the intrafollicular control of ovarian function, progress has been made in the practical application of metabolic modification of reproductive function. Metabolic manipulation of follicles is most successful in follicles of intermediate development, which have the potential for atresia or ovulation, depending on metabolic and gonadotrophin support. Progress in control of follicular development and ovulation rate will depend on a better understanding of these factors as well as the influence of metabolic state when metabolic hormones are administered. It can be concluded that metabolic modifiers may be harmful if body condition is below average and may

Table 3. Summary of representative experiments to affect reproductive function of gilts and sows with porcine somatotrophin (pST)

Dose/duration	Reproductive period or age		Effect of pST[a]	Reference
	Prepubertal administration			
70 µg kg^{-1} for 30 or 65 days	72 kg weight at start	+	Follicle IGF-I, progesterone, number of follicles	Bryan *et al.*, 1989
70 µg kg^{-1} day^{-1}, 21 or 42 days	Age 159 day to puberty	0	Corpora lutea	Bryan *et al.*, 1990
35 or 70 µg kg^{-1} day^{-1} for 50 days	From age 140 day	0 0	Age at puberty Corpora lutea	Andres *et al.*, 1993
implant; 2 or 4 mg day^{-1} for 6 weeks	From age 160 day	+ +	Medium follicles (4 mg day^{-1}) follicle IGF-I	Echternkamp *et al.*, 1994
	Oestrous cycle administration			
90 µg kg^{-1} day^{-1}; day 14 of cycle to 24 h after oestrus	First cycle	+ –	Corpora lutea (12.4 versus 14.3) % in oestrus (55 versus 100%)	Kirkwood *et al.*, 1988
90 µg kg^{-1} day^{-1}; day 14–17 or 14–22	First cycle	0	Corpora lutea (15.1 versus 14.3 and 14.0 for day 14–17, day 14–22 and control)	Gilbertson *et al.*, 1991
	Postweaning follicular period			
6 mg day^{-1}; daily or alternate	–2 to +4 days from weaning	0 0	Farrowing rate Litter size	Kirkwood *et al.*, 1993
40 µg kg^{-1} day^{-1}	Days 1 to 5 after weaning	+ +	Follicular IGF-I Follicular progesterone, oestradiol, testosterone	Whitley *et al.*, 1997b

[a] +, – or 0 denote positive, negative or absence of effects of pST, respectively.

be unnecessary if body condition is above average. Enhancing ovulation rate by manipulating the preovulatory follicle is possible in gilts and appears to be possible in sows, although less reliable. Gilts can be managed such that ovulation rate is not limiting to litter size. However, follicle development in sows may benefit from anabolic signals applied for a short enough time to enhance reproduction without causing deleterious peripheral metabolic effects. Thus it is more practical to consider application of metabolic modifiers to produce anabolic signals during preovulatory follicle growth in sows. Future progress in this area will involve assessing effects of metabolic manipulation on embryo quality, as well as integration of research on intrafollicular factors which control fertility.

The author gratefully acknowledges the excellent assistance in interpretation of data and preparation of the manuscript provided by A. B. Moore, J. Purvis, B. F. Stevens, N. C. Whitley and H. Zhang and the data generously contributed by H.D. Guthrie, W. M. Garrett and M. C. Lucy.

References

Andres CJ, Green ML, Clapper JA, Cline TR, Diekman MA and Thomford PJ (1993) Influence of porcine somatotropin on endocrine and histological variables in gilts *Journal of Animal Science* **71** 1552–1560

Angell CA, Tubbs RC, Moore AB, Barb CR and Cox NM (1996) Depressed luteinizing hormone response to estradiol *in vivo* and gonadotropin-releasing hormone *in vitro* in experimentally diabetic swine *Domestic Animal Endocrinology* **13** 453–463

Beltranena E, Aherne FX, Foxcroft, GR and Kirkwood RN (1991) Effects of pre- and post-pubertal feeding on production traits at first and second estrus in gilts *Journal of Animal Science* **69** 886–893

Besnard N, Pisselet C, Monniaux D and Monget P (1997) Proteolytic activity degrading insulin-like growth factor-binding protein-2,-3,-4, and -5 in healthy growing and atretic follicles in the pig ovary *Biology of Reproduction* **56** 1050–1058

Booth PJ (1990) Metabolic influences on hypothalamic–pituitary–ovarian function in the pig *Journal of Reproduction and Fertility Supplement* **40** 89–100

Britt JH, Armstrong JD, Cox NM and Esbenshade KL (1985) Control of follicular development during and after lactation in sows *Journal of Reproduction and Fertility Supplement* **33** 37–54

Britt JH, Armstrong JD and Cox NM (1988) Metabolic interfaces between nutrition and reproduction in pigs *Proceedings of the 11th International Congress on Animal Reproduction and Artificial Insemination* **5** 117–125

Bryan KA, Hammond JM, Canning S, Mondschein J, Carbaugh DE, Clark AM and Hagen DR (1989) Reproductive and growth responses of gilts to exogenous porcine pituitary growth hormone *Journal of Animal Science* **67** 196–205

Bryan KA, Clark AM and Hagen DR (1990) Effect of treatment with and subsequent withdrawal of exogenous porcine somatotropin on growth and reproductive characteristics of gilts *Journal of Animal Science* **68** 2357–2361

Cardenas H and Pope WF (1997) Administration of testosterone on day 13 of the estrous cycle to estrus increased the number of corpora lutea and conceptus survival in gilts *Journal of Animal Science* **75** 202–207

Charlton ST, Cosgrove JR, Glimm DR, and Foxcroft GR (1993) Ovarian and hepatic insulin-like growth factor-I gene expression and associated metabolic responses in prepubertal gilts subjected to feed restriction and refeeding *Journal of Endocrinology* **139** 143–152

Chun S, Billig H, Tilly JL, Furuta I, Tsafriri A and Hsueh AJW (1994) Gonadotropin suppression of apoptosis in cultured preovulatory follicles: mediatory role of endogenous insulin-like growth factor I *Endocrinology* **135** 1845–1853

Chun S, Eisenhauer KM, Minami S, Billig H, Perlas E and Hsueh AJW (1996) Hormonal regulation of apoptosis in early antral follicles: follicle-stimulating hormone as a major survival factor *Endocrinology* **137** 1447–1456

Clowes EJ, Aherne FX and Foxcroft GR (1994) Effect of delayed breeding on the endocrinology and fecundity of sows *Journal of Animal Science* **72** 283–291

Cosgrove JR and Foxcroft GR (1996) Nutrition and reproduction in the pig: ovarian aetiology *Animal Reproduction Science* **42** 131–141

Cox NM, Stuart MJ, Althen TG, Bennet WA and Miller HW (1987) Enhancement of ovulation rate in gilts by increasing dietary energy and administering insulin during follicular growth *Journal of Animal Science* **64** 507–516

Cox NM, Meurer KA, Carlton CA, Tubbs RC and Mannis DP (1994) Diabetes during the luteal phase of the estrous cycle affects preovulatory follicular function and prevents ovulation but does not impair gonadotropins *Journal of Reproduction and Fertility* **101** 77–86

Cox NM and Qiu Y (1995) Effect of duration of insulin withdrawal on apoptosis and estradiol in preovulatory follicles of swine with insulin-dependent diabetes mellitus *Journal of Animal Science* **73** (Supplement 1) 218

Cox NM, Quirk MN, Moore AB and Ramirez JL (1995) Influences of parity and level of food intake on reproductive responses to insulin administered to sows after weaning *Journal of Animal Science* **73** (Supplement 1) 218

Echternkamp SE, Spicer LJ, Klindt J, Vernon RK, Yen JT and Buonomo FC (1994) Administration of porcine somatotropin by a sustained-release implant: effect on follicular growth, concentrations of steroids and insulin-like growth factor I, and insulin-like growth factor binding protein activity in follicular fluid of control, lean, and obese gilts *Journal of Animal Science* **72** 2431–2440

Edwards JL, Hughey TC, Moore AB and Cox NM (1996) Depletion of insulin in streptozocin diabetic pigs alters estradiol, IGF-I and its binding proteins in cultured ovarian follicles *Biology of Reproduction* **55** 775–781

Einarsson S and Rojkittikhun (1993) Effects of nutrition on pregnant and lactating sows *Journal of Reproduction and Fertility* **48** 229–239

Findlay JK, Drummond AE and Fry RC (1996) Intragonadal regulation of follicular development and ovulation *Animal Reproduction Science* **42** 321–331

Flowers B, Martin MJ, Cantley TC and Day BN (1989) The effect of pregnant mare serum gonadotropin on follicle stimulating hormone and estradiol secretion in the prepuberal gilt *Animal Reproduction Science* **21** 93–100

Flowers WL and Turner ZA (1996) *In vitro* growth of preantral follicles isolated from prepuberal gilts *Journal of Animal Science* **74** (Supplement 1) 220

Foxcroft GR, Shaw HJ, Hunter MG, Booth PJ and Lancaster RT (1987) Relationships between luteinizing hormone, follicle-stimulating hormone and prolactin secretion and ovarian follicular development in the weaned sow *Biology of Reproduction* **36** 175–191

Fricke PM, Ford JJ, Reynolds LP and Redmer DA (1996) Growth and cellular proliferation of antral follicles throughout the follicular phase of the estrous cycle in Meishan gilts *Biology of Reproduction* **54** 879–887

Garrett WM and Guthrie HD (1996) Expression of androgen receptors and steroidogenic enzymes in relation to follicular growth and atresia following ovulation in pigs *Biology of Reproduction* **55** 949–955

Gilbertson J, Kirkwood RN and Thacker PA (1991) Timing of growth hormone injections and reproduction in gilts *Canadian Journal of Animal Science* **71** 717–723

Guthrie HD and Cooper BS (1996) Follicular atresia, follicular fluid hormones, and circulating hormones during the midluteal phase of the estrous cycle in pigs *Biology of Reproduction* **55** 543–547

Guthrie HD, Bolt DJ and Cooper BS (1993) Changes in follicular estradiol-17β, progesterone and inhibin immunoactivity in healthy and atretic follicles during preovulatory maturation in the pig *Domestic Animal Endocrinology* **10** 127–140

Guthrie HD, Welch GR, Cooper BS, Zakaria AD and Johnson LA (1994) Flow cytometric determination of degraded deoxyribonucleic acid in granulosa cells to identify atretic follicles during preovulatory maturation in the pig *Biology of Reproduction* **50** 1303–1311

Guthrie HD, Grimes RW, Cooper BS and Hammond JM (1995a) Follicular atresia in pigs: measurement and physiology *Journal of Animal Science* **73** 2834–2844

Guthrie HD, Grimes RW and Hammond JM (1995b) Changes in insulin-like growth factor-binding protein-2 and -3 in follicular fluid during atresia of follicles grown after ovulation in pigs *Journal of Reproduction and Fertility* **104** 225–230

Hammond JM, Samaras SE, Grimes R, Leighton J, Barber J, Canning SF and Guthrie HD (1993) The role of insulin-like growth factors and epidermal growth factor-related peptides in intraovarian regulation in the pig ovary *Journal of Reproduction and Fertility Supplement* **48** 117–125

Homburg R, Eshel A, Abdalla HI and Jacobs HS (1988) Growth hormone facilitates ovulation induction by gonadotropins *Clinical Endocrinology* **29** 113–117

Howard HJ and Ford JJ (1992) Relationships among concentrations of steroids, inhibin, insulin-like growth factor-1 (IGF-1), and IGF-binding proteins during follicular development in weaned sows *Biology of Reproduction* **47** 193–201

Howell JL, Hughey TB, Cox NM, Lehman FD, Armstrong JD, Campbell RM and Heimer EP (1993) Effects of the absence of insulin and immunoneutralization of growth hormone releasing factor (GRF) on intrafollicular hormones in preovulatory follicles *Journal of Animal Science* **71** (Supplement 1) 217

Hsueh AJW, Billig H and Tsafriri A (1994) Ovarian follicle atresia: a hormonally controlled apoptotic process *Endocrine Reviews* **15** 707–724

Hughey TB, Howell JL, Lehman FD, Cox NM and Moore AB (1993) Effects of exogenous insulin on intrafollicular hormones of preovulatory follicles during the estrous cycle *Journal of Animal Science* **71** (Supplement 1) 218

Johnston LJ, Pettigrew JE, Wilson ME, Walker RD and Crooker BA (1994) Influence of dietary molasses on reproductive performance of lactating sows *Journal of Animal Science* **72** Supplement 1 334

Kemp B, Soede NM, Helmond FA and Bosch MW (1995) Effects of energy source in the diet on reproductive hormones and insulin during lactation and subsequent estrus in multiparous sows *Journal of Animal Science* **73** 3022–3029

King BF, Britt JH, Esbenshade KL, Flowers WL, Sesti LAC, Martin TL and Ireland JJ (1993) Ovulatory and endocrine responses after active immunization of gilts against a synthetic fragment of bovine inhibin *Journal of Animal Science* **71** 975–982

King BF, Britt JH, Esbenshade KL, Flowers WL and Ireland JJ (1995) Evidence for a local role of inhibin or inhibin a subunits in compensatory ovarian hypertrophy *Journal of Reproduction and Fertility* **104** 291–295

Kirkwood RN and Thacker PA (1991) The influence of premating feeding level and exogenous insulin on the reproductive performance of sows *Canadian Journal of Animal Science* **71** 249–251

Kirkwood RN, Thacker PA, Gooneratne AD, Guedo BL and Laarveld B (1988) The influence of exogenous growth hormone on ovulation rate in gilts *Canadian Journal of Animal Science* **68** 1097–1103

Kirkwood RN, Thacker PA, Guedo BL and Laarveld B (1989) The effect of exogenous growth hormone on the endocrine status and the occurrence of estrus in gilts *Canadian Journal of Animal Science* **69** 931–937

Kirkwood RN, Peacock AJ and Thacker PA (1993) The influence of growth hormone injections either pre- or post-breeding on the reproductive performance of sows and gilts *Canadian Journal of Animal Science* **73** 259–265

Koketsu Y, Dial GD, Pettigrew JE, Marsh WE and King VL (1996) Influence of imposed feed intake patterns during lactation on reproductive performance and on circulating concentrations of glucose, insulin, and luteinizing hormone in primiparous sows *Journal of Animal Science* **74** 1036–1046

Matamoros IA, Cox NM and Moore AB (1990) Exogenous insulin and additional dietary energy affect follicular distribution, follicular steroid concentrations and granulosa cell human chorionic gonadotropin binding in swine *Biology of Reproduction* **43** 1–7

Matamoros IA, Moore AB and Cox NM (1991) Effects of exogenous insulin and body condition on metabolic hormones and gonadotropin-induced follicular development in prepuberal gilts *Journal of Animal Science* **69** 2081–2091

McKinnie MR, Britt JH and Esbenshade KL (1988) Ovarian function and hormone secretion of gilts actively immunized against androstenedione *Journal of Animal Science* **66** 3131–3143

Meurer KA, Cox NM, Matamoros IA and Tubbs RC (1991) Decreased follicular steroids and insulin-like growth factor-I and increased atresia in diabetic gilts during follicular growth stimulated with PMSG *Journal of Reproduction and Fertility* **91** 187–196

Purvis J, Skelton JO, Quirk MN, Moore AB, Whitley NC and Cox NM (1997) Influence of insulin and insulin-like growth factor-I (IGF-I) on the function of porcine ovarian follicles in culture *Journal of Animal Science* **75** (Supplement 1) 1

Ramirez JL (1994) *Use of Exogenous Insulin to Influence Fertility in Swine* PhD Dissertation. Mississippi State University

Ramirez JL, Moore AB and Cox NM (1997) Influence of insulin administration to primiparous sows after weaning on subsequent reproductive performance *Journal of Animal Science* (in press)

Rojkittikhun T (1992) *Clinical and Endocrinological Studies in Lactating and Post-Weaning Primiparous Sows* PhD Dissertation. Swedish University of Agricultural Science, Uppsala

Spicer LJ and Echternkamp SE (1995) The ovarian insulin and insulin-like growth factor system with an emphasis on domestic animals *Domestic Animal Endocrinology* **12** 223–245

Tilton JE, Plaine KJ, Weigel RM and Swantek PM (1996) Insulin-like growth factor and luteinizing hormone concentrations in the insulin treated weaned sow *Proceedings of the 13th International Congress on Animal Reproduction* **3** 16-6

Tokach MD, Pettigrew JE, Dial GD, Wheaton JE, Crooker BA and Johnston LJ (1992a) Characterization of luteinizing hormone secretion in primiparous lactating sows: relationship to blood metabolites and return-to-estrus interval *Journal of Animal Science* **70** 2195–2201

Tokach MD, Pettigrew JE, Dial GD, Wheaton JE, Crooker BA and Koketsu Y (1992b) Influence of glucose infusions on luteinizing hormone secretion in the energy-restricted, primiparous, lactating sow *Journal of Animal Science* **70** 2202–2206

Tonetta SA and diZerega GS (1990) Local regulatory factors controlling folliculogenesis in pigs *Journal of Reproduction and Fertility Supplement* **40** 151–161

Vesseur PC, Kemp B and den Hartog LA (1994) The effect of the weaning to estrus interval on litter size, live born piglets and farrowing rate in sows *Journal of Animal Physiology and Animal Nutrition* **71** 30–38

Whitley NC, Quirk MN, Skelton JO, Moore AB, Purvis J, Qiu Y and Cox NM (1997a) Influence of insulin on follicular development and the intrafollicular IGF-I system in sows *Journal of Reproduction and Fertility* (in press)

Whitley NC, Moore AB and Cox NM (1997b) Comparative effects of insulin and porcine somatotropin (pST) on post-weaning follicular development in sows *Journal of Animal Science* (in press)

Wu MC, Hentzel MD and Dziuk PJ (1987) Relationships between uterine length and number of fetuses and prenatal mortality in pigs *Journal of Animal Science* **65** 762–770

Yuan W, Lucy MC and Smith MF (1996) Messenger ribonucleic acid for insulin-like growth factors-I and -II, insulin-like growth factor binding protein-2, gonadotropin receptors, and steroidogenic enzymes in porcine follicles *Biology of Reproduction* **55** 1045–1054

Zak LJ, Cosgrove JR, Aherne FX and Foxcroft GR (1997) Pattern of feed intake and associated metabolic and endocrine changes differentially affect postweaning fertility in primiparous lactating sows *Journal of Animal Science* **75** 208–216

Zhou J, Adesanya OO, Vatzias G, Hammond JM and Bondy CA (1996) Selective expression of insulin-like growth factor system components during porcine ovary follicular selection *Endocrinology* **137** 4893–4901

Journal of Reproduction and Fertility Supplement **52**, 47–61

Mechanisms mediating nutritional effects on embryonic survival in pigs

G. R. Foxcroft

Alberta Pork Research Centre, Department of Agricultural, Food and Nutritional Science, 310-C Agriculture-Forestry Centre, University of Alberta, Edmonton, Alberta, T6G 2P5, Canada

The inconsistency of data from experiments designed to show nutritional effects on embryonic survival is perplexing. However, a number of experimental models have provided some insight into the mechanisms that potentially mediate interactions between nutrition, metabolic state and embryonic survival. The developing ovarian follicle provides the maturational environment for the oocyte, and differences in follicular maturation are associated with differences in the ability of these follicles to support oocyte maturation. In turn, the rate of oocyte maturation and the maturational state of the oocyte immediately before ovulation may contribute to differences in embryonic survival. Therefore, evidence that nutritionally induced differences in metabolic state affect follicular development and the maturational state of the oocyte in the late follicular phase may constitute a mechanism by which nutrition affects the very early stages of conceptus development. Once ovulation occurs, the process of fertilization and early cleavage occurs in the environment of the oviduct. Nutritional state might affect the secretory and motile activity of the oviduct both directly, by influencing the physiology of the oviductal cells, or indirectly by affecting the secretion of key regulatory hormones. Thus evidence for nutritionally dependent effects on plasma progesterone concentrations in early pregnancy and associations with differences in embryonic survival may be partly mediated at the oviductal stage of development. Nutritional effects on circulating progesterone concentrations may also affect the uterine environment. However, the metabolic state of the gilt or sow, or specific nutrients in the diet, may directly affect the integrity of the endometrium and thus affect embryonic survival at this stage of development.

Introduction

An adequate understanding of the physiology of gamete production and of early pregnancy in pigs is essential as background information for a review of the mechanisms affecting embryo survival. The presentation of such background information is beyond the scope of this review but an excellent summary of most of the relevant information was presented by Pope (1994). In the same publication Ashworth (1994) described much of the data linking nutritional state to differences in embryo survival and provided some insight into the mechanisms involved. Such reviews of the literature should leave little doubt that various nutritional factors can affect embryo survival in pigs. The purpose of this review is to identify some of the key stages in the reproductive cycle at which nutrition may affect embryo survival and extend the discussion of the possible mechanisms that might mediate these effects. Initially, however, it seems appropriate to discuss the reasons for the inconsistency in much of the literature regarding the effects of nutrition on embryo survival. This inconsistency has been a major problem in developing good experimental models with which to explore the mechanistic aspects of nutritional effects. By combining a critical approach to the better design of experiments with information already available on the putative mechanisms involved, it should be possible to achieve greater progress in describing the physiological mechanisms by which nutrition affects embryonic survival.

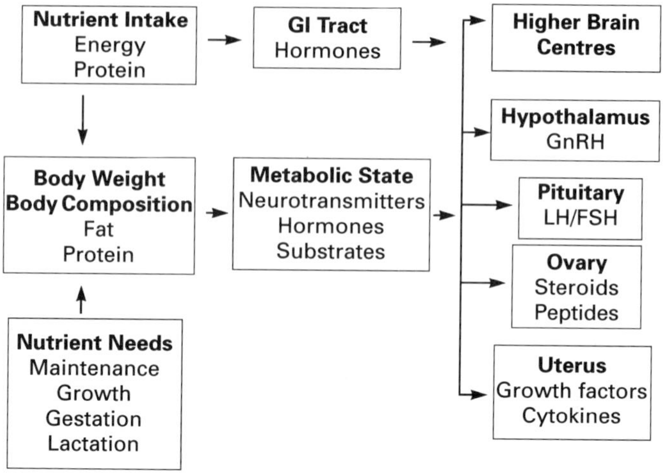

Fig. 1. Schematic representation of the interactions between metabolic state of the gilt or sow and the reproductive axis. Nutrient intake can affect reproduction via gut-mediated mechanisms throughout the gastro-intestinal (GI) tract or, depending on the nutrient demands of the animal, may affect metabolic state and therefore mechanisms at all levels of the hypothalamic–pituitary–ovarian–uterine system.

Problems in Consistently Demonstrating Effects of Nutrition on Embryonic Survival

How to define nutritional effects

Much of the confusion that has arisen in defining the effects of nutrition on reproductive function comes from the reluctance to move from an empirical to a mechanistic approach to this problem. The advantages of adopting a mechanistic approach have been discussed in detail by I'Anson *et al.* (1991) but some of the key issues are worth considering. Firstly, if the metabolic theory of nutritional effects on reproduction is accepted, then even long-term effects of nutritional state on reproduction must be explained on the basis of metabolically mediated mechanisms. Experiments that simply explore the relationships between changes in body composition and reproductive function, without information being provided on the metabolic and endocrine status of the animal, are unlikely to make a significant contribution to our understanding of the problem. In many instances the mechanisms that mediate the effects of nutrition on reproduction will operate in an additive or even synergistic manner. Thus for instance, the point at which nutritional manipulations result in measurable changes in protein or adipose tissue will be relatively late in a cascade of events that will already have involved very rapid responses to changes in energy balance. It is equally likely that the same 'nutritional' change will exert its effects at different levels of the hypothalamo–hypophysial–ovarian axis (Fig. 1).

If nutritional treatment involves changes in total feed intake, it is likely that different components of the diet will be exerting many effects on the reproductive system. Thus, although studies in gilts suggest that changes in energy intake may be a key regulator of follicular development and hence ovulation rate (see Cox, 1997), in sheep, responses to the protein component of the diet may be equally, if not more, effective when used during 'flush feeding' to increase ovulation rate (see Ashworth, 1994). Furthermore, while there may be effects of nutrition on embryo survival that reflect something as general as energy balance mediated through the actions of, for example, insulin and insulin-like growth factors, other nutritional effects may relate to an effect of specific nutrients, such as minerals or vitamins, that regulate essential mechanisms for successful embryo development. Given all these alternatives, it is clear that only a critical and reductionist approach to dissecting the mechanisms mediating effects of nutrition on embryonic survival is likely to be successful. A further advantage of embracing the metabolic approach to considering nutritional effects on reproduction is the realization that nutrient intake *per se* is only one component of the complex interactions (Fig. 1) that determine the final outcome in terms of fertility. By relating the pool of available nutrients to the needs of the gilt or sow for maintenance and productive functions like growth, gestation and lactation, it should be possible to develop greater comparability between different experimental paradigms.

A further problem in defining nutritional effects is that the response to nutrition at one stage of the reproductive cycle may confound responses at a later stage. This problem was addressed in the review of Aherne and Kirkwood (1985) in relation to the fertility of gilts maintained on a high plane of nutrition. Although high feed or energy intakes (flushing) may be beneficial in maximizing ovulation rate, the negative impact of continued high planes of nutrition after mating, discussed later, may negate this effect. Therefore, the continuous application of dietary treatments throughout the preovulatory, periovulatory and postovulatory periods does not allow us to define the stage of the reproductive cycle that is being critically affected.

The general problem of achieving comparability among nutritional experiments was highlighted by Schultz *et al.* (1966) and in more recent studies that have sought to establish effects of nutrient intake during the postmating period on embryo survival in the gilt; the inconsistency of the results may be related to the precise time in relation to ovulation and fertilization at which nutritional changes were implemented (Jindal *et al.*, 1996). Therefore, as suggested by Pope (1994), a consideration of the potential effects of nutrition on embryo mortality should try to differentiate clearly between effects at each stage of oocyte and embryonic development.

How to optimize the chance of defining effects at a mechanistic level

One very obvious feature of any set of data on embryonic survival in gilts or sows is the variability involved. Populations of gilts that show greater overall embryonic survival also tend to show less variability in survival (Jindal *et al.*, 1996). Interestingly, in populations of gilts and sows subjected to adverse nutritional regimens, some animals invariably show 100% survival to the postimplantation stage. Given this variability, it is frustrating that many studies are still conducted with inadequate numbers of animals. Assuming the variance in embryonic survival that we have encountered in populations of sows (Zak *et al.*, 1997a; L.J. Zak, R. Jindal and G.R. Foxcroft, unpublished; J. Mao, L.J. Zak, J.R. Cosgrove and G.R. Foxcroft, unpublished) in which embryo survival was compromised (mean ± SD embryo survival of 70.7 ± 14.0), we calculate that to observe a difference in embryo survival in response to treatment of 5% or 10% would require 65 and 16 observations, respectively. Comparable data from our recent experiments in gilts (Pharazyn *et al.*, 1991a; Pharazyn, 1992; Jindal *et al.*, 1996; Jindal *et al.*, 1997) in which mean embryo survival was 84% indicate a need for 11 and 17 animals, respectively, to demonstrate a 5% and 10% improvement in embryo survival in response to experimental treatments. In many reports the level of possible response may be below 5%, yet limited numbers of sows have been used to estimate treatment effects. Experimental paradigms that seem to provide a level of response that may be conducive to effective mechanistic studies include comparisons between prolific and less prolific genotypes and inbred lines (see Pope, 1994), variants of postmating nutritional manipulation in gilts (see Ashworth, 1994; Jindal *et al.*, 1996) and nutritional manipulation in lactating and weaned sows (Zak *et al.*, 1997a).

Inconsistent responses to nutritional manipulation may also be related to differences in embryo survival in the 'control' population used for comparison. This may be particularly true where research is conducted in large commercial units and careful selection of the population of sows under study may be beneficial, even though large numbers of animals may be available. A comparison between the data of Tremblay *et al.* (1989a) and Zak *et al.* (1997a; shown in Table 1) illustrates this point, in that the highest embryonic survival in any treatment group in each of these experiments was 67.4% and 87.5%, respectively. Thus the baseline against which treatment effects were being assessed was very different.

Table 1. Effects of pattern of feed intake in lactation on postweaning fertility in primiparous sows*

Parameter	Treatment 1	Treatment 2	Treatment 3
Weight loss in lactation (kg)	11.00[a]	21.12[b]	24.75[b]
Backfat loss in lactation (mm)	2.19[c]	4.61[b]	5.38[b]
Ovulation rate	19.86[c]	15.44[b]	15.43[b]
Embryo survival (%)	87.53[a]	64.43[b]	86.50[a]
Weaning to oestrus interval (h)	88.71[a]	122.33[b]	134.71[b]

* This experiment involved three patterns of feeding during lactation: treatment 1 sows were fed *ad libitum* (100%) throughout lactation; treatment 2 sows were fed *ad libitum* from day 0 to day 21 and then restrict fed to 50% of *ad libitum* feed intake from day 22 to day 28 of lactation: Treatment 3 sows were restrict fed from day 0 to day 21 and then *ad libitum* from day 22 to day 28 of lactation. All sows suckled litters of six piglets to reduce the direct inhibitory effects of suckling *per se* and thereby allow the effects of nutrition to be more clearly demonstrated (from Zak *et al.*, 1997a)

[a] and [b] denote treatment differences at $P < 0.002$, and [c] denotes differences at $P < 0.05$.

Given the reported association between nutritionally induced increases in ovulation rate and reduced embryo survival in pigs (see review of Anderson and Melampy, 1972), nutritionally induced changes in ovulation rate may confound the study of embryo survival. In a number of studies in cyclic gilts, this problem was avoided by imposing nutritional treatments only after the expected time of ovulation (Pharazyn *et al.*, 1991a; Liao and Veum, 1994). However, the later studies of Pharazyn (1992) and Jindal *et al.* (1996, 1997) suggest that this may have been unnecessary and indeed, as discussed later, the critical period for an effect on embryonic survival may have been missed. If the confounding effect of nutrition on ovulation rate cannot be avoided, then as in the study of Liao and Veum (1994), it may need to be removed by covariance analysis. A better solution would be to develop experimental paradigms in which nutrition differentially affects ovulation rate and embryo survival. The recent studies of Zak *et al.* (1997a) and J. Mao, L.J. Zak and G.R. Foxcroft (unpublished) suggest that nutritional manipulation during lactation provides this opportunity, in that different patterns of feed restriction resulted in a similar reduction in ovulation rate compared with sows fed to appetite, but substantial differences in embryo survival (Table 1).

The confounding effect of inducing variable ovulation rate is also clearly demonstrated in the experiment of Tremblay *et al.* (1989a) in which PMSG was used as a deliberate strategy to increase ovulation rate and thereby the chance of showing a beneficial effect of folic acid treatment on embryo survival (discussed later). In their control (no folic acid treatment) sows given PMSG, an ovulation rate of 28.6 was reported, compared with 19.2 and 21.5 ovulations, respectively, in the two control groups not treated with PMSG. These ovulation rates were associated with estimated embryonic survival rates of 41.8% in the PMSG-treated control sows compared with 60.8 and 63.0%, respectively, in the other groups. Thus, as in other studies, the increase in ovulation rate was completely offset by the increase in embryo loss and the estimated number of live fetuses at day 31–36 after mating was not different (12.2, 11.4 and 13.4, respectively). Personal discussions with breeding companies in recent months also raise another potentially important confounding effect that needs consideration when interpreting experiments on embryo survival. The advent of prolific dam lines for use with crosses to terminal line boars means that very different animals are used compared with those used in earlier experiments. Litter size in these prolific dam lines shows some increase with parity (for example 11 increasing to 13 born live) but this is apparently associated with an even greater parity effect on ovulation rate. In cyclic gilts that have not been mated, ovulation rates of 14 to 16 are typical, increasing to 18 to 20 ovulations in parity 2. However, in higher parity sows it appears that ovulation rates of 25 or greater are not uncommon. This discrepancy between the magnitude of the increase in litter size and ovulation rate suggests that prenatal survival over different parities is very different, probably decreasing from about 70% at parity 1 to 50% or less at higher parities. Whether this relates to differences in embryo survival in the preimplantation period, or to fetal losses due to problems with uterine capacity in later gestation needs to be clarified. The data of Harper *et al.* (1996) suggest that the former is the case, as in their control groups, an increase in ovulation rate from 14.4 in parity 1 to 19.5 in parity 3 was associated with a decrease in embryo survival from 77.8% to 59.3% at day 40–48. However, on the basis of an expected litter size born alive of 10.9 in first parity sows in our herd fed to appetite during a 28 day lactation and mated at first oestrus (Clowes *et al.*, 1994), the data in Table 1 suggest that relatively low embryonic mortality (13%) in sows fed to appetite (Treatment 1) would be associated with losses in later gestation of the order of 35%. Parity and genotype may therefore result in very different patterns of prenatal loss and these differences should be appreciated when selecting populations of animals in which to study potential nutritional effects on embryo survival, and when the results of different experiments are compared.

Finally the contribution of the semen used and the pattern of breeding management to observed differences in embryo survival merits consideration. The standardization of any 'boar' effect among different experimental treatments is often difficult using natural matings, and yet an unbalanced use of boars will probably contribute unwanted non-treatment effects to the results of studies on embryo mortality. The logical and usually very practical solution is to use artificial insemination. If this is combined with the use of pooled semen from a group of boars designated to a particular experiment, then it is usually possible to inseminate gilts from all treatments with semen from the same collections and even to equalize the use of day 1 and day 2 semen among treatments.

Potential Mechanisms Mediating Nutritional Effects at Different Reproductive Stages

The preovulatory period

A range of follicle maturational stages has been reported in both gilts (Grant *et al.*, 1989; Hunter *et al.*, 1989) and sows (Foxcroft *et al.*, 1987) at the time of recruitment into the ovulatory population. This prompted speculation about the consequences of this follicular 'heterogeneity' for oocyte maturation, luteal function and subsequent embryo survival (Foxcroft and Hunter, 1985; Hunter and Wiesak, 1990) The essential stages of maturation through which the follicle and oocyte progress during the oestrous cycle were summarized by Foxcroft and Hunter (1985) and Pope (1994). These reviews included reference to a number of studies that indicate that nutritional effects on the size of the preovulatory pool of follicles is one factor determining ovulation rate in swine (Clark *et al.*, 1972; Dailey *et al.*, 1972; Clark *et al.*, 1973; Dailey *et al.*, 1975a,b). As the estimated time for an antral follicle to mature to the preovulatory stage is 19 days (Morbeck *et al.*, 1992), differences in nutritional and metabolic state could exert a protracted effect on the quality of the follicle and oocyte.

The essential role of the follicle in supporting the maturation of mammalian oocytes is well established and the hormonal microenvironment of the follicle has associations with oocyte maturation (Xie *et al.*, 1990a). Direct evidence for a functional relationship between follicular status and oocyte maturation came from studies of Ding and Foxcroft (1992, 1994) in which the origin of the follicle, or follicle conditioned medium, used to support the *in vitro* maturation of standardized pools of immature oocytes affected the outcome of the maturation process. More recently, as part of a series of studies comparing the endocrine and follicular characteristics of prolific Meishan to European commercial White-type gilts, Faillace and Hunter (1994) reported that oocyte maturation to metaphase II of the second meiotic division occurred earlier in the Meishan females. This raises the possibility that differences in the preovulatory development of the follicle and its enclosed oocyte could already be a contributory factor to differences in embryo survival. It seems likely that nutritional state could exert similar effects.

The results of an earlier study in pigs (Bazer *et al.*, 1968) and more recent studies in sheep (McKelvey *et al.*, 1988) led Ashworth (1994) to conclude that "...these data suggest that nutritional modifications affect embryo survival by altering maternal function rather than by influencing embryo viability per se". However, the series of experiments reported by McEvoy *et al.* (1995)

provided evidence for effects of nutritionally mediated differences in circulating progesterone in the preovulatory period in ewes on the developmental potential of the early embryo *in vitro*.

The extensive studies of Pope and his colleagues have provided compelling evidence that in pigs the preovulatory development of the follicle and oocyte has consequences for subsequent embryo survival (see Pope *et al.,* 1990). Variability in follicular development within the recruited pool of follicles affects the sequence of ovulation and use of electrocautery to change the population of ovulatory follicles also affected the variability in embryo development at day 11 of pregnancy (Pope *et al.,* 1988). Subsequently it was shown that the maturational state of oocytes in the periovulatory period was associated with their development as early zygotes (Xie *et al.* 1990b) and that later ovulating oocytes developed into the smallest embryos *in utero* and were less effective in stimulating uterine secretory activity (Xie *et al.,* 1990c). We therefore used the lactating and weaned sow paradigm of Zak *et al.* (1997a) described earlier to demonstrate that nutritionally induced differences in metabolic state in late lactation affect the state of nuclear maturation in oocytes recovered in the late follicular phase preceding the first post-weaning oestrus (Zak *et al.,* 1997b). Furthermore, follicular fluid recovered from the larger follicles of sows that were previously the most catabolic were less able to support nuclear maturation of oocytes in a standard *in vitro* maturation system. Although further evidence is required to relate such differences in preovulatory oocyte maturation to development after fertilization either *in vivo* or *in vitro* in pigs, the potential impact of nutritionally mediated effects on follicular development and hence on oocyte maturation cannot be ignored. The laborious nature of such studies in lactating and weaned sows has prompted the development of a comparable experimental paradigm involving nutritional manipulation within the 21 day oestrous cycle.

The periovulatory period

Clearly, any effects on the preovulatory development of the oocyte will carry over into the periovulatory period and the same nutritional effects could continue to influence pre- and post-fertilization events within the oviduct. Although it has been suggested that nutritional effects on the fertilization process probably contribute little to reproductive failure in pigs (Ashworth, 1994; Hunter, 1994), definitive information as to whether nutrition might influence oviductal function is lacking. Only a brief discussion of potential mechanisms that might be involved will therefore be presented.

As nutrition can affect both the number of follicles in the preovulatory pool and their steroidogenic activity, as measured by the concentration of follicular fluid oestradiol (Zak *et al.,* 1997b), it seems increasingly likely that there will be carry-over effects on the pattern of luteinization of these follicles. Therefore, assuming that oviductal secretions are functionally important during the periovulatory period (see Buhi, 1997) and that this secretory activity will be controlled by the prevailing steroidogenic stimulation at the level of the oviductal circulation, the timing of the switch from oestrogen to progesterone dominance could be critical. Hunter *et al.* (1988, 1996) described local mechanisms that appear to allow steroids to control periovulatory events, and evidence for the existence of effective local countercurrent exchange mechanisms in the sub-ovarian plexus was extensively reviewed by Krzymoski *et al.* (1990). The effectiveness of these mechanisms in concentrating progesterone in the oviductal circulation was demonstrated by Pharazyn *et al.* (1991b). Therefore, if a high plane of nutrition in the immediate postmating period results in a delayed rise in peripheral plasma progesterone concentration (Pharazyn, 1992; Jindal *et al.,* 1997), this will be reflected in an even earlier difference at the oviduct and there is a potential mechanism by which nutrition can exert a critical effect on oviductal function in the periovulatory period. If oviductal secretions affect the development of the cleaving embryo, nutrition could exert a direct effect via changes in the steroid environment of the Fallopian tube. Alternatively, nutritionally dependent changes in the steroid environment could affect the rate of tubal transport and hence the precise synchrony between the development of the uterus and the preimplantation embryo.

In the context of the earlier discussion of the need to define clearly the stage of development at which embryo mortality might be affected by nutrition, it is important to realise that when changes

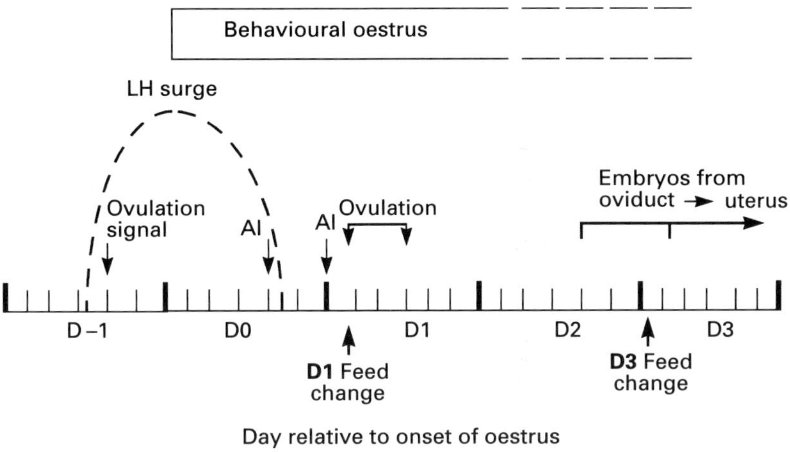

Fig. 2. Representation of the relationship between key events in the periovulatory period in the gilt and sow and key elements in the experimental design of the study of Jindal *et al.* (1996). By implementing a reduction of feed intake at day 1, effects on progesterone concentrations could exert effects at both the oviductal and uterine stages of development. By delaying the feed change to day 3, only uterine stages of development could be affected.

in nutrition are made in the immediate post-mating period there is the possibility that both the oviductal and uterine stages of development are affected (Fig. 2). However, if changes in nutritional state are delayed by only 1 or 2 days, only uterine effects would be involved. If future experiments do provide convincing evidence for nutritionally mediated effects at the oviductal stage of development, from a mechanistic viewpoint this may also help to explain why differences in the timing of nutritional treatments have produced such inconsistent effects on embryo survival.

The uterine phase of development

The literature on the physiology of the embryo and uterus in the preimplantation stages of pregnancy is already vast and has been reviewed by Bazer *et al.* (1982). A key feature of this period of development is the complex interaction between the developing blastocyst and the uterus (see Geisert *et al.*, 1982; Roberts *et al.*, 1993; Geisert and Yelich, 1997). For successful implantation to occur it is essential that the development of the embryo and uterus is synchronous (Pope, 1988). In some of the earliest embryo transfer experiments in pigs that addressed this question, Bazer *et al.* (1969) and Webel *et al.* (1970) established that a lack of synchrony of only 12 h affected blastocyst survival in the recipient uterus. Compared with monotocous species, in litter bearing species like the pig, asynchrony among embryos within a litter is an important factor in embryo survival. The relationships between follicular maturation, the time of ovulation and subsequent diversity of early embryo development has already been discussed and leads to functionally important differences in embryonic development at the crucial transitional stage of development (see Pope, 1994; Fig. 3). Therefore, if nutrition affects the developmental competence of the early embryo, or the variance in embryo development within a litter, this will compromise the ability of embryos to mature appropriately at the transitional stage and therefore survive the implantation process.

On its part, the uterus is required to progress through a sequence of developmental changes that prepare the endometrium for implantation and provide essential support to the developing blastocyst. Nutritional factors may affect this process indirectly by effects on progesterone secretion, or directly at the uterus.

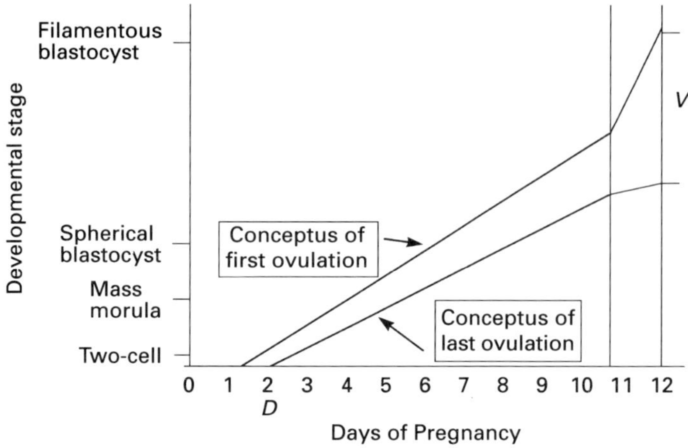

Fig. 3. Diagrammatic representation of the relationship between the duration of ovulation (*D*), representing the time of ovulation of the first to the last ovulating ovarian follicle, and the variance (*VT*) in embryo diversity during the transitional period of development when blastocysts undergo rapid elongation. Although the immediate impact of the spread in ovulation time may appear limited, the complex interactions between embryos of different maturity with the uterine environment greatly amplifies developmental differences in this transitional stage of development on day 11 to 12, and places the smaller embryos at risk. (Modified from Pope, 1994.)

Progesterone-mediated effects. Effects of progesterone-induced changes in uterine secretory activity on embryo development were first reported by Knight *et al.* (1974a) and the dose dependency of this secretory effect of progesterone by Knight *et al.* (1974b). A more comprehensive review of the regulation and functions of uterine secretions was presented by Roberts and Bazer (1988). There is both indirect (Ashworth, 1991; Pharazyn, 1992; Jindal *et al*, 1996) and direct (Ashworth, 1991; Jindal *et al.*, 1997) evidence that reduced plasma progesterone concentrations in early pregnancy in gilts may mediate the detrimental effects of high planes of nutrition on embryo survival. A positive association between plasma progesterone concentrations 72 h after the onset of oestrus and embryo survival has also been observed in primiparous weaned sows in which embryo survival was depressed by nutritionally induced catabolism in late lactation (L.J. Zak, R. Jindal and G.R. Foxcroft, unpublished); furthermore, Clowes *et al.* (1994) reported that the initial increase in plasma progesterone concentrations was earlier in primiparous sows mated at the second compared with the first oestrus after weaning and was associated with a significant increase in litter size born. Finally, in a study of the endocrinology of the prolific Meishan breed, an earlier increase in plasma progesterone concentrations was observed compared with European commercial type gilts (Hunter *et al.*, 1996). On the strength of these observations, it seems reasonable to suggest that effects on circulating progesterone concentrations may be an important mechanism by which nutrition and metabolic state can influence embryo survival, and similar effects have been clearly demonstrated in sheep (Parr *et al.*, 1982, 1987).

Several mechanisms may contribute to nutritionally induced changes in progesterone secretion. Prime and Symonds (1993) reported effects of the plane of nutrition on hepatic portal blood flow and the metabolic clearance rate of progesterone in ovariectomized gilts, again supporting comparable observations in sheep (Parr *et al.*, 1993a,b). However, it is also possible that the marked changes in splanchnic circulation generally as a consequence of changes in feed intake may divert blood away from the ovarian circulation. Even if progesterone secretion rates remain constant, changes in the perfusion rate of the utero–ovarian vasculature may markedly affect the efficiency of the

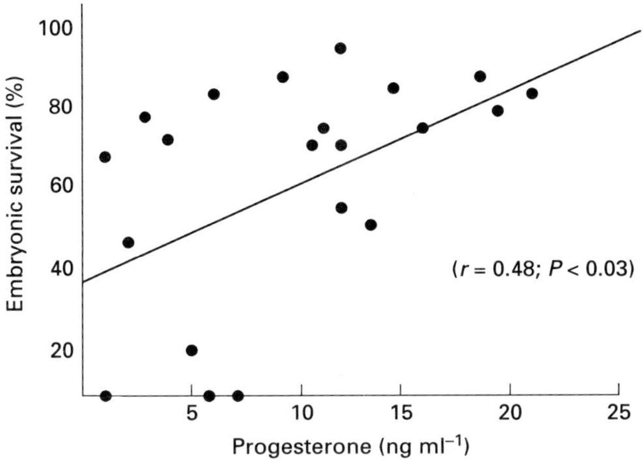

Fig. 4. Relationship between plasma progesterone at 72 h after onset of oestrus and embryonic survival (ES) at day 28 of pregnancy in primiparous sows subjected to severe feed restriction during the last week of a 28 day lactation and mated at the first postweaning oestrus (L.J. Zak, R. Jindal and G.R. Foxcroft, unpublished data).

countercurrent exchange mechanism and thus the distribution of steroids between the local and peripheral circulation. Finally, there is also the possibility, alluded to previously, that the prior effects of nutrition on the steroidogenic competence of the periovulatory follicle may translate into important differences in the pattern of luteinization of these follicles in the periovulatory period. Again, the time at which critical thresholds for progesterone actions were reached, either locally or peripherally, may be functionally much more important than the eventual peak concentrations of progesterone in the circulation.

Returning to the need for effective synchrony between embryonic and uterine development, the above studies may indicate the essential timing involved in demonstrating these effects of nutrition on embryo survival. By delaying the time at which plasma progesterone reaches the threshold for stimulating the normal cascade of endometrial changes, high planes of nutrition may effectively deprive less developed embryos of essential support and thus compromise their ability to survive the transitional stage of development. Furthermore, the ability to demonstrate that exogenous progesterone can reverse this effect may rely to a great extent on the fact that the nutritionally induced increase in embryo survival is indeed progesterone mediated. In other instances in which progesterone therapy has failed to improve embryo survival this situation may not apply. Because the pig is polyovulatory, the likelihood of plasma progesterone concentrations being below some essential threshold in established pregnancy is probably low; in contrast in monovulatory species like the ewe and cow, and in women, luteal insufficiency is more likely to be a cause of infertility in early pregnancy. Thus in pigs, the critical window of time during which progesterone-mediated effects of nutrition (or any other factor) will be observed is probably limited to the first 3–4 days after ovulation. If progesterone mediated effects of nutrition on early embryo survival can be consistently demonstrated, these experimental paradigms need to be exploited to determine the changes in uterine function involved.

Direct nutritional effects on the conceptus or uterus. Although a multitude of nutritional effects on embryo survival have been reported that might exert their effects at the uterus, the effects of folic acid, β-carotene and vitamin A provide some insight into the mechanisms that might be involved. The discussion of the effect of specific nutrient effects at the uterus will therefore be limited to these effects.

Although an effect of folic acid treatment on litter size and embryonic survival has not been consistently established using accepted statistical criteria, in summarizing published data from previous experiments, Lindemann (1993) concluded that "The likelihood of 11 positive responses in 11 trials, if in fact the response is not real, is 1/2,048…" and therefore that folic acid treatment effects on embryo survival were real. After some initial experiments which identified the potential for obtaining beneficial effects with folic acid treatment, the physiological basis for effects of folic acid was explored in a series of studies by Matte, Tremblay and their colleagues. A decline in serum folate concentrations after weaning and through the breeding and early gestation period in sows (Matte *et al.*, 1982) suggested that the demand for folic acid was not being met from dietary sources in this critical period of the reproductive cycle. This deficiency could be corrected with folic acid injections (Matte *et al.* 1984a), and treatment at weaning, mating and periodically up to week 12 of gestation resulted in a significant improvement in litter size born (Matte *et al.*, 1984b). Later studies demonstrated the effectiveness of dietary supplementation with folic acid in maintaining serum folate concentrations (Tremblay *et al.*, 1986) but variable responses in both litter size and embryonic mortality to such dietary supplementation have been reported (Lindemann and Kornegay, 1989; Thaler *et al.*, 1989; Tremblay *et al.*, 1989a; Harper *et al.*, 1994, 1996). In general, the greatest responses to folic acid supplementation were observed in situations in which potential litter size was increased by an initial increase in ovulation rate, due to 'flush' feeding in the preovulatory period (Matte *et al.*, 1984b), treatment with PMSG (Tremblay *et al.*, 1989a) or increased parity of the sows studied (Harper *et al.*, 1996). The potential for folic acid supplementation to be more effective when the number of implanting embryos increases is consistent with the suggestion that folic acid may act by supporting DNA synthesis at critical stages of embryonic development (see Lindemann, 1993). Even if embryonic survival is not improved by folic acid treatment, an increase in fetal protein content (Tremblay *et al.*, 1989a; Harper *et al.*, 1996) and a trend for increased placental growth (Harper *et al.*, 1996) suggest that postimplantation survival may also contribute to overall effects on litter size at term. Beyond the suggestion that folic acid supplementation may prevent deficiencies in DNA synthesis in the early conceptus, little else is known about its mechanism of action, although Tremblay *et al.* (1989b) reported interactions with serum zinc concentrations during early gestation.

The role of vitamin A and its provitamin, β-carotene, in pig reproduction were reviewed by Chew (1993). Because of reported effects of retinoic acid *in vitro* on progesterone secretion from porcine luteal cells (Talavera and Chew, 1988), effects of vitamin A might be mediated indirectly through differences in circulating progesterone, with implications for effects on any progesterone-dependent changes in endometrial function. However, in the light of the existence of retinol-binding proteins in both the endometrium (Adams *et al.*, 1981) and conceptus (Harney *et al.*, 1990) and cellular mechanisms by which retinoic acid can affect cellular metabolism and gene expression by binding to nuclear receptors (see Chew, 1993), the retinoids can potentially have very direct effects on uterine function, and conceptus growth and differentiation.

Consistent with some of these possible mechanisms of action, Chew *et al.* (1982) reported an increase in uterine-specific proteins on day 15 of gestation after injecting gilts daily with β-carotene from the day of mating. This work was followed by the study of Brief and Chew (1985) in which various combinations of dietary and injected vitamin A and β-carotene were administered to gilts for one oestrous cycle before mating until weaning at day 21. Although there is extensive reference in the results to significant effects of both vitamin A and β-carotene injections on 'embryonic survival', only an increase in final litter size born was actually recorded and clear evidence that effects on embryonic development was an important component of the response was not provided. Actual improvements in embryo survival *per se* were later reported by Britt *et al.* (1992) in gilts in response to injections of retinol palmitate on day 15 of the oestrous cycle before breeding, and this treatment was also able to counteract the detrimental effects of high plane feeding after mating. There was also a suggestion in this report that the effectiveness of the treatment might have been related to reduced variance in embryo development at day 11–12 of gestation. A subsequent study (Coffey and Britt, 1993) involved large numbers of primiparous and multiparous sows already provided with diets supplemented with vitamin A. In Expt 1, sows received single injections of increasing doses of β-carotene (n = > 166 sows per dose) on the day of weaning. Although an overall effect of β-carotene

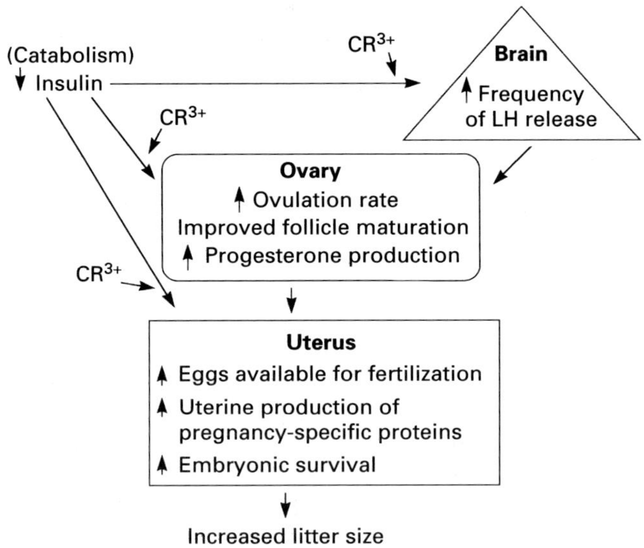

Fig. 5. Modification of the conceptual diagram of Trout (1995) indicating the mechanisms by which chromium (Cr^{3+}) may potentiate the effects of insulin and therefore improve reproductive function in insulin-deficient sows. The critical point is that chromium alone has no direct beneficial effects on reproduction; rather it exerts its effects by altering the overall metabolic state of the sow.

on litter size born failed to reach significance, there was evidence for a treatment by parity interaction, and in multiparous sows a linear increase in litter size born alive was recorded over the dose range of 0–200 mg. These effects were associated with an increase in plasma β-carotene on day 6 and 13 after injection but not at day 18, and no increase in plasma vitamin A compared with controls. No effect of treatment on plasma progesterone was recorded, although as samples were taken at approximately day 14 of pregnancy, this may not be very meaningful. On the assumption that these effects of β-carotene were direct (i.e. not mediated by changes in plasma vitamin A concentrations), and that increased plasma concentrations of β-carotene may need to be maintained beyond day 13 of treatment, in Expt 2 sows received repeated injections of either vitamin A, β-carotene or control vehicle (n = 140 per treatment) on the day of weaning, the day of mating and on day 7 after mating. A marginal effect of both vitamin A and β-carotene treatments on litter size born alive (10.6 in both treated groups compared with 10.0 in controls; $P < 0.10$) was reported. In the light of these generally positive responses to treatment with either vitamin A or β-carotene, it was suggested that effects on endometrial glandular secretion might be involved.

As with the studies on folic acid, there therefore appears to be evidence that there may be a beneficial response to vitamin A and β-carotene on embryo survival, but that in this case an inherent deficiency in either vitamin A or β-carotene may not be needed to show a response to supplementation. In the different studies reviewed, it is not always certain whether vitamin A and β-carotene are exerting independent effects and it will be important to clarify this in future studies. It is also uncertain whether treatment in the preovulatory period is essential for producing consistent responses. Therefore the mechanism of action may include effects on follicle and oocyte maturation, as well as effects on the uterus.

Indirect nutritional effects on the conceptus or uterus. A good example of how this might occur is seen in recent literature related to potential effects of chromium on reproductive function in sows.

The mechanisms mediating effects of chromium supplementation on growth performance in swine was recently extensively reviewed by White *et al.* (1993) and these authors stressed a number of important points. Firstly, chromium is a nutrient and not a therapeutic agent, and therefore its mechanism of action must be explained in terms of its metabolic effects. Secondly, effects of chromium were seen only in situations of deficiency but there was evidence that traditional diets used in swine production could be chromium deficient. Finally, at a mechanistic level chromium acted as an essential cofactor for insulin and therefore many of the clinical responses to chromium deficiency were analogous to insulin-dependent diabetes.

Given this scenario, and the discussion of insulin-mediated effects on reproductive function reviewed above, it is not unexpected that evidence for effects of chromium deficiency on sow fertility have been identified (Lindemann *et al.*, 1995). In addition, given the proven and potential effects of insulin at all levels of the reproductive system, it is difficult to identify the precise mechanism by which chromium supplementation would affect embryo survival *per se*. The various possibilities were summarized by Trout (1995) and a modification of his conceptual outline of these mechanisms is shown in Fig. 5. The importance of this particular example is that it illustrates that by understanding that the mechanism of action of chromium is to potentiate insulin-mediated effects on all tissues, we can also appreciate why it may have very diverse effects on reproductive function. In some of the examples of specific nutrient effects discussed earlier, it appears that the mechanism of action may be much more specific.

Conclusion

Building on previous reviews, this paper specifically addressed the question of nutritional effects on embryo survival at a mechanistic level. There seems to be convincing evidence that there are effects of nutrition and metabolic state on embryo survival in gilts and sows. Not surprisingly, the clearest responses to changing nutritional state occur when nutrition is limiting and in this situation many effects at all levels of the hypothalamo–hypophysial–ovarian axis can be anticipated. Furthermore, in situations in which embryo survival is limited by inadequate nutrition, a number of specific nutrients, as well as an overall increase in total dietary intake may produce positive responses.

From a general perspective, there seems to be increasing evidence that a number of the effects of nutrition and metabolic state on embryonic survival are mediated by progesterone. However, the precise mechanisms involved need to be elucidated and a concerted effort is needed to determine the timing and the location of progesterone-mediated effects.

A final comment relates to the conceptual outlines shown in Figs 1 and 3. The hypothalamic–hypophysial release of gonadotrophins, and particularly LH, is the essential driver of ovarian development. In many situations in which the nutrition of the gilt or sow is compromised, episodic release of LH is suppressed (see review of Foxcroft *et al.*, 1995). In these situations the lack of tonic LH secretion may have consequences for follicular development that will carry over to the time of ovulation and affect oocyte quality and luteal function. However, although there has been speculation that differences in LH secretion may affect subsequent fertility (for instance in the catabolic, first parity sow bred at first oestrus after weaning), there is no convincing evidence that these reproductive characteristics are functionally linked. It is possible that the pattern of LH secretion merely reflects the energy or protein status of the sow and that effects of metabolic state at the ovary or uterus mediate the decrease in subsequent fertility. A clearer definition of these possibilities is therefore needed. We are presently using GnRH-induced increases in LH secretion during periods of feed restriction in late lactation in sows to address this question.

The author wishes to express appreciation to present and former members of his research group for their input to the preparation of this review and particularly to Jody Barry and Shirley Shostak for preparation of the final manuscript and to Drs Frank Aherne, Jim Pettigrew and Louisa Zak for their comments and input. Financial and other support of the National Sciences and Engineering Research Council, The Alberta Pork Producers Development Corporation, The Alberta Agriculture Research Institute, Alberta Swine Genetics Inc., Pig Improvement (Canada) Ltd, and the University of Alberta for research reported in this review is also gratefully acknowledged.

References

Adams KL, Bazer FW and Roberts RM (1981) Progesterone-induced secretion of a retinol binding protein in the pig uterus *Journal of Reproduction and Fertility* **62** 39–47

Aherne FX and Kirkwood RN (1985) Nutrition and sow prolificacy *Journal of Reproduction and Fertility Supplement* **33** 169–183

Anderson LL and Melampy RM (1972) Factors affecting ovulation rate in the pig. In *Pig Production* Ed. DJA Cole. Butterworths, London

Ashworth CJ (1991) Effect of pre-mating nutritional status and post-mating progesterone supplementation on embryo survival and conceptus growth in gilts *Animal Reproduction Science* **26** 311–321

Ashworth CJ (1994) Nutritional factors related to embryonic mortality in the domestic species. In *Embryonic Mortality in Domestic Species* pp 179–194 Eds MT Zavy and RD Geisert. Boca Raton, Florida

Bazer FW, Clawson AJ, Robinson OW, Vincent CK and Ulberg LC (1968) Explanation for embryo death in gilts fed a high energy intake *Journal of Animal Science* **27** 1021–1026

Bazer FW, Clawson AJ, Robinson OW and Ulberg LC (1969) Uterine capacity in gilts *Journal of Reproduction and Fertility* **18** 121–124

Bazer FW, Geisert RD, Thatcher WW and Roberts RM (1982) The establishment and maintenance of pregnancy. In *Control of Pig Reproduction* pp 227–252 Eds DJA Cole and GR Foxcroft. Butterworths, London

Brief S and Chew BP (1985) Effects of vitamin A and β-carotene on reproductive performance in gilts *Journal of Animal Science* **60** 998–1004

Britt JH, Whaley SL and Hedgpeth VS (1992) Improvement of embryo survival by injection of vitamin A in gilts fed normal or high-energy diets before and after mating *Journal of Animal Science* **70** (Supplement 1) 271

Buhi WC (1997) Oviductal regulation of fertilization and early embryonic development *Journal of Reproduction and Fertility Supplement* **52** 285–300

Chew BP (1993) Effects of supplemental β-carotene and vitamin A on reproduction in swine *Journal of Animal Science* **71** 247–252

Chew BP, Rasmussen H, Pubols MH and Preston RL (1982) Effects of vitamin A and β-carotene on plasma progesterone and uterine protein secretions in gilts *Theriogenology* **18** 643–654

Clark JR, Daily RA, First NL, Chapman AB and Casida LE (1972) Effect of feed level and parity on ovulation rate in three genetic groups of swine *Journal of Animal Science* **35** 1216–1222

Clark JR, Edey TN, First NL, Chapman AB and Casida LE (1973) The effects of four genetic groups and two levels of feeding on ovulation rate and follicular development in pubertal gilts *Journal of Animal Science* **36** 1164–1169

Clowes EJ, Aherne FX and Foxcroft GR (1994) Effect of delayed breeding on the endocrinology and fecundity of sows *Journal of Animal Science* **72** 283–291

Coffey MT and Britt JH (1993) Enhancement of sow reproductive performance by β-carotene or vitamin A *Journal of Animal Science* **71** 1198–1202

Cox NM (1997) Regulation of follicular development and ovulation rate *Journal of Reproduction and Fertility Supplement* **52** 31–46

Dailey RA, Clark JR, First NL, Chapman AB and Casida LE (1972) Effects of high and low feeding at two stages of the oestrous cycle on follicular development in gilts from four genetic groups *Journal of Animal Science* **35** 1210–1222

Dailey RA, Clark JR, First NL, Chapman AB and Casida LE (1975a) Loss of follicles during the follicular phase of the estrous cycle of swine as effected by genetic group and level of feed intake *Journal of Animal Science* **41** 835–841

Dailey RA, Clark JR, First NL, Chapman AB and Casida LE (1975b) Effect of short-term 'flushing' on follicular development at estrus and ovulation rate of gilts of different genetic groups *Journal of Animal Science* **41** 842–848

Ding J and Foxcroft GR (1992) Follicular heterogeneity and oocyte maturation *in vitro* in pigs *Biology of Reproduction* **47** 648–655

Ding J and Foxcroft GR (1994) Conditioned media produced by follicular shells of different maturity affect maturation of pig oocytes *Biology of Reproduction* **50** 1377–1384

Faillace LS and Hunter MG (1994) Follicle development and oocyte maturation during the immediate preovulatory period in Meishan and white hybrid gilts *Journal of Reproduction and Fertility* **101** 571–576

Foxcroft GR and Hunter MG (1985) Basic physiology of follicular maturation in the pig *Journal of Reproduction and Fertility Supplement* **33** 1–19

Foxcroft GR, Shaw HJ, Hunter MG, Booth PJ and Lancaster RT (1987) Relationship between luteinizing hormone, follicle-stimulating hormone, prolactin secretion and ovarian follicular development in weaned sows *Biology of Reproduction* **36** 175–191

Foxcroft GR, Aherne FX, Clowes EJ, Miller HM and Zak LJ (1995) Sow fertility: the role of suckling inhibition and metabolic status. In *Animal Science Research and Development: Moving Towards a New Century* pp 377–391 Ed. M Ivan. Center for Food and Animal Research, Ottawa

Geisert RD and Yelich J (1997) Regulation of conceptus development and attachment *Journal of Reproduction and Fertility Supplement* **52** 133–149

Geisert RD, Renegar RH, Thatcher WW, Roberts RM and Bazer FW (1982) Establishment of pregnancy in the pig. I Interrelationship between preimplantation development of the pig blastocyst and uterine endometrial secretions *Biology of Reproduction* **27** 925–939

Grant SA, Hunter MG and Foxcroft GR (1989) Morphological and biochemical characteristics during ovarian follicular development in the pig *Journal of Reproduction and Fertility* **86** 171–183

Harney JP, Mirando MA, Smith LC and Bazer FW (1990) Retinol binding protein: a major secretory product of the pig conceptus *Biology of Reproduction* **42** 523–532

Harper AF, Lindemann MD, Chiba LI, Combs GE, Handlin DL, Kornegay ET and Southern LL (1994) An assessment of dietary folic acid levels during gestation and lactation on reproductive and lactational performance of sows: a cooperative study *Journal of Animal Science* **72** 2338–2344

Harper AF, Lindemann MD and Kornegay ET (1996) Fetal survival and conceptus development after 42 days of gestation in gilts and sows in response to folic acid supplementation *Canadian Journal of Animal Science* **76** 157–160

Hunter MG and Wiesak T (1990) Evidence for and implications of follicular heterogeneity in pigs *Journal of Reproduction and Fertility Supplement* **40** 163–171

Hunter MG, Southee JA and Lamming GE (1988) Function of abnormal corpora lutea *in vitro* after GnRH-induced ovulation in the anoestrous ewe *Journal of Reproduction and Fertility* **84** 139–148

Hunter MG, Grant SA and Foxcroft GR (1989) Histological evidence for heterogeneity in the development of preovulatory pig follicles *Journal of Reproduction and Fertility* **86** 165–170

Hunter MG, Picton HM, Biggs C, Mann GE, McNeilly AS and Foxcroft GR (1996) Periovulatory endocrinology in high ovulating Meishan sows *Journal of Endocrinology* **150** 141–147

Hunter RHF (1994) Causes for failure of fertilization in domestic species. In *Embryonic Mortality in Domestic Species* pp 1–22 Eds MT Zavy and RD Geisert. Boca Raton, Florida

I'Anson H, Foster DL, Foxcroft GR and Booth PJ (1991) Nutrition and reproduction *Oxford Reviews of Reproductive Biology* **13** 239–311

Jindal R, Cosgrove JR, Aherne FX and Foxcroft GR (1996) Effect of nutrition on embryonal mortality in gilts: association with progesterone *Journal of Animal Science* **74** 620–624

Jindal R, Cosgrove JR and Foxcroft GR (1997) Progesterone mediates nutritionally induced effects on embryonic survival in gilts *Journal of Animal Science* **75** 1063–1070

Knight JW, Bazer FW and Wallace HD (1974a) Effect of progesterone induced increase in uterine secretory activity on development of porcine conceptus *Journal of Animal Science* **39** 743–746

Knight JW, Bazer FW, Wallace HD and Wilcox CJ (1974b) Dose–response relationship between exogenous progesterone and estradiol and porcine uterine protein secretions *Journal of Animal Science* **39** 747–751

Krzymowski T, Kotwica J and Stefanczyk–Krzymowska S (1990) Uterine and ovarian countercurrent pathways in the control of ovarian function in the pig *Journal of Reproduction and Fertility Supplement* **40** 179–191

Liao CW and Veum TL (1994) Effects of dietary energy intake by gilts and heat stress from days 3 to 24 or 30 after mating on embryo survival and nitrogen and energy balance *Journal of Animal Science* **72** 2369–2377

Lindemann MD (1993) Supplemental folic acid: a requirement for optimizing swine reproduction *Journal of Animal Science* **71** 239–246

Lindemann MD and Kornegay ET (1989) Folic acid supplementation to diets of gestating lactating swine over multiple parities *Journal of Animal Science* **67** 459–464

Lindemann MD, Wood CM, Harper AF, Kornegay ET and Anderson RA (1995) Dietary chromium picolinate additions improve gain: feed and carcass characteristics in growing–finishing pigs and increase litter size in reproducing sows *Journal of Animal Science* **73** 457–465

McEvoy TG, Robinson JJ, Aitken RP, Findlay PA, Palmer RM and Robertson IS (1995) Dietary-induced suppression of pre-ovulatory progesterone concentrations in superovulated ewes impairs the subsequent *in vivo* and *in vitro* development of their ova *Animal Reproduction Science* **39** 89–107

McKelvey WAC, Robinson JJ and Aitken RP (1988) The use of reciprocal embryo transfer to separate the effects of pre- and post-mating nutrition on embryo survival and growth of the ovine conceptus *Proceedings of the 11th International Congress for Animal Reproduction and Artificial Insemination, Dublin* (Abstract 176)

Matte JJ, Girard CL and Brisson GJ (1982) Serum folates levels during the reproductive cycle of sows *Journal of Animal Science* **55** (Supplement 1) 284

Matte JJ, Girard CL and Brisson GJ (1984a) Folic acid and reproductive performance of sows *Journal of Animal Science* **59** 1020–1025

Matte JJ, Girard CL and Brisson GJ (1984b) Serum folates during the reproductive cycle of sows *Journal of Animal Science* **59** 158–163

Morbeck DE, Esbenshade KL, Flowers WL and Britt JH (1992) Kinetics of follicle growth in the prepubertal gilt *Biology of Reproduction* **47** 485–491

Parr RA, Cumming IA and Clark IJ (1982) Effects of maternal nutrition and plasma progesterone concentrations on survival and growth of the sheep embryo in early gestation *Journal of Agricultural Science* **98** 39–46

Parr RA, Davis IF, Fairclough RJ and Moss MA (1987) Overfeeding during early pregnancy reduces peripheral progesterone concentration and pregnancy rate in sheep *Journal of Reproduction and Fertility* **80** 317–320

Parr RA, Davis IF, Miles MA and Squires TJ (1993a) Feed intake affects metabolic clearance rate of progesterone in sheep *Research in Veterinary Science* **55** 306–310

Parr RA, Davis IF, Miles MA and Squires TJ (1993b) Liver blood flow and metabolic clearance rate of progesterone in sheep *Research in Veterinary Science* **55** 311–316

Pharazyn A (1992), *Edmonton, Canada Nutritional Effects on Embryo Survival in the Gilt* PhD Thesis, University of Alberta, Edmonton

Pharazyn A, Den Hartog LA, Foxcroft GR and Aherne FX (1991a) The influence of dietary energy and protein intake during early pregnancy on plasma progesterone and embryo survival *Canadian Journal of Animal Science* **71** 949–952

Pharazyn A, Foxcroft GR and Aherne FX (1991b) Temporal relationship between plasma progesterone concentrations in the utero–ovarian and jugular veins during early pregnancy in the pig *Animal Reproduction Science* **26** 323–332

Pope WF (1988) Uterine asynchrony: a cause of embryonic loss *Biology of Reproduction* **39** 999–1003

Pope WF (1994) Embryonic mortality in swine. In *Embryonic Mortality in Domestic Species* pp 53–77 Eds MT Zavy and RD Geisert. Boca Raton, Florida

Pope WF, Wilde MH and Xie S (1988) Effect of electro-cautery of nonovulated day 1 follicles on subsequent morphological variation among day 11 porcine embryos *Biology of Reproduction* **39** 882–887

Pope WF, Xie S, Broermann DM and Nephew KP (1990) Causes and consequences of early embryonic diversity *Journal of Reproduction and Fertility Supplement* **40** 250–260

Prime GR and Symonds HW (1993) The influence of plane of nutrition on portal blood flow and the metabolic clearance rate of progesterone on ovariectomized gilts *Journal of Agricultural Science* **121** 389–397

Roberts RM and Bazer FW (1988) The function of uterine secretions *Journal of Reproduction and Fertility* **82** 875–892

Roberts RM, Xie S and Trout WE (1993) Embryo–uterine interactions in pigs during week 2 of pregnancy *Journal of Reproduction and Fertility Supplement* **48** 171–186

Schultz JR, Speer VC, Hays VW and Melampy RM (1966) Influence of feed intake and progesterone on reproductive performances in swine *Journal of Animal Science* **25** 157–160

Talavera F and Chew BP (1988) Comparative role of retinol, retinoic acid and β-carotene on progesterone secretion by pig corpus luteum *in vitro*. *Journal of Reproduction and Fertility* **82** 611–615

Thaler RC, Nelssen JL, Goodband RD and Allee GL (1989) Effect of dietary folic acid supplementation on sow performance through two parities *Journal of Animal Science* **67** 3360–3369

Tremblay GF, Matte JJ, Lemieux L and Brisson GJ (1986) Serum folates in gestating swine after folic acid addition *Journal of Animal Science* **63** 1173–1178

Tremblay GF, Matte JJ, Dufour JJ and Brisson GJ (1989a) Survival rate and development of fetuses during the first 30 days of gestation after folic acid addition to a swine diet *Journal of Animal Science* **67** 724–732

Tremblay GF, Matte JJ, Girard CL and Brisson GJ (1989b) Serum zinc, iron and copper status during early gestation in sows fed a folic acid supplement diet *Journal of Animal Science* **67** 733–737

Trout WE (1995) Hypothesis provides possible explanation as to chromium's effect on reproductive efficiency in swine *Feedstuffs* **67** 12–23

Webel SK, Peters JB and Anderson LL (1970) Synchronous and asynchronous transfer of embryos in the pig *Journal of Animal Science* **30** 565–568

White M, Pettigrew J, Zollitsch-Stelzl J and Crooker B (1993) Chromium in swine diets. In *Proceedings of the 54th Minnesota Nutrition Conference and National Renderers Technical Symposium* pp 251–261

Xie S, Broermann DM, Nephew KP, Ottobre JS, Day ML and Pope WF (1990a) Changes in follicular endocrinology during final maturation of porcine oocytes *Domestic Animal Endocrinology* **7** 75–82

Xie S, Broermann DM, Nephew KP, Bishop MD and Pope WF (1990b) Relationship between oocyte maturation and fertilization of zygotic diversity in swine *Journal of Animal Science* **68** 2027–2033

Xie S, Broermann DM, Nephew KP, Geisert RD and Pope WF (1990c) Ovulation and early embryogenesis in swine *Biology of Reproduction* **43** 236–240

Zak LJ, Cosgrove JR, Aherne FX and Foxcroft GR (1997a) Pattern of feed intake and associated metabolic and endocrine changes differentially affect postweaning fertility in primiparous lactating sows *Journal of Animal Science* **75** 208–216

Zak LJ, Xu X, Hardin RT and Foxcroft GR (1997b) Impact of different patterns of feed intake during lactation in the primiparous sow on follicular development and oocyte maturation *Journal of Reproduction and Fertility* **110** 99–106

Poster titles

Quesnel H, Pasquier A Mounier AM and A Prunier
Restricting feed intake during lactation inhibits LH pulsatility and follicular development around weaning in sows

Messias de Braganca M, Mounier AM and A Prunier
Comparison of the effect of undernutrition and elevated ambient temperature on reproductive function of primiparous sows

Prunier A, Peintre J and AM Mounier
Role of the nutrient deficit in the lactational inhibition of the reproductive function in multiparous sows

Brand H van den, Kemp B, Soede NM and JW Schrama
Effects of a carbohydrate and fat rich diet at two feeding levels during lactation of first litter sows on subsequent reproduction parameters

Dawson A, Reynolds TS, Boulton M and AR Peters
Expression of type 1 insulin-like growth factor receptor mRNA in the ovary of the sow throughout the oestrous cycle

Parmeggiani A, Galeati G, Govoni N, Spinaci M and E Seren
2-Deoxy-D-glucose effect on pulsatile luteinizing hormone secretion in prepubertal gilts

Swanchara KW, Armstrong JD, Farin CE and WL Flowers
Immunization against growth hormone releasing factor (GrFi) in prepubertal gilts delays normal ovarian development and alters the concentrations of insulin-like growth factor binding proteins (IGFBP)-2, -3, -4 and -5 in follicular fluid

Feng SM and GW Almond
Affinities and concentrations of PGE receptors on porcine luteal cells

Barb CR, Barrett JB, Kraeling RR, Rampacek GB, Yan X and TG Ramsay
Leptin modulation of luteinizing hormone (LH) secretion by pig pituitary cells in culture

Gregoraszczuk EL and M Slomczynska
Effect of ß-endorphin on progesterone and oestradiol secretion by porcine granulosa cells during follicular development

Brussow KP, Schneider F and W Kanitz
How the gonadotropin-releasing hormone antagonist antarelix influences the reproductive hormone levels in cyclic gilts

Kemp B, Soede NM and AMB ten Vergert
Effect of timing of ovulation induction after PMSG on the ovulation process and early luteal and embryonic development in sows

Jayes FCL and JH Britt
Oestradiol-17ß decreases basal and GnRH-stimulated release of LH and FSH from perifused porcine pituitary cells

FACTORS AFFECTING BREEDING EFFICIENCY IN THE PIG

Chair
T. van der Lende

Journal of Reproduction and Fertility Supplement **52**, 67–78

Management of boars for efficient semen production

W. L. Flowers

Department of Animal Science, North Carolina State University, Raleigh, NC 27695–7621, USA

Consistent production of large quantities of fertile semen is a primary concern of boar management programmes. Common visual tests such as motility are poor indicators of the fertilizing capacity of an ejaculate. Research studies have demonstrated that, in general, if motility is 60% or greater, there are no relationships among the percentage of motile spermatozoa, *in vitro* sperm penetration rates, farrowing rates and litter size. Biological and environmental factors that influence the production of total numbers of spermatozoa are either inhibitory or stimulatory. Chronic undernutrition and heat stress inhibit spermatogenesis, while photoperiod and genetic factors are stimulatory. A consistent stimulatory effect of photoperiod does not appear to occur under all environmental conditions. In contrast, selection for increased testis size consistently results in greater sperm production. In addition, libido is decreased by heat stress and undernutrition. However, failure to allow boars to interact socially with other pigs during pubertal development appears to be the most damaging factor. Effective boar management programmes must minimize the occurrence of factors that adversely affect spermatogenesis and simultaneously accentuate those with a stimulatory role.

Introduction

Each time a mating occurs, the boar and sow are generally viewed as contributing equally towards its outcome. Fertilization requires placement of adequate numbers of fertile spermatozoa in the female reproductive tract before ovulation which is closely linked to reproductive processes and behaviour in boars. Establishment of pregnancy involves maintenance of a uterine environment conducive for embryonic and fetal development which is predominantly a biological function of sows. However, when the entire breeding herd is considered, a single boar has a significantly greater impact on efficiency and productivity than an individual sow. This is because a female normally produces two litters per year, whereas a male can sire from 50 to 3000 depending on the type of mating system that is used. Consequently, management strategies for boars tend to influence herd productivity in an exponential fashion and are a major component of reproductive efficiency of swine operations.

Regardless of whether matings are natural or artificial, production of fertile spermatozoa is an absolute requirement for boars and the measure by which their reproductive merit should be evaluated. Therefore, the primary intent of this review is to examine recent developments in terms of our current understanding of biological and environmental factors that influence qualitative and quantitative aspects of sperm production in boars. In addition to spermatogenesis, mating behaviour of boars also has important ramifications with regard to the efficient production of semen especially in breeding programmes using artificial insemination. Consequently, a secondary objective is to consider strategies that have been implemented to enhance the sex drive of boars. The end result should be a summary of relevant reproductive and management data from which informed decisions can be made with regard to managing semen production in boars.

Semen Production

Consistent production of large numbers of fertile spermatozoa is the ultimate goal of a successful boar management programme. As a result, qualitative and quantitative aspects of semen production have been the primary focus of research for improving boar fertility. For the purposes of this review, quality will refer to fertility, defined as the ability of semen to produce live pigs, and quantity will be defined as the number of spermatozoa produced. It is important to note that there is not an obligatory association between these two aspects of semen production. Thus, each will be considered separately.

Production of fertile semen

The ability to assess the fertility of semen is a necessary prerequisite for studying the effects of various management strategies designed to improve its quality. In both the scientific and commercial sectors of the swine industry, the most commonly used tests for evaluating sperm quality involve visual estimations of the percentage of motile or morphologically normal spermatozoa (Colenbrander and Kemp, 1990; Woelders, 1991). Furthermore, a common assumption is that ejaculates with a high percentage of motile or morphologically normal sperm cells are of higher quality and more fertile than those with low estimates of the same parameters. In fact, a common practice within commercial boar studs is to adjust the total number of spermatozoa in insemination doses based on sperm motility, or morphology or both factors (Colenbrander *et al.*, 1993). While it is true that, during fertilization, pig spermatozoa rely on their forward motion to move through the cumulus layers and penetrate the zona pellucida (Polge, 1978), the value of using the motility of a raw ejaculate as an indication of its ability to fertilize eggs upon insemination is less clear (See Holt and Medrano, this supplement). In other species, the percentage of motile sperm cells is generally considered to be a poor measure of semen fertility (Linford *et al.*, 1976; Amann, 1988).

Results from a recent study (Flowers and Turner, in press) that examined the relationship between motility and fertility are presented in Table 1. In this experiment, spermatozoa were obtained once a week from 12 mature boars for 26 weeks. Evaluations for motility were conducted on each ejaculate. A portion of each collection was processed for *in vitro* fertilization, while the remainder was extended and used to inseminate at least five sows. Semen characteristics were divided into ten groups, for example 90 to 100%, 80 to 89%, and treated as an independent variable. Special precautions were taken to ensure that all insemination doses administered to an individual sow during oestrus and the corresponding subsample used *in vitro* were from the same motility classification. Consequently, the experimental design provided both *in vitro* and *in vivo* estimates of fertility for ejaculates exhibiting a wide range of semen motility.

These data provide several important observations concerning the use of the percentage of motile spermatozoa as an estimate of semen fertility in swine. First, motility, at best, can be used to establish the lower limit of acceptability for semen use, but is of little analytical value for ejaculates above this value. This is evident in that sperm penetration rates, farrowing rates and number of pigs born alive were not different among ejaculates estimated to contain more than 60% motile spermatozoa. In contrast, reproductive performance was reduced in sows inseminated with semen the motility of which was less than 60%. Second, there appear to be individual boars that deviate from this general relationship between motility and fertility. When motility groups were considered in conjunction with individual boar effects, a significant interaction was present. This interaction was attributed to boars whose fertility tended to be low when semen parameters were high or vice versa (Table 2). Finally, inclusion of additional visual criteria such as morphology and acrosome integrity do not improve the predictive value of motility. Adjustments for morphology and acrosome integrity within each motility group did not alter the relationship between the percentage of motile spermatozoa and fertility estimates (data not shown). In summary, it appears that use of the percentage of motile spermatozoa is a qualitative rather than a quantitative measure of semen fertility.

This conclusion should not be surprising. In addition to the mechanical requirements associated with egg penetration, spermatozoa must also undergo biochemical, metabolic and molecular

Table 1. Relationships among motility, sperm penetration rates, farrowing rates and number of pigs born alive for boar semen

Motility[a] (%)	Sperm penetration rate[b] (%)	Farrowing rate[c] (%)	Number born alive
94.7	89.5[v] (58)	86.9[v] (460)	10.6[v]
82.3	81.7[v,w] (55)	87.1[v] (330)	10.5[v]
76.1	84.3[v,w] (50)	84.5[v] (300)	10.5[v]
66.2	74.7[w] (44)	86.1[v] (264)	10.1[v]
52.4	55.5[x] (40)	72.4[w] (201)	9.2[w]
44.2	34.7[y] (28)	72.3[w] (168)	9.2[w]
32.6	21.3[z] (17)	51.7[x] (85)	7.8[x]
SEM[D]	4.8	5.8	0.3

[a]Motility is expressed as the average number of motile spermatozoa within the following classes: >90; 80–89; 70–79; 60–69; 50–59; 40–49; and 30–39.
[b]Sperm penetration rate is defined as the percentage of eggs that were fertilized. Numbers in parentheses represent the number of ejaculates within a motility category.
[c]Numbers in parentheses represent the number of sows inseminated with semen within a motility category.
[d]Standard error of the mean.
[v,w,x,y,z]Means within the same column with different superscripts are significantly different ($P < 0.05$).

changes before they acquire fertilizational competence. A variety of different tests based on one or more of these events has been or is in the process of being developed (Table 3). For some of these tests there are estimates of their relationship to sperm fertility. The most promising of these attempt to quantify specific aspects of sperm binding to egg membranes or structural changes within sperm membranes or DNA. Obviously, the ideal fertility test for efficient production of semen needs to be proactive, accurate, quick and economical. While many of these new tests appear to fulfil the first two requirements, most require extended analytical periods or sophisticated equipment. As a result, their acceptability and use within the commercial sector, at the present time, remains to be determined.

In addition to spermatozoa, there is an increasing abundance of evidence that indicates that other components in semen participate in physiological events associated with fertilization (Claus, 1990; Flowers and Esbenshade, 1993; Waberski *et al.*, 1995; Waberski, this supplement). Consequently, it is reasonable to speculate that male fertility may be related, in part, to concentrations of these compounds normally present in semen. At the present time, there is little direct evidence for this speculation in swine. However, consideration of the results from several related experiments provide indirect support for this contention.

In a recent study, DNA fingerprinting techniques were used to determine the paternity of piglets within a litter when several boars were used for breeding (Flowers *et al.*, 1996). The conclusions were that a single boar sired the majority of pigs within a litter (>80%) and that this boar was usually the one used for mating 12 or 24 h after the first detected oestrus. This was expected because previous studies have suggested that the mating that occurs closest to ovulation yields most of the piglets within a litter (Vesseur *et al.*, 1996) and, in this particular herd, the average oestrus-to-ovulation interval was determined by ultrasonography to be 32.6±4.5 h. However, there was one boar that consistently sired the majority of pigs within a litter when he mated sows at 0 and 12, but not 24, h after the onset of oestrus. These data are presented in Table 4 and this boar was referred to as being

Table 2. Relationships among semen motility, sperm penetration rates, farrowing rates and number of pigs born alive for four individual boars

Boar number	Motility[a] (%)	Sperm penetration rate[b] (%)	Farrowing rate[c] (%)	Number of pigs born alive
1	87.3 ± 5.3^x (26)	84.6 ± 4.8^x	83.2 ± 6.3^x (147)	10.9 ± 0.3^x
2	78.3 ± 6.1^x (26)	62.3 ± 5.1^y	67.3 ± 5.8^y (133)	9.5 ± 0.4^y
3	55.3 ± 5.7^y (26)	81.4 ± 6.7^x	84.1 ± 6.1^x (130)	10.7 ± 0.3^x
4	51.3 ± 4.1^y (26)	52.3 ± 5.4^y	70.3 ± 4.8^y	8.7 ± 0.3^y

[a]Motility is the average percentage of motile spermatozoa. Numbers in parentheses indicate the number of ejaculates that were evaluated.
[b]Sperm penetration rate is the average percentage of eggs that were fertilized.
[c]Numbers in parentheses indicate the number of sows that were mated per boar.
[x,y]Means within the same column with different superscripts are significantly different ($P < 0.05$).

dominant because if he was used for the first or second mating, then he sired the majority of the litter. One of the explanations for this observation was that, for whatever reason, spermatozoa from the dominant boar were retained longer in the female reproductive tract than were those of other males used in the study (designated as non-dominant). Variation among boars in the retention time of their spermatozoa in the female reproductive tract has been described by Mburu *et al.* (1996) and, in this particular case, was confirmed by subsequent studies as numbers of oviductal spermatozoa from the dominant boar at 24 h after mating were five times higher than those from non-dominant males (W. L. Flowers, unpublished). An equally feasible explanation that was not explored was that the dominant boar induced an earlier ovulation compared with non-dominant boars. Both retention of spermatozoa in the female reproductive tract and advancement of ovulation have been attributed to seminal plasma (Claus, 1990; Waberski *et al.*, 1995).

Another study was initiated in which heterogeneous inseminations were made by mixing the sperm-rich and seminal plasma fractions of the dominant boar with those of a non-dominant male to determine whether this advantage in sperm retention was a function of spermatozoa or seminal plasma (W. L. Flowers, D. G. McLaren and K. Siggens, unpublished). An outline of the experimental design is shown (Fig. 1). The various combinations of spermatozoa and seminal plasma were used for matings at first detected oestrus, whereas additional boars were used for matings 12 and 24 h later. Paternity results from these mating treatments are shown in Table 5.

These results indicate that at least part of the dominance effect resides in the seminal plasma. This is evident by the fact that the proportion of piglets sired by the dominant boar was lower when his spermatozoa were mixed with seminal plasma from a non-dominant male, whereas the converse was true with the reciprocal combination, non-dominant semen with dominant seminal plasma. With regards to semen fertility, the relevant finding in this study was that the paternity pattern for piglets was altered by the removal and addition of seminal plasma to spermatozoa. Thus, the relative importance of seminal plasma and its composition deserves consideration when attempting to evaluate the fertility of pig semen. Further credence for this statement is provided by recent studies with cattle. In dairy bulls, the presence of four seminal plasma proteins has been shown to be highly correlated with fertility (Killian *et al.*, 1993), and there is a strong relationship between calving rates and concentrations of heparin-binding proteins in semen of beef bulls (Bellin *et al.*, 1994).

In summary, to produce fertile pig semen, it is necessary to evaluate its fertility accurately. At the present time, most of the visual criteria used to assess semen quality are, at best, crude indicators of its fertilizing properties. Consequently, as new tests with greater accuracy are developed, it will be important to use these for evaluation of semen not only on commercial farms, but also in the

Table 3. Summary of various measurements used to estimate fertilizing ability of boar spermatozoa

Measurement	Rationale	References
Percoll gradients	Velocity of spermatozoa is positively correlated with fertility. Spermatozoa with increased velocities can be separated via Percoll gradients	Grant *et al.*, 1994
Fluorescent stains	Some fluorescent stains (Hoechst 33258) enter sperm cell if membrane is damaged and stain only dead cells. Others (SYBR-14) enter spermatozoa with a membrane potential and stain only live cells	Johnson *et al.*, 1996
Sperm chromatin structure	When subjected to acidic conditions and stained with metachromatic dye acridine orange, double-stranded DNA give a green fluorescence, while single-stranded DNA (damaged) give a red fluorescence. Ratio of green to red spermatozoa is related to fertility	Evenson *et al.*, 1994
Sperm plasma membrane protein profile	Plasma membrane proteins were separated and quantified. Three proteins were positively correlated with sperm binding to egg membranes ($r = 0.38$ to 0.53). Two proteins were negatively correlated with sperm binding ($r = -0.42$ and -0.37)	Ash *et al.*, 1994
Oocyte membrane binding assay	Fertile spermatozoa bind to egg membranes. Correlation between egg binding and fertility *in vivo* from heterospermic inseminations was 0.80	Berger *et al.*, 1996
Hemizona binding assay	Oocytes are bisected and the number of spermatozoa from different boars binding to each half of the zona pellucida are counted. Within a hemizona pair, increased binding equates to increased fertility	Fazeli *et al.*, 1995

research community as strategies for enhancing male fertility are investigated. Moreover, the composition of seminal plasma may also need to be included in future assessments of semen fertility.

Production of total number of spermatozoa

In general, management programmes designed to optimize production of the total number of spermatozoa can be divided into two general categories: (i) prevention of detrimental situations and (ii) enhancement of stimulatory influences. In essence, techniques in the first group can be thought of as removal of situations that reduce spermatogenesis, while those in the latter classification are attempts to enhance male gamete formation. Although most of the techniques discussed are dependent upon the age of the boar, there are a few instances where prepubertal males appear to respond in a different way from mature males. When appropriate, these situations will be acknowledged and discussed.

Table 4. Paternity distribution (%±SEM) of pigs from sows mated to three different boars during oestrus

Variable[a]	First sire[b]	Second sire[b]	Third sire[b]
Litters in which the dominant boar was the first sire (10)	89.3 ± 7.5^{x}	4.5 ± 1.2^{x}	6.2 ± 1.3^{x}
Litters in which the dominant boar was the second sire (10)	7.7 ± 4.3^{y}	88.2 ± 8.6^{y}	4.3 ± 1.2^{x}
Litters in which the dominant boar was the third sire (10)	25.4 ± 7.5^{z}	35.7 ± 8.2^{z}	38.1 ± 9.1^{z}
Litters in which non-dominant boars were used as sires (70)	25.7 ± 3.3^{z}	41.5 ± 3.7^{z}	32.8 ± 4.5^{z}

[a]Numbers in parentheses indicate number of litters that were evaluated.
[b]First, second and third sire refer to boars that inseminated sows at 0, 12 and 24 h after the onset of oestrus, respectively.
[x,y,z]Means with different superscripts within the same column are significantly different ($P < 0.05$).

Prevention of detrimental situations. Nutritional influences on sperm production in mature boars have been reviewed by Colenbrander and Kemp (1990) and Kemp and Verstegen (1991). The general consensus from studies in this area is that prolonged periods of undernutrition reduce the total number of spermatozoa produced. Specifically, the magnitude of protein and energy restrictions reported varied from 25–60% and 15–67%, respectively, and the restriction period had to be at least 6–8 weeks before significant differences were observed. The negative effect of undernutrition on spermatogenesis is more pronounced when protein rather than energy is restricted. In fact, Louis and coworkers (1994a) did not observe a reduction in sperm output in boars on a restricted energy diet. However, when dietary levels of only protein or both protein and energy were lowered, the total numbers of spermatozoa were reduced by 18% and 27%, respectively. Similar studies conducted in prepubertal boars revealed that moderate restrictions in energy and protein during sexual maturation did not affect spermatogenesis (Hacker *et al.*, 1994).

Several different groups have attempted to estimate protein and energy requirements for male reproduction (Kemp *et al.*, 1991; Close and Roberts, 1993). For mature boars (226 kg), 37 MJ of metabolizable energy and 230 g of crude protein per day appear to be adequate for meeting maintenance, libido and spermatogenic requirements. It is important to note that maintenance requirements for boars are influenced by a variety of environmental factors including health status, ambient temperature and frequency of use. Thus, it is probable that maintenance needs for boars differ among farms. As a result, variability in studies investigating the effect of energy and protein manipulations on reproductive function in boars should be expected.

Perhaps a more relevant point of discussion concerning boar nutrition is the early identification of suboptimal nutritional regimens. Because spermatogenesis requires between 39 and 50 days in boars (Swierstra, 1968), numbers and morphology of spermatozoa would not be expected to reflect nutritional deficiencies immediately. Consequently, a problem could remain undetected for several months. In studies that used chronic restriction regimens, aberrant mating behaviour consistently preceded measurable changes in ejaculate characteristics (Louis *et al.*, 1994a, b). These included extended recognition times, reduced durations of ejaculation and increased frequencies of mating refusals. On the basis of these observations, changes in normal mating behaviour may have potential for use in diagnosing marginal protein and energy deficiencies in boars.

In contrast to the effects of nutrition, exposure to high ambient temperatures has immediate, severe and consistent effects on spermatogenesis (Christenson *et al.*, 1972; Wettemann and Bazer, 1985). These include ejaculation of a large proportion of spermatozoa with morphological abnormalities, fewer

Table 5. Paternity distribution (%+SEM) of pigs from sows mated with various combinations of seminal plasma and spermatozoa from dominant and non-dominant boars at the onset of oestrus

Treatment[a]	First mating[b]	Second mating[c]	Third mating[c]
Dominant spermatozoa + dominant seminal plasma (10)	80.3±6.1[x]	8.1±3.2[x]	11.6±3.9[x]
Dominant spermatozoa + non-dominant seminal plasma (10)	62.2±8.3[y]	17.4±6.9[x]	20.4±5.1[x]
Non-dominant spermatozoa + dominant seminal plasma (10)	48.8±9.1[y]	30.1±5.4[y]	18.7±6.2[x]
Non-dominant spermatozoa + non-dominant seminal plasma (10)	22.4±8.3[z]	44.1±5.9[z]	33.5±6.1[y]

[a]Numbers in parentheses indicate number of litters that were evaluated.
[b]First mating refers to combinations of seminal plasma and spermatozoa from dominant and non-dominant boars that were administered at the onset of oestrus.
[c]Second and third matings were administered at 12 and 24 h after the onset of oestrus, respectively, by non-dominant boars not used for the first mating.
[x,y,z]Means with different superscripts within the same column are significantly different ($P < 0.05$).

spermatozoa and reduced fertility. The minimum exposure time and critical air temperature above which sperm production is impaired have been reported to be 30°C (Stone, 1982) and 72 h (McNitt and First, 1970), respectively. However, most of the studies conducted to date have used longer periods of exposure (1–4 weeks) and higher temperatures (33–37°C).

Obviously, an important aspect of efficient semen production is to prevent boars from being exposed to acute periods of heat stress. However, few studies have examined alterations in sperm output from boars exposed for extended periods to temperatures in the upper range of their thermal comfort zone, 26–29°C. Chronic exposure to temperatures within this range probably occurs routinely during the summer months in mechanically ventilated facilities located in temperate and semi-tropical climates. Field data from seven commercial boar studs in southeastern North Carolina from June through September is shown in Fig. 2. Each stud houses an average of 200 boars and has a weekly production goal of 2000 insemination doses. The average weekly high temperature in these facilities never increased above 29°C during the recording period. However, a significant increase in the number of ejaculates that were rejected due to poor quality and a corresponding decrease in the number of insemination doses per ejaculate were observed. It is interesting to note that the downward trend in the number of doses per ejaculate began 5 to 6 weeks after the weekly high temperatures stabilized at 27.5°C. In acute studies, a lag period of about 2 weeks is commonly observed between the initiation of heat stress and the first indications of abnormal sperm production (Wettemann *et al.*, 1979). It is possible that these data provide indirect evidence that boars may also be sensitive to chronic periods of moderately high temperatures not normally associated with heat stress conditions. The longer interval between the temperature increase and the first signs of abnormal sperm production may reflect a gradual reduction in the ability of boars to dissipate heat or subtle adjustments in energy partitioning to compensate for the increased demands for maintaining thermal neutrality over an extended period. Regardless of the mechanism, these field data indicate that the idea of chronic heat stress resulting from long periods of exposure to moderately high temperatures merits further investigation.

Enhancement of stimulatory influences. From a physiological perspective, an increase in the total number of spermatozoa produced can be accomplished in one of two ways: either by increasing the number of type A spermatogonia that initiate mitosis at a given time or by increasing the frequency

Seminal plasma \ Spermatozoa		
	Dominant + Dominant	Non-dominant + Dominant
	Dominant + Non-dominant	Non-dominant + Non-dominant

Fig. 1. Experimental treatments consisting of combinations of spermatozoa and seminal plasma. Sperm rich fractions of semen were collected separately from seminal plasma of dominant and non-dominant boars and recombined in a factorial arrangement of treatments as indicated (sperm origin plus seminal plasma origin).

at which this occurs. Both processes are regulated by complex endocrine and paracrine interactions. As a result, most of the management strategies that have attempted to increase sperm production are believed to work through these regulatory pathways.

The effect of photoperiod on sperm production in boars still awaits resolution. Studies investigating the influence of changes in photoperiod on sperm production are equivocal. Mazzari *et al.* (1968) reported that mature boars exposed to long daylengths exhibited a reduction in total numbers of spermatozoa. In contrast, other studies have reported increased spermatozoa output by increasing light when natural daylength is decreasing or by decreasing light when the natural daylength is increasing (Claus and Weiler, 1985). Brandt and Diekman (1985) did not observe any effects of supplemental lighting on spermatogenesis or concentrations of LH and testosterone. A similar situation is found for the influence of photoperiod on sperm production in prepubertal boars. Additional lighting may hasten the onset of mature testicular function (Hoagland and Diekman, 1982), but does not appear to yield consistent effects on cellular and endocrine aspects of spermatogenesis (Brandt and Diekman, 1985; Minton and Wettemann, 1987).

Thus, it is difficult to formulate definitive recommendations regarding the optimal lighting regimen for spermatogenesis. In addition, failure to measure changes in endocrine activity associated with photoperiodic responses suggests that while in unique situations the light : dark ratio may temporarily influence spermatogenesis, it is probably not a major avenue through which sperm output can be manipulated with consistent results. This statement is supported by the observation that in most males, including wild boars, in which spermatogenesis is linked to photoperiod, there is a corresponding change in the secretion of gonadotrophins and testosterone (Mauget, 1982).

In contrast to the influence of photoperiod, there appears to be a high positive correlation between testis size and sperm production in both prepubertal and mature boars (Rathje *et al.*, 1995). Boars selected for testis size produced 14.6×10^9 more spermatozoa than did randomly selected control males (Huang and Johnson, 1996). On the basis of the industry standard of $3–4 \times 10^9$ spermatozoa per insemination, these values would translate into an additional four to five doses per week. Furthermore, repeatabilities for sperm production traits in this study were reported to be between 0.26 and 0.40 (Huang and Johnson, 1996). Collectively, these data indicate that selection for increased testis size can be used to enhance the reproductive capacity of boars.

As stated earlier, management strategies that enhance sperm production, including selection for increased testicular size probably have a strong endocrine component. This is because most major testicular functions such as establishment of normal populations of Sertoli and Leydig cells and regulation of the timing and frequency of spermatogenic waves are controlled primarily by testosterone, gonadotrophins and growth hormone (Sharpe, 1995). Consequently, hormonal manipulation may offer potential for increasing sperm production. However, for this approach to be practical, identification of critical periods during which brief periods of administration would have long-lasting effects on

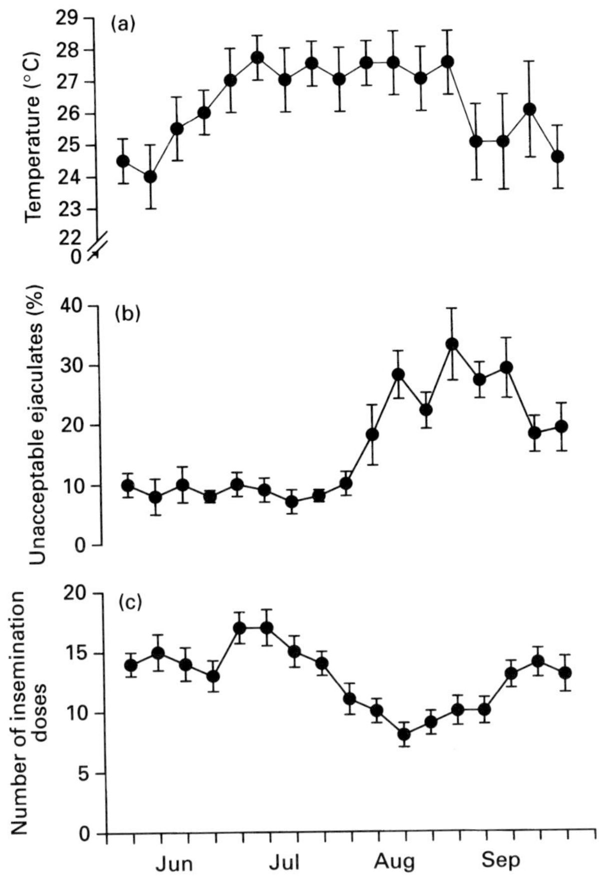

Fig. 2. Mean weekly high ambient temperature (a), percentage of ejaculates rejected for further processing (b) and average number of insemination doses per ejaculate (c) for commercial boar studs in North Carolina between June and September.

spermatogenesis is necessary. At the present time, several studies indicate that this may be possible. Swanlund *et al.* (1995) administered FSH or growth hormone to boars between 8 and 40 days of age. Growth hormone promoted tubular and Sertoli cell maturation, whereas FSH increased the number of Sertoli cells. In another study, administration of a reversible goitrogen, 6-propyl-2-thiouracil, for the first 8, 16 or 24 days after birth increased testis size and daily sperm production in adult rats (Cooke *et al.*, 1992). Obviously, considerably more research is needed to determine whether sperm production in boars can be manipulated effectively via hormonal regimens. However, it is important to remember that the use of exogenous gonadotrophins and prostaglandins at specific times during the production cycle has proved to be beneficial for enhancing reproductive efficiency in female swine.

Libido and Mating Behaviour

In boar management programmes, maintenance of high levels of libido probably deserves equal consideration as increasing sperm output. It could be argued that production of large amounts of fertile semen is of limited value if the desire to mate is not present. Unfortunately, our current understanding of factors affecting the sex drive of boars is more anecdotal than scientific. This may be

Table 6. Summary of management situations that influence libido and mating behaviour in boars

Situation	Age	Effect	Reference
Protein restriction	Mature (> 1 year)	Increased time to initiate ejaculation Decreased duration of ejaculation	Louis *et al.*, 1994a,b
Ad libitum feeding	Prepubertal (between 1 and 8 months of age)	Decreased chomping and salivation	Hacker *et al.*, 1994
Individual housing	Prepubertal (between 1 and 8 months of age)	Increased recognition time Increased incorrect mounts Decreased mating scores	Hemsworth *et al.*, 1977 Hacker *et al.*, 1994
High temperatures	Mature (> 1 year)	Decreased mounting attempts Decreased duration of ejaculation	Christenson *et al.*, 1972 Claus and Weiler, 1985
Decreased light when natural photoperiod is increasing	Mature (> 1 year)	Decreased reaction time	Claus and Weiler, 1985
Supplemental light	Prepubertal (between 15 and 32 weeks of age)	Increased libido scores	Berger *et al.*, 1980; Hoagland and Diekman, 1982

because behavioural responses are often more variable and less repeatable than physiological responses.

The results of several studies that have examined the influence of various aspects of the production environment of sexual behaviour in boars are summarized (Table 6). Collectively, these data indicate that isolation during pubertal development, chronic nutritional restriction and exposure to high ambient temperatures reduce libido or the performance of male mating behaviours. The effect of isolation during pubertal development appears to be related to the lack of experience in practising or participating in mounting behaviours during adolescence (Hemsworth *et al.*, 1977), while nutritional and environmental effects are related, at least in part, to a reduction in the endocrine activity of the hypothalamo–hypophyseal–testis axis (Wetteman and Desjardins, 1979; Louis *et al.*, 1994b). It is interesting to note that most of the detrimental influences outlined in Table 6 are chronic as opposed to acute deficiencies in the production environment.

Reports of strategies with acute effects on libido in boars are scarce. Intuitively, administration of testosterone would seem to be a logical approach. Although testosterone does produce short-term improvements in sex drive and mounting activity, it also results in a suppression in spermatogenesis if used for extended periods owing to its negative effect on LH secretion (Bishop *et al.*, 1992). Prostaglandins are involved in a variety of male sexual behaviours (Miesel and Sacks, 1994) and, thus, are commonly administered in an attempt to stimulate sexual interest and mounting behaviour in boars. Definitive studies are still needed to establish the effectiveness of this practice and any repercussions it may have on sperm production.

Conclusion

Most visual tests commonly used to evaluate semen quality in boars are poor indicators of fertility. As a result, our current understanding of the manner in which nutrition, environment and other management parameters affect sperm fertility from studies that have used these tests may not be as accurate as once believed. Consequently, as tests more closely linked to the fertilization process are developed, it may be necessary to revisit some of these areas of research. In addition, it appears that most of the situations that are detrimental to sperm production and male sexual behaviour, with the obvious exception of exposure to acute heat stress conditions, result from chronic or long-term inadequacies in the production environment. Given the relatively short period during which boars are normally used within operations for breeding, 7 months to 1.5 years, management of boars during pubertal development should be given as much, if not more, consideration as the manner in which they are handled as adults.

References

Amann RP (1988) Relationships between computerized evaluations of spermatozoal motion and competitive fertility index *Proceedings of the 11th International Congress on Animal Reproduction and Artificial Insemination* Dublin pp 38–44

Ash KL, Berger T, Horner CM and Famula TR (1994) Boar sperm plasma membrane protein profile: correlation with the zona-free hamster ova assay *Theriogenology* **42** 1217–1226

Bellin ME, Hawkins HE and Ax RL (1994) Fertility of range beef bulls grouped according to presence or absence of heparin-binding proteins in sperm membranes and seminal fluid *Journal of Animal Science* **74** 173–182

Berger T, Mahone JP, Syoboda GS, Metz KW and Clegg ED (1980) Sexual maturation of boars and growth of swine exposed to extended photoperiod during decreasing natural photoperiod *Journal of Animal Science* **51** 672–678

Berger T, Anderson DL and Penedo MCT (1996) Porcine sperm fertilizing potential in relationship to sperm functional capacities *Animal Reproduction Science* **44** 231–239

Bishop DK, Wettemann RP and Maxwell CV (1992) Prepuberal testosterone treatment and testes development, spermatogenesis and sexual behavior of boars. In *Animal Science Research Report* pp 383–388 Agricultural Experiment Station, Oklahoma State University, Stillwater, OK

Brandt KE and Diekman MA (1985) Influence of supplemental lighting on serum LH, testosterone and semen quality in prepubertal and postpubertal boars *Animal Reproduction Science* **8** 287–294

Christenson RK, Teague HS, Grifo AP, Jr and Roller WL (1972) The effect of high environmental temperature on the boar. In *Ohio Swine Research and Information Report* pp 19–23 Ohio Agricultural Research and Development Center, Wooster, Ohio

Claus R (1990) Physiological role of seminal components in the reproductive tract of the female pig *Journal of Reproduction and Fertility Supplement* **40** 117–131

Claus R and Weiler U (1985) Influence of light and photoperiodicity on pig prolificacy *Journal of Reproduction and Fertility Supplement* **33** 185–197

Close WH and Roberts FG (1993) Nutrition of the working boar. In *Recent Developments in Pig Nutrition 2* pp 347–368 Eds DJA Cole, W Haresign and PC Garnsworthy. Nottingham University Press, Nottingham

Colenbrander B and Kemp B (1990) Factors influencing semen quality in pigs *Journal of Reproduction and Fertility Supplement* **40** 105–115

Colenbrander B, Feitsma H and Grooten HJ (1993) Optimizing semen production for artificial insemination in swine *Journal of Reproduction and Fertility Supplement* **48** 207–215

Cooke PS, Porcelli J and Hess RA (1992) Induction of increased testis growth and sperm production in adult rats by neonatal administration of the goitrogen propylthiouracil (PTU): the critical period *Biology of Reproduction* **46** 146–154

Evenson DP, Thompson L and Jost L (1994) Flow cytometric evaluation of boar semen by the sperm chromatin structure assay as related to cryopreservation and fertility *Theriogenology* **41** 637–651

Fazeli AR, Holt C, Steenweg W, Bevers MM, Holt WV and Colenbrander B (1995) Development of a sperm hemizona binding assay for boar semen *Theriogenology* **44** 17–27

Flowers WL and Esbenshade KL (1993) Optimizing management of natural and artificial matings in swine *Journal of Reproduction and Fertility Supplement* **48** 217–228

Flowers WL and Turner ZA Relationships among motility, morphology and fertility estimates for boar semen *Journal of Animal Science* (in press)

Flowers WL, Siggens K and McLaren DG (1996) Paternity of pigs from sows bred with combinations of natural service (NS) and AI *Journal of Animal Science 74 (Supplement 1)* p 223 (Abstract)

Grant SA, Long SE and Parkinson TJ (1994) Fertilizability and structural properties of boar spermatozoa prepared by Percoll gradient centrifugation *Journal of Reproduction and Fertility* **100** 477–483

Hacker RR, Du Z and D'arcy CJ (1994) Influence of penning type and feeding level on sexual behavior and feet and leg soundness in boars *Journal of Animal Science* **72** 2531–2537

Hemsworth PE, Beilharz RG and Galloway DB (1977) Influence of social conditions during rearing on the sexual behaviour of the domestic boar *Animal Production* **27** 201–212

Hoagland TA and Diekman MA (1982) Influence of supplemental lighting during increasing daylength on libido and reproductive hormones in prepubertal boars *Journal of Animal Science* **55** 1483–1489

Huang Y-T and Johnson RK (1996) Effect of selection for size of testes in boars on semen and testis traits *Journal of Animal Science* **74** 750–760

Johnson LA, Maxwell WMC, Dobrinsky JR and Welch GR (1996) Staining sperm for viability assessment *Reproduction in Domestic Animals* **31** 37–47

Kemp B and Verstegen MWA (1991) Nutrition and sperm production *Reproduction in Domestic Animals Supplement* **1** 287–298

Kemp B, Bakker GCM, den Hartog LA and Verstegen MWA (1991) The effect of semen collection frequency and food intake on semen production in breeding boars *Animal Production* **52** 355–364

Killian GJ, Chapman DA and Rogowski LA (1993) Fertility-associated proteins in holstein bull seminal plasma *Biology of Reproduction* **49** 1202–1207

Linford E, Glover FA, Bishop C and Stewart DL (1976) The relationship between semen evaluation methods and fertility in the bull *Journal of Reproduction and Fertility* **47** 283–291

Louis GF, Lewis AJ, Weldon WC, Ermer PM, Miller PS, Kittock RJ and Stroup WW (1994a) The effect of energy and protein intakes on boar libido, semen characteristics, and plasma hormone concentrations *Journal of Animal Science* **72** 2051–2060

Louis GF, Lewis AJ, Weldon WC, Miller PS, Kittock RJ and Stroup WW (1994b) The effect of protein intake on boar libido, semen characteristics, and plasma hormone concentrations *Journal of Animal Science* **72** 2038–2050

McNitt JI and First NL (1970) Effect of 72-hour heat stress on semen quality in boars *International Journal of Biometrics* **14** 373–380

Mauget R (1982) Seasonality of reproduction in the wild boar. In *Control of Pig Reproduction* pp 509–526 Eds DJA Cole and GR Foxcroft. Butterworth Scientific, London

Mazzari G, du Mesnil du Buisson F and Ortavant R (1968) Action de la temperature et de la lumiere sur la spermatogenese, la production et le pouvoir lecondant de sperme chez le verrat *Proceedings of the 6th International Congress on Animal Reproduction and Artificial Insemination* **1** 305–308

Mburu JN, Einarsson S, Lundeheim N and Rodriquez-Martinez H (1996) Distribution, number and membrane integrity of spermatozoa in the pig oviduct in relation to spontaneous ovulation *Animal Reproduction Science* **45** 109–121

Meisel RL and Sachs BD (1993) The physiology of male sexual behavior. In *The Physiology of Reproduction* Vol. 2 pp 3–106 Eds E Knobil and JD Neill. Raven Press, New York

Minton JE and Wettemann RP (1987) Growth, testicular and endocrine function of boars exposed to 8, 16 or 24 hours of light daily before puberty *Animal Reproduction Science* **14** 239–244

Polge C (1978) Fertilization in the pig and horse *Journal of Reproduction and Fertility* **54** 461–474

Rathje TA, Johnson RK and Lunstra DD (1995) Sperm production in boars after nine generations of selection for increased weight of testis *Journal of Animal Science* **73** 2177–2188

Sharpe RM (1993) Regulation of spermatogenesis. In *The Physiology of Reproduction* Vol. 1 pp 1363–1434 Eds E Knobil and JD Neill. Raven Press, New York

Stone BA (1982) Heat induced infertility of boars: the inter-relationship between depressed sperm output and fertility and an estimation of the critical air temperature above which sperm output is impaired *Animal Reproduction Science* **4** 283–299

Swanlund DJ, N'Diaye MR, Loseth KJ, Pryor JL and Crabo BG (1995) Diverse testicular responses to exogenous growth hormone and follicle-stimulating hormone in prepubertal boars *Biology of Reproduction* **53** 749–757

Swiestra EE (1968) Cytology and duration of the cycle of the seminiferous epithelium of the boar. Duration of the spermatozoa transit through the epididymis *Anatomical Record* **161** 171–186

Vesseur PC, Kemp B and den Hartog LA (1996) Factors influencing the proportion of offspring from a second insemination in sows *Animal Reproduction Science* **41** 255–265

Waberski D, Sudhoff H, Hahn T, Jungblut PW, Kallweit E, Calvete JJ, Ensslin M, Hoppen H-O, Wintergalen N, Weitze KF and Toper-Peterson E (1995) Advanced ovulation in gilts by the intrauterine application of a low molecular mass pronase-sensitive fraction of boar seminal plasma *Journal of Reproduction and Fertility* **105** 247–252

Wettemann RP and Bazer FW (1985) Influence of environmental temperature on prolificacy of pigs *Journal of Reproduction and Fertility Supplement* **33** 199–208

Wettemann RP and Desjardins C (1979) Testicular function in boars exposed to elevated ambient temperature *Biology of Reproduction* **20** 235–241

Wettemann RP, Wells ME and Johnson RK (1979) Reproductive characteristics of boars during and after exposure to increased ambient temperature *Journal of Animal Science* **49** 1501–1505

Woelders H (1991) Overview of *in vitro* methods for evaluation of semen quality *Reproduction in Domestic Animals Supplement* **1** 145–164

Journal of Reproduction and Fertility Supplement **52**, 79–89

Consequences of variation in interval from insemination to ovulation on fertilization in pigs

B. Kemp and N. M. Soede

Agricultural University Wageningen, Wageningen Institute of Animal Sciences (WIAS), Marijkeweg 40, 6709 PG Wageningen, The Netherlands

This review describes effects of variation in the interval between insemination and ovulation on the fertilization process in the sow. Inseminations performed too early or too late relative to ovulation decrease litter size and especially farrowing rate. This effect can be explained to a large extent by the increase in the percentage of non-fertilized eggs, resulting in partial fertilization or no fertilization at all. No effects of variation in the interval from insemination to ovulation are found on the percentage of degenerate embryos. Only moderate effects are found on mean embryonic development and variation in embryonic development at day 5 after insemination. In general, insemination between 0 and 24 h before ovulation gives good fertilization results. Factors influencing the optimal interval from insemination to ovulation, such as number of sperm cells used for insemination, storage time of liquid semen and use of frozen semen, and sow factors such as parity and breed are discussed.

Introduction

Many reviews have been published about the distribution of spermatozoa within the genital tract after insemination and the fertilization process in the sow (for example Hancock, 1962; Baker and Polge, 1976; Hunter, 1982; Einarsson, 1985; Hunter, 1990; Flowers and Esbenshade, 1993). From these reviews it has become clear that eggs that enter the oviduct after ovulation have a short fertile lifespan and should therefore be fertilized quickly after ovulation by a population of capacitated sperm cells. Sperm cells seem to survive for extended periods after insemination when stored in the isthmus of the oviduct near the utero–tubal junction. The lifespan of eggs after ovulation and the lifespan of a sufficient number of sperm cells capable of fertilization within the oviduct therefore define the time during which inseminations can lead to successful fertilization relative to ovulation.

During recent years (especially with the development of ultrasound techniques to study the timing and process of ovulation) more quantitative data have become available on consequences of variation in the interval from insemination to ovulation for reproductive parameters. This review describes effects of variation in the interval from insemination to ovulation on fertilization in the sow. In addition, factors that influence the duration of the optimal interval from insemination to ovulation (in terms of fertilization) are discussed.

Knowledge of the range of interval from insemination to ovulation that leads to successful fertilization is important because it will set the accuracy with which insemination should be timed relative to ovulation. Together with the possibility for predicting ovulation during oestrus, this can lead to greater knowledge of the effects of insemination strategies on reproductive performance.

In spontaneously ovulating sows, the duration of ovulation ranges from 1 to 3 h (Soede *et al.*, 1992); since this is a rather short period, in this review we refer to it as the moment of ovulation rather than the duration of ovulation.

Short Description of the Transport of Gametes to the Site of Fertilization

Ovulation occurs about 30–35 h after the peak of the preovulatory LH surge, which is about 44 h after the onset of the preovulatory LH surge (Soede *et al.*, 1994; Mburu *et al.*, 1995). Eggs are transported to the site of fertilization at the ampullary–isthmic junction in 30–45 min or less (Hunter, 1974) and remain there for fertilization to occur.

During oestrus, sperm cells are deposited by AI or natural insemination at the utero–cervical junction and are transported to the utero–tubal junction due mainly to the contractile activity of the uterus (for review, see Einarsson, 1985). Although a large quantity of semen and a great number of spermatozoa are deposited into the uterus, only a relatively small number reach the oviducts (First *et al.*, 1968; Hunter, 1982). The uterus is a hostile environment for sperm cells. The motility of sperm cells in the uterus decreases quickly after insemination and many sperm cells are removed from the uterus within a few hours after insemination by back flow and phagocytosis (Baker *et al.*, 1968; Lovell and Getty, 1968; Pursel *et al.*, 1978; Viring and Einarsson, 1980). Results from Viring and Einarsson (1981) show that within 2 h after insemination, approximately one-third of the inseminated sperm cells were found in back flow from the vagina. In our laboratory similar results were found (D. W. B. Steverink, unpublished) and although the amount of back flow showed a large variation between animals, this was not related to the fertilization results of the sows.

A sufficient number of sperm cells to effect fertilization have entered the oviduct within 15–30 min after insemination or mating. This population of sperm cells forms a sperm reservoir in the caudal 1–2 cm of the oviductal isthmus (Hunter, 1984). This reservoir has favourable storage conditions for sperm cells and regulates the release of capacitated, activated sperm cells to the site of fertilization.

The lifespan of eggs and the lifespan of a sufficient number of sperm cells in the sperm reservoir capable of fertilization define the time course during which insemination can be successful relative to ovulation.

Limits of the Interval from Insemination to Ovulation

Studies from the 1960s showed that sows inseminated early or late relative to oestrus showed a significant reduction in fertilization rate (Hancock and Hovell, 1962) and litter size and farrowing rate (Willemse and Boender, 1967). Since the time of ovulation relative to onset of oestrus varies considerably (for review, see Soede and Kemp 1997), these experiments do not lead to an accurate estimate of the effects of the interval from insemination to ovulation on reproduction. However, these studies did lead to the understanding that during the period of oestrus there were suboptimal and optimal times for insemination and that these were probably related to the moment of ovulation.

A summary of the results of experiments in which ovulation was controlled by hCG, or monitored by the progesterone rise or ultrasound, to study effects of variable intervals from insemination to ovulation on reproductive parameters given in Table 1.

Dziuk (1970) carried out a study in which double inseminations were performed at certain times relative to ovulation that was induced with hCG. He concluded that the optimal time for insemination was 12 h before ovulation and that a range of 6–18 h would probably have only moderately adverse effects on litter size. Hunter (1967a) studied effects of postovulatory inseminations on fertilization rate and concluded that eggs are optimally fertilized up to 8 h after ovulation and undergo normal development at least to the four- to eight-cell stage. Results of this early work should be interpreted with some caution since ovulation was induced with hCG and expected to occur at 40–42 h after induction. Experiments in which timing of ovulation after hCG injection was checked by slaughter, laparotomy or ultrasound show that ovulation time after hCG administration may vary between 35 h and 48 h (Hunter, 1967a, 1972; Pope *et al.*, 1988; Brüssow *et al.*, 1990; Soede and Kemp, 1993). Therefore, such data only allow rough estimates of the optimal interval from insemination to ovulation.

Helmond *et al.* (1986) assumed that ovulation occurred when progesterone increased by 1 ng above basal values. However, Soede *et al.* (1994) showed that progesterone concentrations above 1 ng occurred 6–19 h (mean 13 h) after ovulation as determined by ultrasound. After correcting for this, the estimate of optimal time of insemination of Helmond *et al.* (1986) would be 0–15 h before

Table 1. Estimates of the optimal time for insemination relative to ovulation in pigs

Time range (h before ovulation)		AI dose	Evaluation time	Ovulation	
Optimal	Studied	(× 10^9)	(embryo age)	assessment	Reference
Gilts					
12 (6–18)	6–30	10–15% of an ejaculate, twice[a]	Day 85–90	40 h after hCG	Dziuk (1970)
6–8	6–20	80–120 ml semen	Day 3	41–42 h after hCG	Hunter (1967a)
13–28[b]	–16–48	AI	Day 3–10	Progesterone rise[c]	Helmond *et al.* (1986)
0–12	0–16	2	Day 2–5 + 28	Ultrasound every 4 h	Waberski *et al.* (1994a)
0–24	0–> 24	2	Day 2–4	Ultrasound every 12 h	Waberski *et al.* (1994b)
Sows					
0–24	–16–48	3	Day 5	Ultrasound every 4 h	Soede *et al.* (1995a)
–8–24	–8–32	3	Day 5	Ultrasound every 4 h	Soede *et al.* (1995b)
–4 –28	–9–41	2	Day 28 + full term	Ultrasound every 6 h	Nissen *et al.* (1997)

[a]33 gilts were inseminated twice and seven gilts were mated twice.
[b]When data of Helmond *et al.* (1986) are corrected for the increase in progesterone after ovulation according to Soede *et al.* (1994), the optimal interval from insemination to ovulation ranges from 15 h to 0 h before ovulation.
[c]Ovulation time was assumed when plasma progesterone concentrations increased by 1 ng above basal values.

ovulation. But again, owing to the variation in rise of progesterone after ovulation, these data should be interpreted with some caution.

Waberski *et al.* (1994a) used transcutaneous ultrasonography to assess ovulation and concluded that good fertilization rates were found in gilts inseminated between 0 and 12 h before ovulation. However, their estimate for the optimal interval from insemination to ovulation may not be very accurate as only a few animals were inseminated more than 12 h before ovulation ($n = 5$) and no animals were inseminated more than 16 h before ovulation.

Studies in which ultrasound was used to monitor the time of ovulation covering a wide range of intervals from insemination to ovulation (Waberski *et al.*, 1994b; Soede *et al.*, 1995a,b; Nissen *et al.*, 1997) led to the conclusion that inseminations can be performed between 0 and 24 h before ovulation with no significant adverse effects on fertilization rate. This interval is significantly longer than previously reported values. Nissen *et al.* (1997) found optimal results in terms of numbers of day 16 embryos and farrowing rate and litter size in sows inseminated between 28 h before ovulation and 4 h after ovulation (as detected by transrectal ultrasound).

Several factors may influence the range of optimal interval from insemination to ovulation as found by various workers and these will be discussed later in this review.

Effects of Interval from Insemination to Ovulation on Partial Fertilization and Embryonic Development

To study the effects of the interval from insemination to ovulation on subsequent fertilization, Soede *et al.* (1995a,b) inseminated sows at various times relative to ovulation (as detected by transrectal ultrasound) and slaughtered the animals at about 120 h after ovulation to study fertilization rate and embryonic development. The morphology of the embryos was evaluated and the number of nuclei (to calculate the number of cell cycles) and of spermatozoa bound to the zona pellucida (accessory sperm count) was determined. An oocyte was classified as unfertilized if the nuclear count was zero or one. Embryos with degenerate morphology and a small number of nuclei were classified as degenerated; the remaining embryos were considered to be normal. Fertilization rate was defined as the percentage of normal embryos relative to all embryos and oocytes recovered. Sows were assigned on the basis of fertilization rate to classes varying from no fertilized eggs to total fertilization of all ovulated eggs. The number of degenerated embryos was about 4% irrespective of the interval between insemination and ovulation.

The percentage of sows within each class of fertilization rate is shown (Fig. 1) for different intervals from insemination to ovulation. The percentage of sows with partial or no fertilization increases when insemination takes place more than 24 h before ovulation or directly after ovulation. However, in these suboptimal periods a large variation between sows can be seen. Even when insemination occurs more than 40 h before ovulation, 15% of the sows showed 100% fertilization.

Mean embryo development in a litter and within litter variation in embryo development of normal embryos are shown in relation to interval from insemination to ovulation (Fig. 2). These data are from three experiments (Soede *et al.*, 1995a, b; Steverink *et al.*, 1997) and are corrected for experiment and age of embryos at slaugher (between 118 h and 130 h). Although not significant, a longer interval from insemination to ovulation and inseminations after ovulation on average result in slightly lower embryonic development and increased variation in embryonic development. In the study of Soede *et al.* (1995a) in each insemination to ovulation class, successful fertilization (>90% normal embryos) was significantly associated with better embryonic development. On average, development (corrected for embryo age) in terms of number of cell cycles was 5.7±0.7 for the 'good fertility' sows and 5.1±0.8 for the 'poor fertility' sows. This effect was completely independent of the interval from insemination to ovulation, which means that embryonic development of 'good fertility sows' inseminated at optimal or suboptimal intervals from insemination to ovulation was similar.

Therefore, these results show that variation in interval from insemination to ovulation induces partial fertilization or no fertilization; furthermore litters with partial fertilization also have slightly retarded embryo development and increased variation in embryonic development at day 5 of pregnancy. Whether the effect on embryo development has an impact on embryonic mortality remains to be investigated.

It should be mentioned that fertilization is often assumed to be an 'all or none' phenomenon; either all oocytes are fertilized or none are (see Hunter, 1994). The above mentioned studies clearly show that partial fertilization frequently occurs, even when timing of insemination relative to ovulation is optimal. This finding is of importance for research on embryo survival since embryo survival is normally defined as the number of embryos relative to the number of corpora lutea and assumes that fertilization rate is 100%.

Possible Causes of Reduced Fertilization at Long Intervals Between Insemination and Ovulation

When insemination is performed early relative to ovulation the fertilization success depends on the lifespan of a sufficient number of sperm cells capable of fertilization in the sperm reservoir.

The functional sperm reservoir in pigs has been reviewed by Hunter (1990). In short, the sperm reservoir in pigs is established within 15–30 min after insemination in the caudal 1–2 cm of the oviduct. Sperm cells can be trapped in this reservoir for more than 36 h while motility of the sperm cells is depressed and the acrosomal membrane remains intact. By the time ovulation approaches, sperm cells are released and become hyperactive and are able to fertilize eggs. It was suggested that this synchronized release at the time of ovulation is co-ordinated at least in part by changing ovarian hormone concentrations after the LH surge influencing the oviduct through local countercurrent transfer from the ovarian vein and the oviductal branch of the ovarian artery.

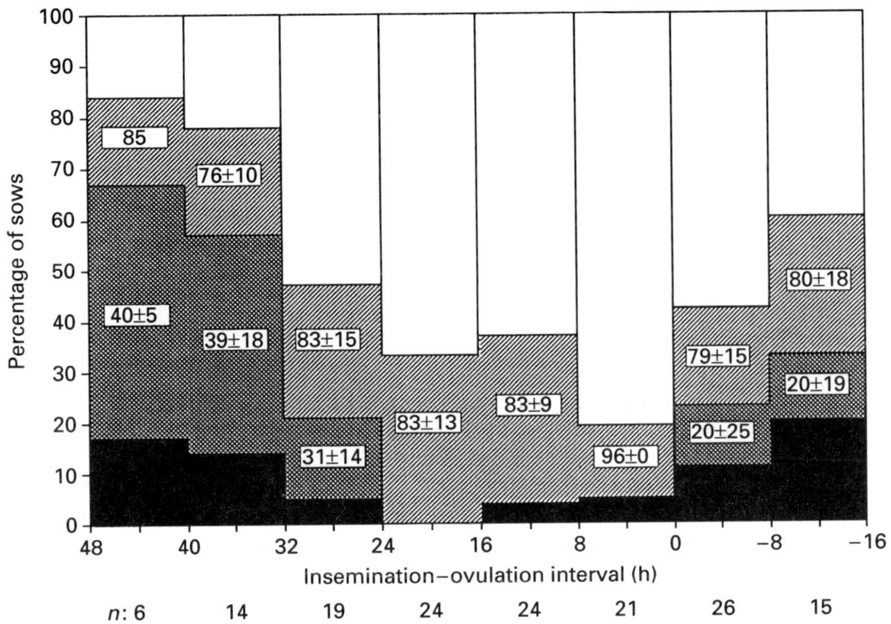

Fig. 1. Percentage of sows with 0% (■), 0%–50% (▩), 50%–100% (▨) and 100% (□) fertilization for each 8 h insemination to ovulation interval ($n = 151$). For sows with partial fertilization, mean fertilization rate (mean ± SD) is given per time interval. (Reproduced from Soede *et al.*, 1995a.)

Suarez *et al.* (1991) and Raychoudhury and Suarez (1991) showed that *in vitro* sperm cells bind to ciliated oviductal cells within minutes after incubation. Binding to the epithelium was improved when oestradiol was added to the medium. Binding of sperm cells to the epithelium was significantly reduced after 24 h. Of those sperm cells that remained attached to the epithelium, 70% remained motile for more than 44 h. Unbound sperm cells were not motile. It was also shown that the oviductal isthmus cells produced a thick mucus which traps sperm cells that are released from the epithelium.

Mburu *et al.* (1996) studied the distribution of spermatozoa in the pig oviduct in relation to spontaneous ovulation and showed that before ovulation most sperm cells were found in the lower isthmus and these sperm cells seemed partly trapped in thick mucus and in the folded endosalpinx. During and after ovulation the endosalpinx folds were much less pronounced, mucus was less viscous and the number of sperm cells in the lower oviduct was reduced. However, the number of sperm cells in the upper part of the oviduct was increased around and after ovulation.

Collectively these results seem to suggest that sperm cells enter the oviduct within 15 min after insemination and form a sperm reservoir at the caudal isthmus, because sperm cells attach to the epithelium and are trapped in a mucus plug and the intense folding of the isthmus. Sperm cells attached to the epithelium can survive for more than 44 h. Around ovulation this reservoir is lost and sperm cells can be released to the oviduct.

It is not known why fertility is lower when inseminations are performed more than 24 h before ovulation if sperm cells are able to survive in the isthmus for more than 44 h. A possible explanation is that the number of capacitated sperm cells capable of fertilization decreases over time and that this is the limiting factor for fertilization success. An indication that the number of capacitated sperm cells capable of attachment to the zona pellucida is lower with a longer interval from insemination to ovulation is the rapid decrease in number of accessory sperm cells in the embryos fertilized at longer intervals between insemination and ovulation as found by various authors (see Fig. 3).

Smith and Yanagimachi (1990, 1991) showed that in hamsters only uncapacitated sperm cells are capable of attachment to the isthmic epithelium. These sperm cells sometimes detach from the

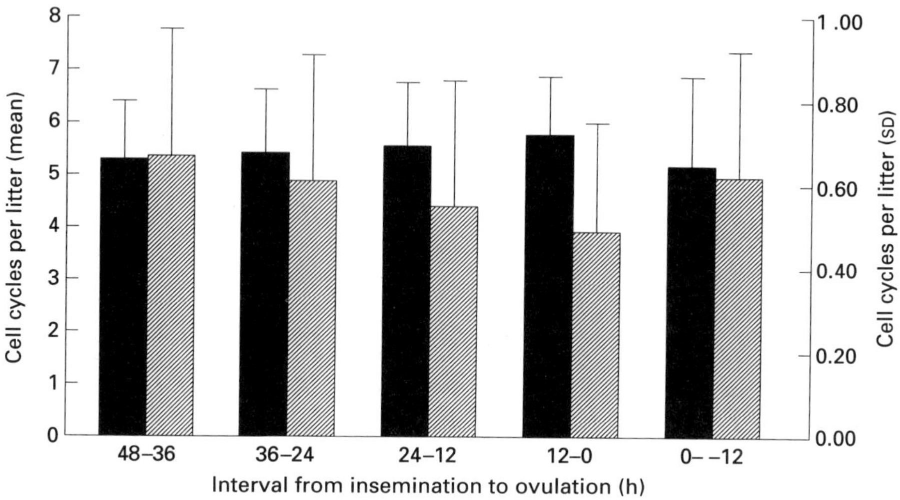

Fig. 2. Mean number of cell cycles (■) and standard deviation in cell cycles (▨) within sows of day 5 embryos (corrected for age) at various 12 h intervals from insemination to ovulation. (Based on data from Soede *et al.*, 1995a,b; and Steverink *et al.*, 1997.)

epithelium for a short period and then re-attach (similar observations in mouse: Demott and Suarez, 1992). However, when sperm cells become capacitated they lose their affinity for oviductal mucosa. Further observations showed that most sperm cells that were not attached to the oviduct but detached before ovulation are dead and do not play a role in the fertilization process. Such observations seem to suggest that uncapacitated sperm cells are stored in the oviduct and that part of the population gradually becomes capacitated and enters the lumen of the oviduct. Populations of these capacitated sperm cells that are released long before ovulation will not survive and therefore the population of remaining sperm cells gradually decreases. Research on the mechanism that regulates the release of capacitated sperm cells from the epithelium of the oviductal reservoir before ovulation might improve our understanding of the causes of reduced fertilization at longer intervals between insemination and ovulation.

Possible Causes of Reduced Fertilization due to Insemination after Ovulation

When insemination takes place after ovulation, fertilization is often poor (as shown in Table 1 and Fig. 1). It is assumed that the decrease in fertilization rate is caused by the limited lifespan of the oocytes in combination with the time needed for sperm cells to capacitate and reach the site of fertilization (Hunter, 1994). Recent studies of oocytes and embryos from postovulatory insemination showed that fertilized oocytes contain large numbers of accessory sperm cells (see Fig. 3). This corresponds to the finding of Hunter (1984) that, after insemination, the regulation of the number of sperm cells passing from the uterus to the site of fertilization is less effective, and therefore increases the chance that oocytes are confronted by two or more spermatozoa simultaneously, increasing the rate of polyspermic fertilization. In pig husbandry, many sows are inseminated two or more times, with an interval between two subsequent inseminations of 12–24 h. A greater number of inseminations increases the chance of postovulatory insemination and consequently may increase the chances of polyspermic fertilization. To study this, Soede *et al.* (1995b) conducted an experiment in which sows were inseminated before ovulation followed by an insemination within 0–5 h after ovulation. The second insemination resulted in a marked increase in the number of accessory sperm cells but did not result in a decrease in the number of well-developed embryos at 120 h after ovulation. In addition, no relationship was found between the number of accessory sperm cells and embryonic

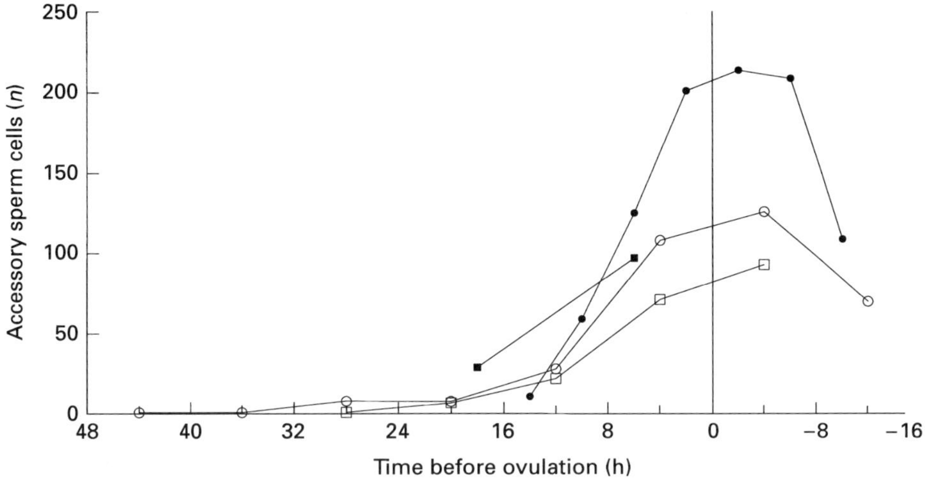

Fig. 3. Mean number of accessory sperm cells on day 2–5 embryos in relation to interval from insemination to ovulation. (Based on data from Waberski *et al.*, 1994a, (●) b (■); and Soede *et al.*, 1995a, (○) b (□).)

development in sows inseminated a second time after insemination. It was concluded therefore that a second insemination after ovulation did not result in a reduction in normal fertilization. It might therefore be suggested that polyspermic fertilization is not a cause of embryonic losses under normal artificial insemination conditions. Polyspermic fertilization is normally assessed in the first hours after fertilization by counting the number of sperm cells or pronuclei in the cytoplasm of the oocyte. Polyspermic embryos are said to degenerate early in development, but may seem normal for up to the four- to eight-cell cycle (Hunter, 1967b). In the experiment of Soede *et al.* (1995b) no assessment of polyspermic embryos was made; however, it can be assumed that (expanded) blastocysts at 120 h after ovulation are viable normal embryos not resulting from polyspermic fertilization.

Estimates of polyspermic fertilization have previously been made under unnatural conditions (progesterone injections, Day and Polge, 1968; hCG injection, Hunter, 1967a; oviductal inseminations, Hunter, 1973). Even polyspermic fertilization at the end of the fertile lifespan of oocytes (Hunter, 1967b) may not occur under natural conditions; reduced fertilization of sows inseminated after ovulation in the experiment of Soede *et al.* (1995a) was due to lack of fertilization and not due to an increased incidence of degenerated embryos. Therefore, the time after ovulation in which successful inseminations can be applied seems to depend on the lifespan of the eggs or the capacitation rate of the sperm cells.

Factors Influencing the Optimal Interval from Insemination to Ovulation in Sows

Parity and breed effects

From the above mentioned studies it has become clear that success of fertilization depends on the interval from insemination to ovulation. However, it is also clear that there are large differences between individual sows. Although the general trend suggests that, if fresh semen is used at an adequate dose, the optimal interval from insemination to ovulation is between 0 h and 24 h before ovulation, it is clear that some sows have good fertilization results outside these limits and that some sows have poor fertilization results inside this time period.

To our knowledge there are no experiments that have studied the effect of parity on the optimal interval from insemination to ovulation. From the experiments listed in Table 1, it could be concluded that experiments performed in gilts result in a shorter optimal interval from insemination to ovulation. However, owing to the limited numbers of gilts used in these experiments and the various techniques used to detect ovulation, these results should be interpreted with caution. Furthermore, Waberski *et al.* (1994b) showed that fertilization in gilts was not affected by an interval from insemination to ovulation ranging from 0 to 24 h which is comparable with studies performed with older parity sows (as listed in Table 1).

In the experiments of Soede *et al.* (1995 a, b), three different genetic lines (one sow line and two boar lines) were used that were parent lines for commercial crossbreeds. Data from these experiments were used to study the possible effect of breed on fertilization rate at different intervals from insemination to ovulation. In Table 2, fertilization results are shown for boar and sow lines inseminated at different intervals from insemination to ovulation. The boar lines showed similar results and are therefore not separated. The data show that inseminations performed after ovulation resulted in marked differences between lines. The sow line showed good fertilization results, whereas boar lines showed a marked decline in fertilization. However, when insemination occurred before ovulation, the sows from the boar lines tended to have better fertilization results. These data suggest that oocyte survival, sperm transport, sperm survival in the oviduct or sperm capacitation may vary between breeds and consequently influence the optimal period for insemination relative to ovulation.

Semen characteristics

Semen characteristics might influence the optimal interval from insemination to ovulation. To our knowledge, no information is available on the effects of semen quality parameters assessed before insemination (like motility characteristics) on the optimal interval from insemination to ovulation. Information on the number of sperm cells inseminated, storage time of liquid semen and insemination with frozen semen is presented below.

Number of sperm cells inseminated. In the early years of AI, insemination dosages of 40×10^9 sperm cells were used (Rigby, 1966; First *et al.*, 1968). In 1968, Baker *et al.* studied the effects of different volumes (20, 100 or 200 ml) and different numbers of sperm cells (1, 5 or 10×10^9 sperm cells) on the fertilization results after AI and concluded that best results were to be expected when using 5 or 10×10^9 sperm cells in a volume of 100 ml. During the following years new extenders and improved quality assessment of semen resulted in the use of dosages of $2.0–3.5 \times 10^9$ sperm cells (of which 70–90% were scored as motile) in 80–100 ml in most countries (Colenbrander, 1991).

The experiments of Soede *et al.* (1995a, b) were performed with a dose of 3×10^9 total sperm cells and the experiments of Waberski *et al.* (1994b) and Nissen (1995) were performed with a dose of 2×10^9 total sperm cells. These data collectively suggest that use of a dose of $2–3 \times 10^9$ sperm cells was compatible with an the optimal interval from insemination to ovulation of about 0–24 h.

Steverink *et al.* (1997) studied whether this optimal interval could be longer when a higher sperm dosage (6×10^9 total sperm cells) was used or shorter when a lower sperm dosage was used (1×10^9 sperm cells). However, results showed that the optimal period for insemination of 0–24 h was hardly influenced by the dosages of spermatozoa applied in this study. These results indicate that the lifespan of the functional sperm reservoir is not very sensitive to variation in the number of sperm cells inseminated, at least not within the range of $1–6 \times 10^9$ total sperm cells.

Liquid semen storage before AI. Liquid semen can be stored after dilution in commercial extenders (for example BTS, Androhep or Kiev) for several days without apparent loss in fertilizing capacity. The general trend in the literature indicates that liquid semen can be stored up to about 3 days without adverse effects on litter size and pregnancy rate (Larsson *et al.*, 1979; Aalbers *et al.*, 1984; Waberski *et al.*, 1990; Sather *et al.*, 1991; Machaty *et al.*, 1992; Baltes, 1993; Alexopoulos *et al.*, 1996).

Waberski *et al.* (1994b) conducted an experiment to study effects of long-term storage of liquid semen in relation to fertilization rate at different intervals from insemination to ovulation. Semen

Table 2. Percentage of normal embryos (mean±SEM) at day 5 after ovulation in sows of sow-lines[a] and boar-lines[b] with a variable interval between insemination and ovulation

Time of insemination relative to ovulation	Sow-lines ($n=46$)	Boar-lines ($n=154$)	P^c
After ovulation			
0–16 h	94±4	62±6	0.007
Before ovulation			
0–24 h	83±6	92±2	0.07
24–48 h	49±12	63±6	0.23

[a]Sow-lines have been selected mostly for fertility parameters. Mothers of commercial crossbreed sows.
[b]Boar-lines have been selected mostly for production parameters. Fathers of fattening pigs.
[c]Based on analyses with arcsine transformed percentages.

diluted in BTS and stored at 17°C for 0–48 h could be inseminated between 0 and 24 h before ovulation without apparent loss in fertilizing capacity (based on the percentage of good embryos at days 2–4). Semen stored for 48 h to 87 h resulted in a decrease in fertilization rate when inseminated more than 12 h before ovulation and semen stored for 87 to 118 h already resulted in a lower fertilization rate when inseminated between 0 and 12 h before ovulation. Postovulatory insemination (between 0 and 4 h after ovulation) with long-term stored liquid semen was carried out in a second trial. Fertilization rate seemed to decrease when semen was stored for more than 24 h in BTS and for more than 48 h in Androhep. Motility and percentage normal apical ridge assessed *in vitro* showed no decrease until after storage for 48 h.

In the experiments of Soede *et al.* (1995 a, b), semen was diluted in BTS and retrospective analyses showed that storage times between 12 h and 38 h did not affect the optimal interval from insemination to ovulation. From these results it seems that storage time of liquid semen affects the optimal interval from insemination to ovulation and effects seem to depend on the type of extender used.

Frozen semen. Reduced fertility associated with frozen semen has been the main reason for its limited use in AI practice (Reed, 1985; Didion and Schoenbeck, 1996). In studies with hCG = treated sows, Larsson (1976) showed that insemination with frozen semen close to the expected time of ovulation resulted in the highest fertilization results. Waberski *et al.* (1994a) studied effects of interval from insemination to ovulation in spontaneously cyclic gilts on fertility using frozen semen. They showed that the percentage of normal day 2–5 embryos significantly decreased when insemination was performed more than 4 h before ovulation or directly after insemination. Fertilization rate in the group of sows inseminated between 0 h and 4 h before ovulation was 88.1%, which was only slightly less than the fertilization expected when using fresh semen. However, a dose of 5×10^9 total sperm cells was used for insemination with frozen semen which is substantially higher than dosages used for fresh semen (Waberski *et al.*, 1994a,b; Soede *et al.*, 1995a,b). It was suggested (Pursel *et al.*, 1978; Saacke, 1982) that cell damage after cryopreservation results in a higher elimination rate of sperm cells in the female tract, resulting in fewer spermatozoa at the site of fertilization. Comparing trials with liquid and frozen semen, Waberski *et al.* (1994a) found that insemination between 0 and 4 h before ovulation resulted in a tenfold reduction in the number of accessory spermatozoa when using frozen semen even though a higher dose of sperm cells (5×10^9 versus 2×10^9) was used in the frozen semen trial.

These results indicate that frozen semen seriously reduces the optimal interval from insemination to ovulation. In the Netherlands, fresh semen is used at a dosage of 2.5×10^6 sperm cells and frozen semen at a dosage of $10–15\times10^6$ sperm cells to obtain comparable pregnancy rates in cattle (van Wagtendonk, personal communication). Such a four- to six-fold increase in dosage is not an option for AI practice in pigs since only a limited number of dosages can be obtained from a boar. Perhaps timing of ovulation with hCG in combination with insemination just before ovulation is a better option.

Conclusion

Variation in the interval from insemination to ovulation affects fertilization rate in sows and results in partial fertilization or no fertilization at all. No effects are found on the percentage of degenerated embryos and only small effects are found on embryonic development.

In general, insemination between 0 and 24 h before ovulation gives good fertilization results; however, there are large differences between sows. Even in suboptimal periods some sows show 100% fertilization. This finding indicates that factors like storage conditions in the sperm reservoir, capacitation rate of semen in the sow and lifespan of the eggs and semen in the sow may vary considerably between individual sows. Part of this variation seems genetic but factors like insemination conditions (natural breeding versus AI) may also play a role. This role remains to be elucidated.

References

Aalbers JG, Johnson, LA, Rademaker JMH and Grooten HJG (1984) Use of boar spermatozoa for AI: fertility and morphology of semen diluted in BTS and used for insemination within 24 h or 25 h after collection. In *Proceedings of the 10th International Congress on Animal Reproduction and Artificial Insemination* Vol. 2. Abstract 180

Alexopoulos C, Boscos C, Saratsis P, Saoulidis C and Kyriakis S (1996) The effect of storage time and number of spermatozoa per insemination dose on semen characteristics and fertilization capacity of boar semen diluted with BTS extender *Animal Science* **62** 599–604

Baker RD and Polge C (1976) Fertilization in swine and cattle *Canadian Journal of Animal Science* **56** 105–119

Baker RD, Dziuk PJ and Norton HW (1968) Effect of volume of semen and drugs on transport of sperm in artificial inseminated gilts *Journal of Animal Science* **27** 88–93

Baltes TJ (1993) Plasma membrane evaluation with fluorescent stains, and computer measured motility as indicators of *in vitro* aging of boar spermatozoa *MSc Thesis, Tierarztlicher Hochschule Hannover*

Brüssow KP, Ratky J, Kanitz W and Becker F (1990) Determination of duration of ovulation in gilts by means of laparoscopy *Reproduction in Domestic Animals* **25** 184–190

Colenbrander B (1991) Commercial use of swine AI worldwide: a roundtable *Reproduction in Domestic Animals Supplement* **1** 298–333

Day BN and Polge C (1968) Effects of progesterone on fertilization and egg transport in the pig *Journal of Reproduction and Fertility* **17** 227–230

Demott RP and Suarez SS (1992) Hyperactivated sperm progress in the mouse oviduct *Biology of Reproduction* **46** 779–785

Didion BA and Schoenbeck RA (1996) Fertility of frozen boar semen used for AI in commercial settings *Reproduction in Domestic Animals* **31** 175–178

Dziuk P (1970) Estimation of optimum time for insemination of gilts and ewes by double-mating at certain times relative to ovulation *Journal of Reproduction and Fertility* **22** 277–282

Einarsson S (1985) Transport of boar sperm in the female reproductive tract. In *Deep Freezing of Boar Semen* pp 180–198 Eds LA Johnson and K Larsson. Swedish University of Agricultural Sciences, Uppsala

First NL, Short RE, Peters JB, Stratmen FW (1968) Transport and loss of boar spermatozoa in the reproductive tract of the sow *Journal of Animal Science* **27** 1037–1040

Flowers WL and Esbenshade KL (1993) Optimizing management of natural and artificial matings in swine *Journal of Reproduction and Fertility Supplement* **48** 217–228

Hancock JL (1962) Fertilization in farm animals *Animal Breeding Abstracts* **30** 285–310

Hancock JL and Hovell GJR (1962) Insemination before and after onset of heat in sows *Animal Production* **4** 91–96

Helmond FA, Aarnink A and Oudenaarden C (1986) Preovulatory hormone profiles in relation to embryonic development and mortality in pigs. In *Embryonic Mortality in Farm Animals*, pp 165–187 Ed. ESE Hafez. Lea and Febiger, Philadelphia

Hunter RHF (1967a) The effect of delayed insemination on fertilization and early cleavage in the pig *Journal of Reproduction and Fertility* **13** 133–147

Hunter RHF (1967b) Polyspermic fertilization in pigs during the luteal phase of the oestrous cycle *Journal of Experimental Zoology* **165** 451–460

Hunter RHF (1972) Ovulation in the pig: timing of the response to injection of hCG *Research in Veterinary Science* **13** 356–361

Hunter RHF (1973) Polyspermic fertilization in pigs after tubal deposition of excessive numbers of spermatozoa *Journal of Experimental Zoology* **183** 57–64

Hunter RHF (1974) Chronological and cytological details of fertilization and early embryonic development in the domestic pig, *Sus scrofa. Anatomical Records* **178** 169–186

Hunter RHF (1982) Interrelationships between spermatozoa, the female reproductive tract, and egg investments. In *Control of Pig Reproduction*, pp. 49–63. Eds DJA Cole and GR Foxcroft. Butterworths, London

Hunter RHF (1984) Pre-ovulatory arrest and peri-ovulatory redistribution of competent spermatozoa in the isthmus of the pig oviduct *Journal of Reproduction and Fertility* **72** 203–211

Hunter RHF (1990) Fertilization of pig eggs *in vivo* and *in vitro*. *Journal of Reproduction and Fertility Supplement* **40** 211–226

Hunter RHF (1994) Causes for failure of fertilization in domestic species. In *Embryonic Mortality in Domestic Species* pp 1–22 Eds MT Zavy and RD Geisert. CRC Press Inc., Boca Raton

Larsson K (1976) Fertility of deep frozen boar spermatozoa at various intervals between insemination and induced ovulation. Influences of boars and thawing diluents *Acta Veterinaria Scandinavica* **17** 63–73

Larsson K, Swenson T and Wass K (1979) A field trial on the fertility of liquid boars semen after utilization of two different diluents *Nordisk Veterinaermedicin* **31** 337–338

Lovell JE and Getty R (1968) Fate of semen in the uterus of the sow: histological study of endometrium during 27 h after natural service *American Journal of Veterinary Research* **29** 609–625

Machaty Z, Takacs T and Gathy I (1992) Fertilizing capacity of boar semen diluted with BTS and Modified Kiev (MK) extenders in relation to storage time and number of spermatozoa per insemination dose *Animal Reproduction Science* **29** 289–295

Mburu JN, Einarsson S, Dalin AM, and Rodriquez-Martinez H (1995) Ovulation as determined by transrectal ultrasonography in multiparous sows: relationships with oestrus symptoms and hormonal profiles *Journal of Veterinary Medicine Series A* **42** 285–292

Mburu JN, Einarsson S, Lundeheim N, and Rodriquez-Martinez H (1996) Distribution, number and membrane integrity of spermatozoa in the pig oviduct in relation to spontaneous ovulation *Animal Reproduction Science* **45** 109–121

Nissen AK, Soede NM, Hyttel P, Schmidt M and D'Hoore L (1997) The influence of time of insemination relative to time of ovulation on farrowing frequency and littersize in sows, as investigated by ultrasonography *Theriogenology* 47 1571–1582

Pope WF, Wilde MH and Xie S (1988) Effect of electrocautery of nonovulated day 1 follicles on subsequent morphological variation amoung day 11 porcine embryos *Biology of Reproduction* **39** 822–887

Pursel VG, Schulmann LL and Johnson LA (1978) Distribution and morphology of fresh and frozen–thawed sperm in the reproductive tract of gilts after artificial insemination *Biology of Reproduction* **19** 69–76

Raychoudhury SS and Suarez SS (1991) Porcine sperm binding to oviductal explants in culture *Theriogenology* **36** 1059–1070

Reed HCB (1985) Current use of frozen boar semen – future needs of frozen semen. In *Deep Freezing of Boar Semen* pp 225–238 Eds LA Johnsen and K Larrson. Swedish University of Agricultural Sciences, Uppsala

Rigby JP (1966) The persistence of spermatozoa at the utero–tubal junction of the sow *Journal of Reproduction and Fertility* **11** 153–155

Saacke RG (1982) Semen quality in relation to semen preservation *Journal of Dairy Science* **66** 2635–2644

Sather AP, Harbison DS and Seth PC (1991) A note on the influence of storage duration of fresh semen on fertilizing capacity and embryo survival in sows *Animal Production* **52** 554–557

Smith TT and Yanagimachi R (1990) The viability of hamster spermatozoa stored in the isthmus of the oviduct: the importance of sperm–epithelium contact for survival *Biology of Reproduction* **42** 450–457

Smith TT and Yanagimachi R (1991) Attachment and release of spermatozoa from the caudal isthmus of the hamster oviduct *Journal of Reproduction and Fertility* **91** 567–573

Soede NM and Kemp B (1993) In synchronized pigs, the duration of ovulation is not affected by insemination and not a determinant for early embryonic diversity *Theriogenology* **39** 1043–1053

Soede NM and Kemp B (1997) Expression of oestrus and timing of ovulation in pigs *Journal of Reproduction and Fertility Supplement* **52** 91–103

Soede NM, Noordhuizen JPTM and Kemp B (1992) The duration of ovulation in pigs, studied by transrectal ultrasonography, is not related to early embryonic diversity *Theriogenology* **38** 653–666

Soede NM, Helmond FA and Kemp B (1994) Periovulatory profiles of oestradiol, LH and progesterone in relation to oestrus and embryo mortality in multiparous sows using transrectal ultrasonography to detect ovulation *Journal of Reproduction and Fertility* **101** 633–641

Soede NM, Wetzels CCH, Zondag W, de Koning MAI and Kemp B (1995a) Effects of time of insemination relative to ovulation, as determined by ultrasonography, on fertilization rate and accesssory sperm count in sows *Journal of Reproduction and Fertility* **104** 99–106

Soede NM, Wetzels CCH, Zondag W, Hazeleger W, and Kemp B (1995b) Effects of a second insemination after ovulation on fertilization rate and accessory sperm count in sows *Journal of Reproduction and Fertility* **105** 135–140

Steverink DWB, Soede NM, Bouwman EG and Kemp B (1997) Influence of insemination–ovulation interval and sperm-cell dose on fertilization in sows *Journal of Reproduction and Fertility* **111** 165–171

Suarez SS, Redfern K, Raynor P, Martin F, Phillips DM (1991) Attachment of boar sperm to mucosal explants of the oviduct: possible role in formation of a sperm reservoir *Biology of Reproduction* **44** 998–1004

Viring S and Einarsson S (1980) Influence of boar seminal plasma on the distribution of spermatozoa in the genital tract of gilts *Acta Veterinaria Scandinavica* **21** 589–606

Viring S and Einarsson S (1981) Sperm distribution within the genital tract of natural inseminated gilts *Nordisk Veterinar Medicin* **33** 145–149

Waberski, D, Dirksen G, Weitze KF, Leidig C and Hahn R (1990) Spermienmotilitat und morphology in ihrer auswirkung af die fruchtbarkeit von besamungsebern in feldversuchen *Tierartzliche Praxis* **18** 591–594

Waberski D, Weitze KF, Gleumes T, Schwartz M, Willmen T and Petzholdt R (1994a) Effects of time of insemination relative to ovulation on fertility with liquid and frozen boar semen *Theriogenology* **42** 831–840

Waberski D, Weitze KF, Lietmann C, Lubbert zur Lage W, Bortolozzo FP, Willmen T, Petzoldt R (1994b) The initial fertilizing capacity of longterm stored liquid semen following pre- and postovulatory insemination *Theriogenology* **41** 1367–1377

Willemse AH and Boender J (1967) The relation between the time of insemination and fertility in gilts *Tijdschrift Diergeneeskunde* **92** 18–34

Journal of Reproduction and Fertility Supplement **52**, 91–103

Expression of oestrus and timing of ovulation in pigs

N. M. Soede and B. Kemp

Wageningen Agricultural University, Wageningen Institute of Animal Sciences (WIAS), Department of Animal Husbandry, PO Box 338, 6700 AH Wageningen, The Netherlands

Oestrus is the period around ovulation in which sows show a standing response for boars, thus allowing the boars to mate with the sows. The duration of oestrus may vary between sows from 24 h up to 96 h, which is not related to systemic oestrogen concentrations. The moment of ovulation after onset of oestrus may vary from 10 h to 85 h. Consequently, onset of oestrus is not a good predictor for the time of ovulation. A reliable prediction of ovulation time would be worthwhile, since fertilization results are highly dependent on the moment of insemination relative to the moment of ovulation. However, at present, the duration of oestrus is the best 'predictor' of ovulation time and is independent of the duration of oestrus; ovulation takes place on average 70% of the way through oestrus. The following factors have been found to influence the duration of oestrus (and consequently the optimal time for insemination): boar-stimuli, stress, interval from weaning to oestrus, season and parity. Further study is needed to determine the causes of the variability in oestrus expression between sows, including the regulation of oestrus behaviour at the level of the brain.

Introduction

The period around and during oestrus is characterized by a combination of externally assessable changes of the vulva and behavioural changes, both of which show considerable variation between sows in intensity as well as in duration. Although in pigs little is known about the regulation of oestrous behaviour at the brain, many factors have been found that influence the duration of oestrus and these will be summarized.

The timing of insemination or mating in pig husbandry is normally based on the onset of (standing) oestrus. Since fertilization results are highly dependent on the time of insemination relative to ovulation (see Kemp and Soede, 1997), an important question is whether or not certain oestrous characteristics can 'predict' the time of ovulation. In this review, available data on the characteristics of oestrus in relation to ovulation time will be summarized.

In spontaneously ovulating sows, the duration of ovulation is only 1–3 h (Soede *et al.*, 1992) or at most 5 h under extreme chronic stress conditions (Soede *et al.*, 1997a) and studies relating oestrous characteristics to ovulation time only assess ovulation time with an accuracy of 4 h or more; in this review ovulation is regarded as taking place at one moment rather than during a period.

This review will not deal with factors affecting or regulating attainment of puberty in gilts or the intervals from weaning to oestrus in sows, which have been reviewed by Hughes *et al.* (1990) and Hemsworth and Barnett (1990) and Prunier *et al.* (1996) and Vesseur (1997), respectively.

Unless specified otherwise, in this review the expression 'oestrus' and 'duration of oestrus' is used for sows that show a standing response in the presence of a boar.

Externally Assessable Characteristics of (approaching) Oestrus

Vulva

The period of increasing follicular development and oestrogen production by the ovaries, during which symptoms of approaching oestrus can be observed, is called pro-oestrus (Willemse and Boender, 1966). During pro-oestrus, the vulva swells and becomes red(dish) under the influence of the higher oestrogen concentrations. Furthermore, the amount of vaginal mucus increases and also changes in fluidity, colour, conductivity, pH and crystallization pattern. During pro-oestrus, the swelling and redness of the vulva differ between animals, not only in intensity but also in duration. In primiparous sows, Sterning *et al.* (1994) showed that the reddening and swelling during pro-oestrus lasted 2.0±1.5 days, with a range of 0 to 8 days in 158 sows during the first oestrus after weaning. The maximum intensity of the reddening and swelling (scored as 0, 1, 2 and 3) was 2.6±0.6 and varied from 0 to 3. Not much is known about factors that influence the duration of pro-oestrus. Sterning *et al.* (1994) found indications of a seasonal influence; pro-oestrus was shorter in spring and summer (March–September; 1.5 days) than in winter (January–March; 2.3 days). In primiparous sows, the duration of pro-oestrus is not related to the duration of standing oestrus (Sterning *et al.*, 1994). It is unclear whether differences in vulval swelling and reddening between sows are related to differences in systemic oestrogen concentrations.

Behaviour

The differentiation of sexual behaviour, resulting in typical female and male sexual behaviour in pigs takes place after birth, as has been reviewed by Ford (1990). However, a very small part of the differentiation seems to take place during pregnancy, as is suggested by the fact that gilts that have been exposed to testosterone *in utero* show a decrease in oestrus expression (testosterone-treated: Ford and Christenson, 1987; gilts between two males *in utero*: Rohde-Parfet *et al.*, 1988).

In female mammals, the behaviour leading to mating is divided in proceptive and receptive behaviour. Proceptive behaviour has been defined as 'various reactions by the female towards the male which constitute her assumption of initiative in establishing or maintaining sexual interaction' and receptive behaviour is defined in terms of 'female responses necessary and sufficient for the male's success in achieving intra-vaginal (or intra-cervical) ejaculation' (Beach, 1976). Oestrus is defined as the period in which the sow shows receptive behaviour or, as defined by Willemse and Boender (1966), the only period in which the female animal permits mating.

Approaching and during oestrus, sows often show a reduced appetite, in general they have a higher level of activity and may show mounting behaviour. Proceptive behaviour in female pigs consists of boar-directed behaviour; given the opportunity, the sow will go to the boar and will stay close to the boar for longer periods of time approaching oestrus and during oestrus (Signoret, 1970; Bressers *et al.*, 1991; de Jonge *et al.*, 1994). Although receptive behaviour is normally the basis for oestrus detection, in group-housing systems, this (proceptive) boar-directed behaviour has been the basis for a system of automated oestrus-detection (Bressers *et al.*, 1991).

Sows show receptive behaviour by the so-called 'standing response'; the sow stands immobile, arches her back and cocks the ears (Signoret, 1970). Olfactory and tactile stimuli of the boar are particularly involved in evoking the standing response (Signoret, 1970). During oestrus, the standing response is not an all or none phenomenon; at the onset of oestrus, many (boar) stimuli may be needed to evoke the standing response, but later in oestrus, fewer stimuli are needed. Willemse and Boender (1966), therefore, divided oestrus into a first boar-period (standing response can be evoked by the boar, but not by the inseminator), an inseminator-period (standing response can also be evoked by the inseminator who performs the Back-Pressure-Test (BPT), and a second boar-period. During the receptive period, not only receptivity but also proceptivity increases reaches a maximum and then decreases again (de Jonge *et al.*, 1994).

The duration of oestrus may vary considerably, both within and between experiments or farms; for example, Weitze *et al.* (1994) found that the duration of oestrus varied between 35 h and 96 h (on average 60±15 h) in 483 sows and Soede *et al.* (1995a) found a variation of between 24 h and 88 h (on average 50±13 h) in 144 sows. Furthermore, when a similar strategy for detecting oestrus (twice daily with a boar) was used on 21 farms in The Netherlands, the average duration of oestrus varied between 38 h and 69 h (G. D. R. Groenland, unpublished).

The duration of the inseminator period (or standing oestrus in the absence of a boar) may also vary considerably; in gilts, Willemse and Boender (1966) found a variation in duration between less than 21 and more than 65 h (on average 43±12 h; excluding the gilts that did not show oestrus in the absence of a boar) and in sows Soede *et al.* (1996) found a variation between 0 h and 64 h (on average 32±14 h; including 8 of 144 sows that did not show oestrus in the absence of a boar). Not all gilts and sows that show a standing response for the boar also show a standing response for man. The proportion of oestrous gilts and sows that do not show oestrus in the absence of a boar was reported to be as high as 46% and 30% (gilts; Schenk, 1967 and Signoret, 1971, respectively)) and 19% (sows; Schenk, 1967). However, in those studies, no account was made of the stage of oestrus of the animals. In studies in which oestrus detection was performed repeatedly at short intervals, the proportion of pigs at oestrus not showing oestrous behaviour in the absence of a boar was only 5% (gilts: Willemse and Boender, 1966; sows: Soede *et al.*, 1996).

From the few experimental data available, it seems that vulval and behavioural characteristics of oestrus are poorly related. For example, a significant difference in the duration of pro-oestrus (vulval swelling and redness) between the second and fourth oestrus of gilts was not accompanied by a change in the duration of standing oestrus (Andersson *et al.*, 1983). Furthermore, both gilts and sows may show vulval changes and ovulate, without showing a standing oestrus (see for example Ford and Christenson, 1987; Eliasson, 1991); not much is known about the cause of this 'silent' oestrus. However, it has been suggested that the occurrence of silent oestrus may be an indication that concentrations of oestrogen needed for initiation of the preovulatory LH peak are less than the concentration that is needed for initiation of oestrous behaviour (see for example Ford and Christenson, 1987).

Regulation of Oestrous Behaviour

Oestrous behaviour, like the preovulatory LH peak, is initiated by oestrogen produced by the growing follicles. From this it might be concluded that expression of oestrus is simply related to differences in systemic oestrogen profiles. In ovariectomized gilts, one injection with oestradiol benzoate can induce oestrous behaviour; higher doses of oestradiol benzoate result in a higher percentage of gilts that show oestrus and also increase the duration of oestrus in those gilts (for example Signoret, 1967; Esbenshade and Huff, 1989). Progesterone priming does not affect oestrous behaviour in pigs (Signoret, 1969) as it does for example in sheep. Furthermore, in pigs, GnRH and circulating LH and FSH do not seem to affect oestrus expression directly. This was shown by Ebenshade and Huff (1989) who eliminated these hormones by active immunization against GnRH and who, after subsequent ovariectomy , did not find any difference in expression of oestrus in these gilts compared with non-immunized ovariectomized gilts in several oestradiol benzoate treatments.

In intact animals, however, there does not appear to be a relationship between systemic oestrogen concentration and the duration of oestrus. Andersson *et al.* (1983) found higher oestrogen concentrations together with a shorter duration of pro-oestrus (but there was no relationship to the duration of oestrus) in gilts during their fourth compared with their second oestrus. In sows, no relationships were found between concentrations of oestrogen, LH or progesterone or their temporal relationships and the duration of oestrus (Soede *et al.*, 1994). Furthermore, in two groups of sows (tethered versus loose-housed) that differed in the duration of oestrus, no differences were found in profiles of oestrogen, LH or progesterone (Soede *et al.*, 1997a). In conclusion, even though oestrogen regulates oestrous behaviour, systemic concentrations of oestrogen do not seem to be related to differences in the duration of oestrus in pigs.

A considerable amount of literature, especially in rats, has shown that sexual behaviour in females is regulated for a major part by the ventromedial nucleus of the hypothalamus (VMH) and other areas of the brain that have a high density of oestrogen receptors (reviewed by

Pfaff *et al.*, 1994). Between and within these areas, there are complicated regulatory mechanisms that may affect oestrous expression depending on internal and external circumstances. In pigs, however, hardly any information is available on the regulation of expression of oestrus at the level of the brain.

Time of Ovulation in Relation to Characteristics of Oestrous Expression

Insemination more than 24 h before ovulation or insemination after ovulation result in a reduced fertilization rate and consequently a reduction in farrowing rate and litter size (reviewed by Kemp and Soede, 1997). Therefore, an important question for optimal detection of oestrus and management of insemination is: which characteristics of oestrus best predict the time of ovulation? In this section, current knowledge regarding this question is evaluated, taking into account both physical changes and behavioural parameters.

Vaginal mucus conductivity. During pro-oestrus and oestrus, the vaginal mucus changes in amount, consistency and conductivity. Several investigations have taken place in which changes in conductivity of the vaginal mucus have been used to determine the optimal time of insemination (e.g. Harbison *et al.*, 1987; Ko *et al.*, 1989). However, the reproductive success of sows inseminated based on vaginal mucus conductivity were at best similar to when inseminations were based on onset of oestrus. A study by Stokhof *et al.* (1996) has related vaginal conductivity to the timing of ovulation. They showed that, although vaginal conductivity increased slightly during oestrus, vaginal conductivity varied considerably between sows and there was no relationship between vaginal conductivity and ovulation time as determined by ultrasonography. In their study, even onset of oestrus was a better predictor for the moment of ovulation than was vaginal conductivity.

Body temperature. In contrast to cattle and humans, in pigs there appears to be no relationship between ovulation time and body temperature (Henne, 1991; Soede *et al.*, 1997b). In eight sows, continuous measurements of vaginal temperature did not show any consistent rises or falls between 4 days before ovulation and 2 days after ovulation (see Fig. 1).

Standing response for the boar. Until the early 1990s, sows were thought to ovulate at a relatively fixed time, at 38 to 40 h after the onset of oestrus. Studies in which ovulation was assessed by ultrasonography have shown that ovulation takes place at very variable moments after the onset of (spontaneous) oestrus in both German, Dutch, Danish and Swedish experiments (see Table 1). Although the average ovulation time is not very variable between the experiments (between 35 h and 48 h), the ovulation time of individual sows varies between 10 h and 85 h after the onset of oestrus. Therefore, although in practice onset of oestrus is the parameter used to assess insemination time, it is not a good predictor for the moment of ovulation.

Standing response for man (in absence of a boar). In 1966, Willemse and Boender found that gilts showed a positive standing response in the absence of a boar during the middle two-thirds of (boar) oestrus, the so-called inseminator period, irrespective of the duration of oestrus. However, in 1996, Soede *et al.* found that the duration of oestrus in the absence of a boar was only partly related to the duration of oestrus in the presence of a boar in 144 sows; time without boar (h) $= 4.8 + 0.53 \times$ time with boar ($R^2 = 0.25$). In both experiments, oestrus was detected in the home-pen of the animals. In Soede *et al.* (1996), ovulation time (determined by ultrasonography) relative to the onset of oestrus as determined in the absence of a boar was found to be very variable, from −2 h to 50 h (on average 26 ± 9 h). Furthermore, eight sows did not show a standing response in the absence of a boar and 22 sows

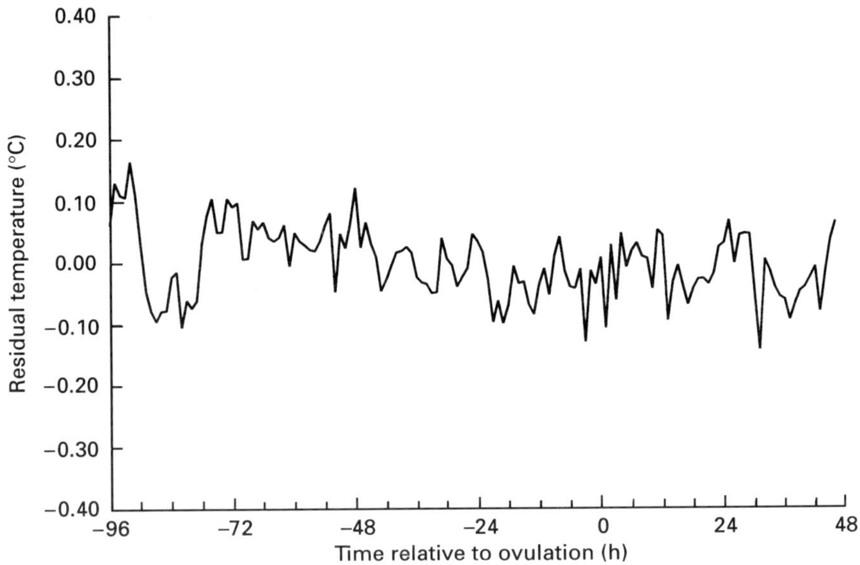

Fig. 1. Residual vaginal temperature (continuously recorded) of eight sows between 96 h before and 48 h after ovulation (checked every 2 h). Since the vaginal temperature is significantly affected by sow and time of day (h), the residual temperature (actual temperature minus estimated temperature) represents the deviation of the expected temperature for a specific sow and a specific time (h) of the day. (Reprinted from Soede *et al.*, 1997b with kind permission from Elsevier Science – NL, Sara Burgerhartstraat 25, 1055 KV Amsterdam, The Netherlands.)

showed a standing response of only 8 or 16 h. Therefore, oestrus as determined in the absence of a boar cannot be used to predict ovulation time.

Duration of oestrus (standing response for the boar). In a number of recent German, Dutch, Swedish and Danish investigations using ultrasonography to assess ovulation in sows, the average timing of ovulation (percentage of oestrus) varies between only 64% and 72% (Table 2). In these experiments, the duration of oestrus explains as much as 50–60% of the variation in ovulation time in oestrus (see Table 2). The variation in ovulation time between individual sows still seems quite large (between 39% and 133%), but this is primarily caused by the sows with oestrus of a short duration (see also Fig. 2). Thus, the relative timing of ovulation during oestrus seems quite constant Weitze *et al.* (1990a) showed that insemination with seminal plasma at the onset of oestrus may advance ovulation by up to 14 h. However, the effects on the relative timing of ovulation are limited (ovulation at 59% of oestrus), because the gilts concomitantly show a reduction in duration of oestrus of on average 8 h. Further studies have shown that the advancement of ovulation is caused by oestrogen and a peptide fraction in the seminal plasma that somehow result in a reduction in the interval from LH to ovulation (see review by Waberski, 1997).

From the above, it is clear that at the moment, the duration of oestrus as determined in the presence of a boar seems the best predictor for ovulation time during oestrus. Unfortunately, the duration of oestrus is highly variable and gives only a retrospective estimate of the time of ovulation. Therefore, research should be aimed at finding better prospective estimators for the timing of ovulation. The ultimate prediction would be to be able to monitor the rise in LH concentrations in a simple way, but for pigs this may still be impractical. Possibly, a combination of parameters (may be both physical and behavioural) can give a better estimate of ovulation time than only one parameter. However, at the moment, no such combination of parameters is known.

Table 1. Timing of ovulation (h after onset of oestrus) in sows and gilts, assessed by ultrasonography

Ovulation time				
Mean ± SD	Range	*n*	Treatment	Reference
Gilts				
38 ± 10	23–48	21	Saline-infusion	Weitze *et al.*, 1990b
45 ± 5	39–48	21	None	Weitze *et al.*, 1990b
Sows				
39 ± 12	20–64	15	None	Dalin *et al.*, 1995
ca 43	17–70	91	AI once	Nissen *et al.*, 1997
37 ± 2	35–43	20	AI once	Mburu *et al.*, 1995
48 ± 6	39–56	13	AI every day	Soede *et al.*, 1992
35 ± 8	10–58	144	AI once	Soede *et al.*, 1995a
41 ± 8	22–58	91	AI once or twice	Soede *et al.*, 1995b
45 ± 13	15–85	427	AI	Weitze *et al.*, 1994

Factors Affecting the Duration of Oestrus

As already shown, a large variation is found in the duration of oestrus between sows and between experiments and farms. Sow factors and management factors that may influence the duration of oestrus are discussed below.

Consistency. The little information that is available shows that the consistency of oestrus duration is not high. In five gilts, the fourth oestrus differed from the second oestrus by – 45 h, + 3 h, + 3 h, + 9 h and + 33 h (Andersson *et al.*, 1983). In 1996, Sterning found no correlation between the duration of oestrus at puberty and the duration of oestrus after weaning (both assessed twice a day) ($r = -0.03$, $n = 161$). Furthermore, in two consecutive parities, the percentage of sows with identical duration of oestrus (checked at 12 h intervals) was 29% (31 of 106 sows) and in 72% of the sows, the duration of oestrus did not differ by more than 12 h (either plus or minus) from the former parity (Weitze, 1996).

Genetic background. In practice, it is well known that differences occur in oestrous expression between the different breeds even though not much research has been done to substantiate this. Within breeds, Willemse and Boender (1966) found that variation in the duration of oestrus within groups of sisters (gilts from 38 litters) was significantly smaller than the variation between groups of sisters, suggesting either a genetic or ontogenic influence on oestrous behaviour. In gilts selected for lean tissue growth rate, Rydhmer *et al.* (1994) found a genetic correlation of –0.49 between the duration of standing oestrus at puberty (1.8 ± 1.7 days) and growth rate to 90 kg (832 ± 97 g) in 600 gilts; the phenotypic correlation was not significant, nor were the genetic and phenotypic correlations with lean percentage. These data suggest that selection for production parameters may result in reduced oestrous expression. However, sows of sow lines (selected mainly for reproduction traits) and boar lines (selected mainly for production traits) were housed under identical conditions and were found to have a similar duration of oestrus (corrected for parity and interval from weaning to oestrus) of 56 h and 57 h, respectively (D. W. B. Steverink, unpublished). During the last 60 years, the performance of the sows (litter size) has changed markedly, while the duration of oestrus does not seem to have changed (in 1930, the duration of oestrus was reported to be on average 40–46 h (McKenzie and Miller (1930 in Gordon, 1997)); therefore, effects of selection for production on oestrous expression have been limited. Nevertheless, the results of Rydhmer *et al.* (1994) suggest that we may be approaching a limit.

Table 2. Timing of ovulation (% of oestrus) in sows, assessed by ultrasonography

Ovulation (% of oestrus)				
Mean ± SD	Range	n	Treatment	Reference
71 ± 14[a]	ND	91	AI once	Nissen *et al.*, 1997
68 ± 8	54–78	20	AI once	Mburu *et al.*, 1995
67 ± 6	58–77	13	AI every day	Soede *et al.*, 1992
72 ± 15[b]	39–133	144	AI once	Soede *et al.*, 1995a
69 ± 1 (± SEM)	ND	60	AI once	Soede *et al.*, 1995b
64 ± 1 (± SEM)	ND	31	AI twice	Soede *et al.*, 1995b
68 ± 10[c]	ND	115	AI once	Steverink *et al.*, 1997
–[d]	ND	427	AI	Weitze *et al.*, 1994

[a]Timing of ovulation (h after onset of oestrus) = 9.9 + 0.53 × duration of oestrus (h) (R^2 = 0.49).
[b]Timing of ovulation (h after onset of oestrus) = 11 + 0.48 × duration of oestrus (h) (R^2 = 0.60).
[c]Timing of ovulation (h after onset of oestrus) = 8.6 + 0.50 × duration of oestrus (h) (R^2 = 0.59).
[d]Timing of ovulation (h after onset of oestrus) = 1.79 + 0.7 × duration of oestrus (h).
ND not determined.

Table 3. Sows classified according to interval from weaning to oestrus (classes of 24 h) and the interval from onset of oestrus to ovulation (classes of 8 h)

Onset of oestrus to ovulation	Weaning to oestrus interval (days)									
	3		4		5		6		Mean	
(h)	n	(%)	n	(%)	n	(%)	n	(%)	n	(%)
0–24	3	(8)	6	(5)	7	(16)	5	(45)	21	(10)
24–32	4	(13)	22	(19)	16	(36)	3	(27)	46	(22)
32–40	10	(27)	39	(34)	11	(25)	2	(18)	62	(31)
40–48	11	(30)	38	(33)	8	(18)	0	(0)	57	(27)
48–56	8	(22)	8	(7)	2	(5)	0	(0)	18	(9)
56–64	0	(0)	3	(2)	0	(0)	1	(9)	4	(2)
Total	36	(18)	116	(56)	44	(21)	11	(5)	208	(100)

From Kemp and Soede, 1996.

Gilts: cycle number. In six successive cycles in crossbred gilts, no influence of cycle number was found on the duration of standing oestrus, but the duration of pro-oestrus became significantly shorter in oestrus 3 to 6 in comparison with oestrus 1 and 2 (Andersson and Einarsson, 1980). Similarly, Andersson *et al.* (1983) studied second and fourth oestrus in crossbred gilts and did not find a difference in the duration of oestrus but did find a reduction in the duration of pro-oestrus; the authors suggested that the latter may have been caused by a change in sensitivity of vulval hormone receptors. In conclusion, in gilts, cycle number does not seem to influence the duration of oestrus.

Parity. The few data that are available are contradictory. Weitze *et al.* (1994) found that, only during summer time, sows of parity 2–4 had a significantly longer oestrus than sows of a higher parity (64 versus 60 h). In contrast, Steverink *et al.* (1997) found that the duration of oestrus (corrected for interval from weaning to oestrus) for sows of parity 1–2 was shorter than that for sows of higher parity (55 ± 11 versus 62 ± 12 h, P < 0.01).

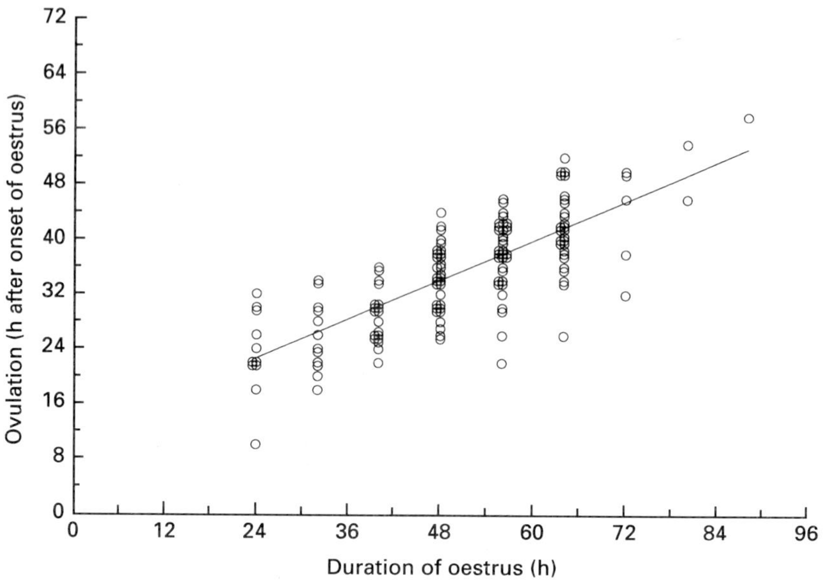

Fig. 2. Time of ovulation (OV) during oestrus in multiparous sows; OV = 11 + 0.48 × Oestrus (R^2 = 0.60, P < 0.001, n = 144) (Soede *et al.*, 1995b).

Season. There are very few experimental data on the effects of season on oestrus expression in sows. In the Netherlands, the combined data from several farms indicate that the duration of oestrus is on average longer in the summer months July, August and September (on average 53, 57 and 60 h) compared with the other months of the year (on average 46–50 h) (G. J. R. Groenland, unpublished). In Germany, Weitze *et al.* (1994) found a small increase in the duration of oestrus in summer compared with spring, whereas in Sweden, in primiparous sows, oestrus was of similar duration in the four seasons (Sterning, 1995). In ovariectomized sows, Cox *et al.* (1987) found that LH concentrations after an injection of oestradiol benzoate remained suppressed for longer in autumn and recovered more slowly (suggesting an increased sensitivity to oestrogen negative feedback). However, in autumn a reduction in oestrus expression was seen which was not related to post-injection concentrations of oestradiol (suggesting a decreased sensitivity to the oestrogen positive feedback). The authors suggested that environmental factors associated with season may alter responsiveness of the brain to oestradiol, consequently affecting sexual behaviour.

Weaning to oestrus interval. An increase in the weaning to oestrus interval (between approximately 3 and 6 days) is related to a decrease in the duration of oestrus (Rojkittikhun *et al.*, 1992; Weitze *et al.*, 1994; Kemp and Soede, 1996). This influence is quite large; an increase in weaning to oestrus interval of 3 days resulted in an average decrease in duration of oestrus of 24 h (Kemp and Soede, 1996). The decrease in duration of oestrus also resulted in a decrease in the interval from onset of oestrus to ovulation (Weitze *et al.*, 1994; Kemp and Soede, 1996). The percentage of sows that ovulate within 32 h from onset of oestrus increases from day 3 to day 6 after weaning from 21%, 24%, 50% to 72% (see Table 3). Rojkittikhun *et al.* (1992) found that the interval from the increase in oestradiol and the onset of oestrus was longer in sows with a prolonged weaning to oestrus interval, suggesting that these sows were less sensitive to oestrogen as a trigger for oestrous behaviour.

Chronic stress. It is generally known that stress conditions may affect several reproductive processes (reviewed by Varley and Stedman, 1994). For example, it is well established that rearing

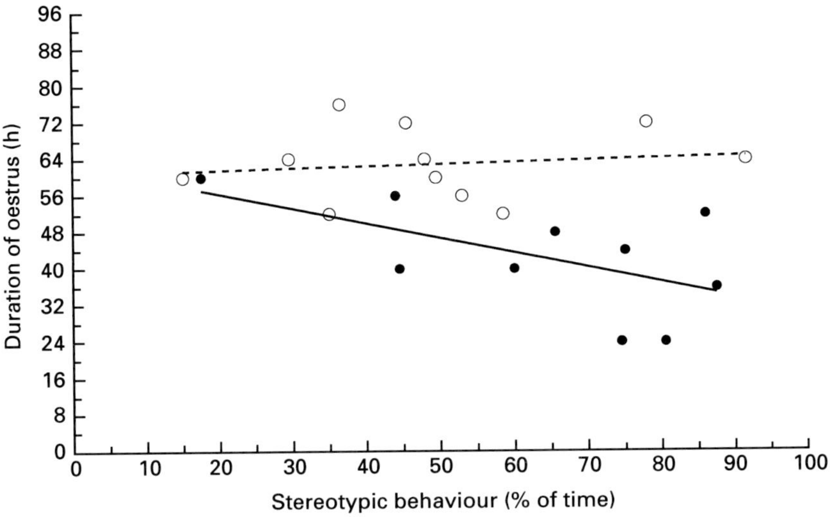

Fig. 3. Level of stereotypic behaviour during 1 h after afternoon feeding in sows between days 12 and 17 before oestrus (% of time) in relation to the duration of oestrus (h) in sows that had been tethered (●) or were housed loose (○) for 2 months after weaning. (Based on data from Soede *et al.*, 1997a.)

conditions (space allowance, group size, boar contact) influence the age of puberty in gilts (Hemsworth and Barnett, 1990). Rearing conditions may also influence oestrous expression; a reduction in space allowance (Hemsworth *et al.*, 1986) resulted in a reduction of the number of oestrous gilts that showed oestrus without the presence of a boar. Apparently, gilts under chronic stress conditions need more stimulation to express oestrus.

In addition, in sows, stressful conditions may influence oestrous expression. Pedersen *et al.* (1993) found that, in sows that were group-housed after weaning, the sows that were lowest in rank had the shortest duration of oestrus. However, no relation with concentrations of cortisol or oestrogen were found. In another study, in which sows were tethered by neck chains for 2 months after weaning, the duration of oestrus was significantly shorter than in sows that were individually housed in a pen of 6 m^2 during the same two months (42 ± 12 versus 63 ± 8 h; $P < 0.001$). In addition, in the tethered sows, the interval from onset of oestrus to ovulation, as detected by ultrasonography, was reduced (28 ± 6 versus 41 ± 6 h; $P < 0.001$). The changes in the duration of oestrus were not related to changes in oestradiol, LH or progesterone profiles of the sows, nor to cortisol concentrations (Soede *et al.*, 1997a). However, in the tethered sows, the duration of oestrus was negatively related to the level of stereotypic behaviour (an indication of chronic stress conditions) of these sows (see Fig. 3).

In general, oestrous behaviour seems to be sensitive to stressful conditions. The involvement of hormones of the hypothalamus–pituitary–adrenal (HPA) axis and endogenous opioid peptides (EOP) in oestrous behaviour have been shown in several experiments. Systemic injections with ACTH, cortisol and morphine (an EOP-agonist) reduced the duration of oestrus in pigs (Liptrap, 1970; Esbenshade *et al.*, 1983; Esbenshade and Huff, 1989). Esbenshade *et al.* (1983) concluded that '…it was not clear whether the competitive effects of cortisol and oestradiol on the expression of oestrus occurs at the level of the systemic circulation, at hypothalamic receptors or at higher brain centres'. This is still unclear. However, hormones of the HPA-axis have been found to affect LH release on all three concentrations. Therefore, oestrus expression may also be a resultant of interactions with hormones of the HPA-axis at the systemic, hypothalamic and higher brain level.

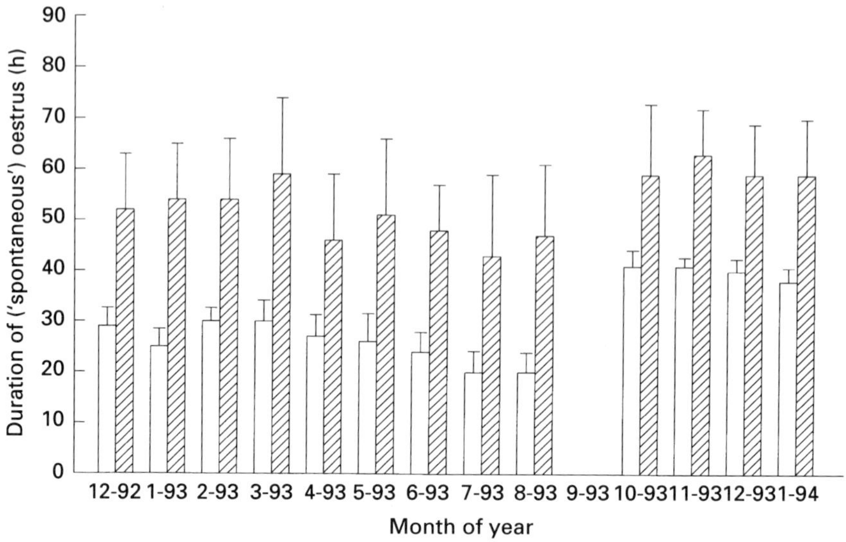

Fig. 4. The average duration of oestrus (□) and of the spontaneous standing response (▨; standing response in presence of the boar but before application of the back pressure test) per month in multiparous sows between 3 and 6 days after weaning in two experiments (Expt 1: December 1992 to August 1993, Expt 2: October 1993 to January 1994) that differed only in the boar that was used for detection of oestrus.

Boar stimulation. Many studies have been performed regarding effects of specific stimuli of the boar (visual, olfactory, auditory and tactile) on the attainment of puberty (reviewed by Hughes *et al.*, 1990), but only a few studies have incorporated the duration of oestrus.

An increase in the amount of boar stimulation during oestrus detection increases the expression of oestrus of sows. Jongman *et al.* (1996) even showed that stimulation by more than one boar (oestrus detection in a detection-mating-area where the sows are surrounded by several boars) increased the duration of oestrus by half a day. Between boars, variability in stimulation value (for a major part related to the olfactory stimuli), affect both proceptive and receptive behaviour (Hughes *et al.*, 1985) and, under multi-sire conditions, there was a certain partner preference (Tanida *et al.*, 1991). Boar variation was also found on our experimental farm, where we performed two 'identical' experiments with a different boar. The boar influence was quite large (Fig. 4). It is obvious that these experiments were not designed to study the influence of the boar and that other factors, such as season, may play a role here, but the results substantiate the importance of boar stimuli in oestrus expression of sows. However, it must be kept in mind that too long and too close contact with the boar has adverse effects on oestrus expression if gilts habituate to the boar stimuli (Hemsworth *et al.*, 1988). This has also been found in sows after weaning (Dyck, 1988).

Part of the influence of boar stimulation on the duration of oestrus may be explained by an involvement of prolactin, since Prunier *et al.* (1987) found that every introduction of a boar during oestrus causes a release of prolactin, and Van de Wiel *et al.* (1981) found that the duration of oestrus was positively associated with concentrations of prolactin.

Consequences of Variation In Expression of Oestrus for Insemination Strategy

As is clear from the outline above, many factors may be responsible for differences in expression of oestrus between farms and within farms, the most important factors (that have been identified as such): interval from weaning to oestrus, chronic stress and boar stimuli. Factors such as stress conditions and boar stimulation will be mostly responsible for differences in duration of oestrus between farms. However, the interval from weaning to oestrus influences the variability in duration of oestrus within farms. Consequently, an optimal insemination strategy on farms (which implies a high farrowing rate and litter size with few inseminations or matings) varies between farms, but should also vary between sows on a farm depending on the interval from weaning to oestrus.

In several studies, an increase in the interval from weaning to oestrus has been found to cause a reduction in subsequent litter size and farrowing rate (see Vesseur, 1997). It has therefore been suggested that sows with a longer interval from weaning to oestrus are less fertile. However, in a study of Kemp and Soede (1996), fertilization results were high for sows that were inseminated between 0 h and 24 h before ovulation irrespective of the weaning to oestrus interval. This finding suggests that adverse effects of a longer weaning to oestrus interval might be a result of a suboptimal timing of insemination and not of poor fertility in these sows.

Concluding Remarks

Ovulation takes place at on average 70% of the way through oestrus in all experiments that have measured this parameter. At the moment, only one factor, insemination with seminal plasma at the onset of oestrus, is known to advance the relative timing of ovulation (to approximately 59%). Since the duration of oestrus is influenced by many factors, the duration of oestrus (and consequently the optimal time for insemination) is very variable within and between farms. However, the influence of some factors is quite fixed, resulting in, for example, a rather stable average duration of oestrus on a specific farm and a rather stable effect of the interval from weaning to oestrus on the duration of oestrus. Consequently, assessment of the average duration of oestrus of sows on a farm, taking into account the interval from weaning to oestrus, can be used for evaluation of the insemination strategy. For most factors that are known to influence the duration of oestrus, the mechanism behind the change in oestrus expression is unclear. Therefore, future studies are needed to determine causes of variability in oestrus expression between sows, including studies concerning the regulation of oestrous behaviour.

References

Andersson AM and Einarsson S (1980) Studies on the oestrus and ovarian activity during five successive oestrous cycles in gilts *Acta Veterinaria Scandinavica* **21** 677–688

Andersson AM, Einarsson S, Edqvist L-E and Lundeheim, N (1983) Endocrine pattern and external oestrous symptoms at second and fourth oestrus in gilts *Animal Reproduction Science* **6** 301–310

Beach FA (1976) Sexual attractivity, proceptivity and receptivity in female mammals *Hormones and Behaviour* **7** 105–138

Bressers HPM, Te Brake JHA and Noordhuizen JPTM (1991) Oestrus detection in group-housed sows by analysis of data on visits to the boar *Applied Animal Behaviour Science* **31** 183–193

Cox NM, Ramirez JL, Matamoros IA, Bennett WA and Britt JH (1987) Influence of season on estrous and luteinizing hormone responses to estradiol benzoate in ovariectomized sows *Theriogenology* **27** 395–405

Dalin AM, Nanda T, Hultén F and Einarsson S (1995) Ovarian activity at naturally attained oestrus in the sow: an ultrasonographic and LH study *Acta Veterinaria Scandinavica* **36** 377–382

Dyck GW (1988) The effect of housing facilities and boar exposure after weaning on the incidence of postlactational anestrus in primiparous sows *Canadian Journal of Animal Science* **68** 983–985

Eliasson L (1991) Relationships between puberty and production traits in the gilt. 2. Oestrous symptoms at puberty *Animal Reproduction Science* **25** 255–264

Esbenshade KL and Huff BG (1989) Involvement of hypothalamic compounds in the expression of estrus in the pig *Animal Reproduction Science* **18** 51–59

Esbenshade KL, Paterson AM and Day BN (1983) Glucocorticoids and estrus in swine. I The relationship between triamcinolone acetonide and estrogen in the expression of estrus in gilts *Journal of Animal Science* **56** 460–465

Ford JJ (1990) Differentiation of sexual behaviour in pigs *Journal of Reproduction and Fertility Supplement* **40** 311–321

Ford JJ and Christenson RK (1987) Influences of pre- and postnatal testosterone treatment on defeminization of sexual receptivity in pigs *Biology of Reproduction* **36** 581–587

Gordon I (1997) The sow's oestrous cycle and associated events. In *Controlled Reproduction in Pigs* pp 60–76 Ed I Gordon. CAB International, New York

Harbison S, Kirkwood RN, Aherne FX and Sather AP (1987) Reproductive performance of sows inseminated with fresh or thawed semen: influence of insemination at a fixed time or as indicated by changes in vaginal mucus conductivity *Canadian Journal of Animal Science* **67** 865–867

Hemsworth PH and Barnett JL (1990) Behavioural responses affecting gilt and sow reproduction *Journal of Reproduction and Fertility Supplement* **40** 343–354

Hemsworth PH, Barnett JL, Hansen C and Winfield CG (1986) Effects of social environment on welfare status and sexual behaviour of female pigs II Effects of space allowance *Applied Animal Behaviour Science* **16** 259–267

Hemsworth PH, Winfield CG, Tilbrook AJ, Hansen C and Barnett JL (1988) Habituation to boar stimuli: possible mechanism responsible for the reduced detection rate of oestrous gilts housed adjacent to boars *Applied Animal Behaviour Science* **19** 255–264

Henne I, (1991) Körpertemperaturen von Jungsauen im ovulationsnahen Zeitraum *Monatshefte Veterinär–Medizin* **46** 674–676

Hughes PE, Hemsworth PH and Hansen C (1985) The effects of supplementary olfactory and auditory stimuli on the stimulus value and mating success of the young boar *Applied Animal Behaviour Science* **14** 245–252

Hughes PE, Pearce, GP and Patterson, AM (1990) Mechanisms mediating the stimulatory effects of the boar on gilt reproduction *Journal of Reproduction and Fertility Supplement* **40** 323–341

Jonge F de, Mekking P, Abbott K and Wiepkema PR (1994) Proceptive and receptive aspects of oestrus behaviour in gilts *Behavioural Processes* **31** 157–166

Jongman EC, Hemsworth PH and Galloway DB (1996) The influence of conditions at the time of mating on the sexual behaviour of male and female pigs *Applied Animal Behaviour Science* **48** 143–150

Kemp B and Soede NM (1996) Weaning to oestrus interval in relation to timing of ovulation and fertilization results in sows *Journal of Animal Science* **74** 944–949

Kemp B and Soede NM (1997) Consequences of variation in interval from insemination to ovulation on fertilisation in pigs *Journal of Reproduction and Fertility Supplement* **52** 79–89

Ko JCH, Evans LE and Hopkins SM (1989) Vaginal conductivity as an indicator for optimum breeding time in the sow after weaning *Theriogenology* **32** 961–968

Liptrap RM (1970) Effect of corticotrophin and corticosteroids on oestrus, ovulation and oestrogen secretion in the sow *Journal of Endocrinology* **47** 197–205

McKenzie FF and Miller JC (1930) Length of the period of heat in gilts and sows and the length of the interval between heat periods *Missouri Agricultural Experiment Station Research Bulletin no. 285*

Mburu JN, Einarsson S, Dalin A-M and Rodriguez-Martinez H (1995) Ovulation as determined by transrectal ultrasonography in multiparous sows: relationships with oestrus symptoms and hormonal profiles. *Journal of Veterinary Medicine Series A* **42** 285–292

Nissen AK, Soede NM, Hyttel P, Schmidt M and D'Hoore L (1997) The influence of time of insemination relative to time of ovulation on farrowing frequency and litter size in sows as investigated by ultrasonography *Theriogenology* **47** 1571–1582

Pedersen LJ, Rojkittikhun T, Einarsson S and Edqvist L-E (1993) Postweaning grouped sows: effects of aggression on hormonal patterns and oestrous behaviour *Applied Animal Behaviour Science* **38** 25–39

Pfaff DW, Schwartz-Giblin S, McCarthy MM and Kow L-M (1994) Cellular and molecular mechanisms of female reproductive behaviours. In *The Physiology of Reproduction* pp 107–220 Eds E Knobil and JD Neill, Raven Press, New York

Prunier A, Martinat-Botte F, Ravault JP and Camous S (1987) Perioestrous patterns of circulating LH, FSH, prolactin and oestradiol-17ß in the gilt *Animal Reproduction Science* **14** 205–218

Prunier A, Quesnel H, Messias de Braganca M and Kermabon AY (1996) Environmental and seasonal influences on the return-to-oestrus after weaning in primiparous sows: a review *Livestock Production Science* **45** 103–110

Rohde-Parfet KA, Ganjam VK, Lamberson WR, Rieke AR, Vom Saal FS and Day BN (1988) Intrauterine position effects in female swine: subsequent reproductive performance, and social and sexual behaviour *Applied Animal Behaviour Science* **26** 349–362

Rojkittikhun T, Sterning M, Rydmer L and Einarsson S (1992) Oestrous symptoms and plasma levels of oestradiol-17ß in relation to the interval from weaning to oestrus in primiparous sows *Proceedings of the 12th Congress of the International Pig Veterinary Society* 485 (Abstract), The Hague

Rydhmer L, Eliasson-Selling L, Johansson K, Stern S and Andersson K (1994) A genetic study of estrus symptoms at puberty and their relationship to growth and leanness in gilts *Journal of Animal Science* **72** 1964–1970

Schenk PM (1967) An investigation into the oestrus symptoms and behaviour of sows *Zeitschrift fur Tierzucht und Zuchtungskunde* **2** 87–110

Signoret JP (1967) Durée du cycle oestrien et de l'oestrus chez la truie. Action de benzoate d'oestradiol chez la femelle ovariectomisée *Annales Biologie Animales Biochemie et Biophysiologie* **7** 407–421

Signoret JP (1969) Action de la progestérone sur le comportement d'oestrus induit par le benzoate d'oestradiol chez la truie ovariectomisée *Annales Biologie Animales Biochemie et Biophysiologie* **9** 361–368

Signoret JP (1970) Reproductive behaviour of pigs *Journal of Reproduction and Fertility Supplement* **11** 105–117

Signoret JP (1971) The mating behaviour of the sow. In *Pig Production* pp 295–313 Ed. DJA Cole. Butterworths, London

Soede NM, Kemp B and Noordhuizen, JPTM (1992) The duration of ovulation in pigs, studied by transrectal ultrasonography, is not related to early embryonic diversity *Theriogenology* **38** 653–666

Soede NM, Helmond FA and Kemp B (1994) Periovulatory profiles of oestradiol, LH and progesterone in relation to oestrus and embryo mortality in multiparous sows using transrectal ultrasonography to detect ovulation *Journal of Reproduction and Fertility* **101** 633–641

Soede NM, Wetzels CCH, Zondag W, de Koning MAI and Kemp B (1995a) Effects of time of insemination relative to ovulation, as determined by ultrasonography, on fertilization rate and accessory sperm count in sows *Journal of Reproduction and Fertility* **104** 99–106

Soede NM, Wetzels, CCH, Zondag W, Hazeleger W and Kemp B (1995b) Effects of a second insemination after ovulation on fertilization rate and accessory sperm count in sows *Journal of Reproduction and Fertility* **105** 135–140

Soede NM, Wetzels CCH and Kemp B (1996) Oestrus (standing response for boar and man) and ovulation in sows *Reproduction in Domestic Animals* **31** 293–294 (Abstract)

Soede NM, Helmond FA, Schouten WPG and Kemp B (1997a) Oestrus, ovulation and peri-ovulatory hormone profiles in tethered and loosehoused sows *Animal Reproduction Science* **46** 133–148

Soede NM, Hazeleger W, Broos J and Kemp B (1997b) Vaginal temperature is not related to the time of ovulation in sows *Animal Reproduction Science* (in press)

Sterning M (1995) Oestrous symptoms in primiparous sows 2. Factors influencing the duration and intensity of external oestrous symptoms *Animal Reproduction Science* **40** 165–174

Sterning M (1996) *Reproductive Performance and Oestrous Symptoms in Primiparous Sows* PhD Thesis Swedish University of Agricultural Sciences

Sterning M, Rydhmer L, Einarsson S and Andersson K (1994) Oestrous symptoms in primiparous sows. 1. Duration and intensity of external oestrous symptoms *Animal Reproduction Science* **36** 305–314

Steverink DWB, Soede NM, Bouwman EG and Kemp B (1997) Fertilisation results as influenced by insemination to ovulation interval and sperm cell dosage in sows *Journal of Reproduction and Fertility* (in press)

Stokhof S, Soede NM and Kemp B (1996) Vaginal mucus conductivity as measured by Walsmeta MKIV does not accurately predict the moment of ovulation or the optimum time for insemination in sows *Animal Reproduction Science* **41** 305–310

Tanida H, Miyazaki N, Tanaka T and Yoshimoto T (1991) Selection of mating partners in boars and sows under multi-sire mating *Applied Animal Behaviour Science* **32** 13–21

van de Wiel DFM, Erkens J, Koops W, Vos E and van Landeghem AAJ (1981) Periestrous and midluteal time courses of circulating LH, FSH, prolactin, estradiol-17ß and progesterone in the domestic pig *Biology of Reproduction* **24** 223–233

Varley M and Stedman R (1994) Stress and Reproduction. In *Principles of Pig Science* pp 277–296 Eds DJA Cole, J Wiseman and MA Varley, Nottingham University Press, Nottingham

Vesseur PC (1997) *Causes and Consequences of Variation in Weaning to Oestrus Interval in the Sow* PhD Thesis Wageningen Agricultural University

Waberski D (1997) Effects of semen components on ovulation and fertilization *Journal of Reproduction and Fertility Supplement* **52** 105–109

Weitze KF (1996) Timing of AI and ovulation in breeding herds I *Reproduction in Domestic Animals* **31** 193–199

Weitze KF, Lotz, JF, Everwand A, Willmen T and Waberski D (1990a) Interaction between inseminate, uterine and ovarian function in the sow. II Investigations into the influencing of ovulation by the use of sperm free media *Reproduction in Domestic Animals* **25** 197–204

Weitze KF, Rath D, Willmen T, Waberski D and Lotz J (1990b) Advancement of ovulation in the sow related to seminal plasma application before insemination *Reproduction in Domestic Animals* **25** 61–67

Weitze KF, Wagner-Rietschel H, Waberski D, Richter L and Krieter J (1994) The onset of heat after weaning, heat duration, and ovulation as major factors in AI timing in sows *Reproduction in Domestic Animals* **29** 433–443

Willemse AH and Boender J (1966) A quantitative and qualitative analysis of oestrus in gilts *Tijdschrift voor Diergeneeskunde* **91** 349–363

Journal of Reproduction and Fertility Supplement **52**, 105–109

Effects of semen components on ovulation and fertilization

D. Waberski

Institute for Reproductive Medicine, Veterinary School of Hannover, Bünteweg 15, D- 30559, Hannover, Germany

In pigs, transcervical infusion of seminal plasma at the onset of oestrus advances ovulation and thus improves the chance of gametes meeting during their full fertilizing competence. An animal model that allows single uterine horn infusion was used in combination with transcutaneous sonographic monitoring of ovaries. Preparative surgery involved the detachment of one uterine horn from the corpus, leaving the caudal end open to the peritoneal cavity but sealing the corpus wound. Transcervical infusion of 100 ml seminal plasma immediately after the detection of oestrus advanced ovulation by between 8 and 14 h on the ipsilateral ovary adjacent to the infused horn compared with the contralateral ovary. In addition, the seminal plasma infusion did not influence the LH profile compared with uterine infusion of PBS. This finding indicates that the effect is mediated by a local mechanism in the female genital tract. The advancement of ovulation depends on the time of infusion early in oestrus and is more pronounced in gilts exhibiting a long interval between the onset of oestrus and spontaneous ovulation compared with early ovulators. At 24 h after the detection of oestrus, seminal plasma was ineffective. Apparently, seminal plasma does not affect maturation and fertilizing competence of oocytes. The activity resides in a low molecular mass protein fraction and, to a far lesser extent, in seminal oestrogens. Further characterization of the active components may allow a practical application in AI.

Induced Ovulation in Spontaneously Ovulating Species

Success of insemination depends on the precise timing of the meeting of fully fertilizing competent gametes in the oviduct. Consequences of ageing of either spermatozoa or eggs in the female genital tract in pigs have been shown extensively (Hunter, 1967; Waberski *et al.*, 1994; Soede *et al.*, 1995). Furthermore, using spermatozoa of low quality or of low number may lead to high fertility when insemination is performed only a few hours before ovulation (Waberski *et al.*, 1994). There is growing evidence that the essential steps that the spermatozoa have to undergo before fertilization, such as capacitation and ascent towards the eggs, are under the control of the periovulatory events in the female genital tract (Hunter, 1995). However, extended storage time of spermatozoa in the lower isthmus invariably leads to irreversible loss of fertilization capacity. This is especially pronounced in so called spontaneously ovulating species, such as pigs and horses, in which intervals between the onset of oestrus and ovulation extend for several days. In these species, repeated inseminations are usually required to ensure high fertility.

Coitus-induced ovulation is a known physiological mechanism that regulates the timing of the fertilization process. Some species known as spontaneous ovulators, such as cattle, sheep and pigs, may become temporarily induced ovulators for the improved coordination of the essential steps of fertilization (Jöchle, 1975). In pigs, natural mating shortens the interval between the onset of oestrus and ovulation (Pitkjanen, 1958; Signoret *et al.*, 1972). This has been explained by the neurohormonal stimulation associated with the presence of a boar and with copulation (Ziecik *et al.*, 1981; Kirsch *et al.*, 1985) and specific components of boar semen (Seglin'sh and Brutgans, 1981; Claus, 1989; Weitze *et al.*, 1990). The present paper summarizes our current knowledge on active substances and mechanisms involved in the advancement of ovulation by boar seminal plasma and the consequences for fertilization.

Local Effect of Seminal Plasma on the Time of Ovulation

Since boar seminal plasma contains a remarkably high oestrogen content (up to 11.5 μg per ejaculate), it was originally considered to be an active component regulating the function of the female genital tract. It was thought that prostaglandins released by the endometrium in response to seminal oestrogen might promote passive sperm transport and induce ovulation (Claus *et al.*, 1987, 1989). A vascular countercurrent pathway between uterine vein and ovarian arteries (Einer-Jensen 1988; Krzymowski *et al.*, 1990) provides a local pathway by which the prostaglandins might travel to the ovary. In addition, some authors (Kirsch *et al.*, 1985; Claus, 1989) suggest that seminal plasma may also influence the time of ovulation systemically by advancing the preovulatory LH surge.

The suspected influence of transcervical infusions of seminal plasma on the time of ovulation has been confirmed by the use of transcutaneous sonography for the detection of ovulation in gilts (Weitze *et al.*, 1990). However, in this study an infused oestrogen solution was less effective than whole seminal plasma in shortening the interval between the onset of oestrus and ovulation. Pigs show a high physiological variation in their intervals from the onset of oestrus to spontaneous ovulation, ranging from 20 to 120 h (Weitze *et al.*, 1994). Therefore, studies on the effect of a given solution (seminal plasma or a component thereof) require the investigation of a large number of oestrous cycles. We tested the hypothesis that the seminal plasma effect is mediated by a local pathway using an animal model described by Jungblut *et al.* (1991). The use of the animal model in combination with the sonographic detection of ovulation also allowed for a reduced number of investigated cycles since it provides a treated uterine horn and an untreated control horn within the same animal. Briefly, one uterine horn was detached from the corpus, leaving the caudal end open to the peritoneal cavity but sealing the corpus wound. Thus, a transcervically infused medium only has access to one uterine horn (Fig. 1) (Waberski *et al.*, 1995). The infusion of 100 ml seminal plasma at the onset of oestrus shortened the interval from the onset of oestrus to ovulation by 8–14 h on the ipsilateral ovary adjacent to the patent horn compared with the contralateral ovary on the non-infused control side (Waberski *et al.*, 1995, 1997). Seminal plasma pools were used in all studies since the concentration of steroid hormones is known to vary widely between ejaculates of different boars and between ejaculates of the same boar at different seasons (Claus *et al.*, 1983). These results provide evidence that seminal plasma advances ovulation by a locally active mechanism. This does not necessarily exclude an additional systemic effect of seminal plasma on the time of ovulation. Therefore, in a recent study using the same animal model, the measurement of LH was included. LH concentrations and LH surge profiles were not influenced by the infusion of seminal plasma compared with PBS infusion into a single uterine horn (Waberski *et al.*, 1997). An influence of the insemination procedure and the infused volume on LH release and, therefore, on ovulation cannot be ruled out. However, this was not considered as our experiments were designed to study the specific effect of seminal plasma. As revealed by infusion of PBS, a local volume effect on the time of ovulation was not detectable in the animal model. In addition, the intervals from the onset of oestrus to ovulation on the control ovary were not any shorter in gilts receiving an infusion of seminal plasma compared with non-infused gilts. We, therefore, concluded that the observed effect of seminal plasma is a local phenomenon.

The advancement of ovulation by seminal plasma via a local mechanism consequently shortens the interval from the LH increase to ovulation. Since LH also triggers the resumption of meiosis in follicular oocytes, the question arises whether seminal plasma would affect oocyte maturation. Recent studies indicate that fertilization competence of oocytes is not diminished after uterine infusion of seminal plasma (D. Waberski, R. Classen, P. W. Jungblut, E. Kallweit and K. F. Weitze, unpublished). This seems plausible since there is no reason to assume that mating early in oestrus followed by an additional mating close to ovulation negatively affects fertility.

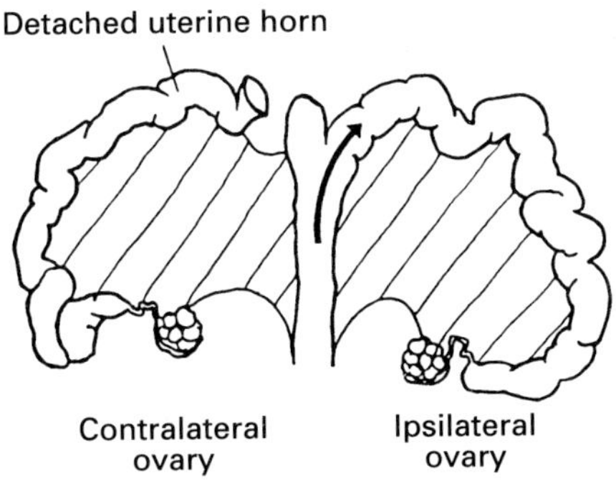

Fig. 1. Schematic design of uterus and ovaries in the animal model. The arrow indicates the flow of an infused solution. (Reproduced from Waberski *et al.*, 1995.)

Time Dependency of the Effect of Seminal Plasma on Ovulation

The advantages of a mating-induced ovulation via the stimulus from copulation and specific effects of seminal plasma are evident. We studied the ovarian response to seminal plasma in relation to the duration from the onset of oestrus to 'spontaneous' ovulation and in relation to different times of infusion in the oestrous cycle (Waberski *et al.*, 1997). In the first experiment, seminal plasma was infused into single uterine horns immediately after the onset of oestrus, which was detected by tolerance to the mounting of a teaser boar. The interval from the onset of oestrus to ovulation on the control ovary was recorded and related to the ovulation-inducing effect of seminal plasma as calculated from the time difference between ipsi- and contralateral ovulation. In the second experiment, using three groups of gilts prepared according to the same animal model, the effect of the time of seminal plasma infusions was tested at the onset of oestrus (0 h), 16 h and 24 h later. These studies showed that the seminal plasma effect is more pronounced in gilts with long oestrus–ovulation intervals and at an early time of infusion in oestrus (Fig. 2). Infusion of seminal plasma at 24 h after the onset of oestrus showed no advancement of ovulation. When calculated retrospectively, it was shown that seminal plasma was ineffective when the interval between infusion and contralateral ovulation was less than 20 h (Waberski *et al.*, 1997). In several farm studies it was shown that irrespective of the duration of oestrus, which varies between 33 and 153 h, ovulation occurs fairly constantly at the beginning of the last third of oestrus (Weitze *et al.*, 1994; Soede *et al.*, 1995). Since seminal plasma is less effective in sows with early spontaneous ovulation compared with late ovulators, seminal plasma seems to have a synchronizing effect on the time of ovulation.

Active Components in Seminal Plasma

Using the animal model, we confirmed our previous result (Weitze *et al.*, 1990) that seminal oestrogens are effective only to some extent. Advancement of ovulation was observed on average only for 3.3 h after infusion of 10 μg oestradiol into a single uterine horn, while infusion of a steroid-depleted seminal plasma fraction resulted in ovulations occurring 7.3 h earlier on the ipsilateral than on the contralateral ovary. Addition of 10 μg oestradiol to the steroid-depleted fraction restored the effect to that seen with whole seminal plasma (10.7 h on average). Further characterization provided evidence that the proteinaceous 1–10 kDa fraction of seminal plasma plays a key role in eliciting ovulation. It is noteworthy that transcervical infusions of 250 μg of either prostaglandin $F_{2\alpha}$ or prostaglandin E_2 in 100 ml PBS were ineffective (Waberski *et al.*, 1995). Boar seminal plasma contains only very low concentrations of prostaglandins (Poyser, 1974, Waberski *et al.*, 1995), which are unlikely to have biological activity. However, prostaglandins released from the endometrium were suspected to induce ovulation (Claus, 1989). The rise of intrafollicular prostaglandins a few hours

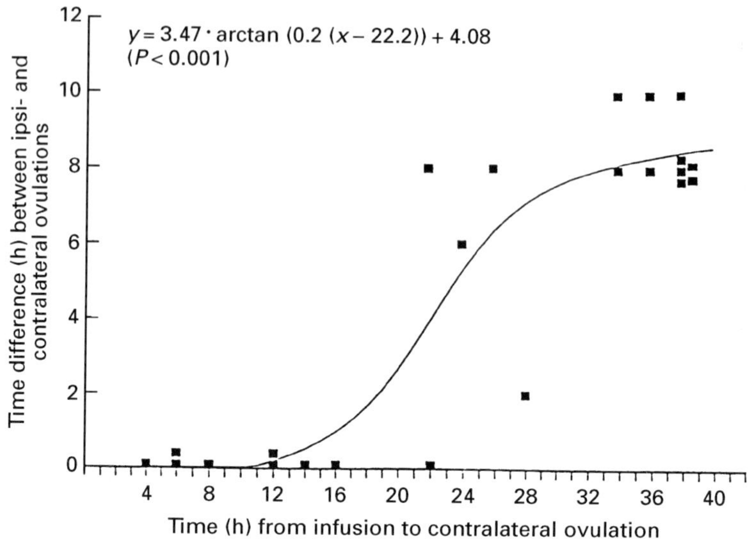

Fig. 2. Time dependency of the seminal plasma effect on the time of ovulation in 22 gilts. (Reproduced from Waberski *et al.*, 1997.)

before ovulation is widely accepted as stimulating the cascade leading to follicular rupture (Espey and Lippner, 1994). However, the rise of intrafollicular prostaglandins in response to the infusion of oestrogens as measured by Weiler and Claus (1991) is likely to occur too early and not in sufficient quantity to contribute to the induction of ovulation. It is still not known whether seminal plasma components act directly on the ovary or whether the effect is mediated by a signal from the female genital tract after the exposure to seminal plasma. As candidates for a direct effect on ovulation, proteins of low molecular mass with a possible ovulation inducing-effect were tested in the animal model. However, neither uterine infusion of relaxin nor GnRH as a substitute for GnRH-like peptides affected ipsilateral ovulations (D. Waberski, R. Classen, P. W. Jungblut, E. Kallweit and K. F. Weitze, unpublished). Further progress in the characterization of the active fraction requires sophisticated separation methods that do not influence the activity of the relevant seminal plasma components.

Practical Application for AI

The question arises as to whether the dilution of a boar ejaculate and division into at least 20 portions for use in AI might diminish the benefits of seminal plasma in a single AI dosage as compared with the whole ejaculate. As mentioned, at 24 h after the onset of oestrus, when the first AI is commonly performed, seminal plasma does not affect the time of ovulation. However, when given at an advanced stage of oestrus, seminal plasma may promote passive sperm transport in the female genital tract (Viring and Einarsson, 1980; Claus *et al.*, 1989). However, in a previous study, we did not find evidence that the seminal plasma plays a major role in the success of fertilization under well balanced reproductive management conditions (Waberski *et al.*, 1996). Since in this study, high fertility was achieved even after a single insemination with a very few spermatozoa (0.5×10^9), the experimental AI condition may not be representative of on-farm conditions. Nevertheless, the major effect in terms of the reproductive biological interaction between male genital secretions and the female genital tract is in the advancement of ovulation after an intrauterine infusion of seminal plasma early in oestrus.

Conclusion

In conclusion, an intrauterine infusion with active components at the onset of oestrus may facilitate the timing of the subsequent insemination close to ovulation and, thereby, enhance the chance of fertilization. Since availability of boar seminal plasma is limited to AI centres, and for hygienic reasons, the use of active substitutes is desirable. As a prerequisite for the application of the effect of seminal plasma, further studies should consider individual variance of boar seminal plasma and further characterization of the active component should be undertaken.

The support of the German Association of Pig Production (ZDS, Bonn) and the Dr. Dr. h.c. Karl Eibl Foundation (Neustadt a.d. Aisch) is gratefully acknowledged.

References

Claus R (1989) Oestrogens of the boar: effects on male and female reproductive functions. In *Reproductive Biology and Medicine* pp 136–147 Eds AF Holstein, KD Voigt and D Grässlin, Diesbach Verlag, Berlin

Claus R, Schopper D and Wagner H-G (1983) Seasonal effects on steroids in blood plasma and seminal plasma of boars *Journal of Steroid Biochemistry* **19** 725–729

Claus R, Hoang-Vu C, Ellendorff F, Meyer HD, Schopper D and Weiler U (1987) Seminal oestrogens in the boar: origin and function in the sow *Journal of Steroid Biochemistry* **27** 331–335

Claus R, Ellendorff F and Hoang-Vu C (1989) Spontaneous electromyographic activity throughout the cycle in the sow and its change by intrauterine oestrogen infusion during oestrus *Journal of Reproduction and Fertility* **87** 543–551

Einer-Jensen N (1988) Countercurrent transfer in the ovarian pedicle and its physiological implications *Oxford Reviews of Reproductive Biology* **10** 348–381

Espey LL and Lippner H (1994) Ovulation. In *The Physiology of Reproduction* 2nd edn Eds E Knobil and JD Neill. Raven Press Ltd., New York

Hunter RHF (1967) The effects of delayed insemination on fertilization and early cleavage in the pig *Journal of Reproduction and Fertility* **13** 133–147

Hunter RHF (1995) Ovarian endocrine control of sperm progression in the fallopian tubes *Oxford Reviews of Reproductive Biology* **17** 85–124

Jöchle W (1975) Current research in coitus-induced ovulation: a review *Journal of Reproduction and Fertility Supplement* **22** 165–207

Jungblut PW, Sierralta WD, Henning M and Kallweit E (1991) Die transzervikale Injektion in das Uteruslumen von kastrierten Schweinen als zell- und organbiologische Versuchsanordnung. Das Mariensee Modell *Journal of Experimental Animal Science* **34** 229–235

Kirsch JD, Tilton JE, Ziecik A, Weigl R, Schaffer T and Williams GL (1985) Effect of various mating stimuli on pituitary release of luteinizing hormone in the gilt *Domestic Animal Endocrinology* **2** 99–104

Krzymowski T, Kotwica J and Stefanczyk-Krzymowska S (1990) Uterine and ovarian countercurrent pathways in the control of ovarian function in the pig *Journal of Reproduction and Fertility Supplement* **40** 179–191

Pitkjanen IG (1958) The time of ovulation in sows *Animal Breeding Abstracts* **27** 212 (Abstract 928)

Poyser NL (1974) Some aspects of prostaglandins in reproduction *Biochemistry Society Transactions* **2** 1196

Seglin'sh A and Brutgans YA (1981) The effect of boar seminal plasma, oxytocin and carbachol on conception *Pig News and Information* **2** 197–198

Signoret JP, Du Mesnil du Buisson D and Mauléon P (1972) Effect of mating on the onset and duration of ovulation in the sow *Journal of Reproduction and Fertility* **31** 327–330

Soede NM, Wetzels CCH, Zondag W, de Koning MAI and Kemp B (1995) Effects of time of insemination relative to ovulation, as determined by ultrasonography, on fertilization rate and accessory sperm count in sows *Journal of Reproduction and Fertility* **104** 99–106

Viring S and Einarsson S (1980) Effect of boar seminal plasma on uterine and oviductal motility in oestrous gilts *Acta Veterinaria Scandinavia* **21** 607–616

Waberski D, Weitze KF, Gleumes T, Schwarz M and Petzoldt R (1994) Effect of time of insemination relative to ovulation on fertility with liquid and frozen boar semen *Theriogenology* **42** 831–840

Waberski D, Südhoff H, Hahn T, Jungblut PW, Kallweit E, Calvete JJ, Ensslin M, Hoppen H-O, Wintergalen N, Weitze KF and Töpfer-Petersen E (1995) Advanced ovulation in gilts by the intrauterine application of a low molecular mass pronase-sensitive fraction of boar seminal plasma *Journal of Reproduction and Fertility* **105** 247–252

Waberski D, Soares JAG, Bandeira de Arruda E and Weitze KF (1996) Effect of a transcervical infusion of seminal plasma prior to insemination on the fertilising competence of low numbers of boar spermatozoa at controlled AI–ovulation intervals *Animal Reproduction Science* **44** 165–173

Waberski D, Claassen R, Hahn T, Jungblut PW, Parvizi N, Kallweit E and Weitze KF (1997) LH profile and advancement of ovulation after transcervical infusion of seminal plasma at different oestrous stages in gilts *Journal of Reproduction and Fertility* **109** 29–34

Weiler U and Claus R (1991) Endocrine aspects of testicular function, especially hormones in the seminal plasma and their fate in the female reproductive tract; testicular steroids and their relevance for male and female reproductive functions *Reproduction in Domestic Animals Supplement* **1** 41–59

Weitze KF, Habeck O, Willmen T and Waberski D (1990) Interaction between inseminate, uterine and ovarian function in the sow. II. Investigations into influencing of ovulation by the use of sperm-free media *Reproduction in Domestic Animals* **29** 433–443

Weitze KF, Wagner-Rietschel H, Waberski D, Richter L and Krieter J (1994) The onset of heat after weaning, heat duration, and ovulation as major factors in AI timing in sows *Reproduction in Domestic Animals* **29** 433–443

Ziecik A, Tilton JE and Williams GL (1981) Effect of mating on the luteinizing hormone surge in the pig *Journal of Animal Science* **53** 434–438

Poster titles

Martinat-Botté F, Guillouet P, Mercat MJ, Landrevie J and M Terqui
Changes in the electrical conductivity of vaginal mucus of sows in different reproductive situations

Mwanza AM, Pettersson A and S Einarsson
A computer assisted method for recording and analysis of oviductal intraluminal pressure in sows

Johansson M, Mburu J and H Rodriguez-Martinez
Ultrastructure of the endosalpinx an demonstration of oviduct luminal mucus in the pig

Soede NM, Bouwman EG and B Kemp
Seminal plasma does not advance ovulation in hCG-treated sows

Tonanon A, Chaitrikal J, Ekawatpanich J, Inthisorn J, Poolperm P and P Udomprasert
Effect of semen temperature on artificial insemination in swine

Borchardt Neto G, Waberski D, Wagner-Rietschel H and KF Weitze
Effect of inseminations relative to ovulation on farrowing rate and litter size of multiparous sows

Vesseur PC and NM Soede
Check of insemination time relative to ovulation on a sow farm

Lamberson WR, Boyle AM and MC Lucy
A simulation model for comparing timing of insemination

Steverink DWB, Bouwman EG, Soede NM and B Kemp
The effect of semen backflow on fertilisation results in sows

Vesseur PC, Kemp B, Hartog LA den and JPTM Noordhuizen
Breeding of first parity sows during second oestrus after weaning

Vesseur PC, Kemp B, Hartog LA den and JPTM Noordhuizen
Effects of split-weaning in first and second parity sows

Mburu JN, Einarsson S, Madej A and H Rodriguez-Martinez
Influence of post-ovulatory food deprivation on accessory sperm count, embryo development and plasma progesterone in sows

Weitze KF, Schmidt A, Waberski D and L Richter
Incidence of ovarian cysts in a sow herd detected by transcutaneous sonography

Hultén F, Lundeheim N, Dalin A-M and S Einarsson
A field study on fertility in sows group-housed during late lactation

Peters AR and RL Pitt
Seasonal breeding pattern in outdoor pigs in the UK

Hoste S, Gommert A, Glossop CE and A Cliff
The use of AI in UK outdoor herds during the summer months

Tantasuparuk W, Lundeheim N, Dalin A-M, Kunavongkrit A and S Einarsson
Seasonal variation in littersize of Landrace, Yorkshire and Crossbred sows in Thailand

EMBRYONIC AND FETAL DEVELOPMENT IN THE PIG

Chair
F. W. Bazer

Journal of Reproduction and Fertility Supplement **52**, 115–131

Role of uterine immune cells in early pregnancy in pigs

H. Engelhardt[1], B. A. Croy[2] and G. J. King[1]

Departments of [1]Animal and Poultry Science and [2]Biomedical Sciences, University of Guelph, Guelph, Ontario, Canada N1G 2W1

The immune system discriminates 'self' from 'non-self', and eliminates that which it determines to be non-self. Mammalian pregnancy appears to represent a failure of self–non-self discrimination, yet it is a highly successful reproductive strategy. We present evidence that the immune system of the female pig responds to the challenges of both mating and the presence of conceptuses. Mating induces an influx of inflammatory leukocytes into the endometrial stroma and uterine lumen. This response, while partially under endocrine control, is amplified by as yet unidentified factors in seminal plasma. In addition to preventing microbial infection, this mating-induced immune response may enhance reproductive performance. During the first month of pregnancy when intimate contact between maternal and fetal tissues is being established, the number of uterine lymphocytes decrease in the luminal epithelium and increase in the endometrial stroma at sites of conceptus attachment. The majority of these lymphocytes express the CD2 and CD8 surface markers, consistent with either T or natural killer (NK) cell lineage. Dispersed endometrial cells obtained during early gestation exhibit a pregnancy-specific increase in NK-like lytic activity. Our ongoing efforts to determine whether these NK-like lymphocytes are the ones that localize to sites of conceptus attachment are discussed. We speculate on how the reactions of pig uterine leukocytes to seminal plasma and conceptuses might contribute to successful pregnancy.

Introduction

Immunology is commonly thought of as the highly complex interactions between cells and molecules that mediate the recognition and elimination of potential pathogens. However, virtually all functions of the immune system can be related to its ability to distinguish 'self' from 'non-self' and to eliminate or otherwise inactivate that which it determines to be non-self. Pregnancy appears to represent a failure of self–non-self discrimination. At mating, the female is exposed to spermatozoa and male-derived proteins in semen. Throughout gestation, she is exposed to the conceptus, which is non-self by virtue of its expression of developmentally regulated, unique embryonic antigens and paternally inherited antigens. Yet mammalian pregnancy is a proven, highly successful reproductive strategy.

Most hypotheses about the role of uterine leukocytes in pregnancy are based on investigations conducted in rodents and humans. In these species, implantation is associated with degeneration of the uterine epithelial layer and invasion of the endometrial stroma. There is direct contact between circulating maternal blood and the trophoblast. In contrast, the placentae of conventional livestock species are characterized by little or no invasion of maternal tissues and no direct contact between maternal and fetal blood. The pig exemplifies the least intimate form of placentation. The uterine epithelium remains intact throughout gestation, invasion is completely prevented and the relationship between maternal and fetal tissues is more appropriately described as 'attachment' rather than 'implantation' (reviewed by King, 1993). The ultimate placental union consists of interlocking microvilli of chorionic and uterine epithelia which begin to form by day 12 or 13 (Keys and King, 1990). By day 30 (term = 114 days), attachment extends over the entire surface of the maternal–fetal interface except at gland openings and the extreme ends of the chorion.

In the haemochorial placenta of rodents and humans, the presence of specialized subsets of immune cells or leukocytes within the pregnant uterus and the activation of these cells by pregnancy are emerging themes (Croy *et al.*, 1997a). The fact that the relationship between maternal and fetal tissues in the epitheliochorial placenta of the pig is so superficial raises the question of whether the pig uses similar strategies for coping with the immunological challenges of pregnancy. In this review, we first provide a brief overview of the immune system with emphasis on immune cells likely to be present in the uterine mucosa. We then present evidence that uterine leukocytes in sows do indeed respond to the challenges of both mating and contact with conceptuses and we consider how these responses might contribute to successful pregnancy.

The Immune System: An Overview

Functional divisions of the immune system

Immune responses can be classified as either antigen specific or innate. The latter responses are also called natural immunity. Natural immune mechanisms are generally effective without prior exposure to the antigen and are available at very short notice. Antigen-specific or acquired immunity requires prior experience, is specific for the stimulating antigen, and develops over several days. The effector cells in natural immunity are phagocytes (macrophages and neutrophils) and natural killer (NK) lymphocytes. The important cell types in antigen-specific immunity are B and T lymphocytes, and antigen-presenting cells (macrophages and dendritic cells). These divisions of immune responses are largely a matter of convenience. In reality, antigen specific and innate immune mechanisms are inextricably linked with one another.

Morphology and function of the major leukocyte classes

Granulocytes. Granulocytes, named for their abundant cytoplasmic granules, are the most numerous leukocytes in peripheral blood. These leukocytes, often referred to as inflammatory cells, play important roles in both natural and specific immunity. Neutrophils, also known as polymorphonuclear leukocytes or polymorphs because of their distinctive multilobed, morphologically diverse nuclei, are the predominant circulating granulocyte. Neutrophils respond rapidly to chemotactic stimuli and are the major cell population in acute inflammatory responses.

Lymphocytes. The next most abundant circulating leukocytes are lymphocytes (B, T and NK cells), which are typically small- to medium-sized cells with large nuclei surrounded by a thin rim of cytoplasm. Their relatively nondescript appearance contrasts with remarkable functional diversity. B lymphocytes are the only cells capable of producing antibodies. In mammals, the early stages of B-cell maturation occur in the bone marrow. Differentiation into specialized forms called plasma cells, found only in lymphoid organs or at sites of immune responses, is the final fate of a B cell.

The second major class of lymphocytes is the T cells. They originate from precursors in the bone marrow and migrate to the thymus for final maturation. T cells are further subdivided into several functionally distinct subpopulations, the best defined of which are helper (T_H) cells and cytotoxic (T_C) cells. In response to antigenic stimulation, T_H cells release cytokines which promote the proliferation and differentiation of other T cells, B cells, NK cells and macrophages. T_C cells lyse cells that express non-self antigens such as cells infected by viruses. 'Suppressor' T cells that inhibit immune responses may also exist. However, there is considerable controversy over whether suppression is mediated by a distinct T-cell subset, or by T_C or T_H cells under certain conditions. The only structures that T_H and T_C cells recognize are peptide antigens attached to proteins that are encoded in the major histocompatibility complex (MHC) and expressed on the surfaces of other cells

(Abbas *et al.*, 1994). There are two different types of MHC gene products, called MHC class I and class II molecules. While T_C cells recognize non-self peptides attached to class I MHC molecules, T_H cells recognize only those peptides associated with class II molecules. In general, class I molecules are present on virtually all nucleated cells, whereas class II molecules are normally expressed only on B cells, antigen-presenting cells and endothelial cells.

The third major class of lymphocyte does not express markers for either T or B cells, and was therefore initially called the null cell population. In humans and mice, most null cells are relatively large lymphocytes with numerous cytoplasmic granules that are capable of spontaneous lysis of a variety of tumour cells and virally infected cells. This lymphocyte population is now known as NK cells. There is mounting evidence that uterine leukocytes of this lineage play important roles in pregnancy in humans and rodents (reviews: Croy *et al.*, 1996; Head, 1996; King *et al.*, 1996).

Mononuclear phagocytes. These cells originate from a common precursor in the bone marrow and, after maturation and subsequent activation, achieve varied morphological forms. Monocytes, large cells with characteristic horseshoe-shaped nuclei, are immature forms that enter the blood from the bone marrow. Once they settle in tissues, these cells mature and become macrophages, which can in turn be activated and assume different forms. Macrophages are central for both specific and natural immunity. They are the principal scavenger cells of the body and recruit other inflammatory cells by secretion of cytokines. In specific immunity, macrophages function as accessory or antigen-presenting cells by presenting foreign antigens to T cells in association with MHC molecules. Dendritic cells, identified morphologically by their spine-like projections, are thought to be related in lineage to mononuclear phagocytes. They are found in the interstitial tissue of most organs, scattered through the epidermis of the skin, and in T cell-rich areas of lymph nodes and spleen.

The CD nomenclature for leukocytes

Although lymphocytes are relatively homogeneous by light microscopy, functionally distinct populations have been found to express different membrane proteins. A uniform nomenclature has been adopted in which a cell-surface marker identifying a particular lineage or differentiation stage that has a defined structure and is recognized by a group ('cluster') of monoclonal antibodies is designated a 'CD' or 'cluster of differentiation', followed by a number indicating its order of discovery. Many of these phenotypic markers have been found to play important roles in the function of these cells. For example, all T cells express CD3, which refers to a group of proteins that form part of the T-cell receptor complex. T_H cells express CD4, while T_C cells express a different marker called CD8. The surface phenotype of these T-cell subsets would thus be designated $CD3^+CD4^+CD8^-$ and $CD3^+CD4^-CD8^+$, respectively.

Although this nomenclature was originally developed for humans, it is now common practice to refer to homologous markers in other species by the same CD designation. The expression of selected CD molecules by pig leukocytes is shown in Table 1. In both species, the T-cell receptor, which is composed of two covalently linked glycoprotein chains, exists as an $\alpha\beta$ dimer or a $\gamma\delta$ dimer. In humans, the CD8 antigen also has two different forms: an $\alpha\beta$ heterodimer and $\alpha\alpha$ homodimer (Moebius *et al.*, 1991). Lymphocytes expressing the $\alpha\alpha$ dimer have been found to be enriched at mucosal surfaces such as the gut (Guy-Grand *et al.*, 1991). When detected using an antibody that recognizes both forms, cell-surface expression of the $\alpha\beta$ dimer is high ($CD8^{bright}$) and that of the $\alpha\alpha$ dimer low ($CD8^{dim}$). It appears that two different forms of CD8 also exist in the pig, in that there are $CD8^{bright}$ and $CD8^{dim}$ lymphocytes. As in the human, T_C cells in the pig are $CD4^-CD8^{bright}$, T_H cells are $CD4^+CD8^-$, and double positive T cells are $CD4^+CD8^{dim}$ (Pescovitz *et al.*, 1985; Yang and Parkhouse, 1996). In the human circulation, virtually all T cells fall into either T_H or T_C categories and $\gamma\delta$, double positive and double negative T cells are rare. In pigs, in contrast, $\gamma\delta$ T cells with a $CD4^-CD8^-$ phenotype and $CD4^+CD8^+$ $\alpha\beta$ T cells constitute major circulating populations (Binns, 1982; Pescovitz *et al.*, 1985; Saalmüller *et al.*, 1987; Saalmüller *et al.*, 1990; Hirt *et al.*, 1990; Yang and Parkhouse, 1996). As in humans, NK cells in pigs are $CD2^+CD4^-CD8^{dim}$ (Pescovitz *et al.*, 1988; Yang and Parkhouse, 1996).

Table 1. Expression of selected CD molecules by pig leukocytes

	Antigen distribution							
	Lymphocytes							
	$\alpha\beta$ T							
Marker	T_H	T_C	$\gamma\delta$ T	NK	B	Gran	Mϕ	Known or proposed function
CD2	+	+	-	+	-	-	-	Adhesion, T-cell activation
CD3	+	+	+	-	-	-	-	Signal transduction (component of T-cell receptor)
CD4	+	-	-	-	-	-	-	Adhesion (binds to class II MHC), signal transduction
CD8	-	+	-	+	-	-	-	Adhesion (binds to class I MHC), signal transduction
CD16	-	-	-	+	-	+	+	Low affinity Fcγ receptor, activation of NK cells
CD44	+	+	+	+	+	+	+	Adhesion
CD45	+	+	+	+	+	+	+	Signal transduction (leukocyte common antigen)

$\alpha\beta$ and $\gamma\delta$: expressing those variants of the T-cell receptor, respectively; T_H and T_C: helper and cytotoxic T cells, respectively; NK: natural killer cells; gran: granulocytes; Mϕ: cells of monocyte/macrophage lineage.

The uterus is a component of the mucosal immune system

The immune system can also be classified anatomically. Leukocytes in the peripheral blood, bone marrow, spleen, thymus, lymph nodes, liver and thoracic duct constitute the systemic immune system. In many textbooks, aggregates of lymphoid cells in the gastrointestinal and respiratory tracts are presented as the mucosal immune system (Abbas *et al.*, 1994). However, both male and female reproductive tracts belong in the mucosal category. A study in mice showing that intravenously injected cells from the mesenteric lymph node migrated to the female reproductive tract along with several other mucosal surfaces (McDermott and Bienenstock, 1979) provided evidence that the uterus is a member of a common mucosal immune system. Mucosal surfaces comprise an epithelium resting on a basement membrane supported by a layer of connective tissue. Because the uterine epithelium is maintained throughout gestation in pigs, the pig uterus retains the structural arrangement of a mucosal surface even in pregnancy. In addition to the mechanisms that normally regulate immune cell function, uterine leukocyte populations are likely to be under endocrine regulation. Oestradiol treatment stimulates the appearance of IgA-positive cells in the rat uterus (Wira *et al.*, 1980) and increases the tissue content of IgA (Wira *et al.*, 1980, 1983). Migration of mesenteric lymph node cells to the female reproductive tract of mice is greater at pro-oestrus and oestrus than at met- and dioestrus, while migration to the small intestine is unchanged (McDermott *et al.*, 1980). In spite of these differences, leukocyte populations of the uterus face challenges similar to those encountered at other mucosal surfaces. Within the gastrointestinal tract, a protective local response against viruses, bacteria and other hazardous agents is necessary but local and systemic antigen-specific tolerance to nutritional components is also required. Similarly, the uterus must be able to eliminate infectious

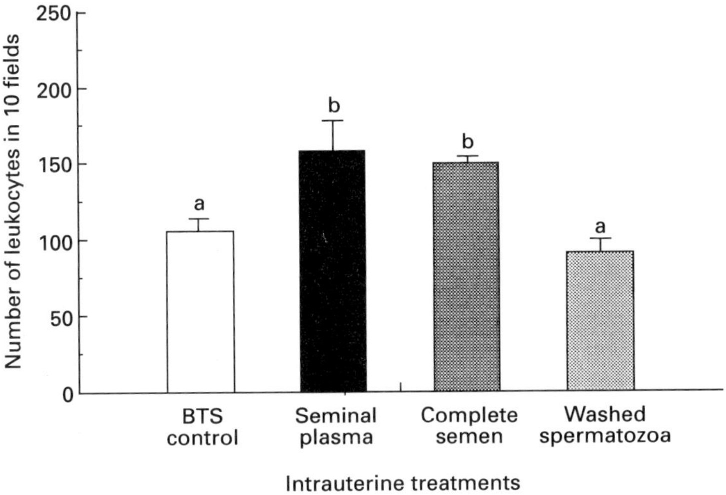

Fig. 1. The effect of intrauterine infusion of complete semen, washed spermatozoa, seminal plasma or semen extender (Beltsville Thawing Solution; BTS) at oestrus on the number of endometrial leukocytes. Uteri were obtained at slaughter on the morning after insemination (approximately 18 h after treatment). In this preliminary trial, leukocytes were identified by their morphology in sections stained with haematoxylin and eosin. Numbers of leukocytes in epithelial and stromal compartments were pooled within animal. Bars with different letters are significantly different ($P < 0.05$).

agents while coping with the introduction of potentially antigenic male factors at mating and intimate contact with the genetically disparate conceptus throughout pregnancy.

The Maternal Immune Response to Mating

The pig is an excellent animal for investigating the immunological effects of semen on the uterus, because mating involves the deposition of large volumes (up to 500 ml) of semen directly into the uterine lumen. Semen consists of mature spermatozoa suspended in seminal plasma, which is composed of the combined secretions of the male accessory sex glands. The major accessory sex glands in the boar are the bulbourethral glands which are the source of the gel fraction of the ejaculate, and the seminal vesicles, which contribute the bulk of the fluid. In natural mating, approximately one third of the ejaculate is lost via backflow through the cervix (Viring and Einarsson, 1981). Of the remainder, most is absorbed by the uterine mucosa within 5 h (du Mesnil du Buisson and Dauzier, 1955). Spermatozoa reach the oviduct within 15 min after mating (Burger, 1952). When added to *in vitro* tests of immune function, seminal plasma in species studied so far has been shown to inhibit activities of most immune cell types, including T cells, B cells, NK cells, macrophages and neutrophils (reviews: Alexander and Anderson, 1987; Thaler, 1989). However, these findings may not reflect events occurring within the female reproductive tract.

The leukocytic response to seminal plasma

Mating in pigs is associated with a rapid influx of inflammatory leukocytes (neutrophils and macrophages) into the endometrial stroma and uterine lumen (Lovell and Getty, 1968). In agreement with Hadjisavas *et al.* (1994), we found that infusion of either seminal plasma or complete semen into the uterine lumen was equally effective in eliciting an influx of leukocytes into the endometrium

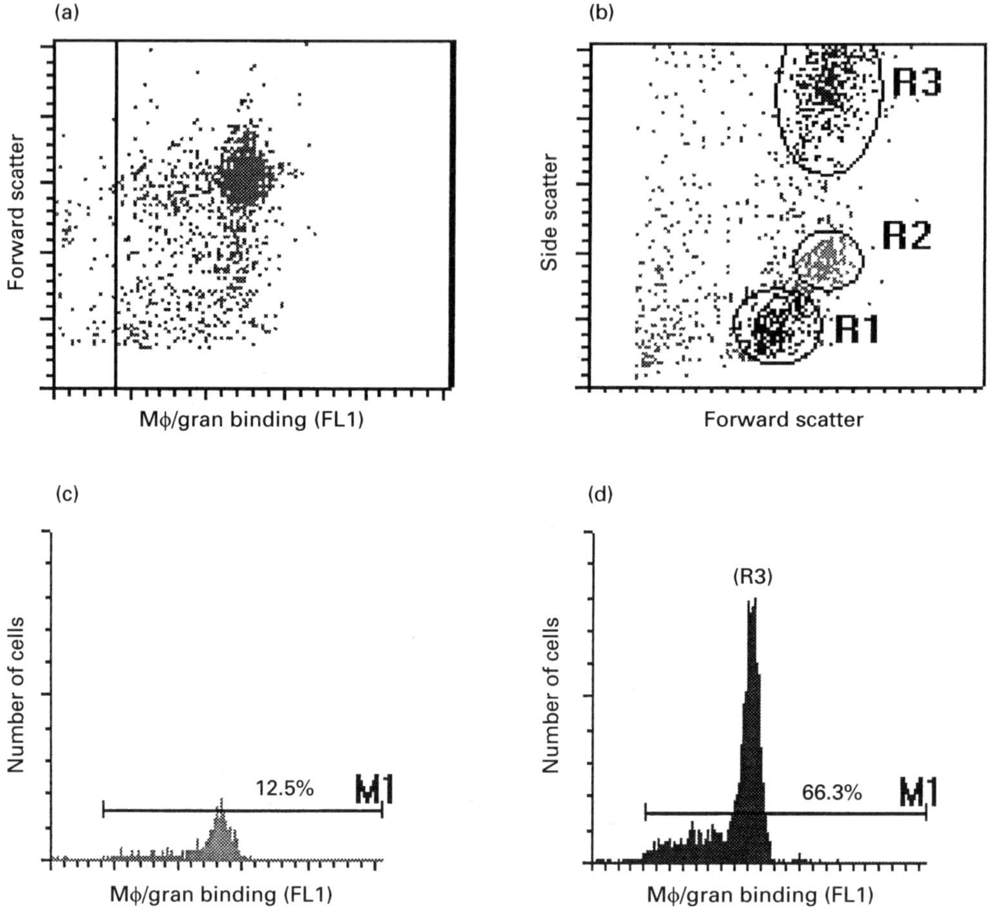

Fig. 2. Flow cytometric characterization of cells flushed from the uterine lumen of a gilt receiving an intrauterine infusion of seminal plasma at oestrus. The uterus was obtained at slaughter approximately 18 h after treatment. Cells were incubated with 74-22-15, an antibody reactive with macrophages (Mϕ) and granulocytes, followed by a goat anti-mouse secondary antibody coupled to fluorescein isothiocyanate (FITC; detected in FL1). The scatter plot (a) (FL1 versus forward scatter) shows the total number of positive cells. Scatter characteristics (forward scatter ~ cell size; side scatter ~ cell granularity) were used to gate on lymphocytes, Mϕ and granulocytes (R1, R2 and R3, respectively; (b)). The histograms indicate the proportion of (c) Mϕ and (d) granulocytes binding to the primary antibody.

(Fig. 1; Engelhardt *et al.*, 1996), indicating that the active factor was associated with seminal plasma rather than spermatozoa. The number of endometrial leukocytes in gilts infused with washed spermatozoa was not different from that of controls infused with a semen extender (Beltsville Thawing Solution (BTS) without antibiotics).

The luminal infiltrate consists primarily of either neutrophils (30 min to 27 h after mating; Lovell and Getty, 1968) or macrophages (9–10 h after mating; Hadjisavas *et al.*, 1994). In our flow cytometric study, granulocytes and macrophages were distinguished from other leukocyte types by binding to a granulocyte–macrophage marker and from each other by scatter characteristics. Neutrophils were the predominant cell type in luminal flushings collected approximately 18 h after exposure to seminal plasma (Fig. 2). Assessment of binding to a pan leukocyte marker (plots not shown) indicated that (i) the cells with the scatter characteristics of lymphocytes (R1 in Fig. 2) were not leukocytes and (ii) granulocytes and macrophages accounted for nearly all the leukocytes in the

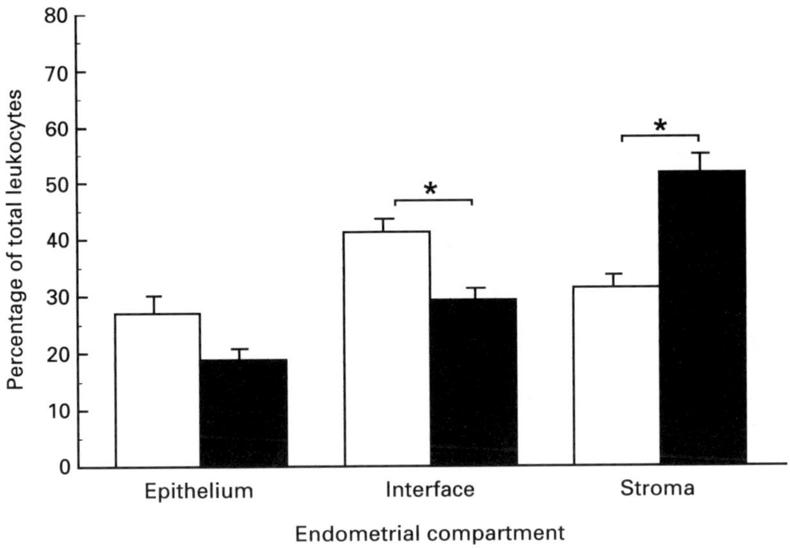

Fig. 3. The effect of stage of cycle on the distribution of leukocytes within endometrial tissue. Uteri were obtained at slaughter on the morning after insemination (approximately 18 h after treatment) at (■) oestrus ($n = 9$) or (□) dioestrus (day 10–12; $n = 6$). Leukocytes were identified in uterine tissue by binding to a pan leukocyte marker (CD44). Positive cells located immediately beneath the luminal epithelium were considered to be in a third endometrial compartment called 'interface'. Data are expressed as percentage of total leukocytes within each animal. Because intrauterine treatments (semen extender control versus seminal plasma) had no significant effect on leukocyte distribution ($P > 0.05$), data for these treatments were pooled. *Pairs of bars are significantly different ($P > 0.05$)

luminal infiltrate. In inflammatory responses to injury or infection, neutrophils are the first phagocytes on the scene, arriving within 30–60 min (Abbas *et al.*, 1994). If the cause of the response persists, the area will be infiltrated by macrophages and lymphocytes within 5–6 h. The fact that lymphocytes were not present in luminal flushings collected 18 h after exposure to seminal plasma suggests that the leukocytic response to mating differs from a conventional inflammatory response.

Endocrine influences on the response to seminal plasma

To assess the influence of endocrine status on the ability of seminal plasma to alter endometrial leukocyte distribution, we compared the proportion of leukocytes in epithelial and stromal compartments of the endometrium in gilts receiving intrauterine infusions of seminal plasma at either oestrus or dioestrus (days 10–12). Control gilts at each stage were infused with the same volume of BTS. Leukocytes were identified in paraffin wax embedded tissue using the then available pig pan leukocyte marker (BAT31A; VMRD Inc., Pullman, WA). This antibody has subsequently been identified as an anti-CD44 antibody which binds to 100% of pig peripheral blood leukocytes (Zuckerman *et al.*, 1994). In gilts infused with both BTS and seminal plasma, large numbers of leukocytes congregated at the epithelial–stromal interface, particularly in oestrous animals. Because of the difficulty in determining their position relative to the basement membrane, leukocytes located immediately beneath the luminal epithelium were considered to be in a third endometrial compartment called the 'interface'. The proportion of leukocytes located in epithelial, interface and stromal compartments was not influenced by intrauterine treatment ($P > 0.05$; Fig. 3). However, stage of cycle affected distribution among these compartments: oestrous gilts had a greater proportion of endometrial leukocytes at the interface and a

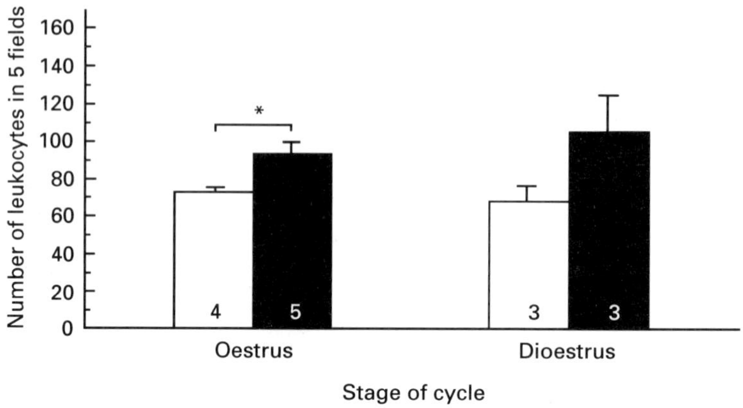

Fig. 4. The effect of intrauterine infusion of (■) seminal plasma or (□) semen extender (Beltsville Thawing Solution; BTS) at oestrus versus dioestrus on the number of endometrial leukocytes. Uteri were obtained at slaughter on the morning after insemination (approximately 18 h after treatment). Leukocytes were identified in uterine tissue by binding to a pan leukocyte marker (CD44). Numbers of leukocytes in epithelial and stromal compartments were pooled within animal. Numbers within bars indicate numbers of animals. Infusion of seminal plasma resulted in a greater number of endometrial leukocytes relative to controls receiving BTS semen extender. This effect was significant ($P < 0.05$) at oestrus but not at dioestrus..

correspondingly lower proportion within the stroma compared with dioestrous gilts ($P < 0.01$; Fig. 3). The fact that stage of cycle strongly influenced uterine leukocyte distribution was consistent with earlier reports describing the infiltration of large numbers of neutrophils into the uterus at oestrus in both unmated and mated pigs (Corner, 1921; Bischof *et al.*, 1994).

Although the shift in leukocyte distribution among endometrial compartments appears to be a response to the endocrine milieu at oestrus, seminal plasma does influence the magnitude of the leukocytic response. The total number of endometrial leukocytes 18–20 h after intrauterine infusion of BTS (control) and seminal plasma at oestrus and dioestrus are shown (Fig. 4). Seminal plasma infusion increased the number of endometrial leukocytes at both stages of the cycle, although this effect was not significant in dioestrous gilts ($P > 0.05$). It must be noted that responses in dioestrous gilts were more variable, probably because a large proportion of these animals developed pyometra in response to treatment. Pyometra (identified by cloudiness of fluid flushed from the uterine lumen) was never observed in gilts infused at oestrus or in three dioestrous gilts killed without treatment. This observation is discussed below.

Contribution of the leukocytic response to successful pregnancy

Prevention of microbial infection. Both the endocrine and seminal plasma-mediated aspects of the leukocytic response to mating may be important in containing the inevitable microbial contamination introduced into the uterus by insemination. The potential importance of this antimicrobial effect should not be underestimated, particularly with respect to the preparation of semen for artificial insemination (AI). Scofield *et al.* (1974) assessed uterine bacterial flora and early embryonic loss rates in gilts 9 and 13 days after natural mating. Of the 50 pigs examined, 46% showed growth on culture. In the study of Scofield *et al.* (1974), although uterine bacterial flora were compatible with pregnancy, early embryonic losses were higher in pigs with infected versus uninfected uteri (sterile: 37.5%; infected: 69.4%; assessed at day 13). Assuming that at least five embryos are necessary for the maintenance of pregnancy in pigs (Polge *et al.*, 1966), four of the six pigs with infected uteri at day 13 in that study would have been expected to return to service, compared with only one of the seven gilts with no infection.

In cattle, it has long been known that resistance to uterine infection is greatest at oestrus and lowest during dioestrus (Black *et al.*, 1953). That this difference was related to circulating oestrogen and progesterone concentrations was demonstrated by Rowson *et al.* (1953), who showed that insemination of luteal phase or ovariectomized, progesterone-treated cows almost invariably resulted in pyometritis, while oestrous or ovariectomized, oestrogen-treated cows were remarkably resistant to infection even using semen known to be microbially contaminated. Addition of antibiotics to semen, which is standard practice in AI in livestock, prevented infection in luteal phase animals.

Both progesterone-mediated immunosuppression and oestrogen-mediated increases in resistance could contribute to the cycle-related differences in susceptibility. Oestrogen could increase resistance through its ability to increase blood flow, vascular permeability and uterine motility. In experimentally infected rabbits, ligation of the uterus at oestrus to impede drainage resulted in an accumulation of leukocytes to form a sterile pyometra similar to that observed in pseudopregnant animals (Broome *et al.*, 1959). In our study, the viability of leukocytes in luminal flushings was high in oestrous gilts (97±3%) but much lower in dioestrous animals (55±4%), which could reflect a lower clearance rate in the progesterone-dominated uterus.

Immunological 'priming' of the uterus for pregnancy. The leukocytic response to mating may have a proactive role in preparing the maternal immune system for pregnancy. Several groups have reported that litter size in pigs can be increased by enhancing the immunological response to mating, by intrauterine infusion of either non-viable semen at the oestrus before mating (Murray *et al.*, 1983; Murray and Grifo, 1986) or peripheral blood leukocytes at the time of insemination (Skjervold *et al.*, 1979; Almlid, 1981). However, others have been unable to reproduce these effects (Blichtfeldt, 1984; Van der Lende *et al.*, 1986; Giles *et al.*, 1990). Interpretation of these studies is difficult because different stimuli were applied at different times, immune responses were not assessed, and 'litter size' was determined at different times ranging from day 8–10 of gestation to 2 weeks post partum. Moreover, no attempt was made to control for exposure to seminal plasma, which is crucial if, as our data suggest, a factor(s) in seminal plasma initiates the response.

The mechanism by which seminal plasma elicits a leukocytic response has been characterized in mice (reviewed by Robertson *et al.*, 1996). Robertson *et al.* (1996) hypothesized that seminal plasma-induced inhibition of potentially harmful cell-mediated immune activity against paternally inherited antigens of the conceptus facilitates implantation and placental development. Although their findings provide a conceptual framework for investigations of effects of seminal plasma in pigs, whether this state of anergy influences the outcome of pregnancy is unknown. The success of AI in the dairy industry in which a single ejaculate, the total volume of which varies from 2 to 10 ml, can be diluted to service hundreds of cows does not support a requirement for seminal-plasma mediated immunological priming in cattle. With respect to AI in pigs, any effect of seminal plasma in preventing uterine infection is met by the addition of antibiotics to semen extenders, but it is possible that variations in exposure to seminal plasma could contribute to differences in litter size. Owing to the relatively low concentrations of spermatozoa in boar semen, the proportion of seminal plasma in a dose of semen diluted for AI can range from 50% to less than 10%. We suggest that subtle influences on uterine leukocytes as a result of exposure to seminal plasma very early in gestation are most likely to have discernible effects on pregnancy outcome when conditions are suboptimal (as has been reported in women undergoing IVF or gamete intrafallopian transfer (GIFT): Bellinge *et al.*, 1986; Marconi *et al.*, 1989), or in litter-bearing species such as pigs in which incremental increases in prolificacy can be realised.

The Maternal Immune Response to Pregnancy

Uterine leukocyte populations in the non-pregnant pig

Before puberty, neutrophils are the predominant immune cells in the pig uterus; MHC class II$^+$ cells (macrophages and dendritic cells) and lymphocytes are also common (Bischof *et al.*, 1994). In the cyclic animal, the most numerous leukocytes are lymphocytes, followed by macrophages and neutrophils. Lymphocytes are located in the stroma and the luminal epithelium, both between epithelial cells and in basal positions (Bischof *et al.*, 1994). The most striking change in uterine leukocyte distribution during the cycle is the infiltration of large numbers of neutrophils at oestrus (Corner, 1921; Bischof *et al.*, 1994). This event is probably mediated, directly or indirectly, by high concentrations of circulating oestrogens. Oestradiol has been shown to increase uterine blood flow (Ford and Christenson, 1979) and endometrial vascular permeability in pigs (Keys and King, 1988). Other stage-dependent changes in leukocyte populations are more debatable. King (1988) noted a modest, but statistically significant increase in the number of lymphocytes (identified morphologically) within the uterine epithelium from days 10 to 19 of the cycle. However, using lymphocyte surface markers (CD2, CD4 and CD8), Bischof *et al.* (1994) reported a marked reduction in the number of lymphocytes at midcycle (days 10–12) relative to earlier in the cycle (days 2–4) in both the epithelial and stromal compartments. Most lymphocytes in the pig uterus express the CD2 surface marker; $CD2^+$ lymphocytes were predominantly $CD8^+$, particularly in the epithelium, with fewer $CD4^+$ cells (Bischof *et al.*, 1994). It must be noted that reliance on surface markers (primarily CD2, CD4, CD8) may result in the failure to detect lymphocytes with the $CD2^-CD4^-CD8^-$ (triple negative) phenotype. In pigs, this phenotype is typical of $\gamma\delta$ T cells (Hirt *et al.*, 1990), which are known to play important roles at mucosal surfaces.

The effect of pregnancy on lymphocytes within the luminal epithelium

Because intraepithelial lymphocytes, usually T cells, are believed to have special functions at other mucosal surfaces, it is possible that uterine intraepithelial lymphocytes play a role in pregnancy. King (1988) found that numbers of lymphocytes within the luminal epithelium at days 10, 13, 16 and 19 of pregnancy were significantly lower than those at comparable stages of the oestrous cycle, and decreased as gestation progressed. Similarly, Bischof *et al.* (1995) reported that there were significantly fewer lymphocytes ($CD2^+$ cells) within the epithelium on days 18–21 of gestation than on days 18–21 of the cycle. Trophoblast may mediate this effect. Bischof *et al.* (1995) noted that at days 28–30 and 45–80 of gestation, intraepithelial lymphocytes were completely absent except in areolar regions, where glands open into the uterine lumen. At these sites, there is no contact between trophoblast and maternal tissues. Although placental structure is somewhat different in ruminants, similar findings have been reported. In cattle and sheep, chorionic binucleate cells migrate across the placental interface and fuse with uterine epithelial cells or other binucleate cells (reviewed by King, 1993). These multinucleate complexes subsequently spread to produce an extensive syncytium on the maternal side. As in pigs, intraepithelial lymphocytes decrease in number during early pregnancy in sheep (Staples *et al.*, 1983; Lee *et al.*, 1992) and cattle (Van der Wielen and King, 1984). Analogous to the situation at areolae in the pig placenta, intraepithelial lymphocytes are completely absent from syncytial regions in cattle (Van der Wielen and King, 1984) and sheep (Staples *et al.*, 1983; Gogolin-Ewens *et al.*, 1989).

Because most of the research in pigs has concentrated on the first third of pregnancy, little is known about uterine leukocyte distribution later in gestation. In sheep (Lee *et al.*, 1992) and possibly other ruminant species (Kellas, 1961; Lee *et al.*, 1995), the decrease in intraepithelial lymphocytes during early pregnancy is overshadowed by a marked increase in the frequency of granulated lymphocytes within the luminal epithelium during mid- to late gestation. These lymphocytes have been found to be an unusual subset of $\gamma\delta$ T cells (Meeusen *et al.*, 1993) that increase in size, number and apparent activation during mid- to late gestation (Gogolin-Ewens *et al.*, 1989; Lee *et al.*, 1992; Meeusen *et al.*, 1993). Using a model in which pregnancy was surgically confined to one uterine horn, Liu *et al.* (1997) showed that in the pregnant horn, a higher proportion of these cells expressed selected activation markers, suggesting that a conceptus-derived molecule induced the activation.

Because pigs are similar to ruminants in that $\gamma\delta$ T cells constitute a major circulating T-cell subpopulation (Hirt *et al.*, 1990; Hein and Mackay, 1991), a possible role for uterine $\gamma\delta$ T cells in pregnancy in pigs warrants further investigation. However, the lack of antibodies to detect the pig $\gamma\delta$ T-cell receptor has hindered investigations. An antibody recognizing a determinant of the δ chain of the pig $\gamma\delta$ T-cell receptor recently became available (VMRD Inc., Pullman, WA). In preliminary

Table 2. Distribution of leukocytes in epithelial and stromal compartments of pig endometrium during early pregnancy (means ± SEM)[a]

Stage of gestation (day)	Position	Epithelium (cells mm^{-1})	Stroma (cells mm^{-2})
15–17	at	0.6±0.2	411±77*
	between	0.5±0.04	127±27
18–19	at	0.8±0.4	272±24*
	between	1.5±0.2	79±1.8
21–22	at	0.2±0.1	290±37*
	between	0.1±0.1	142±24
25–27	at	0.4±0.1	295±17*
	between	0	135±42

[a]Tissue was collected from the mesometrial aspect of the uterus from sites in contact with and equidistant from attached conceptuses (designated 'at' and 'between' attachment sites, respectively). Leukocytes were localized by indirect immunohistochemistry using a pan leukocyte marker (CD44) as the primary antibody. $CD44^+$ cells were quantified in luminal epithelium and in the associated area of endometrial stroma to a depth of approximately 150 μm using image analysis software (Mocha Verson 1.2, Jandel Scientific, San Rafael, CA).

*Significantly greater than in tissue collected 'between' attachment sites at the same stage of gestation ($P < 0.05$).

investigations using enzymatically dispersed, gradient-purified endometrial cells from day 21–22 of gestation, we have found that this anti-δ chain antibody binds to a subpopulation of uterine leukocytes (about 7%) and to a somewhat higher proportion of peripheral blood mononuclear cells (about 23%). We are currently examining sections of uteri of pregnant pigs using this antibody.

Conceptus-associated leukocytes in the endometrial stroma during early pregnancy

It is generally necessary to use surface markers to identify accurately leukocytes in the heterogeneous mixture of cells making up the endometrial stroma. Bischof *et al.* (1995) detected stromal lymphocytes in pig endometrial tissue by their expression of CD2 (T cells and NK cells), and further characterized them by their expression of CD4 and CD8 antigens. They noted a significant increase in numbers of stromal lymphocytes from days 10–14 to 18–21 of gestation. At days 28–30, lymphocytes were frequently seen closely packed within blood vessels immediately beneath the luminal epithelium, around glands and blood vessels, and also scattered through the stroma, occasionally in large groups. MHC class II^+ macrophages and dendritic cells were also seen in the stroma. By days 45–80, macrophages and dendritic cells were the predominant stromal leukocytes.

We have taken a somewhat different approach to sample selection for histology. We collected tissues from the mesometrial aspect of the pregnant uterus at sites immediately adjacent to, and equidistant from conceptuses (designated 'at' and 'between' attachment sites, respectively) during early pregnancy (days 15–17, 18–19, 21–22 and 25–27; Swatman *et al.*, 1996). By comparing these regions with respect to the number of leukocytes in epithelial and stromal compartments, we hoped to distinguish between local influences of conceptuses and the systemic effects of pregnancy. Leukocytes were identified using the CD44 pan leukocyte marker. Proximity to the conceptus had a marked effect on stromal leukocyte distribution, in that leukocytes were 2.0–3.5 times more numerous in regions in direct contact with conceptuses than in regions between attachment sites at all stages ($P = 0.0001$; Table 2). Numbers of intraepithelial leukocytes were low and were not influenced by proximity to attachment sites ($P = 0.55$).

Chemotactic factors secreted by the pig conceptus. The mechanism by which the pig conceptus attracts leukocytes to the endometrial stroma is unknown. Although leukocytes are passively transported by the vascular system, their entry into tissues involves a coordinated series of events, namely activation of leukocyte integrins, adhesion to the vascular endothelium, and migration

(Dunon *et al.*, 1996). Expression of selected integrin subunits on luminal epithelium and trophoblast has been investigated during early pregnancy in pigs with respect to the timing of blastocyst attachment and regulation by steroid hormones (Bowen *et al.*, 1996). In sheep, the presence of a conceptus is associated with increased expression of L-selectin by γδ T cells within the luminal epithelium (Liu *et al.*, 1997). However, no information has been reported on the expression of these molecules in the subepithelial stroma in either species.

Through their induction of subset-specific adhesion of leukocytes to endothelial cells, cytokines are believed to be key players in leukocyte entry into tissues (Tanaka *et al.*, 1993; Dunon *et al.*, 1996). CSF-1 expression by pig trophectoderm has been reported, but the fact that expression was low up to day 20 (Tuo *et al.*, 1995) makes it unlikely to mediate the shift in leukocyte distribution, which has already occurred by days 15–17. The pig conceptus secretes two types of interferons (IFNs) during the attachment period: IFNγ and spI-IFN, a novel type I IFN (Lefèvre *et al.*, 1990; 1993). Although the timing and site of their synthesis has been well documented (Cross and Roberts, 1989; Mirando *et al.*, 1990; LaBonnardière *et al.*, 1991), their role in pregnancy is still unknown. IFNs are known to modulate leukocyte function, and IFNγ in particular is a potent activator of NK cells (Reiter, 1993). It is therefore possible that one or both of these conceptus-derived IFNs is involved in the attraction of leukocytes to attachment sites.

Although not normally considered a chemoattractant molecule for leukocytes, the fact that oestrogen is secreted by the pig trophoblast around the time of attachment (reviewed by Geisert *et al.*, 1990) suggests that it may participate in the attraction of lymphocytes to attachment sites. Oestrogen may indirectly promote lymphocyte accumulation through its effects on endometrial vasculature. Blood flow to uterine segments in contact with conceptuses is higher than that in segments not in contact (Ford *et al.*, 1982). Initial contact between the conceptus and the endometrium (Dantzer, 1984), and, after a delay of 1–2 days, the first increase in capillary density (Dantzer and Leiser, 1994) occur at the embryonic disc regions which have the highest rates of oestrogen synthesis (Bate and King, 1988). Localized alterations in subepithelial capillaries resembling those occurring in early pregnancy can be simulated by surgical insertion of oestrogen-secreting Silastic beads into the uterine lumen (Keys and King, 1992). It is also possible that conceptus-derived oestrogen plays a more direct role in lymphocyte homing to the endometrium. Oestrogen receptors are expressed on vascular endothelial cells of several species (Farhat *et al.*, 1996) and on specialized lymphocytes that localize to attachment sites in the mouse uterus (Van den Heuvel *et al.*, 1996).

Pregnancy-associated NK cells in the pig uterus

Evidence from functional studies suggests that NK-like cells are present in the pig endometrium (Croy *et al.*, 1988; Yu *et al.*, 1993). Lysis of conventional NK target cells by dispersed endometrial cells was high at days 10 and 20 of pregnancy and undetectable in cyclic animals (Yu *et al.*, 1993). The fact that this increase in lytic activity was not detected at comparable stages of oestrogen-induced pseudopregnancy suggests that it was a response to the presence of conceptuses rather than the endocrine milieu of pregnancy (Yu *et al.*, 1993). Because NK cell-specific markers have not been widely available in pigs, the distribution of uterine NK cells relative to the maternal–fetal interface is not currently known. Lymphocytes in pig peripheral blood exhibiting NK-cell-like activity have the phenotype $CD2^{+}CD4^{-}CD8^{dim}$ (Yang and Parkhouse, 1996), such that the only feature that distinguishes them from subsets of both αβ and γδ T cells is their lack of CD3 expression. Functionally, the distinction between NK cells and γδ T cells in pigs is somewhat unclear. Cells with the precise phenotype and activity of NK cells as defined in mice may not exist in livestock species (reviewed by Engelhardt and King, 1996). However, typical NK functions may be carried out by other lymphocyte subsets. In cattle, for instance, γδ T cells exhibit non MHC-restricted cytolytic activity against NK-sensitive targets (Brown *et al.*, 1994).

Investigations into the identity of lytic cells from early pregnant pig endometrium have continued using the monoclonal antibody G7 (generously provided by Y. B. Kim, University of North Chicago). This antibody was raised in mice immunized with a pig peripheral blood fraction

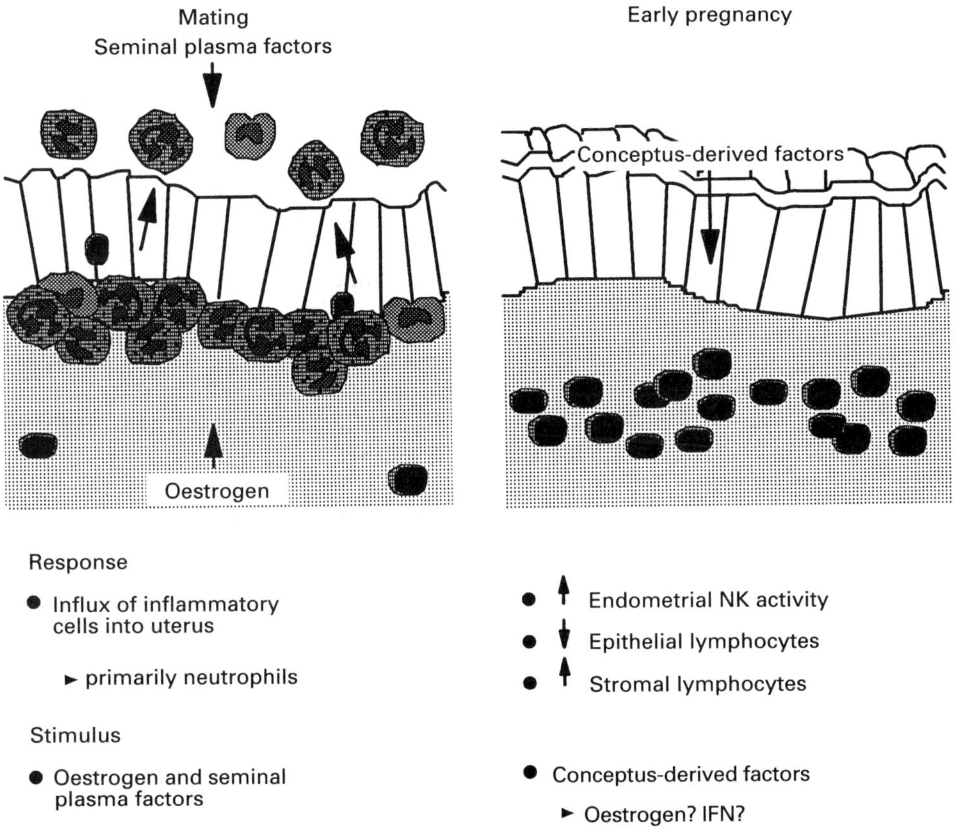

Fig. 5. Summary of the responses of pig uterine immune cells to mating and pregnancy. The influx of inflammatory leukocytes at mating is largely hormonally regulated, but is amplified by factors in seminal plasma. In addition to preventing infection, the leukocytic response to mating may have subtle effects on the maternal immune system that enhance implantation or embryo survival. Marked changes occur in the distribution of uterine immune cells during early pregnancy (days 15–27). The factors mediating these responses have not yet been identified, but conceptus-derived oestrogen and interferon are likely candidates. These conceptus-associated leukocytes may be mediators of communication between conceptuses and other leukocytes, non-immune cells or vascular elements within the uterus during pregnancy.

enriched for NK cells (Dato *et al.*, 1992). In addition to NK cells, G7 reacts with granulocytes and macrophages. Using flow cytometry, we have determined the proportion of uterine leukocytes ($CD44^+$ cells) binding to the G7 antibody in dispersed, gradient-purified endometrial cells obtained both at and between attachment sites during early pregnancy (Engelhardt *et al.*, 1996; Engelhardt and King, 1996). However, after accounting for granulocyte and macrophage reactivity, the frequency of $G7^+$ lymphocytes at versus between attachment sites was found to be similar.

It is currently not known whether the $G7^+$ uterine lymphocytes are the same population mediating the pregnancy-associated lytic activity. The cell surface antigen recognized by G7 has been shown to be the pig homologue of a form of the low-affinity Fc receptor for IgG (FcγRIIIAα), designated CD16 (Halloran *et al.*, 1994). In humans, while circulating NK cells are typically $CD16^+$, the subset populating the pregnant uterus are $CD16^-$ (King *et al.*, 1996). Because uterine lymphocytes in the pig placenta are never in contact with trophoblast, regulation of triggering molecules such as CD16 on uterine NK cells may be quite different from that in humans. Pig uterine NK cells may all be $CD16^+$, as in the circulation, or positive and negative subpopulations may be present, with the $CD16^-$ subset being activated by pregnancy. Efforts are currently underway to assess the lytic activity of isolated G7$^+$ uterine lymphocytes, and to localize putative NK cells in tissue sections using G7 along with other surface markers.

Speculations on the role of uterine NK cells in pregnancy in pigs. The presence of potentially lytic cells in the endometrium of the pregnant uterus raises the question of whether trophoblast cells are susceptible to lysis by NK cells. Reports in humans (King *et al.*, 1989; King and Loke, 1990) and rodents (Zuckerman and Head, 1988; Drake and Head, 1989) that NK cells were unable to lyse trophoblast unless first activated by IL-2 led to the hypothesis that preventing IL-2 activation of NK cells might be involved in the protection of trophoblast. In pigs, however, endometrial cells obtained at early post-attachment times (days 17–20) recognize and kill both conventional NK target cells and trophoblast cells without prior activation of IL-2 (Croy *et al.*, 1994; Yu *et al.*, 1994). Because pig uterine lymphocytes are not likely to encounter trophoblast directly *in vivo*, the evolution of mechanisms to block IL-2 activation of uterine NK cells in this species may not have been necessary.

Given the architecture of the pig placenta, the cytokine-secreting abilities of uterine NK cells are likely to be more important than lytic functions that require cell to cell contact. In response to a conceptus-derived signal, for example IFN, uterine NK cells may communicate with leukocytes or non-immune cells within the uterus via secretion of cytokines, effecting alterations in levels of activity. The nature of these interactions and the identity of the target cells are undefined. Recent observations in three strains of mice genetically lacking NK cells have provided new insights on the possible role of uterine NK cells (Croy *et al.*, 1997b; Guimond *et al.*, 1997). In these mice, placentae of viable conceptuses were less than half of normal size at mid-gestation, and pathological changes in placental vasculature were evident (Guimond, 1997). Because placental size and development in four strains of mice lacking T cells are normal (Croy and Kiso, 1993; Croy *et al.*, 1997c; Guimond, 1997), these effects were attributed to the lack of uterine NK cells. This was supported by the findings that placental vasculature and size were partially restored by the adoptive transfer of bone marrow from *scid* mice (T$^-$B$^-$NK$^+$) into TgE26 females (T$^-$B$^+$NK$^-$; Guimond, 1997). These findings have led to the hypothesis that uterine NK cells promote trophoblast growth or viability indirectly by regulating blood flow to the implantation site. This notion is also supported by evidence that in the uterus of normal mice, uterine NK cells are the major source of nitric oxide synthase, an enzyme responsible for the release of the vasoactive compound nitric oxide (Hunt *et al.*, 1997). In pigs, as early as day 15 when interdigitating microvilli form between maternal and fetal layers, there is a remarkable increase in capillary density and size in the maternal subepithelial bed on the mesometrial side (Dantzer and Leiser, 1994). The fact that extensive angiogenesis occurs during the period of high NK-like activity in the pig suggests that secretory products of NK cells or other leukocytes might be involved in the development or regulation of endometrial vasculature in this species.

Conclusion

Clearly, uterine leukocytes in sows do respond to the immunological challenges of both mating and pregnancy (summarized in Fig. 5). The leukocytic response to mating is partially under endocrine control, but is amplified by exposure of the uterine mucosa to as yet unidentified factors in the seminal plasma. In addition to preventing microbial infection, we propose that the mating response exerts subtle influences on uterine leukocytes that may enhance reproductive performance in litter-bearing species such as pigs. Whether exposure to seminal plasma in pigs produces a state of hypo-responsiveness to paternal antigens comparable with that described in mice by Robertson *et al.* (1996), and whether this state affects pregnancy outcome remains to be determined. These findings may have important implications for the AI industry, in that variations in the proportion of seminal plasma in semen prepared for AI may contribute to between and within boar variations in litter size.

The first month of pregnancy in pigs is interesting both immunologically, because intimate contact between maternal and fetal tissues is established, and economically, because a relatively large proportion of each litter is lost during this time. The accumulation of lymphocytes in the endometrial stroma at sites of conceptus attachment, together with the pregnancy-specific increase in activity of NK cells isolated from the endometrium indicate not only that the maternal immune system is aware of the genetically disparate conceptus, but also that the conceptus seems to be directing the activities of uterine leukocytes to its own advantage. Characterizing the pregnancy-associated lymphocytes, particularly with respect to whether the NK-like lymphocytes are the ones that localize to sites of conceptus attachment, and identifying the conceptus-derived factors that attract or activate them are areas of ongoing research. The role of these uterine leukocytes in pregnancy is still open to speculation. We propose that successful pregnancy requires a dialogue between the conceptus and leukocyte and non-leukocyte populations within the endometrium via soluble factors, and that interfering with this signalling may increase early pregnancy losses. Understanding the immunological events during this critical period of the establishment of maternal–fetal contact would allow the development of strategies to increase embryo survival, through either direct immunological interventions or the modification of husbandry practices shown to affect maternal–fetal signalling.

References

Abbas AK, Lichtman AH and Pober JS (1994) *Cellular and Molecular Immunology* (2nd Edn) WB Saunders Company, Philadelphia, PA

Alexander NJ and Anderson DJ (1987) Immunology of semen *Fertility and Sterility* **47** 192–205

Almlid T (1981) Does enhanced antigenicity of semen increase the litter size in pigs? *Zeitschrift für Tierzüchtung und Züchtungsbiologie* **98** 1–10

Bate LA and King GJ (1988) Production of oestrone and oestradiol-17β by different regions of the filamentous pig blastocyst *Journal of Reproduction and Fertility* **84** 163–169

Bellinge BS, Copeland CM, Thomas TD, Mazzucchelli RE, O'Neil G and Cohen MJ (1986) The influence of patient insemination of the implantation rate in an *in vitro* fertilization and embryo transfer program *Fertility and Sterility* **46** 2523–2526

Binns RM (1982) Organisation of the lymphoreticular system and lymphocyte markers in the pig *Veterinary Immunology and Immunopathology* **3** 95–146

Bischof RJ, Brandon MR and Lee C-S (1994) Studies on the distribution of immune cells in the uteri of prepubertal and cycling gilts *Journal of Reproductive Immunology* **26** 111–129

Bischof RJ, Brandon MR and Lee C-S (1995) Cellular immune response in the pig uterus during pregnancy *Journal of Reproductive Immunology* **29** 161–178

Black WG, Simon J, McNutt SH and Casida LE (1953) Investigations on the physiological basis for the differential response of estrous and pseudopregnant rabbit uteri to induced infection *American Journal Of Veterinary Research* **14** 318–323

Blichtfeldt T (1984) Effect of addition of antigen, adjuvant or mitogen to semen on embryonic survival in artificially inseminated gilts *Zeitschrift für Tierzüchtung und Züchtungsbiologie* **101** 298–304

Bowen JA, Bazer FW and Burghardt RC (1996) Spatial and temporal analyses of integrin and muc-1 expression in porcine uterine epithelium and trophectoderm *in vivo* *Biology of Reproduction* **55** 1098–1106

Broome AWJ, Lamming GE and Smith W (1959) Studies on the relationship between ovarian hormones and uterine infection IV: the role of the granulocyte system in uterine defence *Journal of Endocrinology* **19** 274–281

Brown WC, Davis WC, Choe SH, Dobbelaere DAE and Splitter GA (1994) Functional and phenotypic characterization of WC1$^{+}\gamma/\delta$ T cell isolated from *Babesia bovis*-stimulated T cell lines *Cellular Immunology* **153** 9–27

Burger JF (1952) Sex physiology of pigs *Onderstepoort Journal of Veterinary Research* **26** (Supplement) 2

Corner GW (1921) Cyclic changes in the ovaries and uterus of the sow, and their relation to the mechanism of implantation *Contributions to Embryology* **13** 117–146

Cross JC and Roberts RM (1989) Porcine conceptuses secrete an interferon during the preattachment period of early pregnancy *Biology of Reproduction* **40** 1109–1118

Croy BA and Kiso Y (1993) Granulated metrial cells: a natural killer cell subset of the pregnant murine uterus *Microscopy Research and Technique* **25** 189–200

Croy BA, Waterfield A, Wood W and King GJ (1988) Normal murine and porcine embryos recruit NK cells to the uterus *Cellular Immunology* **115** 471–480

Croy BA, Yu Z-M and King GJ (1994) A review of the natural killer cell lineage of the mouse and of the pig *Journal of Animal Science* **72** (Supplement 2) 9–15

Croy BA, Luross JA, Guimond MJ and Hunt JS (1996) Uterine natural killer cells: insights into lineage relationships and functions from studies of pregnancies in mutant and transgenic mice *Natural Immunity* **15** 22–33

Croy BA, Whitelaw PF and Engelhardt H (1997a) The influences of immune cells on the success of pregnancy. In *The Endocrinology of Pregnancy* Ed. F W Bazer. Humana Press Inc., Totowa N.J. (in press)

Croy BA, Ashkar AA, Foster RA, DiSanto JP, Magram J, Carson D, Gendler SJ, Grusky MJ, Wagner N, Muller W and Guimond M-J (1997b) Further evidence from histological assessment of implantation sites from gene-ablated mice that natural killer cells have important functions in the uterus during pregnancy *Journal of Reproductive Immunology* (in press)

Croy BA, Guimond M-J, Luross J, Hahnel A, Wang B and Van den Heuvel M (1997c) Uterine natural killer cells do not require interleukin-2 for their differentiation or maturation *American Journal of Reproductive Immunology* **37** 463–470

Dantzer V (1984) Scanning electron microscopy of exposed surfaces of the porcine placenta *Acta Anatomica* **118** 96–106

Dantzer V and Leiser R (1994) Initial vascularisation in the pig placenta: I. Demonstration of nonglandular areas by histology and corrosion casts *The Anatomical Record* **238** 177–190

Dato ME, Wierda WG and Kim YB (1992) A triggering structure recognized by G7 monoclonal antibody on porcine lymphocytes and granulocytes *Cellular Immunology* **140** 468–477

Drake BL and Head JR (1989) Murine trophoblast can be killed by lymphokine-activated killer cells *Journal of Immunology* **143** 9–14

du Mesnil du Buisson F and Dauzier L (1955) La remontée des spermatozoides du verrat dans le tractus génital de la truie en oestrus *Comptes Rendus des Société Biologie* (Paris) **149** 76–79

Dunon D, Piali L and Imhof BA (1996) To stick or not to stick: the new leukocyte homing paradigm *Current Opinion in Cell Biology* **8** 714–723

Engelhardt H and King GJ (1996) Uterine natural killer cells in species with epitheliochorial placentation *Natural Immunity* **15** 53–69

Engelhardt H, Small MR and King GJ (1996) Exposure of pigs to seminal plasma at estrus recruits leukocytes to the uterine epithelium *Biology of Reproduction* **54** (Supplement 1) Abstract 476

Farhat MY, Lavigne MC and Ramwell PW (1996) The vascular protective effects of estrogen *FASEB Journal* **10** 615–624

Ford SP and Christenson RK (1979) Blood flow to uteri of sows during the estrous cycle and early pregnancy: local effect of the conceptus on the uterine blood supply *Biology of Reproduction* **21** 617–624

Ford SP, Reynolds LP and Magness RR (1982) Blood flow to the uterine and ovarian vascular beds of gilts during the estrous cycle or early pregnancy *Biology of Reproduction* **27** 878–885

Geisert RD, Zavy MT, Moffatt RJ, Blair RM and Yellin T (1990) Embryonic steroids and the establishment of pregnancy in pigs *Journal of Reproduction and Fertility Supplement* **40** 293–305

Giles JR, Thompson LH, Arkins S, Camacho T and Eichen PA (1990) Effects of uterine infusion of non-viable semen, seminal plasma or egg albumin prior to breeding on the reproductive efficiency of gilts or sows *Canadian Journal of Animal Science* **70** 129–133

Gogolin-Ewens KJ, Lee CS, Mercer WR and Brandon MR (1989) Site-directed differences in the immune response to the fetus *Immunology* **66** 312–317

Guimond M-J (1997) *Evaluations of functions of uterine natural killer cells during pregnancy in mice* PhD Thesis, University of Guelph

Guimond M-J, Luross JA, Wang B, Terhorst C, Danial S and Croy BA (1997) Absence of natural killer cells during murine pregnancy in associated with reproductive compromise in TgE26 mice *Biology of Reproduction* **56** 169–179

Guy-Grand D, Cerf-Bensussan N, Malissen B, Malassis-Seris ML, Briottet C and Vassalli P (1991) Two gut intraepithelial CD8+ lymphocyte populations with different T cell receptors: a role for the gut epithelium in T cell differentiation *Journal of Experimental Medicine* **173** 471–481

Hadjisavas M Laurenz JC and Bazer FW (1994) Seminal plasma (SPL): a potential mediator of inflammation in the uterus following mating in the pig *Biology of Reproduction* **50** (Supplement 1) Abstract 76

Halloran PJ, Sweeney SE, Strohmeier CM and Kim YB (1994) Molecular cloning and identification of the porcine cytolytic trigger molecule G7 as a FcγRIIIα (CD16) homologue *Journal of Immunology* **153** 2631–2641

Head JR (1996) Uterine natural killer cells during pregnancy in rodents *Natural Immunity* **15** 7–21

Hein WR and Mackay CR (1991) Prominence of gamma–delta T cells in the ruminant immune system *Immunology Today* **12** 30–34

Hirt W, Saalmüller A and Reddehase MJ (1990) Distinct γ/δ T cell receptors define two subsets of circulating porcine $CD2^-CD4^-CD8^-$ T lymphocytes *European Journal of Immunology* **20** 265–269

Hunt JS, Miller L, Vassmer D and Croy BA (1997) Expression of the inducible nitric oxide synthase gene in mouse uterine leukocytes and potential relationships with uterine function during pregnancy *Biology of Reproduction* (in press)

Kellas LM (1961) An intra-epithelial granular cell in the uterine epithelium of some ruminant species during the pregnancy cycle *Acta Anatomica* **44** 109–130

Keys JL and King GJ (1988) Morphological evidence for increased uterine vascular permeability at the time of embryonic attachment in the pig *Biology of Reproduction* **39** 473–487

Keys JL and King GJ (1990) Microscopic examination of porcine conceptus–maternal interface between days 10 and 19 of pregnancy *American Journal of Anatomy* **188** 221–238

Keys JL and King GJ (1992) Effects of topical and systemic estrogen on morphology of porcine uterine luminal epithelia *Biology of Reproduction* **46** 1165–1175

King A and Loke YW (1990) Human trophoblast and JEG choriocarcinoma cells are sensitive to lysis by IL-2-stimulated decidual NK cells *Cellular Immunology* **127** 230–237

King A, Birkby A and Loke YW (1989) Early human decidual cells exhibit NK activity against the K562 cell line but not first trimester trophoblast *Cellular Immunology* **118** 337–344

King A, Burrows T and Loke YW (1996) Human uterine natural killer cells *Natural Immunity* **15** 41–52

King GJ (1988) Reduction in uterine intra-epithelial lymphocytes during early gestation in pigs *Journal of Reproductive Immunology* **14** 41–46

King GJ (1993) Comparative placentation in ungulates *Journal of Experimental Zoology* **266** 588–602

LaBonnardière C, Martinat-Botté F, Terqui M, Lefèvre F, Zouari K, Martal J and Bazer FW (1991) Production of two species of interferon by Large White and Meishan pig conceptuses during the peri-attachment period *Journal of Reproduction and Fertility* **91** 469–478

Lee CS, Meeusen E, Gogolin-Ewens K and Brandon MR (1992) Quantitative and qualitative changes in the intraepithelial lymphocyte population in the uterus of nonpregnant and pregnant sheep *American Journal of Reproductive Immunology* **28** 90–96

Lee CS, Wooding FPB and Morgan G (1995) Quantitative analysis of large granular lymphocyte distribution and maternofetal interactions in the synepitheliochorial placenta of the deer *Journal of Anatomy* **187** 445–460

Lefèvre F and Boulay V (1993) A novel and atypical type one interferon gene expressed by trophoblast during early pregnancy *Journal of Biological Chemistry* **268** 19 760–19 768

Lefèvre F, Martinat-Botté F, Guillomot M, Zouari K, Charley B and La Bonnardière C (1990) Interferon-gamma gene and protein are spontaneously expressed by the trophectoderm early in gestation *European Journal of Immunology* **20** 2485–2490

Liu WJ, Gottshall SL and Hansen PJ (1997) Increased expression of cell surface markers on endometrial γδ T-cell receptor+ intraepithelial lymphocytes induced by the local presence of the sheep conceptus *American Journal of Reproductive Immunology* **37** 1–7

Lovell JE and Getty R (1968) Fate of semen in the uterus of the sow: histologic study of endometrium during the 27 hours after natural service *American Journal of Veterinary Research* **29** 609–624

McDermott MR and Bienenstock J (1979) Evidence for a common mucosal immunologic system: I. migration of B immunoblasts into intestinal, respiratory, and genital tissues *Journal of Immunology* **122** 1892–1898

McDermott MR, Clark DA and Bienenstock J (1980) Evidence for a common mucosal immunologic system: II. influence of the estrous cycle on B immunoblast migration into genital and intestinal tissues *Journal of Immunology* **124** 2536–2539

Marconi G, Auge L, Oses R, Qunitana R Raffo F and Young E (1989) Does sexual intercourse improve pregnancy rates in gamete intrafallopian tube transfer? *Fertility and Sterility* **51** 357–359

Meeusen E, Fox A, Brandon M and Lee CS (1993) Activation of uterine intraepithelial gamma–delta T cell receptor-positive lymphocytes during pregnancy *European Journal of Immunology* **23** 1112–1117

Mirando MA, Harney JP, Beers S, Pontzer CH, Torres BA, Johnson HM and Bazer FW (1990) Onset of secretion of proteins with antiviral activity by pig conceptuses *Journal of Reproduction and Fertility* **88** 197–203

Moebius U, Kober G, Griscelli AL, Hercend T and Meuer SC (1991) Expression of different CD8 isoforms on distinct human lymphocyte subpopulations *European Journal of Immunology* **21** 1793–1800

Murray FA and Grifo APJ (1986) Intrauterine infusion of killer semen to increased litter size in gilts *Journal of Animal Science* **62** 187–190

Murray FA, Grifo AP and Parker CF (1983) Increased litter size in gilts by intrauterine infusion of seminal plasma and sperm antigens before breeding *Journal of Animal Science* **56** 895–900

Pescovitz MD, Lunney JK and Sachs DH (1985) Murine anti-swine T4 and T8 monoclonal antibodies: distribution and effects on proliferative and cytotoxic T cells *Journal of Immunology* **134** 37–44

Pescovitz MD, Lowman MA and Sachs DH (1988) Expression of T-cell associated antigens by porcine natural killer cells *Immunology* **65** 267–271

Polge C, Rowson LEA and Chang MC (1966) The effect of reducing the number of embryos during early stages of gestation on the maintenance of pregnancy in the pig *Journal of Reproduction and Fertility* **12** 395–397

Reiter ZVI (1993) Interferon – a major regulator of natural killer cell-mediated cytotoxicity *Journal of Interferon Research* **13** 247–257

Robertson SA, Mau VJ, Hudson SN and Trellen KP (1997) Cytokine–leukocyte networks and the establishment of pregnancy *American Journal of Reproductive Immunology* **37** 438–442

Rowson LEA, Lamming GE and Fry RM (1953) The relationship between ovarian hormones and uterine infection *The Veterinary Record* **65** 335–340

Saalmüller A, Reddehase MJ, Bühring H-J, Jonjic S and Koszinowski UH (1987) Simultaneous expression of CD4 and CD8 antigens by a substantial proportion of resting porcine T lymphocytes *European Journal of Immunology* **17** 1297–1301

Saalmüller A, Hirt W and Reddehase MJ (1990) Porcine γ/δ T lymphocyte subsets differing in their propensity to home to lymphoid tissue *European Journal of Immunology* **20** 2343–2346

Scofield AM, Clegg FC and Lamming GE (1974) Embryonic mortality and uterine infection in the pig *Journal of Reproduction and Fertility* **36** 353–361

Skjervold H, Amlid T, Onstad O and Fossum K (1979) Evidence of immunological influence on the number of live embryos in pigs *Zeitschrift für Tierzüchtung und Züchtungsbiologie* **96** 235–236

Staples LD, Heap RB, Wodding FBP and King GJ (1983) Migration of leukocytes into the uterus after acute removal of ovarian progesterone duing early pregnancy in the sheep *Placenta* **4** 339–350

Swatman J, King GJ and Engelhardt H (1996) Leukocyte distribution in the pig uterus relative to sites of conceptus attachment during early pregnancy *Biology of Reproduction* **54** (Supplement 1) Abstract 475

Tanaka Y, Adams DH and Shaw S (1993) Proteoglycans on endothelial cells present adhesion-inducing cytokines to leukocytes *Immunology Today* **14** 111–115

Thaler CJ (1989) Immunological role for seminal plasma in insemination and pregnancy *American Journal of Reproductive Immunology* **21** 147–150

Tuo W, Harney JP and Bazer FW (1995) Colony-stimulating factor-1 in conceptus and uterine tissues in pigs *Biology of Reproduction* **53** 133–142

Van den Heuvel M, McBey B-A, Hahnel AC and Croy BA (1996) An analysis of the uterine lymphocyte-derived hybridoma cell line GWM 1-2 for expression of receptors for estrogen, progesterone and interleukin 2 *Journal of Reproductive Immunology* **31** 37–50

Van der Lende T, Donker R and Hazeleger W (1986) Embryonic survival in the pig after insemination with antigen-enriched semen *Theriogenology* **26** 391–396

Van der Wielen AL and King GJ (1984) Intraepithelial lymphocytes in the bovine uterus during the oestrous cycle and early gestation *Journal of Reproduction and Fertility* **70** 457–462

Viring S and Einarsson S (1981) Sperm distribution within the genital tract of naturally inseminated gilts *Nordisk Veterinär Medicin* **33** 145–149

Wira CR, Hyde E, Sandoe CP, Sullivan D and Spencer S (1980) Cellular aspects of the rat uterine IgA response to estradiol and progesterone *Journal of Steroid Biochemistry* **12** 451–459

Wira CR, Sullivan DA and Sandoe CP (1983) Epithelial cell involvement in the estrogen stimulated accumulation of IgA in the rat uterus *Journal of Steroid Biochemistry* **19** 469–474

Yang H and Parkhouse RME (1996) Phenotypic classification of porcine lymphocyte subpopulations in blood and lymphoid tissues *Immunology* **89** 76–83

Yu Z, Croy BA, Chapeau C and King GJ (1993) Elevated endometrial natural killer cell activity during early porcine pregnancy is conceptus-mediated *Journal of Reproduction and Fertility* **24** 153–164

Yu Z, Croy BA and King GJ (1994) Lysis of porcine trophoblast cells by endometrial natural killer-like effector cells *in vitro* does not require interleukin-2 *Biology of Reproduction* **51** 1279–1284

Zuckerman FA and Head JR (1988) Murine trophoblast resists cell-mediated lysis. II. Resistance to natural cell-mediated cytotoxicity *Cellular Immunology* **116** 274–286

Zuckerman FA, Binns RM, Husmann R, Yang H, Carr MM, Kim YB, Davis WC, Misfeldt M and Lunney JK (1994) Analyses of monoclonal antibodies reactive with porcine CD44 and CD45 *Veterinary Immunology and Immunopathology* **43** 293–305

Journal of Reproduction and Fertility Supplement **52**, 133–149

Regulation of conceptus development and attachment in pigs

R. D. Geisert[1] and J. V. Yelich[2]

[1]*Animal Science Department, Oklahoma State University, Stillwater, OK 74078, USA; and*
[2]*Department of Animal Science, University of Florida, Gainesville, FL 3261, USA*

Implantation/placentation in domestic pigs is preceded by synthesis of oestrogen by the conceptus to maintain functional corpora lutea throughout pregnancy and a rapid morphological transformation of conceptuses from spherical to long filamentous thread-like structures. Initial conceptus expansion, reaching a metre in length, not only delineates the surface area for placental attachment, but also provides the mechanism for delivery of oestrogen to signal events necessary for placentation throughout the uterine horn. Timing for conceptus gene expression to induce trophoblast expansion and attachment in pigs is temporally associated with downregulation of progesterone receptors and increase in oestrogen receptors within the uterine epithelium. Within the confines of the uterine lumen, pig conceptuses normally do not erode or invade through the uterine epithelial surface. However, the pig conceptus possesses extensive proteolytic activity as it is highly invasive outside the uterine lumen of the pig. Initial release of oestrogen by the elongating pig conceptus induces endometrial release of cytokines and a variety of protease inhibitors. Recently, endometrial expression for the inter-trypsin inhibitor (IαI) family of protease inhibitors has been detected in the pig endometrium during conceptus elongation and attachment. It is possible that IαIs may function to inhibit trophoblast invasion and also serve as targets for adhesion molecules, such as integrins and heparin, to aid in placental attachment to the uterine epithelium.

Introduction

There are few mammalian species in which early embryonic development can compare with the rapid alteration in conceptus morphology that occurs during establishment of pregnancy in pigs. During the period of maternal recognition of pregnancy, pig conceptuses undergo a phenomenal morphological change from 10 mm spherical to tubular (20–40 mm) and finally filamentous (100 mm in length) shapes in less than 3–4 h (Anderson, 1978; Geisert *et al.*, 1982; Stroband and Van der Lende, 1990). Transformation from spherical to a filamentous thread-like morphology occurs through cytoskeletal reorganization which induces cellular modifications in shape and migration, rather than an increase in mitotic activity (Geisert *et al.*, 1982; Mattson *et al.*, 1990). Initial expansion of the pig trophoblast establishes boundaries for placental attachment and initial allotment of uterine space available to each conceptus to compete for nutrients necessary for growth and survival to term. It not only delineates the surface area for placental attachment, but it also provides the mechanism for delivery of conceptus oestrogen throughout the uterus to maintain functional corpora lutea during pregnancy (see reviews Bazer *et al.*, 1984; Geisert *et al.*, 1994a).

Inclusive of maternal recognition on days 10–12 of pregnancy, events critical for early pig embryonic survival include rapid trophoblast elongation, conceptus attachment to the uterine epithelial surface and inhibition of immune rejection by the maternal system. Uterine and conceptus factors involved with inducing rapid trophoblast elongation are of critical importance for embryonic survival. Trophoblast elongation involves many embryonic factors as well as presentation of uterine adhesion factors on the apical border of the endometrial surface epithelium. Initially, uterine

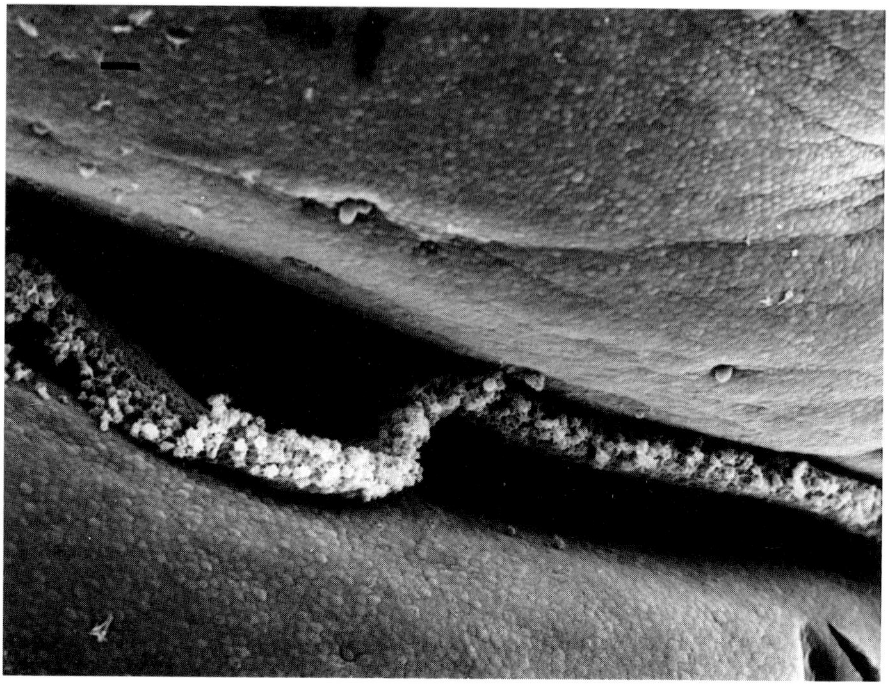

Fig. 1. Scanning electron micrograph showing contact of the elongated pig conceptus apposed to the uterine surface epithelium on day 14 of pregnancy. Scale bar represents 50 μm.

adhesion to the surface epithelium permits trophoblast elongation throughout the uterine lumen (Fig. 1) and then continuous adhesive attachment of the placenta throughout gestation (King *et al.*, 1982). The alterations within the uterine surface glycocalyx that allow trophoblast elongation and then permanent placental attachment are as critical to conceptus survival as the conceptus factors stimulating the morphological changes in the conceptus. Resolving the conceptus and uterine factors involved with trophoblast elongation and attachment within the confines of the pig uterine lumen would clarify our understanding of embryonic survival in pigs. The present paper will review recent literature concerning conceptus and uterine factors involved with trophoblast development and initial attachment to the uterine surface epithelium.

Conceptus Elongation within the Uterus

A critical period of embryonic loss occurs when the peri-implantation conceptus undergoes rapid differentiation and expansion of its trophoblastic membrane between days 11 and 12 of gestation (Geisert *et al.*, 1982; Barends *et al.*, 1989; Pope, 1994). Thus, trophoblast elongation is sensitive to changes in the uterine luminal environment that can be influenced by littermate embryos (Pope, 1994). On day 7–8 of pregnancy, the pig blastocyst is composed of an outer layer of polarized trophectoderm, embryonic disc and an inner layer of endodermal cells (see Stroband and Van der Lende, 1990). By day 10, the pig conceptus expands to a 1–3 mm sphere before enlarging to a 3–8 mm ovoid shape at a rate of approximately 0.3 mm h^{-1} over the next 30 h. Once conceptus diameter reaches 9–10 mm, the conceptus rapidly (30–45 mm h^{-1}) undergoes a transition to a tubular (12–30 mm) and finally thin filamentous form measuring more than 100 mm in length (Geisert *et al.*, 1982). Perry (1981) compared elongation by the ovoid conceptus to rolling a ball of plasticene under your hand. He suggested that the rapidity of elongation would more likely be explained by deformation than by cell division. As predicted by Perry, rapid morphological changes in the pig conceptus are not the result of cellular hyperplasia as proliferative activity, measured by DNA and mitotic index, significantly declines during transition from spherical to filamentous morphology (Geisert *et al.*, 1982; Pusateri *et al.*, 1990). Rather, trophoblastic elongation occurs through massive cellular remodelling of the trophectoderm and endoderm.

Conceptus elongation requires a mechanical force to generate the cellular restructuring necessary to transform its morphology rapidly as the trophectoderm expands. Mattson *et al.* (1990) proposed that actin filaments may be involved with rearrangement of the trophectoderm plasma membrane and initiate the force necessary for conceptus elongation. Modifications of filamentous-actin (f-actin) are consistent with alterations in trophectoderm morphology for axial elongation and narrowing of the conceptus diameter. Polygonal trophectodermal cells of tubular conceptuses vary in the distribution of f-actin in regions proximal to and distal to the embryonic disc. Numerous constricted regions have been observed along the length of the filamentous conceptus (Mattson *et al.*, 1990). Morphology of trophectoderm cells within the constricted region and the distinct arrangement of f-actin suggest that these cells may generate a contractile force to elongate. Although changes in endodermal ultrastructure have been documented (Geisert *et al.*, 1982; Mattson *et al.* 1987), regulation of extraembryonic endoderm migration and function during trophoblast expansion have not been determined.

Currently, the factor(s) that initiates elongation and the cytoskeletal processing that dictates the morphological changes in the pig conceptus is coming under intensive study. It is clear that the force necessary for trophoblast elongation originates within the individual conceptus as a mixture of spherical, tubular and elongating conceptuses which can be found within the same litter (Heuser and Streeter, 1929; Anderson, 1978; Geisert *et al.*, 1982). Moreover, when oestrogen is administered to advance the uterine secretory response before conceptus elongation, conceptuses do not elongate until they have reached the 9–10 mm ovoid morphology (Morgan *et al.*, 1987). Conceptus elongation is, therefore, programmed through developmental cues rather than through direct stimulation by uterine secretions; however, uterine secretions do play a significant role in conceptus growth and survival. The importance of the 10 mm ovoid stage in conceptus development could be related to differentiation and expansion of the extraembryonic mesoderm (Patten, 1931; Geisert *et al.*, 1982; Gupta *et al.*, 1996). First detection of mesoderm is temporally associated with the initial capacity of the 5–6 mm conceptuses to synthesize oestrogens (Fischer *et al.*, 1985; Pusateri *et al.*, 1990; Wilson and Ford, 1997). Protein and gene expression for steroidogenic enzymes such as P450 17α-hydroxylase and aromatase are consistent with the increase in conceptus oestrogen synthesis (Conley *et al.*, 1994; Ko *et al.*, 1994; Green *et al.*, 1995; Yelich *et al.*, 1997a; Wilson and Ford 1997). Later, expansion of the extraembryonic mesoderm between the trophectoderm and endoderm at the 10 mm stage of development may direct cellular remodelling for trophoblast elongation. Certainly, there are many uterine and conceptus factors that contribute to the growth, differentiation and elongation of pig conceptuses.

Endometrial Contributions to Conceptus Development

Although rapid trophoblast elongation is programmed through developmental cues by the conceptus (Morgan *et al.*, 1987), uterine secretions play a significant role in conceptus growth and survival. A number of the major components of progesterone-stimulated uterine secretions during the oestrous cycle and following conceptus oestrogen release have been reviewed by Davis and Blair (1993) and Roberts *et al.* (1993a,b), and will only be briefly addressed here. Notably, endometrial synthesis of uteroferrin and retinol-binding protein (RBP) have been studied extensively in pigs. Uteroferrin was one of the earliest uterine proteins identified within the pig uterus (see Roberts *et al.*, 1993a). Although uteroferrin plays a role in iron transport to the fetus throughout pregnancy, evidence has indicated that uteroferrin may also serve as a haematopoietic stem cell growth factor during early conceptus development (Bazer *et al.*, 1991; Michel *et al.*, 1992).

Endometrial secretion of retinol-binding protein (RBP) may function in the transport and delivery of maternal plasma retinol to the developing conceptus (Trout *et al.*, 1992; Harney *et al.*,

1994a,b). Uterine secretion of retinol bound to RBP permits the uptake and cellular delivery of retinol to the conceptus through cellular RBP (CRBP) (Napoli *et al.*, 1991). Cytosolic retinol can then be metabolized to retinal and the most biologically active metabolite, retinoic acid (Ross, 1991). Retinoic acid binds directly to retinoic acid receptors (RAR) in the nucleus where it effects gene transcription (Chambon *et al.*, 1991). Harney *et al.* (1994a) indicated that the components for retinol transport, metabolism and receptor activation are present in endometrial and conceptus tissues during early pregnancy. Pig endometrial and conceptus tissues synthesize RBP, CRBP, RARα and RARγ during early conceptus development (Trout *et al.*, 1992; Harney *et al.*, 1994a). In fact, RBP is one of the earliest secretory proteins detected in the spherical conceptus on days 10–11 of gestation (Harney *et al.*, 1990). Both uterine and conceptus RBP deliver retinol to tissues, and buffer tissues from the teratogenic (Lammer *et al.*, 1985) and embryotoxic (Thompson *et al.*, 1993) effects of retinoids. Vallet *et al.* (1996) also proposed a role for RBP in protecting uterine and developing conceptus tissues from the lipid oxidizing activity of uteroferrin. Effects of retinoic acid on components of the extracellular matrix (see De Luca, 1991), cell surface adhesive molecules (Agura *et al.*, 1992), and expression of growth factors and their receptors (Roberts and Sporn, 1988) provide an attractive model for its possible involvement with rapid trophoblast elongation and conceptus development in pigs.

In addition to uteroferrin and RBP, the pig endometrium contains many enzymes such as lysozyme, leucine-aminopeptidase (Roberts *et al.*, 1976; Hansen *et al.*, 1985), β-hexosaminidase (Hansen *et al.*, 1985), cathepsins B_1, D, E (Roberts *et al*, 1976) and L (Geisert *et al.*, 1997). Secretion of enzymes such as lysozyme can serve a bactericidal function and for selective proteolysis of proteins for conceptus uptake (Roberts *et al.*, 1993a). Cathepsins are lysosomal cysteine proteases that have been implicated as modulators of invasive implantation of rats (Elangovan and Moulton, 1980) and cats (Li *et al.*, 1992). Cathepsin L activity in the pig uterus is induced by progesterone and increases at the time of trophoblast elongation with peak activity on day 15 of pregnancy (Geisert *et al.*, 1997). Although the pig forms a diffuse, epitheliochorial type of placental attachment (King *et al.*, 1982), the high affinity of cathepsin L for collagen (Kirschke *et al.*, 1982) and elastin (Mason *et al.*, 1982) suggests that it may play a role in placental attachment on day 13–18 of gestation through limited proteolysis of the uterine epithelial surface glycocalyx. Uterine growth and expansion during early pregnancy involves elastase activity and collagen remodelling (Renegar, 1982) in which cathepsin L could play a role in both uterine and placental development.

The pig conceptus is normally noninvasive within the confines of the uterine lumen. However, the pig conceptus possesses extensive invasive activity outside the uterine lumen (Samuel and Perry, 1972). Trophoblast secretion of plasminogen activator at the time of elongation can stimulate the release of plasmin through cleavage of plasminogen which is present in the uterine lumen at the time of implantation/placentation (Fazleabas *et al.*, 1983). Generation of plasmin by the pig conceptuses can activate latent forms of other proteases involved with regulation of the cell basement membrane and extracellular matrix (Werb *et al.*, 1980). Association of plasminogen activator with cellular migration and remodelling of various tissues (Bode and Dziadek, 1979) suggests a primary role for plasmin in the remodelling of the conceptus (Fazleabas *et al.*, 1983). The release of plasmin into the uterine lumen would represent a by-product of conceptus development which must be controlled to avoid damage to the uterine luminal epithelium.

Endometrial secretion of a variety of protease inhibitors regulates the microenvironment during placental attachment in pigs (see Roberts *et al.*, 1993a). The pig endometrium secretes a protease inhibitor specific for plasmin, chymotrypsin and trypsin during the period of trophoblast elongation and placentation (Fazleabas *et al.*, 1983). This inhibitor contains a kunitz domain which provides inhibitory activity against serine proteases (Stallings-Mann *et al.*, 1994). Expression of a uterine elastase/cathepsin G protease inhibitor, antileukoprotinease, during pregnancy in pigs may also support and maintain the noninvasive, epitheliochorial placenta throughout pregnancy (Simmen *et al.*, 1992a). In addition, the pig uterus also secretes a group of low molecular mass, basic proteins that are related to the 'serpin family' of protease inhibitors (Murray *et al.*, 1989; Malathy *et al.*, 1990).

Recently, expression of the inter-trypsin inhibitor (IαI) family of protease inhibitors was detected in the pig endometrium during conceptus elongation and attachment (Geisert *et al.*, 1996; Diederich *et al.*, 1997). Inter-trypsin inhibitors are plasma serine protease inhibitors that have been described during the acute phase reaction to cardiogenic shock (see Salier *et al.*, 1996). The family of inter-trypsin inhibitors are interesting in that they are synthesized as precursor polypeptides that give rise to mature chains with distinct functions and form inter-chain glycosamino–glycan bonds with various molecules. The inter-trypsin inhibitor family of serine protease inhibitors can consist of either a combination of two heavy chains IαIH1, IαIH2 and a single light chain known as bikunin, IαIH3 and bikunin or IαIH2 and bikunin (see review Salier *et al.*, 1996). Heavy chains of IαIH1, IαIH2 and IαIH3 form the various complexes with bikunin through binding to a chrondroitin sulfate chain (Enghild *et al.*, 1993). All the serine protease inhibitory activity is attributed to bikunin, as it contains two Kunitz-type serine protease inhibitor domains (Hochstrasser *et al.*, 1981). Bikunin originates from a separate mRNA in which α-1-microglobulin and bikunin are synthesized as single proteins which undergo proteolytic cleavage (Kaumeyer *et al.*, 1986). However, the IαI heavy chains are translated from separate mRNAs (Diarra-Mehrpour *et al.*, 1989). Through its tandemly arranged kunitz domains, bikunin can target inhibition of trypsin, cathepsin G, elastase and plasmin. The biological role of the IαI family as protease inhibitors is under investigation (Salier *et al.*, 1996). Previously, it was proposed that synthesis of the inter-trypsin inhibitors was restricted to the liver (Saguchi *et al.*, 1995). Uterine synthesis of inter-trypsin inhibitors could regulate conceptus attachment and limit proteolysis. Certainly, bikunin can assist with regulation of endometrial invasion by the pig trophoblast as endometrial bikunin gene and protein expression are detected from day 12 to day 18 of pregnancy (Diedrich *et al.*, 1997). However, it is possible that the heavy chains of IαI may function in the initial attachment of the conceptus to the uterine epithelial surface as will be addressed later.

Continuous growth, development and differentiation of pig conceptuses are highly dependent upon the timing and quantitative amounts of growth factors secreted into the uterine lumen. Insulin-like growth factor I (IGF-I) is one of the earliest and most completely characterized of the growth factors identified in uterine secretions of pigs (Simmen *et al.*, 1993). As endometrial IGF-I expression reaches peak values during conceptus elongation and oestrogen release on day 12 of gestation (Letcher *et al.*, 1989; Simmen *et al.*, 1992b), uterine secretion of IGF-I may enhance conceptus oestrogen synthesis (Hofig *et al.*, 1991), possibly through the enhancement of P450 aromatase gene expression (Ko *et al.*, 1994; Green *et al.*, 1995). The IGF-I receptor mRNA is constitutively expressed during early conceptus development (Green *et al.*, 1995); however, trophectoderm IGF-I receptor protein content is low (Chastant *et al.*, 1994). Uterine secretion of IGF-I may play more of an autocrine role in uterine growth and development as the endometrium contains an abundance of IGF-I receptors (Simmen *et al.*, 1992b; Chastant *et al.*, 1994). Trophectoderm of developing conceptuses expresses IGF-II/mannose 6-phosphate receptor which may stimulate conceptus growth and differentiation (Chastant *et al.*, 1994). However, expression of endometrial IGF-II mRNA increases on day 15 (Simmen *et al.*, 1992b), which is several days after conceptus elongation. It is possible that uterine IGF-I effects early conceptus growth and development by binding to the IGF-II receptor (Czech, 1986). Targeted mutagenesis of the IGF-II/mannose-6-phosphate receptor gene in the mouse has demonstrated its importance in early embryonic survival (Barlow *et al.*, 1991). Since mutation of either IGF-I, IGF-II or IGF-I receptor only reduce prenatal growth (DeChiara *et al.*, 1990; Liu *et al.*, 1993), it is possible that one receptor can substitute for loss of another. Newton *et al.* (1994) indicated that the oestrogen receptor can be transcriptionally activated through IGF-I intracellular signalling suggesting that the oestrogen receptor is one of the nuclear factors involved with growth stimulation. It is possible that conceptus oestrogen and uterine IGF-I coordinate uterine growth and development during trophoblast elongation.

Interest in the role of growth factors in pig conceptus and uterine development has led to identification of a host of growth factors within the uterus (Table 1). Epidermal growth factor (EGF), heparin-binding EGF (HB-EGF), transforming growth factor α (TGF-α) and amphiregulin, all of which bind and activate the EGF receptor (Prigent and Lemoine, 1992), are expressed by the pig endometrium during early pregnancy (Brigstock *et al.*, 1990; Kennedy *et al.*, 1994; Kim *et al.*, 1995; Brigstock *et al.*, 1996a,b). Both EGF and HB-EGF have been immunolocalized to the surface and glandular epithelium of the endometrium (Kennedy *et al.*, 1994; Kim *et al.*, 1995) and the luminal

Table 1. Growth factor and receptor expression by the pig endometrium during early conceptus development and placentation

Expression	Developmental process affected	Reference
Insulin-like growth factors (IGF)	Regulation of cellular proliferation and differentiation	Letcher *et al.*, 1989 Green *et al.*, 1995 Simmen *et al.*, 1992
Insulin-like growth factor receptor	Receptor activation for cellular proliferation and differentiation	Green *et al.*, 1995
Insulin-like growth factor binding protein 2	Regulation of IGF-I and IGF-II activity	Simmen *et al.*, 1992
Epidermal growth factor (EGF)	Regulation of cellular proliferation and differentiation	Kennedy *et al.*, 1994
Heparin binding-epidermal growth factor (HB-EGF)	Regulation of cellular proliferation and differentiation	Brigstock *et al.*, 1990 Kim *et al.*, 1995
Epidermal growth factor receptor	Receptor activation by EGF, HB-EGF, TGF and amphiregulin	Zhang *et al.*, 1992 Kennedy *et al.*, 1994
Transforming growth factor-α (TGFα)	Regulation of cellular proliferation and differentiation	Vaughan *et al.*, 1992 Kennedy *et al.*, 1994
Transforming growth factor-β3	Chemoattractant regulates cellular differentiation and morphogenesis Collagen and fibronectin gene expression	Sun *et al.*, unpublished results
Keratinocyte growth factor	Heparin-binding growth factor Mediates effects of progesterone and androgen on epithelium	Liu *et al.*, 1995
Keratinocyte growth factor receptor	Receptor activation for stromal function	Tuo *et al.*, 1996
Retinoic acid receptor-α, -β, -γ	Retinoic acid receptor for uterine development	Harney *et al.*, 1994 Sun *et al.*, unpublished results
Interleukin 6	Uterine immunological stimulation Uterine hyperaemic response	Mathialagan *et al.*, 1992 Anegon *et al.*, 1994
Fibroblast growth factors	Regulation of cellular proliferation and differentiation	Brigstock *et al.*, 1989
Leukaemia inhibitory factor	Regulation of embryo differentiation	Anegon *et al.*, 1994
Colony stimulating factor 1	Haematopoietic growth factor Cellular proliferation and differentiation	Tuo *et al.* 1995
Pleiotrophin	Neurotrophic factor	Brigstock *et al.*, 1996

content of EGF in uterine secretions is increased on day 12 of pregnancy followed by a decline to day 16 (Diehl *et al.*, 1994). Endometrial and conceptus tissues express EGF receptor (Zhang *et al.*, 1992a,b; Kennedy *et al.*, 1994) which indicates that uterine secretion of EGF and other ligands for EGF receptor could regulate uterine and conceptus development. Although expression of these growth factors, as well as basic fibroblast growth factor (Brigstock *et al.*, 1989) and pleiotrophin (Brigstock *et al.*, 1996a), has been demonstrated, their relationships to conceptus development have not been defined.

In addition to pleiotrophin, several haematopoietic cytokines have been detected within the pig endometrium and uterine secretions during early conceptus development. Given its role as a haematopoietic regulator involved in cellular differentiation and cellular growth, leukaemia inhibitory factor (LIF) is one of the most notable cytokines secreted by pig endometrium at the time of conceptus elongation (Anegon *et al.*, 1994). Endometrial gene expression and secretion of LIF into the uterine lumen are maximal on days 11–12 of pregnancy in pigs (Anegon *et al.*, 1994). The essential role of LIF in blastocyst growth and implantation in mice (see Stewart, 1994) implies that LIF may serve a vital function in conceptus development and implantation in pigs. The importance of conceptus growth and differentiation, and possibly of expansion of the extraembryonic mesoderm, provides an attractive model for studies of endometrial LIF regulation of trophoblast elongation and implantation in pigs.

LIF is related to and binds to a receptor common to interleukin 6 (IL-6), and colony-stimulating factor 1 (CSF-1) (Bazan, 1991). Pig endometrium contains mRNA for IL-6 and IL-6 is present within the uterine secretions during the oestrous cycle and early pregnancy (Anegon *et al.*, 1994). However, changes in mRNA and protein are not associated with changes in conceptus development on days 11–12 of pregnancy as observed for LIF. Gene expression and tissue production of CSF-1 have been detected in the pig endometrium and conceptus (Tuo *et al.*, 1995). Immunoreactive CSF-1 is localized in the uterine surface and glandular epithelium on day 10 of pregnancy, but the greatest tissue content and mRNA expression occur after day 30 of gestation. Temporal changes in CSF-1 mRNA and protein in the endometrium and placenta of pigs suggest that CSF-1 may influence placental and fetal growth following implantation/placentation (Tuo *et al.*, 1995).

The pig endometrium has recently been shown to be a source of relaxin (Knox *et al.*, 1994) and oxytocin (Trout *et al.*, 1995). Although the pig corpus luteum has long been known to be an endocrine source of relaxin (Sherwood, 1994), relaxin has now been localized in the uterine surface and glandular epithelium of gilts during the oestrous cycle and early pregnancy (Knox *et al.*, 1994). Intensity of relaxin immunostaining is weak on day 12 but increases to day 16 and expression is maintained to day 20 in pregnant gilts. Detection of mRNA encoding relaxin demonstrated that the endometrium is the source of relaxin in the uterine epithelium (Knox *et al.*, 1994). Relaxin has a definite uterotrophic effect (Galvin *et al.*, 1991) which could stimulate growth of the uterus to accommodate expansion of the allantochorionic membranes from days 18–30 of pregnancy (see Bazer *et al.*, 1981). Oestrogen stimulation of uterine secretions by the elongating conceptus at the time of maternal recognition of pregnancy initiates release of endometrial oxytocin into the uterine lumen (Trout *et al.*, 1995). Endometrial oxytocin mRNA reaches peak values on day 12 of the oestrous cycle or pregnancy (S. Sun, T. Yelich and R. Geisert, unpublished results). Lack of oxytocin receptor mRNA in day 10–12 conceptus tissue (Yelich *et al.*, 1997a) suggests that uterine oxytocin release may have an autocrine function in uterine contraction and closure of the uterine lumen surrounding the developing conceptus, rather than a direct effect on the developing conceptus. High luminal content of oxytocin may also play a role in regulating the movement of prostaglandin $F_{2\alpha}$ during maternal recognition of pregnancy (see Bazer *et al.*, 1984).

As described previously, there is a great deal of information on the presence of uterine growth factors which may effect conceptus development, many of which are steroid regulated (IGF-I, IGF-II, LIF, RBP). Timing of uterine growth factor expression and subsequent protein release relative to conceptus development needs to be tightly regulated as the uterine environment is not tolerant of asynchrony in conceptus development (see Pope, 1994). The increase in concentrations of plasma progesterone from the corpora lutea is certainly the primary director of uterine development and secretion. However, the marked changes in uterine epithelial secretions and conceptus elongation in pigs are actually temporally related to downregulation of progesterone receptors in the surface and glandular epithelium from day 10 of the oestrous cycle or pregnancy (Geisert *et al.*, 1994b) and the increase in epithelial oestrogen receptors (Geisert *et al.*, 1993). Thus, the loss of progesterone receptors from the uterine epithelium may play an essential role in the timing of growth factor secretion for the developing conceptus and the responsiveness of uterine epithelium to conceptus oestrogens during trophoblastic elongation and implantation. Stromal cell progesterone receptors are maintained during this period and may stimulate the uterine epithelium by secretion of keratinocyte growth factor (KGF) in response to progesterone (Liu *et al.*, 1995). KGF receptors are expressed constitutively by the uterine epithelium (Liu *et al.*, 1996). Although there are now clues to the regulation of uterine growth factor secretion during early conceptus development, definitive roles for each growth factor in conceptus development need to be established.

Table 2. Growth factor and receptor expression by the pig conceptus before and during elongation

Expression	Developmental process affected	Reference
Brachyury	Mesoderm formation	Yelich *et al.*, 1997a
Cytochromes $P45017_{\alpha}$ $P450_{arom}$ $P450_{scc}$	Steroidogenesis	Conley *et al.*, 1994 Green *et al.*, 1995 Corbin *et al.*, 1996
Insulin-like growth factor I	Regulation of cellular proliferation and differentiation	Letcher *et al.*, 1989 Green *et al.*, 1995
Insulin-like growth factor I receptor	Regulation of cellular proliferation and differentiation	Green *et al.*, 1995
Insulin-like growth factor II	Regulation of cellular proliferation and differentiation	Simmen *et al.*, 1992
Insulin-like growth factor II receptor	Regulation of cellular proliferation and differentiation	Chastant *et al.*, 1994
Epidermal growth factor	Regulation of cellular proliferation and differentiation	Vaughan *et al.*, 1992
EGF receptor	Regulation of cellular proliferation and differentiation	Vaughan *et al.*, 1992 Corps *et al.*, 1990 Zhang *et al.*, 1992
Transforming growth factor α	Regulation of cellular morphogenesis	Vaughan *et al.*, 1992
Transforming growth factor β-1, β-2, β-3	Chemoattractant. Regulation of cellular differentiation and morphogenesis. Collagen and fibronectin expression	Yelich *et al.*, 1997a Gupta *et al.*, 1996
Retinol-binding protein	Retinol transport to conceptus	Harney *et al.*, 1990, 1994b Trout *et al.*, 1992 Yelich *et al.*, 1997b
Retinoic acid receptor-α, -β, -γ	Receptor for retinoic acid for embryo morphogenesis	Harney *et al.*, 1994a Yelich *et al.*, 1997b
Leukaemia inhibiting factor receptor	Regulation of cellular proliferation and differentiation	Yelich *et al.*, 1997
Interleukin 1β	Uterine immunological stimulation Haematopoiesis	Tuo *et al.*, 1996
Interleukin 6	Uterine immunological stimulation	Mathialagan *et al.*, 1992 Anegon *et al.*, 1994
Colony stimulating factor 1	Cellular proliferation and differentiation	Tuo *et al.*, 1995
Interferons	Immunological regulation	Lefevre and Boulay, 1993 Cross and Roberts, 1989

Conceptus Regulation of Trophoblastic Elongation

As previously discussed, rapid elongation of the conceptus on days 10–12 of gestation appears to be programmed by endogenous developmental cues. Since conceptuses elongate only after they have reached the 10 mm morphology (Morgan *et al.*, 1987), the critical period of development may prepare the early spherical (1–3 mm) and late-spherical (9–10 mm) conceptuses for elongation. This period is characterized by enhanced conceptus gene expression that orchestrates the sequence of cellular events necessary to initiate and allow trophoblast elongation to occur (Table 2).

Synthesis and secretion of oestrogen by the developing pig conceptus is the fundamental marker of trophoblast elongation (see Geisert *et al.*, 1990). Two key enzymes involved in the steroidogenic pathway for oestrogen synthesis in the conceptus are the cytochromes P450 17α-hydroxylase ($P450_{17\alpha}$) and aromatase ($P450_{arom}$). Although low, initial expression of $P450_{17\alpha}$ is detected in conceptuses of less than 6 mm diameter, and increases as conceptuses approach the large spherical stages (10 mm) just before elongation. Conceptus $P450_{arom}$ gene expression follows a similar pattern that is greatly enhanced at the time of elongation (Conley *et al.*, 1992, 1994; Ko *et al.*, 1994; Green *et al.*, 1995; Yelich *et al.*, 1997a). Immediately after elongation of the trophoblast, gene expression for both enzymes decreases markedly (King and Ackerley, 1985; Conley *et al.*, 1992; Ko *et al.*, 1994; Green *et al.*, 1995). Oestrogen production by the conceptus has a direct effect on uterine function as evidenced by an abundance of oestrogen receptors in the uterine epithelium at the time of trophoblast elongation (Geisert *et al.*, 1993) which, when activated, result in the secretion of numerous uterine proteins as previously discussed in this review. Whether there is a direct effect of oestrogen produced by the conceptus on trophoblast elongation is unclear, although it appears unlikely since oestrogen receptors cannot be detected in the early conceptus by either immunocytochemistry or RT-PCR (R.D. Geisert, unpublished results; Yellich *et al.*, 1997a).

Initial synthesis of oestrogen and elongation of the conceptus is closely associated with first detection of mesoderm differentiation in 5 mm spherical conceptuses that may be the key to the programming of trophoblast elongation (see reviews by Geisert *et al.*, 1990; Stroband and Van der Lende, 1990). The appearance of mesoderm within the embryoblast corresponds to initial expression of the brachyury gene in conceptuses (Yelich *et al.*, 1997a). Brachyury gene expression encodes a transcription factor necessary for mesodermal differentiation in mice (Herrmann *et al.*, 1990), as homozygous mutants for the gene are lethal (Conlon *et al.*, 1995). Brachyury gene expression coincides with and parallels $P450_{17\alpha}$ and $P450_{arom}$ gene expression in early spherical conceptuses (Yelich *et al.*, 1997a). Whether or not expression of brachyury, or the subsequent differentiation of mesoderm, initiates steroidogenesis is debatable. Absence of $P450_{17\alpha}$ and $P450_{arom}$ gene expression in the mesoderm (Conley *et al.*, 1994) suggests that effects of the mesoderm, if any, on oestrogen synthesis are indirect.

Expression of receptors for LIF, IGF-II and EGF have been described in early developing conceptuses (see Table 2). As previously discussed, the uterus secretes IGF I, EGF, HB-EGF and CSF-1 during conceptus elongation, but these growth factors appear to play more of a supportive role in growth and differentiation rather than in cellular remodelling. Initial expression of LIF receptor is observed in 2 mm spherical conceptuses, with a marked increase to a peak at the 7 mm spherical conceptus stage, which is maintained throughout trophoblast elongation (Yelich *et al.*, 1997a). Uterine secretion of LIF at the time of conceptus elongation suggests that LIF may serve a vital function in conceptus development, possibly expansion of the extraembryonic mesoderm, which initiates signalling for remodelling of the trophectoderm. LIF could direct remodelling by enhancing production of conceptus proteases and this is supported by the observation that LIF regulates protease activity in the expanding mouse blastocyst (Harvey *et al.*, 1995). Proteases serve as modifiers of the extracellular matrix (Brenner *et al.*, 1989; Alexander and Werb, 1991).

Expression of mRNA for the EGF receptor has also been detected equally across all stages of pre-elongated conceptuses (Vaughan *et al.*, 1992). Specific binding of EGF to its receptor was also detected (Zhang *et al.*, 1992b), although binding increased temporally from day 10 to day 13 and increased sixfold by day 15 after elongation. The EGF receptor also binds transforming growth factor α (TGF-α). This may be significant because expression of conceptus TGF-α mRNA is maximal before and during elongation, and declines during the post-elongation period (Vaughan *et al.*, 1992). Vaughan *et al.* (1992) speculated that TGF-α plays a role in fluid transport during elongation, and this is supported by the observation that exogenous TGF-α enhances fluid uptake and subsequent blastocoel expansion in the peri-implantation mouse conceptus (Dardick and Schultz, 1991). Whatever role TGF-α may have in conceptus elongation, increased expression of TGF-α could initiate events necessary for rapid trophoblastic elongation, while conceptus EGF gene expression may regulate

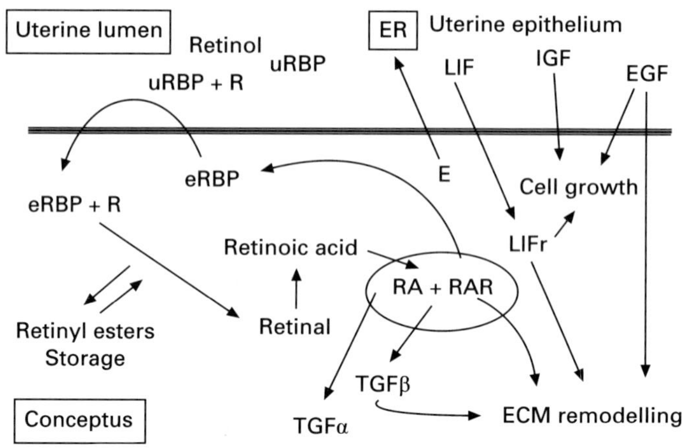

Fig. 2. Proposed simplified model of events associated with rapid trophoblastic elongation in pigs. Oestrogen (E), synthesized by the conceptus, stimulates endometrial oestrogen receptors (ER) which initiate the release of uterine retinol-binding protein (uRBP). As uRBP accumulates in the uterine lumen it transfers retinol (R) to the conceptus either directly or indirectly by exchange with embryonic retinol-binding protein (eRBP). Free R in the conceptus cytoplasm is converted to retinoic acid (RA) which activates retinoic acid receptors (RAR). The RAR may serve as initiators of the extracellular matrix (ECM) remodelling necessary for cell migration and trophoblast elongation either directly or indirectly through activation of the transforming growth factor βs (TGFβ) and other morphogens. Uterine secretion of leukaemia inhibitory factor (LIF) would stimulate conceptus LIF receptors (LIFr) which may initiate the production of proteases that assist in ECM remodelling. Conceptus expression of transforming growth factor α (TGFα) is maximal before and during elongation and it may play an active role in fluid transport during elongation. Other cytokines like the insulin-like growth factors (IGFs) and epidermal growth factor (EGF) may regulate cell proliferation through receptors associated with the trophoblast before, during and following the period of elongation.

development of the embryo during early placentation. Recently, several members of the transforming growth factor β (TGF-β) superfamily and their receptors have been detected in peri-implantation pig conceptuses. Yelich *et al.* (1997b) observed increased expression of mRNA encoding TGFβ-3 during the period of conceptus elongation, but did not detect expression of TGFβ-2. In a comprehensive study Gupta *et al.* (1996) found that TGFβ-1, TGFβ-2 and TGFβ-3 were immunohistochemically localized in peri-implantation pig conceptuses as were type I and type II TGFβ receptors. Gupta *et al.* (1996) concluded that the TGFβs may be involved in induction or advancement of mesoderm migration in developing conceptuses, possibly through regulation of expression of extracellular matrix proteins (Ignotz and Massague, 1986). Laminin and fibronectin are both produced in the early developing pig conceptus (Richoux *et al.*, 1989; Tuo and Bazer, 1996; Bowen *et al.*, 1997).

Several conceptus cytokines may regulate immunofunction within the pig uterus. Interleukin 1β (IL-1β) expression by the early pig conceptus is temporally associated with maternal recognition of pregnancy (Tuo *et al.*, 1996). Conceptus synthesis of IL-1β may play a role in the interplay of the trophoblast and uterus for the establishment of pregnancy through its influence on conceptus remodelling and stimulation of prostaglandin E release. There may be an interaction between oestrogen, PGE, IL-1β and interferon-γ (INF-γ), as conceptus INF-γ increases on day 13 (see La Bonnardiere, 1993) when IL-1β expression has declined (Tuo *et al.*, 1996).

One of the most promising candidates for activation of trophectoderm remodelling is the cellular morphogen, retinoic acid (Harney *et al.*, 1990). Transport of retinol by RBP into the uterine lumen, as described previously, provides the substrate for retinoic acid to the developing conceptus. Transcripts for RBP have been detected in day 11 pig conceptuses (Trout *et al.*, 1991; Harney *et al.*, 1994b; Yelich *et al.*, 1997b) consistent with the presence of immunoreactive RBP in early conceptuses (Harney *et al.*, 1990). Gene expression for RBP is initially observed in 2 mm conceptuses and increases temporally with peak expression just before the initiation of trophoblast elongation (Yelich *et al.*, 1997b). In contrast, uterine expression of mRNA encoding RBP is very low on days 10 and 11 of pregnancy (Harney *et al.*, 1994b), which is consistent with low uterine content of retinol during this period (Trout *et al.*, 1992). The initial rise in conceptus RBP gene expression could serve to protect the developing conceptus from excessive concentrations of retinol released into the uterine lumen in response to conceptus oestrogen secretion (Harney *et al.*, 1990), and to assist in the distribution of retinol to target tissues in the conceptus (Trout *et al.*, 1991) to stimulate cellular remodelling and elongation. Retinoic acid alters embryonic development in mammals (Tickle *et al.*, 1985), affects gene transcription (Chiocca *et al.*, 1989), influences production of extracellular matrix components (see De Luca, 1991) and cell surface adhesive molecules (Agura *et al.*, 1992), and can induce the expression of several peptide growth factors (Roberts and Sporn, 1988) and their receptors (Jetten, 1980). Retinoids exert their biological effects through RAR within the cell. Harney *et al.* (1994a) first described the presence of RARα and RARγ mRNA and protein in day 15 conceptuses. Yelich *et al.* (1997b) demonstrated the presence of RARα, -β and -γ mRNA during both the pre- and post-elongation periods of conceptus development. Gene expression for RARα was activated before and during conceptus elongation, in a similar way to TGFβ-3 gene expression (Yelich *et al.*, 1997b), whereas RARγ expression was more pronounced at the time of trophoblast elongation. It would be expected that RAR would be localized in the trophoblast where RBP is concentrated (Harney *et al.*, 1990; Trout *et al.*, 1992). It is evident that both RBP and RAR may have a role in the initiation or eventual remodelling of the trophoblast during elongation (see model, Fig. 2). Retinoic acid activates expression of laminin β1 (Ross *et al.*, 1994), and integrin β-1 expression (Ross *et al.*, 1994) while it also stimulates gene expression of TGF-βs (Roberts and Sporn, 1988). The TGF-βs are major modifiers of the ECM, particularly in the case of integrins which bind to numerous ECM proteins (Ignotz and Massague, 1986). The ECM may also be modulated through retinoic acid induced activation of conceptus proteases. Retinoic acid influences the production of the protease urokinase-type plasminogen activator and its respective inhibitor (Tienari *et al.*, 1991), both of which are involved in ECM breakdown. In addition, Adler *et al.* (1990) suggested that the expression of ECM-degrading metalloproteinases and their inhibitors are developmentally regulated during the differentiation and spreading of the endoderm in embryonal carcinoma cells stimulated to differentiate with retinoic acid.

Kallikrein – a Key to Conceptus–Endometrial Interaction

Since the time that conceptus oestrogen synthesis was first shown to be involved in establishment of pregnancy in pigs, many alterations in protein secretion, prostaglandin movement, blood flow and changes in uterine morphology have been described (see Geisert *et al.*, 1990). However, the factor(s) induced by conceptus oestrogen to orchestrate these biological changes has not been elucidated. The discovery of the IαI family and a novel inter-trypsin inhibitor H4 produced by the pig endometrium has provided insights to the mechanism whereby conceptus oestrogens regulate uterine function during placentation. Geisert *et al.* (1995) described changes in an endometrial glycoprotein (GP30) which is associated with the time of conceptus attachment to the uterine surface and conceptus survival. GP30 is homologous to the C-terminal region of a larger (120 kDa) pig plasma glycoprotein, IαIH4 (Geisert *et al.*, 1996; Hashimoto *et al.*, 1996). Human and pig IαIH4 are different from other inter-trypsin inhibitor heavy chains, as the consensus DPHFII sequence for binding to bikunin is absent (Saguchi *et al.*, 1995; Hashimoto *et al.*, 1996). Because IαIH4 does not contain protease inhibitory activity or bind other IαIH chains, IαIH4 must have a biological

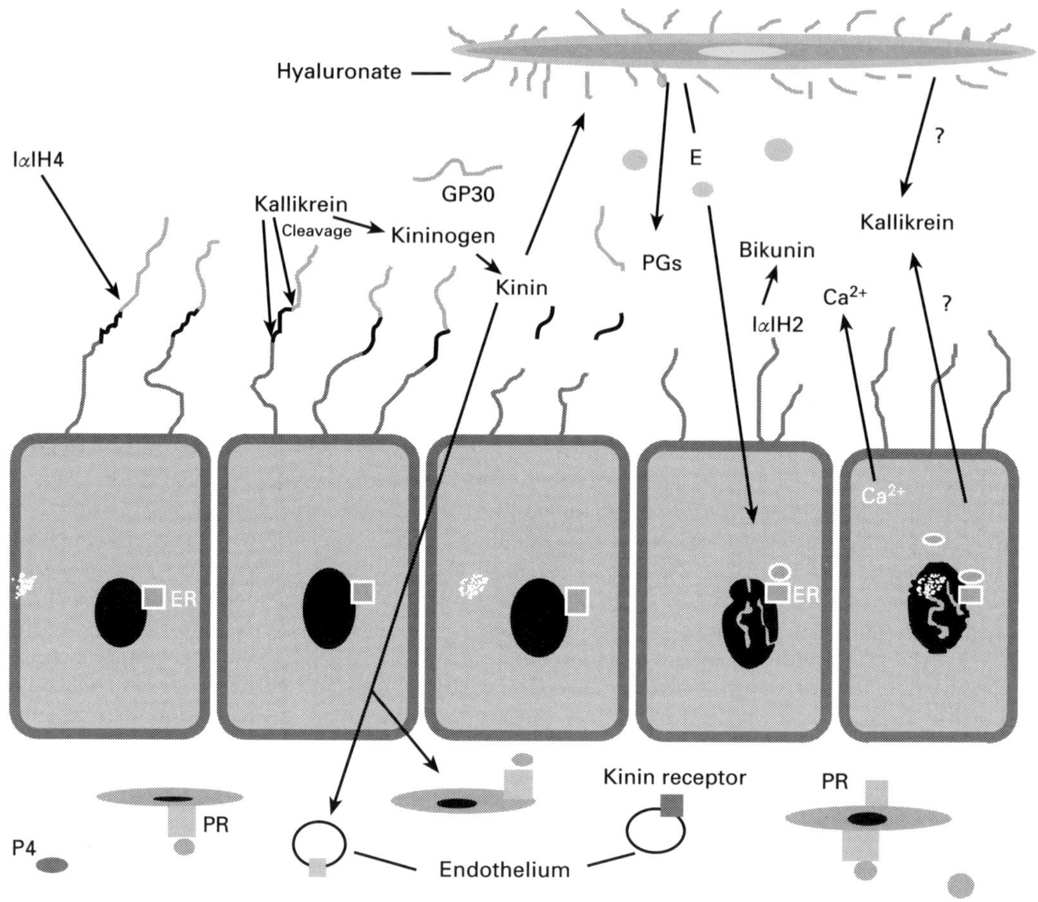

Fig. 3. Proposed model for initial attachment of the conceptus to the uterine epithelium. Release of conceptus oestrogen (E) may stimulate release of uterine kallikrein which acts to cleave inter-trypsin inhibitor heavy chain 4 (IαIH4) present on the uterine epithelial surface. Cleavage of IaIH4 releases a 30 kDa glycoprotein (GP30) the function of which is unknown and a smaller polypeptide fragment. The N-terminal region of IαIH4, which contains binding sites for hyaluronate and integrins, remains on the uterine epithelial surface to bind to the conceptus. Presence of kallikrein may also activate cleavage of kininogen to kinin for stimulation of uterine bloodflow, calcium release and prostaglandin synthesis. Detection of IαIH2 and bikunin in the endometrium indicates that the IαI family of acute phase proteins may regulate uterine inflammation and conceptus invasion through the uterine epithelial surface.

function unrelated to protease inhibition. In contrast to the other IαI heavy chains, plasma human and pig IαIH4 serve as substrates for the plasma serine protease, kallikrein (Nishimura *et al.*, 1995). Plasma kallikrein first cleaves the 120 kDa IαIH4 glycoprotein into 100 and 35 kDa fragments, and then further cleaves the 100 kDa protein to a 70 kDa fragment which can release GP30, as observed in the pig endometrium (Geisert *et al.*, 1995).

Heavy chains of inter-trypsin inhibitors contain calcium-binding sites, potential reactive sites as thiol-protease inhibitors (Salier *et al.*, 1987, 1996) and most importantly, associate with hyaluronan (Zhao *et al.*, 1995). Numerous studies have established the relevance of hyaluronic acid (HA) binding to IαI heavy chains with cell types that display an HA-containing coat (see Salier *et al.*, 1996). All IαI heavy chains possess a von Willebrand type A domain that functions as a target for adhesion molecules such as integrins, collagen, proteoglycans and heparin (see Salier *et al.*, 1996). It has been proposed that inter-trypsin inhibitor heavy chains stabilize the extracellular matrix (Chen *et al.*, 1994;

Jessen *et al.*, 1994). In arthritis, binding of inter-trypsin inhibitor heavy chains to hyaluronate may protect the joint from inflammatory damage possibly caused by free oxygen radicals (Hutadilok *et al.*, 1988). Hyaluronan, a ubiquitous structure within the tissue extracellular matrix serves many roles in tissue morphogenesis, cell proliferation and cell migration. Therefore, in addition to the possible role of pig endometrial IαIH4 in trophoblast attachment during pregnancy, the multipolypeptide chain of the pig IαIH4 could also serve to stabilize the epithelial glycocalyx and protect it from free radical damage. Alteration in IαIH4 may not be the only factor involved with trophoblast attachment; however, cleavage of IαIH4 could induce local alterations in receptivity to the conceptus that permits the conceptus to contact integrins for firm attachment to the uterine epithelium (Bowen *et al.*, 1997). Kallikrein cleaves IαIH4 to allow interaction of the 55 kDa fragment of IαIH4 with integrins and HA for placental attachment.

Kallikrein protease activity increases during conceptus elongation and oestrogen secretion (Vonnahme and Geisert, unpublished results), which is associated with alterations in IαIH4 during early pregnancy (Geisert *et al.*, 1995). The presence of kallikrein in the pig uterus also suggests that the pig uterus may have a functional kallikrein–kininogen–kinin system. A number of physiological responses that occur during early pregnancy in pigs are consistent with functions of bioactive kinins (Margolius, 1996). The pig uterus is responsive to oestrogen during the period of placental attachment and oestrogen stimulates increases in uterine blood flow, vasodilatation of the capillaries surrounding the elongated conceptus, contraction of the myometrium, and endometrial release of calcium, protein and histamine (see Geisert *et al.*, 1990) Kinins have a high affinity for specific membrane receptors on diverse cell types that regulate tissue blood flow, ion transport and smooth muscle contractions (see Bhoola *et al.*, 1992; Rusiniak and Black, 1995; Margolius, 1996). Several studies have demonstrated that the presence of tissue kallikrein in the rat uterus is steroid regulated and correlated with the timing of implantation (Corthorn and Valdes, 1994; Brann *et al.*, 1995; Valdes *et al.*, 1996). These studies suggest that the kallikrein–kininogen–kinin system plays a major role in the induction of the prostaglandin and histamine cascade involved with endometrial permeability and decidual transformation in rats following oestrogen stimulation.

Studies demonstrating a biologically active kallikrein–kininogen–kinin system in the pig uterus during early pregnancy have not been completed; however, we propose that conceptus elongation and oestrogen release initiates a cascade of uterine changes through kallikrein that influence implantation/placentation (see model, Fig. 3). Activation of kallikrein within the uterine lumen would cleave endometrial IαIH4 and permit conceptus trophoblast adhesion to the uterine surface epithelium during and following conceptus elongation. In addition, kallikrein cleaves kininogen into kinin which functions to increase calcium, protein and proteinase inhibitor release locally into the uterine lumen as well as to increase uterine blood flow and production of postaglandins by the conceptus during trophoblast elongation. The localized stimulation of IαIH4 and the short half-life of kinins would permit regulation of the uterine environment by each developing conceptus within the uterine horn.

References

Adler RR, Brenner CA and Werb Z (1990) Expression of extracellular matrix-degrading metalloproteinase and metalloproteinase inhibitors is developmentally regulated during endoderm differentiation of embryonal carcinoma cells *Development* **110** 211–220

Agura ED, Howard M and Collins SJ (1992) Identification and sequence analysis of the promoter for the leukocyte integrin β-subunit (CD18): a retinoic acid-inducible gene *Blood* **79** 602–609

Alexender CM and Werb Z (1991) Extracellular matrix degradation. In *Cell Biology of Extracellular Matrix*, pp 255–302 Ed. ED Hay. Plenum Publishing Corp., New York

Anderson LL (1978) Growth, protein content and distribution of early pig embryos *Anatomical Record* **190** 143–154

Anegon I, Cuturi MC, Godard A, Moreau M, Terqui M, Martinat-Botte F and Soulillou JP (1994) Presence of leukaemia inhibitory factor and interleukin 6 in porcine uterine secretions prior to conceptus attachment *Cytokine* **6** 493–499

Barends PMG, Stroband HWJ, Taverne N, te Kronnie G, Leen MPJM and Blommers PCJ (1989) Integrity of the preimplantation pig blastocyst during expansion and loss of polar trophoblast (Rauber cells) and the morphology of the embryoblast as an indicator for developmental stage *Journal of Reproduction and Fertility* **87** 715–726

Barlow DP, Stoger R, Herrmann BG, Saito K and Schweifer N (1991) The mouse insulin-like growth factor type 2 receptor is imprinted and closely related to the Tme locus *Nature* **349** 84–87

Bazan JF (1991) Neuropoietic cytokines in the hematopoietic fold *Neuron* **7** 197–208

Bazer FW, Goldstein MH and Barron DH (1981) Water and electrolyte transport by the pig chorioallantois. In *Fertilization and Embryonic Development In Vivo*, pp 299–321 Eds L Mastroianni, Jr and JD Biggers. Plenum Publishing Corp., New York

Bazer FW, Marengo SR, Geisert RD and Thatcher WW (1984) Exocrine versus endocrine secretion of prostaglandin F in the control of pregnancy in swine *Animal Reproduction Science* **7** 115–132

Bazer FW, Worthington-White D, Fliss MF and Gross S (1991) Uteroferrin: a progesterone-induced hematopoietic growth factor of uterine origin *Experimental Hematology* **19** 910–915

Bhoola KD, Figueroa CD and Worthy K (1992) Bioregulation of kinins: kallikreins, kininogens, and kininases *Pharmacological Reviews* **44** 1–80

Bode VC and Dziadek MA (1979) Plasminogen activator secretion during mouse embryogenesis *Developmental Biology* **73** 272–289

Bowen JA, Bazer FW and Burghardt RC (1997) Spatial and temporal analyses of integrin and muc-1 expression in porcine uterine epithelium and trophectoderm *in vitro*. *Biology of Reproduction* **56** 409–415

Brann DW, Greenbaum L, Mahesh VB and Gao X (1995) Changes in kininogens and kallikrein in the plasma, brain, and uterus during pregnancy in the rat *Endocrinology* **136** 46–51

Brenner CA, Adler RR, Rappolee DA, Pederson RA and Werb Z (1989) Genes for extracellular matrix-degrading metalloproteinases and their inhibitor, TIMP, are expressed during early mammalian development *Genes and Development* **3** 848–859

Brigstock DR, Heap RB and Brown KD (1989) Polypeptide growth factors in uterine tissues and secretions *Journal of Reproduction and Fertility* **85** 747–758

Brigstock DR, Heap RB, Baker PJ and Brown KD (1990) Purification and characterization of heparin-binding growth factors from porcine uterus *Biochemical Journal* **266** 273–282

Brigstock DR, Kim GY and Steffin CL (1996a) Pig uterine fluid contains the developmentally-regulated neurotrophic factor, pleiotrophin *Journal of Endocrinology* **148** 103–111

Brigstock DR, Kim GY, Steffin CL, Liu A, Vegunta RK and Ismail NH (1996b) High molecular mass forms of epidermal growth factor in pig uterine secretions *Journal of Reproduction and Fertility* **108** 313–320

Chambon P, Zelent A, Petkovich M, Mendelsohn D, Leroy P, Krust A, Kastner P and Brand N (1991) The family of retinoic acid nuclear receptors. In *Retinoids: 10 Years On*, pp 10–27 Ed. JH Saurat, Karger, Basel

Chastant S, Monget P and Terqui M (1994) Localization and quantification of insulin-like growth factor-I (IGF-I) and IGF-II/mannose-6-phosphate (IGF-II/M6P) receptors in pig embryos during early pregnancy *Biology of Reproduction* **51** 588–596

Chen L, Mao SJT, McLean LR, Powers RW and Larsen WJ (1994) Proteins of the inter-α-trypsin inhibitor family stabilize the cumulus extracellular matrix through their direct binding with hyaluronic acid *Journal of Biological Chemistry* **269** 28282–28287

Chiocca EA, Davies PJ and Stein JP (1989) Regulation of tissue transglutaminase gene expression as a molecular model for retinoid effects on proliferation and differentiation *Journal of Cell Biochemistry* **39** 293–304

Conley AJ, Christenson RK, Ford SP, Geisert RD and Mason JI (1992) Steroidogenic enzyme expression in porcine conceptuses during and after elongation *Endocrinology* **131** 896–902

Conley AJ, Christenson RK, Ford SP and Christenson RK (1994) Immunocytochemical localization of cytochrome P450 17α-hydroxylase and aromatase in embryonic cell layers of elongating porcine blastocysts *Endocrinology* **135** 2248–2254

Conlon FL, Wright CV and Robertson EJ (1995) Effects of the TWis mutation on notochord formation and mesodermal patterning *Mechanisms of Development* **49** 201–209

Corthorn J and Valdes G (1994) Variations in uterine kallikrein during cycle and early pregnancy in the rat *Biology of Reproduction* **50** 1261–1264

Cross JC and Roberts RM (1989) Porcine conceptuses secrete an interferon during the preattachment period of early pregnancy *Biology of Reproduction* **40** 1109–1118

Czech MP (1986) Structures and functions of the receptors for insulin and the insulin-like growth factors *Journal of Animal Science* **63** (Supplement 2) 27–38

Dardick A and Schultz RM (1991) Blastocoel expansion in the preimplantation mouse embryo: stimulatory effect of TGF-α and EGF *Development* **113** 919–930

Davis DL and Blair RM (1993) Studies of uterine secretions and products of primary cultures of endometrial cells in pigs *Journal of Reproduction and Fertility Supplement* **48** 143–155

DeChiara TM, Efstratiadis A and Robertson EJ (1990) A growth-deficiency phenotype in heterozygous mice carrying an insulin-like growth factor II gene disrupted by targeting *Nature* **345** 78–80

De Luca LM (1991) Retinoids and their receptors in differentiation, embryogenesis, and neoplasia. *FASEB* **5** 2924–2933

Diarra-Mehrpour M, Bourguignon J, Sesboue R, Mattei MG, Passage E, Salier JP and Martin JP (1989) Human plasma inter-alpha-trypsin inhibitor is encoded by four genes on three chromosomes *European Journal of Biochemisty* **179** 147–154

Diederich MJ, Yelich JV and Geisert RD (1997) Gene and protein expression of the inter-trypsin inhibitor, bikunin, in the porcine conceptus and endometrium *Journal of Animal Science* **75** (Supplement 1) 83

Diehl JR, Henricks DM and Gray SL (1994) EGF and IGF-1 in the uterine and oviductal fluids of pregnant and nonpregnant pigs from day 10 to day 18 *Biology of Reproduction* **50** (Supplement 1) 122

Elangovan S and Moulton BC (1980) Blastocyst implantation in the rat and the immunohistochemical distribution and rate of synthesis of uterine lysosomal cathepsin D *Biology of Reproduction* **23** 663–668

Enghild JJ, Salvesen G, Thogersen IB, Valnickova Z, Pizzo SV and Hefta SA (1993) Presence of the protein–glycosaminoglycan–protein covalent crosslink in the inter-alpha-inhibitor-related proteinase inhibitor heavy chain 2/bikunin *Journal of Biological Chemistry* **268** 8711–8716

Fazleabas AT, Geisert RD, Bazer FW and Roberts RM (1983) Relationship between release of plasminogen activator and estrogen by blastocysts and secretion of plasmin inhibitor by uterine endometrium in the pregnant pig *Biology of Reproduction* **29** 225–238

Fischer HE, Bazer FW and Fields MJ (1985) Steroid metabolism by endometrial and conceptus tissues during early pregnancy and pseudopregnancy in gilts *Journal of Reproduction and Fertility* **75** 69–78

Galvin JM, Anthony RV and Day BN (1991) Effects of purified porcine relaxin on the uterus of gilts during early pregnancy *Animal Reproduction Science* **26** 293–301

Geisert RD, Renegar RH, Thatcher WW, Roberts RM and Bazer FW (1982) Establishment of pregnancy in the pig. I. Interrelationships between peri-implantation development of the pig blastocyst and the uterine endometrial secretions *Biology of Reproduction* **27** 925–939

Geisert RD, Zavy MT, Moffatt RJ, Blair RM and Yellin T (1990) Embryonic steroids and the establishment of pregnancy *Journal of Reproduction and Fertility Supplement* **40** 293–305

Geisert RD, Brenner RM, Moffatt RJ, Harney JP, Yellin T and Bazer FW (1993) Changes in oestrogen receptor protein, mRNA expression and localization in the endometrium of cyclic and pregnant gilts *Reproduction Fertility and Development* **5** 247–260

Geisert RD, Short EC and Morgan GL (1994a) Establishment of pregnancy in domestic farm species. In *Embryonic Mortality in Domestic Species,* pp 23–52 Eds MT Zavy and RD Geisert. CRC Press, Boca Raton

Geisert RD, Pratt T, Bazer FW, Mayes JS and Watson GH (1994b) Immunocytochemical localization and changes in endometrial progestin receptor protein during the porcine oestrous cycle and early pregnancy *Reproduction Fertility and Development* **6** 749–760

Geisert RD, MJ Dixon, T Pratt, RAM Schmitt, BA Lessley and JP McCann (1995) Isolation and characterization of a 30 kDa endometrial glycoprotein synthesized during the estrous cycle and early pregnancy in the pig *Biology of Reproduction* **53** 942–954

Geisert RD, Yelich JV, Pratt T and Pomp D (1996) Identification of porcine endometrial GP30 as inter-trypsin inhibitor family heavy chain-related protein *Biology of Reproduction* **54** (Supplement 1) 78

Geisert RD, Blair RM, Pratt T and Zavy MT (1997) Characterization and proteolytic activity of a cathepsin L-like polypeptide in endometrium and uterine flushings of cycling, pregnant and steroid-treated ovariectomized gilts *Reproduction Fertility and Development* **9** (in press)

Green ML, Simmen RCM and Simmen FA (1995) Developmental regulation of steroidogenic enzyme gene expression in the peri-implantation porcine conceptus: a paracrine role for insulin-like growth factor-I *Endocrinology* **136** 3961–3970

Gupta A, Bazer FW and Jaeger LA (1996) Differential expression of beta transforming growth factors (TGFβ1, TGFβ2, and TGFβ3) and their receptors (Type I and Type II) in peri-implantation porcine conceptuses *Biology of Reproduction* **55** 796–802

Hansen PJ, Bazer FW and Roberts RM (1985) Appearance of β-hexosaminidase and other lysosomal-like enzymes in the uterine lumen of gilts, ewes, and mares in response to progesterone and oestrogens *Journal of Reproduction and Fertility* **73** 411–424

Harney JP, Mirando MA, Smith LC and Bazer FW (1990) Retinol-binding protein: a major secretory product of the pig conceptus *Biology of Reproduction* **42** 523–532

Harney JP, Ali M, Vedeckis WV and Bazer FW (1994a) Porcine conceptus and endometrial retinoid-binding proteins *Reproduction Fertility and Development* **6** 211–219

Harney JP, Smith LC, Simmen RCM, Fliss AE and Bazer FW (1994b) Retinol-binding protein: immunolocalization of protein and abundance of messenger ribonucleic acid in conceptus and maternal tissues during pregnancy in pigs *Biology of Reproduction* **50** 1126–1135

Harvey MB, Leco KJ, Arcellana-Panlilio MY, Zhang X, Edwards DR and Schultz GA (1995) Roles of growth factors during peri-implantation development *Molecular Human Reproduction* **10** 712–718

Hashimoto K, Tobe T, Sumiya J, Sano Y, Choi-Miura N, Ozawa A, Yasue H and Tomita M (1996) Primary structure of the pig homologue of human IHRP: inter-trypsin inhibitor family heavy chain-related protein *Journal of Biochemistry* **119** 577–584

Herrmann BG, Labiet S, Poustka A, King TR and Lehrach H (1990) Cloning of the T gene required in mesoderm formation in the mouse *Nature* **343** 617–622

Heuser CH and Streeter GL (1929) Early stages in the development of pig embryos, from the period of initial cleavage to the time of the appearance of limb-buds *Contributions to Embryology Carnegie Institute* **20** 3–29

Hochstrasser K, Schonberger OL, Rossmanith I and Wachter E (1981) Kunitz-type proteinase inhibitors derived by limited proteolysis of the inter-trypsin inhibitor *Hoppe-Seyler's Zeitschrift für Physiologische Chemie* **362** 1357–1362

Hofig A, Simmen FA, Bazer FW and Simmen RCM (1991) Effects of insulin-like growth factor-I on aromatase cytochrome P450 activity and oestradiol biosynthesis in pre-implantation porcine conceptuses *in vitro. Journal of Endocrinology* **130** 245–250

Hutadilok N, Ghosh P and Brooks PM (1988) Binding of haptoglobin, inter-trypsin inhibitor, and α1 proteinase inhibitor to synovial fluid hyaluronate and the influence of these proteins on its degradation by oxygen derived free radicals *Annals of Rheumatoid Disease* **47** 377–385

Ignotz RA, and Massague J (1986) Transforming growth factor-β stimulates the expression of fibronectin and collagen and their incorporation into the extracellular matrix *Journal of Biological Chemistry* **261** 4337–4345

Jessen TE, L Odum and AH Johnsen (1994) *In vivo* binding of human inter-trypsin inhibitor free heavy chains to hyaluronic acid *Biological Chemistry Hoppe-Seyler* **375** 521–526

Jetten AM (1980) Retinoids specifically enhance the number of epidermal growth factor receptors *Nature* **284** 626–629

Kaumeyer JF, Polazzi JO and Kotick MP (1986) The mRNA for a proteinase inhibitor related to the HI-30 domain of inter-alpha-trypsin inhibitor also encodes alpha-1-microglobulin (protein HC) *Nucleic Acids Research* **14** 7839–7850

Kennedy TG, Brown KD and Vaughan TJ (1994) Expression of the genes for the epidermal growth factor receptor and its ligands in porcine oviduct and endometrium *Biology of Reproduction* **50** 751–756

Kim GY, Besner GE, Steffen CL, McCarthy DW, Downing MT, Luquette MH Abad MS and Brigstock DR (1995) Purification of heparin-binding EGF-like growth factor from pig uterine flushings and its production by endometrial tissues *Biology of Reproduction* **52** 561–571

King GJ and Ackerley CA (1985) Demonstrations of oestrogen in developing pig trophectoderm and yolk sac endoderm between days 10 and 16 *Journal of Reproduction and Fertility* **73** 361–367

King G J, Atkinson BA and Robertson HA (1982) Implantation and early placentation in domestic ungulates *Journal of Reproduction and Fertility Supplement* **31** 17–30

Kirschke H, Kembhavi A A, Bohely P and Barrett A J (1982) Action of rat liver cathepsin L on collagen and other substrates *Biochemical Journal* **201** 367–372

Knox RV, Zhang Z, Day BN and Anthony RV (1994) Identification of relaxin gene expression and protein localization in the uterine endometrium during early pregnancy in the pig *Endocrinology* **135** 2517–2525

Ko Y, Choi I, Green ML, Simmen FA and Simmen RCM (1994) Transient expression of the cytochrome P450 aromatase gene in elongating porcine blastocysts is correlated with uterine insulin-like growth factor levels during peri-implantation development *Molecular Reproduction and Development* **37** 1–11

La Bonnardiere (1993) Nature and possible functions of interferons secreted by the preimplantation pig blastocyst *Journal of Reproduction and Fertility Supplement* **48** 157–170

Lammer EJ, Chen DT, Hoar RM, Angish ND, Benke PJ, Braun JT, Curry CJ, Fernoff PM, Grix AW, Lott IT, Richard JM and Sun SC (1985) Retinoic acid and embryopathy *New England Journal of Medicine* **313** 837–841

Lefevre F and Boulay V (1993) A novel and atypical type one interferon gene expressed by trophoblast during early pregnancy *Journal of Biological Chemistry* **268** 19760–19768

Letcher R, Simmen RCM, Bazer FW and Simmen FA (1989) Insulin-like growth factor-1 expression during early conceptus development in the pig *Biology of Reproduction* **41** 1143–1151

Li W, Jaffe RC and Verhage HG (1992) Immunocytochemical localization and messenger ribonucleic acid levels of a progesterone-dependent endometrial secretory protein (cathepsin L) in the pregnant cat uterus *Biology of Reproduction* **47** 21–28

Liu J-P, Baker J, Perkins AS, Robertson EJ and Efstratiadis A (1993) Mice carrying null mutations of the genes encoding insulin-like growth factor I (IGF-1) and Type 1 IGF receptor (IGF1r) *Cell* **75** 59–72

Liu SH, Bartol FF, Tuo W, Spencer TE and Bazer FW (1995) Expression of keratinocyte growth factor mRNA in porcine endometrium *Biology of Reproduction* **52** (Supplement 1) 87

Liu SH, Tuo W and Bazer FW (1996) Expression of keratinocyte growth factor receptor mRNA in porcine endometrium *Biology of Reproduction* **54** (Supplement 1) 76

Malathy P-V, Imakawa K, Simmen RCM and Roberts RM (1990) Molecular cloning of the uteroferrin-associated protein, a major progesterone-induced serpin secreted by the porcine uterus, and the expression of its mRNA during pregnancy *Molecular Endocrinology* **4** 428–440

Margolius HS (1996) Kallikreins and kinins: molecular characteristics and cellular and tissue responses *Diabetes* **46** S14–S19

Mason RW, Johnson DA, Barrett AJ and Chapman HA (1982) Elastinolytic activity of human cathepsin L *Biochemical Journal* **233** 925–927

Mattson BA, Overstrom EW and Albertini DF (1987) Endodermal cytoskeletal rearrangements during preimplantation pig morphogenesis *Journal of Cell Biology* **105** 257a

Mattson BA, Overstrom EW and Albertini DF (1990) Transitions in trophectoderm cellular shape and cytoskeletal organization in the elongating pig blastocyst *Biology of Reproduction* **42** 195–205

Michel FJ, Fliss MF, Bazer FW and Simmen RC (1992) Characterization and developmental expression of binding sites for the transplacental iron transport protein, uteroferrin, in fetal hematopoietic tissues *Biology of the Neonate* **61** 82–91

Morgan GL, Geisert RD, Zavy MT, Shawley RV and Fazleabas AT (1987) Development of porcine blastocysts in a uterine environment advanced by exogenous oestrogen *Journal of Reproduction and Fertility* **80** 125–131

Murray MK, Malanthy PV, Bazer FW and Roberts RM (1989) Structural relationship, biosynthesis and immuno-cytochemical localization of uteroferrin-associated basic glycoproteins *Journal of Biological Chemistry* **264** 4143–4150

Napoli JL, Posch KP, Fiorella PD and Boerman MHEM (1991) Physiological occurrence, biosynthesis and metabolism of retinoic acid: evidence for roles of cellular retinol-binding protein (CRBP) and cellular retinoic acid-binding protein (CRABP) in the pathway of retinoic acid homeostasis *Biomedicine Pharmacotherapy* **45** 131–143

Newton CJ, Buric R, Trapp T, Brockmeier, Pagotto U and Stalla GK (1994) The unliganded estrogen receptor (ER) tranduces growth factor signals *Journal of Steroid Biochemistry and Molecular Biology* **48** 481–486

Nishimura H, Kakizaki I, Muta T, Sasaki N, Pu PX, Yamashita T and Nagasawa S (1995) cDNA and deduced amino acid sequence of human PK-120, a plasma kallikrein-sensitive glycoprotein *FEBS Letters* **357** 207–211

Patten BM (1931) *The Embryology of the Pig* pp 37–59 P. Blakiston's Son and Co., Philadelphia

Perry JS (1981) The mammalian fetal membranes *Journal of Reproduction and Fertility* **62** 321–335

Pope WF (1994) Embryonic mortality in swine. In *Embryonic Mortality in Domestic Species*, pp 53–78 Eds RD Geisert and MT Zavy. CRC Press, Boca Raton

Prigent SA and Lemoine NR (1992) The type I (EGF-related) family of growth factor receptors and their ligands *Progress in Growth Factor Research* **4** 1–24

Pusateri AE, Rothschild MF, Warner CM and Ford SP (1990) Changes in morphology, cell number, cell size and cellular estrogen content or individual littermate pig conceptuses on days 9 to 13 of gestation *Journal of Animal Science* **68** 3727–3735

Renegar RH (1982) *An Ultrastructural and Cytochemical Investigation of Endometrium from Pregnant And Nonpregnant Gilts* PhD Dissertation, University of Florida, Gainesville

Richoux V, Darribere T, Boucaut J-C, Flechon J-E and Thiery J-P (1989) Distribution of fibronectins and laminin in the early pig embryo *Anatomical Record* **223** 72–81

Roberts AB and Sporn MB (1988) Transforming growth factor-betas: a large family of multifunctional regulatory proteins *Journal of Animal Science* **66** (Supplement 3) 67–75

Roberts RM, Bazer FW, Baldwin N and Pollard WE (1976) Progesterone induction of lysozyme and peptidase activities in the porcine uterus *Archives of Biochemistry and Biophysics* **177** 499–507

Roberts RM, Trout WE, Mathialagan N, Stallings-Mann M and Ling P (1993a) Uterine secretory activity and embryo development. In *Preimplantation Embryo Development*, pp 229–243 Ed. BD Bavister. Springer-Verlag, New York

Roberts RM, Xie S and Trout WE (1993b) Embryo–uterine interactions in pigs during week 2 of pregnancy *Journal of Reproduction and Fertility Supplement* **48** 171–186

Ross AC (1991) Vitamin A: current understanding of the mechanisms of action *Nutrition Today* **1** 6–12

Ross SA, Ahrens RA and De Luca LM (1994) Retinoic acid enhances adhesiveness, laminin and integrin β1 synthesis, and retinoic acid receptor expression in F9 teratocarcinoma cells *Journal of Cell Physiology* **159** 263–273

Rusiniak ME and Black N (1995) Kallikrein–kininogen–kinin system. In *Molecular Biology and Biotechnology*, pp 483–490 Ed. RE Meyers, VCH Publishers, New York, NY

Saguchi K, Tobe T, Hashimoto K, Sano Y, Nakano Y, Miura N-H and Tomita M (1995) Cloning and characterization of cDNA for inter-trypsin inhibitor family heavy chain-related protein (IHRP), a novel human plasma glycoprotein *Journal of Biochemistry* **117** 14–18

Salier J-P, Darra-Mehrpour M, Sesboue R, Bourguignon J, Benarous R, Ohkubo I, Kurachi S, Kurachi K and Martin JP (1987) Isolation and characterization of cDNAs encoding the heavy chain of human inter-trypsin inhibitor: unambiguous evidence for multipolypeptide chain structure of ITI *Proceedings of the National Academy of Sciences USA* **84** 8272–8276

Salier J-P, Rouet P, Raguenez G and Daveau M (1996) The inter-inhibitor family: from structure to regulation *Biochemical Journal* **315** 1–9

Samuel CA and Perry JS (1972) The ultrastructure of the pig trophoblast transplanted to an ectopic site in the uterine wall *Journal of Anatomy* **113** 139–149

Sherwood OD (1994) Relaxin. In *The Physiology of Reproduction*, pp. 861–1009 Eds E Knobil and J Neill, Raven Press, New York, NY

Simmen RCM, Simmen FJ, Fliss AE, Smith LC and Fliss MFV (1992a) Ontogeny, immunocytochemical localization, and biochemical properties of the pregnancy-associated uterine elastase/cathepsin-G protease inhibitor, antileukoproteinase (ALP): monospecific antibodies to a synthetic peptide recognize native ALP *Endocrinology* **130** 1957–1965

Simmen FA, Simmen RCM, Geisert RD, Martinat-Botte F, Bazer FW and Terqui M (1992b) Differential expression, during the estrous cycle and pre- and postimplantation conceptus development, of messenger ribonucleic acids encoding components of the pig uterine insulin-like growth factor system *Endocrinology* **130** 1547–1556

Simmen RCM, Ko Y and Simmen FA (1993) Insulin-like growth factors and blastocyst development *Theriogenology* **39** 163–175

Stallings-Mann ML, Burke MG, Trout WE and Roberts RM (1994) Purification, characterization, and cDNA cloning of a kuntz-type proteinase inhibitor secreted by the porcine uterus *Journal of Biological Chemistry* **269** 24090–24094

Stewart CL (1994) Leukaemia inhibitory factor and the regulation of pre-implantation development of the mammalian embryo *Molecular Reproduction and Development* **39** 233–238

Stroband HWJ and Van der Lende T (1990) Embryonic and uterine development during early pregnancy *Journal of Reproduction and Fertility Supplement* **40** 261–277

Thompson JN, Howell J and Pitt GAJ (1993) Vitamin A and reproduction in rats *Proceedings of the Royal Society of London Series* B **159** 510–535

Tickle C, Lee J and Eichele G (1985) A quantitative analysis of the effect of all-trans-retinoic acid on the pattern of chick wing development *Developmental Biology* **109** 82–95

Tienari J, Alanko T, Lehtonen E and Saksela O (1991) The expression and localization of urokinase-type plasminogen activator and its type 1 inhibitor are regulated by retinoic acid and fibroblastic growth factor in human teratocarcinoma cells *Cell Regulation* **2** 285–297

Trout WE, McDonnell JJ, Kramer KK, Baumbach GA and Roberts RM (1991) The retinol binding protein of the expanding pig blastocyst: molecular cloning and expression in trophectoderm and embryonic disc *Molecular Endocrinology* **5** 1533–1540

Trout WE, McDonnell JJ, Kramer KK, Baumbach GA and Roberts RM (1992) Steroid regulation of the synthesis and secretion of retinol-binding protein by the uterus of the pig *Endocrinology* **130** 2557–2564

Trout WE, Smith GW, Gentry PC, Galvin JM and Keisler DH (1995) Oxytocin secretion by the endometrium of the pig during maternal recognition of pregnancy *Biology of Reproduction* **52** (Supplement 1) 188

Tuo W and Bazer FW (1996) Expression of fetal fibronectin in porcine conceptuses and uterus throughout gestation *Reproduction Fertility and Development* **8** 1207–1213

Tuo W, Harney JP and Bazer FW (1995) Colony-stimulating factor-1 in conceptus and uterine tissues in pigs *Biology of Reproduction* **53** 133–142

Tuo W, Harney JP and Bazer FW (1996) Developmentally regulated expression of interleukin-1β by peri-implantation conceptuses in swine *Journal of Reproductive Immunology* **31** 185–198

Valdes G, Figueroa CD and Corthorn J (1996) Temperospatial changes of kallikrein-like enzymes during the estrous cycle and pregnancy in the rat uterus *Biology of Reproduction* **55** 236–245

Vallet JL, Christenson RK and McGuire WJ (1996) Association between uteroferrin, retinol-binding protein, and transferrin within the uterine and conceptus compartments during pregnancy in swine *Biology of Reproduction* **55** 1172–1178

Vaughan TJ, James PS, Pascall JC and Brown KD (1992) Expression of the genes for TGFα, EGF and the EGF receptor during early pig development *Development* **116** 663–669

Werb Z, Banda MJ and Jones PA (1980) Degradation of connective tissue matrices by macrophages. I. Proteolysis of elastin, glycoproteins and collagen by proteinases isolated from macrophages *Journal of Experimental Medicine* **152** 1340–1357

Wilson EW and Ford SP (1997) Differences in trophectoderm mitotic rate and P450 17α-hydroxylase expression between late preimplantation Meishan and Yorkshire conceptuses *Biology of Reproduction* **56** 380–385

Yelich JV, Pomp D and Geisert RD (1997a) Ontogeny of elongation and gene expression in the early developing porcine conceptuses *Biology of Reproduction* **57** 1256–1264

Yelich JV, Pomp D and Geisert RD (1997b) Detection of transcripts for retinoic acid receptors, retinol binding protein, and transforming growth factors during rapid trophoblastic elongation in the porcine conceptus *Biology of Reproduction* **57** 126–134

Zhang Z, Krause M and Davis DL (1992a) Epidermal growth factor receptors in porcine endometrium: binding characteristics and the regulation of prostaglandin E and $F_{2\alpha}$ production *Biology of Reproduction* **46** 932–936

Zhang Z, Paria BC, Dey SK and Davis DL (1992b) Characterization of the epidermal growth factor receptor in preimplantation pig conceptuses *Developmental Biology* **151** 617–621

Zhao M, Yoneda M, Ohashi Y, Kurono S, Iwata H, Ohnuki Y and Kimata K (1995) Evidence for the covalent binding of SHAP, heavy chains of inter-trypsin inhibitor, to hyaluronan *Journal of Biological Chemistry* **270** 26657–26663

Journal of Reproduction and Fertility Supplement **52**, 151–164

Extracellular matrix and the implantation cascade in pigs

R. C. Burghardt,[1,3] J. A. Bowen,[1,3] G. R. Newton,[3,4] and F. W. Bazer[2,3]

Departments of [1]Veterinary Anatomy & Public Health, [2]Animal Science, [3]Center for Animal Biotechnology, Institute of Biosciences and Technology, Texas A & M University, College Station, TX 77843, and [4]Cooperative Agricultural Research Center, Prairie View A & M University, Prairie View, TX 77446,USA

The structural and functional alterations of uterine epithelial cells that permit the apical–apical union of conceptus and uterine epithelium are complex and are likely to involve many different adhesion molecules with distinct but inter-related functions. A number of changes in the molecular composition at the apical surface of uterine epithelial cells associated with the transition from the pre-receptive to the receptive state in the pig uterus are reviewed. Molecules that function in the adhesion cascade resulting in implantation are represented by a variety of adhesion systems. However, integrins are probably the dominant adhesion systems because their capacity to mediate adhesion is linked to their activation by engaging other surface molecules.

Introduction

The process of implantation in pigs follows a prolonged preimplantation period (the pre-receptive stage) upon arrival of the embryo in the uterus. The preimplantation period is characterized by migration, spacing and elongation of conceptuses and secretion of significant amounts of oestrogen, the signal for maternal recognition of pregnancy, during initial elongation (Geisert *et al.*, 1982a,b). The transition to the receptive state is accompanied by structural and functional changes in endometrial surface and glandular epithelial secretions (Bjorkman *et al.*, 1981; Dantzer, 1985; Keys and King, 1990; Stroband and van der Lende, 1990; Roberts *et al.*, 1993). Receptivity is permissive to the uncommon interaction between apical plasma membranes of the genetically distinct endometrial epithelium and embryonic trophectoderm about 7 days after hatching of the embryo from the zona pellucida. Blastocyst invasion of endometrial stroma and decidualization does not occur in pigs; rather the intimate association between uterine epithelium and conceptus tissues is retained (epitheliochorial placentation). Consequently, the trophoblast and its supporting layer of extra-embryonic mesoderm is simply apposed to the uterine epithelium (Fig. 1) and the conceptus remains within the uterine lumen throughout gestation (Burton, 1992). The implantation process is complete with the formation of a placenta as a means to support fetal development to the end of pregnancy.

Although the pig blastocyst does not normally invade the uterine endometrial epithelium during implantation, it will invade and undergo syncytium formation when transplanted to an ectopic site (Samuel and Perry, 1972). The barrier to implantation in the pre-receptive stage and the refractory stage that develops in the absence of pregnancy or failure of pregnancy recognition is a maternal property resident in the uterine epithelium until it is cyclically transformed to respond to embryonic signals and permit contact with conceptus epithelial tissues in the receptive phase. While our understanding of the implantation process in pigs remains incomplete, opportunities to investigate early stages of the process including conceptus–uterine epithelium apposition and adhesion or attachment has the advantage that the absence of conceptus invasion of the

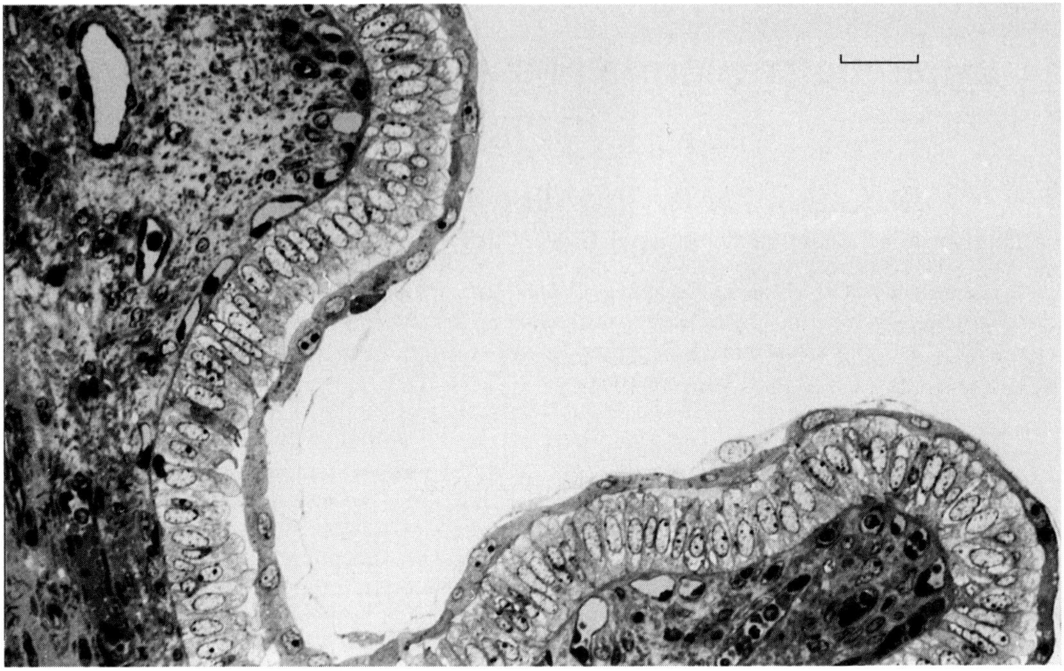

Fig. 1. Light micrograph of conceptus–uterine epithelial interface on day 15 of pregnancy. Scale bar represents 25 μm.

endometrium (as in haemochorial and endotheliochorial placentation) may not obscure or erase earlier events.

The restricted adhesive properties of a blastocyst to a uterine epithelium that is appropriately prepared by ovarian steroid hormones suggests that, under certain conditions, the apical surfaces of uterine epithelial cells express molecules that do not permit (or prevent) blastocyst attachment. Experimental evidence is accumulating that the uterine epithelium responds to a variety of regulatory agents (growth factors, cytokines) in addition to the action of ovarian steroids that modulate structural and functional properties of the uterine epithelium during the transformation from the non-receptive to the receptive state. Insight into the specialized characteristics of uterine epithelial and stromal cells and their interactions has also provided improvements in culture systems that permit the discrimination of functions that are under direct hormonal control from those that are controlled through hormonal effects on other uterine cell types or more distant tissues. Procedures for culturing uterine epithelium in a polarized fashion (Glasser *et al.*, 1988) have stimulated renewed analysis of epithelial receptivity *in vitro*. Because polarized epithelial cells regain their non-receptive state *in vitro* and mimic attributes of these cells *in vivo* (see Glasser and Mullholland, 1993), the utility of these model systems to support the search for, and analysis of, potential markers of pre-receptive and receptive stages is increasing. This review examines several changes in the molecular composition at the apical surface of uterine epithelial cells that are associated with the transition from the pre-receptive to the receptive state in a variety of species, several of which may be relevant in pigs. Particular attention is directed to endpoints that can be reproduced in polarized endometrial epithelial cells. A consideration of extracellular matrix molecules, their receptors and adhesion systems as potential contributors to the receptive state is also included.

Although the following discussion focuses primarily on the epithelial cell component of the endometrial compartment during the peri-implantation period, the potential contributions of other cells to the generation of the non-receptive and receptive status of the epithelium must be mentioned. Despite the fact that the principal cells of the endometrium are separated by a basal lamina, the classic experiments of Cunha *et al.* (1983) define the importance of epithelial–stromal cell interactions in the production of stage-specific responses of the epithelium to endocrine signals. When the uterus enters the receptive phase, circulating progesterone concentrations are high and these conditions result in a marked decrease in immunologically detectable steroid receptors in the luminal epithelium (Geisert *et al.*, 1993, 1994). Other cells including stroma, myometrium and vascular cells are likely to mediate the action of steroid hormones (for example oestromedins, progestomedins) on the surface epithelium. For example, keratinocyte growth factor (KGF, also known as FGF-7) has been shown to be a stroma-derived, progesterone-induced growth factor that may function as a progestomedin on uterine epithelial cell proliferation in primates (Koji *et al.*, 1994). The selective expression of KGF in stroma and the KGF receptor in uterine epithelial cells has recently been reported in pigs (Liu *et al.*, 1996) and indicates the paracrine function of this growth factor in regulating the properties of pig uterine epithelial cells. Similarly, insulin-like growth factor I (IGF-I) and epidermal growth factor (EGF) may also be mediators of oestrogen action in the uterus (Guidice *et al.*, 1993; Hana and Murphy, 1994).

Pre-receptive Stage

During the pre-receptive stage, uterine epithelial cells exhibit an abundant negatively charged apical glycocalyx which diminishes during the receptive phase in most mammals (see Denker, 1993). Lectin-binding characteristics of the pig blastocyst surface change just before attachment (Whyte and Robson, 1984) and the glycocalyx of the uterine epithelium is reduced in thickness at the implantation site during initial attachment (Dantzer, 1985; Blair *et al.*, 1991). Although few candidate glycoprotein markers for the pre-receptive stage have been investigated in detail, Muc-1 and C-CAM, a member of the immunoglobulin superfamily of glycoproteins (Etzioni, 1996) have been evaluated as potential contributors to maintenance of the pre-receptive state (Svalander *et al.*, 1990; Valdizan *et al.*, 1992; Braga and Gendler, 1993). Of these, only Muc-1 has been investigated in pigs (Bowen *et al.*, 1996a; 1997).

The large (200–1000 kDa) heavily glycosylated integral transmembrane glycoprotein, Muc-1, is predominantly associated with the apical domain of epithelial cells, including uterine epithelium, in a variety of species (Pemberton *et al.*, 1992) and may function as an anti-adhesive molecule in both cadherin- and integrin-mediated cell adhesion events *in vitro* (Ligtenberg *et al.*, 1992; Wesseling *et al.*, 1995). Muc-1 mRNA and protein are downregulated during the implantation window in rodents (Braga and Gendler, 1993) and thus Muc-1 may function as an anti-adhesive molecule during the peri-implantation period in rodents by masking embryo attachment sites (Surveyor *et al.*, 1995).

Similarly, the expression of Muc-1 during the oestrous cycle and early pregnancy in pigs is consistent with its postulated role as an anti-adhesive molecule (Bowen *et al.*, 1996a). Antibodies directed to the cytoplasmic domain of Muc-1 revealed intense staining on the apical surface of luminal and glandular uterine epithelium on days 0 and 4. Staining was diminished on day 8 and was not detected on days 10–15 of the oestrous cycle or pregnancy. Therefore, increasing progesterone concentrations correlated with reduced Muc-1 staining in both cyclic and pregnant gilts. These data were further confirmed in ovariectomized gilts treated with exogenous steroids. Immunoreactive Muc-1 on the apical surface of uterine epithelium was present in ovariectomized gilts treated with either vehicle or oestradiol, whereas Muc-1 staining was virtually absent in uterine epithelium of gilts treated with progesterone or oestradiol plus progesterone. These results support the suggestion that reduction or loss of Muc-1 could contribute to the acquisition of endometrial receptivity by exposing constitutively expressed or hormone-induced adhesion molecules or their ligands or both.

The modulation of Muc-1 expression observed in cyclic, pregnant, and ovariectomized gilts is a property that has recently been reproduced in endometrial epithelial cells *in vitro* (Bowen *et al.*, 1997), if conditions for development and maintenance of epithelial cell polarity (Bowen *et al.*, 1996b) were provided. Uniformly low Muc-1 staining was observed at the apical surface of polarized uterine epithelial cells incubated in medium containing charcoal-stripped fetal bovine serum, progesterone, or oestrogen plus progesterone, whereas cells incubated with oestradiol exhibited a marked increase in Muc-1 staining. However, it remains to be determined whether the upregulation of Muc-1 causes polarized epithelial cells to regain their non-receptive state to blastocysts and trophoblastic vesicles *in vitro*.

Receptive Stage

The identification of glycoconjugates that may contribute to the receptive state in mammals has been motivated largely by an interest in the development of techniques for assisted human reproduction in infertile couples. The search for markers that identify or predict receptivity to implantation and the development of noninvasive approaches to detect uterine receptivity is ongoing. While comparative animal studies are contributing greatly to an understanding of implantation, details of the maternal–fetal interactions and adhesion systems in epitheliochorial implantation are not as well developed as in species exhibiting invasive implantation.

Integrins and extracellular matrix

The integrins are a family of glycoconjugates containing members that are expressed on the uterine luminal epithelium and that have spatial and temporal relationships with blastocyst development and implantation. These cation-dependent heterodimeric intrinsic membrane glycoproteins, comprising non-covalently bound α and β subunits, participate in cell–cell and cell–substratum adhesion by virtue of their large extracellular domains (Hynes, 1992). At least 23 heterodimer combinations are known to be formed from the currently identified 17 α and eight β subunits (Luscinskas and Lawler, 1994) with ligand specificity determined by the α and β subunit combinations.

Although integrins are expressed on virtually all cells, the spatial and temporal distribution and modulation of several subunits in the human female reproductive tract during the reproductive cycle and early pregnancy have led to the notion that they are directly involved in the process of implantation (Tabibzadeh, 1990; Lessey *et al.*, 1992). The endometrium is presently the only tissue identified that demonstrates hormone-dependent integrin expression (Lessey *et al.*, 1996). The developmental regulation of their expression on the invasive mouse blastocyst during invasion is also consistent with their potential involvement in this process (Sutherland *et al.*, 1993). Their role in adhesion, migration, invasion and a multitude of intracellular effects on organization of the cytoskeleton as well as bi-directional signalling makes integrins attractive as participants in the complex events of implantation. The bi-directional signalling processes (Humphries, 1993) can involve ligand binding through an integrin receptor leading to specific effects on cytoskeletal organization, activation of signalling pathways, and to the eventual regulation of gene expression (i.e. outside–in signalling). In addition, integrin cytoplasmic domains transduce cell type-specific signals that modulate ligand binding affinity (i.e., inside–out signalling). The involvement of these central functions of integrin receptors in trophectoderm–uterine epithelial cell interactions has been suggested by Damsky *et al.* (1993).

Much of the data on uterine integrin receptors have come from clinical studies and experiments in rodents; only recently have analyses been conducted in an epitheliochorial animal model (Bowen *et al.*, 1996a). Therefore, comparative information relative to invasive and non-invasive modes of implantation is only beginning to emerge. At least 14 different integrin subunits (nine α and five β integrin subunits) have been identified in human endometrium (Lessey *et al.*, 1996) and seven subunits (five α and two β integrin subunits) in the pig uterine endometrium to date (Bowen *et al.*, 1996a). The uterine epithelial cell expression of several integrin subunits in both species is temporally and spatially regulated during the reproductive (oestrus or menstrual) cycle while others are constitutively expressed. The species-specific differences may reflect the major differences in the mode of implantation, although several common subunits have been identified.

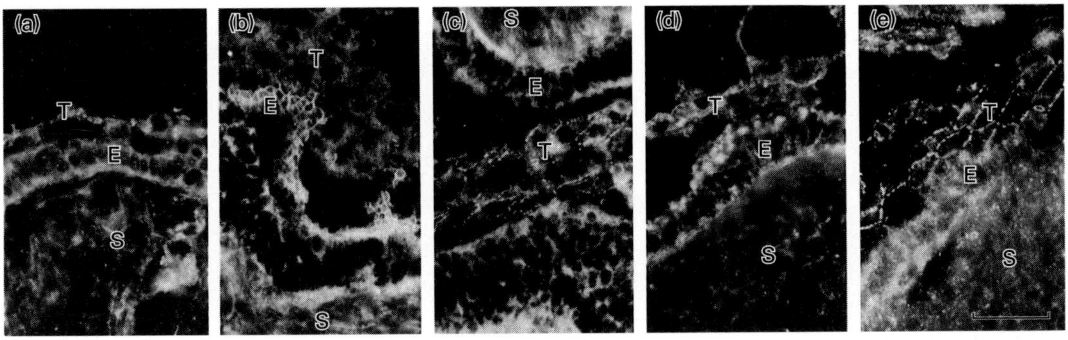

Fig. 2. Immunofluorescence analysis of integrin subunit expression at implantation sites at day 15. Individual subunits shown include (a) α4, (b) α5, (c) αv, (d) β1 and (e) β3 staining in stroma (S), uterine epithelium (E) and trophectoderm (T). Scale bar represents 50 μm.

In humans, considerable attention has been given to αvβ3 as a marker of uterine receptivity (Lessey *et al.*, 1992, 1996). The expression of αvβ3, which was localized along apical microvilli and secretory granules of endometrial epithelial cells, is regulated during the menstrual cycle with maximal expression coincident with the rise in progesterone during the period of maternal recognition of pregnancy; aberrant expression of αvβ3 is associated with infertility (Lessey *et al.*, 1996).

Pig uterine luminal epithelial cell expression of integrin subunits α1, α3, α4, α5, αv, β1 and β3, monitored during the oestrous cycle and early pregnancy (including sites of conceptus attachment) (Bowen *et al.*, 1996a), revealed three distinct patterns of expression. These patterns were classified as: (1) low or no expression (α1 and α3); (2) high intensity constitutive expression (αv and β3); and (3) high intensity, modulated expression (α4, α5, β1). Modulation of α4, α5, β1 resulted in maximal expression during the period corresponding to maternal recognition of pregnancy (i.e. days 10 to 15). In the same study, experimental evidence for steroid modulation of integrin subunits was obtained in ovariectomized gilts. Integrin subunits classified as the high intensity, modulated subunits (α4, α5, β1) were expressed at a higher intensity in gilts that were treated with progesterone (or progesterone in combination with oestrogen) than in oestrogen-treated animals.

Analysis of implantation sites indicated that integrin subunits α4, α5, αv, β1 and β3 were expressed at sites of trophectoderm contact with the uterine epithelium (Fig. 2). The integrin receptors potentially formed from the heterodimers identified in the study included α4β1, α5β1, αvβ1 and αvβ3. These receptors, like those identified in human studies during the initial events of attachment are all members of the fibronectin–vitronectin family of integrin receptors (Hynes, 1992).

The pig conceptus was also found to express the same integrin subunits (α4, α5, αv, β1 and β3) as the uterine epithelium. The significance of the integrin receptors and their modulation must be considered in the context of the available receptors and ligands present on both conceptus and uterine epithelial cells. Apical expression of the extracellular matrix constituents fibronectin and vitronectin is present on the pig conceptus, whereas vitronectin is present on uterine epithelium during early implantation (Fig. 3) (Bowen *et al.*, 1996a). Human studies suggest the involvement of fibronectin in mediating attachment, trophoblast spreading and syncytial formation (Kao *et al.*, 1988; Kliman and Feinberg 1990; Bischof *et al.*, 1995).

Further characterization of the nature of the fibronectin in the pig conceptus and uterus has revealed the constitutive expression of oncofetal fibronectin by both pig conceptus and uterus throughout gestation (Tuo and Bazer, 1996). This fibronectin variant, which exhibits two distinct epitopes, one of which is created by glycosylation of a single amino acid, is absent from normal adult tissues (see Feinberg *et al.*, 1995). The presence of oncofetal fibronectin and vitronectin at the fetal–maternal interface suggests that it plays a role in conceptus interactions, attachment to the endometrium or both processes. This interpretation is strengthened by the fact that the distribution and type of ECM components correspond to the integrin subunits expressed on uterine epithelial

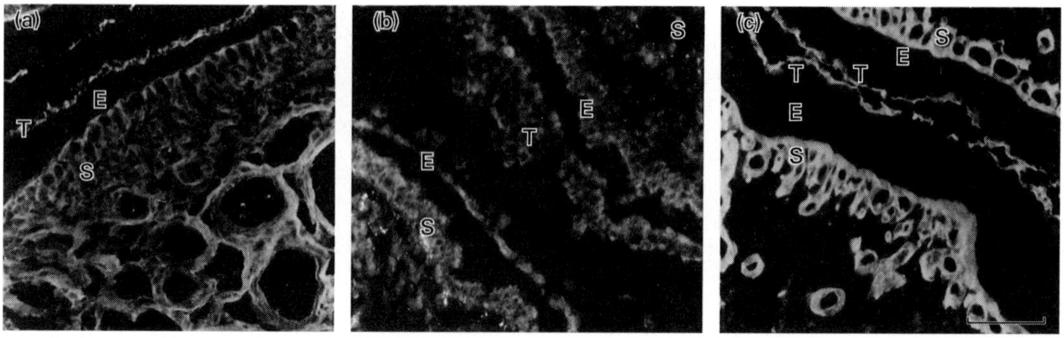

Fig. 3. Immunofluorescence analysis of extracellular matrix expression at implantation sites at day 15. Staining of (a) fibronectin, (b) vitronectin and (c) type IV collagen are shown. Stroma (S), uterine epithelium (E) and trophectoderm (T) are identified. Fibronectin was present on trophectoderm but not epithelial cells whereas vitronectin was present on both. Type IV collagen staining was restricted to basal lamina of epithelium and trophectoderm and was common in stromal vessels. Scale bar represents 50 μm.

cells during implantation in pigs. It is also relevant that in human trophoblasts, transforming growth factor-β (TGF-β) stimulates oncofetal fibronectin synthesis (Feinberg *et al.*, 1994) and three TGF-β isoforms (TGF-β1, TGF-β2, and TGF-β3) and type I and II receptors have been localized in peri-implantation pig conceptuses (Gupta *et al.*, 1996).

Integrins α1β1 and α3β1 are members of the collagen–laminin receptor family. The restricted expression of these integrins at the basal aspect of pig uterine epithelium is consistent with cellular attachment to the basement membrane and their unlikely involvement in interactions between the trophectoderm and uterine epithelium. Integrins α4β1, α5β1 and αvβ1 are members of the large family of fibronectin receptors, which makes them possible candidates for trophectoderm–uterine epithelium attachment based upon the finding that fibronectin and vitronectin are present on the contacting surfaces between the conceptus and the uterine epithelium. In addition to its ability to bind vitronectin, αvβ3 is capable of binding to oncofetal fibronectin and osteopontin. The expression of several different fibronectin receptors on the uterine epithelium and the trophectoderm allows the integrins to bind to fibronectin at two different locations, α4β1 integrin binds to the amino acid sequence, EILDV (Glu-Ile-Leu-Asp-Val), within the IIICS region of fibronectin, while the remaining integrins will bind to the RGD sequence of fibronectin (Hynes, 1992). The αvβ1 (weak affinity vitronectin receptor) and αvβ3 (vitronectin receptor) can also bind to the RGD sequence found in fibronectin. Thus, the four integrin heterodimers and two ECM proteins (fibronectin and vitronectin) are available for participation in implantation events of the pig conceptus. Although not yet described in pigs, osteopontin has been shown to be a ligand for αvβ3 in primate endometrium (Fazleabas *et al.*, 1997).

The redundancy in integrin receptors may ensure the ability of the conceptus to attach to the uterine epithelium. However, it is possible that each of these integrins may carry out different functions related to attachment and cell signalling (Meredith *et al.*, 1996). For example, α5β1 interacts with cytoskeleton, mediates cell spreading on fibronectin, and plays a role in matrix assembly (Lafrenie and Yamada, 1996). In addition to providing a direct link between ECM and cytoskeleton, these integrins also regulate the production of second messengers within cells. A common feature of integrin-dependent cell spreading or integrin receptor clustering is tyrosine phosphorylation of cytoplasmic proteins and the activation of the Na^+–H^+ antiporter, producing high intracellular pH (Schwartz *et al.*, 1991). However, the integrin-dependent adhesion and spreading that results in intracellular Ca^{2+} spikes and oscillations (Schwartz, 1993) seems to be specific for a particular integrin (αvβ3) (Schwartz and and Denninghoff, 1994). These observations suggest that different receptors within the same cell can generate unique signals. Integrin ligand binding and aggregation leads to a transmembrane hierarchy of molecular responses characterized by the prompt transmembrane accumulation of a large class of at least eight cytoskeletal and over 20 signal transduction molecules, and upregulation of cytoplasmic protein tyrosine phosphorylation (Miyamoto *et al.*, 1995). Among the signalling molecules are focal adhesion kinase (pp125^{FAK}), c-Src, cFyn, RhoA, Rac1, Ras, and several MAP kinases.

Regardless of the mode of implantation, it appears that the expression of integrin receptors and their ligands are important elements of adhesion, attachment and cell signalling between the conceptus and uterine epithelium. Integrin receptors expressed on the conceptus appear to be largely conserved among species examined to date, and consist of integrins that are also expressed on the uterine epithelium. This strategy may represent a conserved mechanism of conceptus attachment. The induction of new integrin receptor populations (specific for collagen and laminin) in species that exhibit invasive implantation occurs later in conceptus development (Sutherland *et al.*, 1993). Evaluation of the functional roles of the integrins expressed on pig uterine epithelium and trophectoderm should provide important insights into the relationship of integrin ligation and signal transduction during adhesion and attachment stages of early implantation with broad relevance to reproductive efficiency in a variety of species.

Analysis of integrin expression in pig uterine epithelial cells and conceptus-derived trophoblastic vesicles *in vitro* suggests that the above referenced integrin subunits identified on conceptus trophectoderm and in uterine epithelium of cyclic and pregnant gilts are expressed *in vitro* (Bowen *et al.*, 1997). However, comparable modulation of integrin subunits by steroid hormones was detected in uterine epithelial cells only if conditions for development and maintenance of epithelial cell polarity were provided. Although polarized uterine epithelial cells and trophoblastic vesicle culture systems require further refinements, they may prove useful as *in vitro* model systems to study conceptus–endometrial interactions during the peri-implantation period and the control of uterine receptivity. At this time, it is not feasible to search for direct uterine epithelial cell–trophectoderm signalling *in vivo*; therefore culture systems that express the hormonally modulated adhesion systems observed *in vivo* provide opportunities for investigating integrin-mediated signal transduction in pigs.

pGP30 and IaIH4

Another glycoprotein the expression of which is correlated with uterine receptivity in pigs is a 30 kDa, basic endometrial glycoprotein (pGP30). Initial antibody studies showed that pGP30 expression changes during the oestrous cycle and early pregnancy (Geisert *et al.*, 1995). Maximal staining intensity was observed in surface epithelium on day 12 of the oestrous cycle and from days 12 to 18 of pregnancy; the appearance of pGP30 was stimulated by progesterone administration to ovariectomized gilts. Staining in the area of trophoblast attachment suggested that the glycoprotein is available to the conceptus at the uterine luminal interface. Conceptus oestrogen secretion appears to alter pGP30 secretion from endometrial cultures (Geisert *et al.*, 1995).

More recent analysis has indicated that pGP30 is a cleavage product of a larger pig 120 kDa multi-functional inter-trypsin inhibitor heavy chain-like (IαIH4) protein (Geisert *et al.*, 1996). It appears, therefore, that pGP30 may be a useful marker of receptivity, while IαIH4 may be the functionally relevant player in conceptus–epithelial interaction. The unique properties of pig IαIH4, its interaction with the extracellular matrix and the potential roles of pig IαIH4 in the initial stages of trophoblastic attachment and regulation of trophoblast invasiveness are detailed by Geisert and Yelich (this supplement). Noteworthy is the observation that all IαI heavy chains contain a von Willebrand type A domain, a recognition site for the αvβ3 integrin receptor and other adhesion molecules.

LNF and related determinants

Changes in carbohydrate expression on the epithelium lining the uterine lumen were initially suggested by the reduction in surface charge and thinning of the glycocalyx near the time of implantation (Enders and Schlafke, 1974). Subsequent studies in rodents identified a specific oligosaccharide carrying an antigen at its nonreducing terminal carbohydrate determinants which may play a role in the implantation process. Monoclonal antibodies developed against a blood group determinant, H-type-1 antigen, carrying the lacto-*N*-fucopentaose-1 epitope (LNF-I, Fucα1-2Galβ1-3GlcNAcβ1-) detected changes in expression of this epitope on the surface of mouse uterine epithelial cells during the pre-implantation period (Kimber *et al.*, 1988; Lindenberg *et al.*, 1988; Kimber and Lindenberg, 1990). LNF-I (but not polysaccharides with similar structures such as LNF-II, LNF-III, lacto-*N*-difucohexaose, lacto-*N*-tetraose, and lactodifucotetraose) (Kimber *et al.*, 1994) and an anti-LNF-I monoclonal antibody were also shown to inhibit mouse embryo attachment significantly using an *in vitro* model for implantation (Lindenberg *et al.*, 1988).

In ovariectomized mice, the LNF-I epitope appeared to be stimulated by oestrogen (Kimber and Lindenberg, 1990) rather than progesterone as in rats (Kimber *et al.*, 1995). In pregnant mice, maximal LNF-I epithelial surface expression occurred on days 1–4 and decreased markedly by days 5 and 6 (Kimber *et al.*, 1988). Since the expression of LNF-I does not span the entire duration of the receptive phase, the potential role for LNF-I may be limited to initial steps of adhesion. Mouse conceptuses are believed to contain specific binding sites for LNF-I, although specific receptors have not been identified.

LNF-I expression in pigs has recently been examined using the antibody used by Kimber *et al.* (1994, 1995) in mouse and rat. Expression was greatest in cyclic gilts from days 0–4 and is weak on days 8–15 (Fig. 4). In ovariectomized gilts, LNF-I was detected in oestradiol- but not in progesterone, oestradiol plus progesterone-, or vehicle-treated animals (Fig. 5). Analysis of the LNF-I expression in the polarized uterine epithelial culture system is in progress.

Another carbohydrate antigen localized by immunostaining on luminal epithelium of pregnant pigs from day 14 to day 30 of pregnancy which was specifically blocked by LNF-III is the Lewis^x (Galβ-1[Fucα1-3]GlcNAcβ1-) determinant (Whyte and Binns, 1994). The oligosaccharide carrying this antigen at its nonreducing terminal can function as a ligand for E-selectin and has also been shown to play a role in homotypic adhesion (Lewis^x–Lewis^x) in mouse compaction and F9 embryonal carcinoma cell adhesion (Hakomori, 1992). The absence of immunoreactive Lewis^x on pig trophectoderm indicates the absence of a role for Lewis^x homotypic adhesion in pig uterine epithelial cell–trophectodermal interaction (Whyte and Binns, 1994).

Selectins

The above referenced identification of Lewis^x on pig uterine epithelium has led to the search for the expression of selectins on uterine epithelial cells and trophectoderm. Selectin family members include E-, P-, and L-selectin. The term selectin refers to the common amino-terminal lectin domain and the selective function and expression of these molecules; the nomenclature is based upon the initial detection in cells (i.e. endothelium, platelet and leukocyte, respectively). All members of the selectin family contain the unique extracellular region composed of an amino-terminal calcium-dependent lectin domain, which is central to the ability of each of the selectins to bind the Lewis^x antigen (Tedder *et al.*, 1995; Etzioni, 1996).

Neither E-selectin nor L-selectin were detected within pig uterine endometrial tissue or on trophectoderm by immunocytochemistry, whereas E-selectin was detected in inflammatory cutaneous sites and L-selectin in high endothelial venule cells in maternal lymph nodes in these pigs (Whyte and Binns, 1994). The significance of Lewis^x and LNF-I expression in uterine epithelium and the possible involvement of carbohydrate–selectin interactions in trophoblast–uterine epithelial cell adhesion remains undefined.

HSPGs and HSPG-binding proteins

A body of evidence suggests that heparan sulfate proteoglycans (HSPG) support rodent blastocyst interactions with a variety of substrates including uterine epithelial cells during early attachment (see Wegner and Carson, 1994). HSPGs are developmentally regulated at the surface of

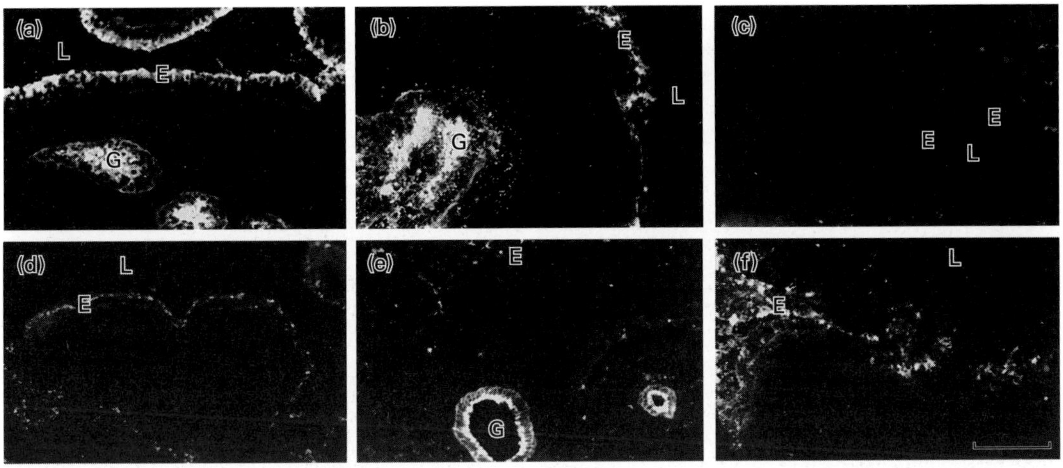

Fig. 4. Immunofluorescence analysis of uterine epithelial lacto-*N*-fucopentaose I (LNF-I) expression in cyclic gilts on days (a)0, (b)4, (c)8, and on days (d)11, (e)13 and (f)15 on luminal epithelial cells (E) and glands (G). The lumen of the uterus (L), where present, is identified in each panel. LNF-I expression was greatest in cyclic gilts from days 0–4 and is weak on days 8–15. Scale bar represents 50 μm.

mouse blastocysts and are detected at sites of embryo–uterine epithelial cell attachment and at the surface of human trophoblastic cell lines (Wegner and Carson, 1994). HSPGs support adhesive interactions of numerous cell types and extracellular matrix constituents and are capable of complexing with a number of growth factors and modulate cellular functions. A heparin-binding epidermal growth factor-like growth factor (HB-EGF) is synthesized as a transmembrane form in mouse uterine epithelium. The mRNA encoding HB-EGF is expressed in uterine luminal epithelium just before implantation and only at the site of blastocyst apposition, and experimental evidence has been provided that HB-EGF can function as a juxtacrine adhesion factor that could be one of the mediators of blastocyst adhesion during early implantation (Raab *et al.*, 1996).

Although comparable analyses of HSPG expression in pig conceptus and uterine epithelial cells have not yet been conducted, the uterine expression of colony-stimulating factor 1 (CSF-1) in the pig exhibits an interesting parallel to HB-EGF. CSF-1 has been detected on the apical side of pig trophectoderm and immunocytochemical analysis suggests that this may be a membrane-bound form of CSF-1 (Tuo *et al.*, 1995). CSF-1 may interact with CSF-1 receptor expressed on both trophectoderm and endometrial cells by a juxtacrine mechanism as the CSF-1 receptor is expressed by both pig conceptus and endometrium (W. Tuo and F.W. Bazer, unpublished).

Implantation as an Adhesion Cascade

Adhesion events that occur during implantation involve a highly coordinated set of sequential interactions that may involve many different adhesion molecules with distinct but inter-related functions. The molecular interactions involved in apposition and attachment of the conceptus and apical uterine luminal epithelial cell surface may constitute an implantation adhesion cascade with core events comparable to other adhesion cascades (Fig. 6). Examples include the cytokine-mediated activation of granulocyte–endothelial interactions and metastasis of malignant cells (see Schweighoffer and Shaw, 1992; Giancotti and Mainiero, 1994). Similarities of early implantation events to the leukocyte–endothelial adhesion cascade and metastatic process have been recognized (Damsky *et al.*, 1993; Kimber *et al.*, 1994).

Several core elements of adhesion cascades have been described (Schweighoffer and Shaw, 1992) including (1) a triggering molecule on the cell surface capable of interacting with an extracellular ligand, (2) a biochemical signal (or series of signals) generated by physical association of the trigger

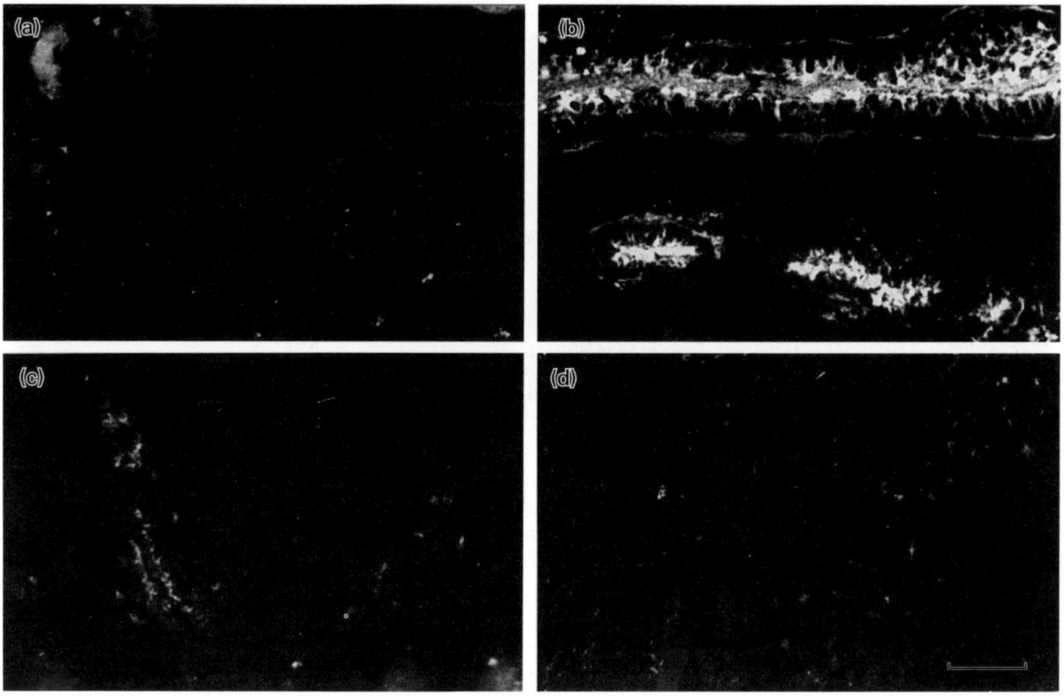

Fig. 5. Immunofluorescence analysis of uterine epithelial lacto-*N*-fucopentaose (LNF-I) expression in ovariectomized gilts receiving vehicle carrier (a), oestradiol (b), progesterone (c) or oestradiol plus progesterone (d). Daily doses of oestradiol valerate (100 μg day^{-1}), progesterone (200 mg day^{-1}), or oestradiol plus progesterone were administered for 10 days. LNF-I staining was intense only in oestradiol-treated gilts. Scale bar represents 50 μm.

molecule to the adhesion molecule, and (3) an adhesion molecule, the adhesive nature of which is altered by the biochemical signal. The molecules that function to trigger the adhesion cascade are diverse and represented by many families of molecules that include cytokine receptors, carbohydrate–selectin interactions, and membrane lipid metabolites that are linked to a variety of signal transduction pathways (Schwartz *et al.*, 1995; Stewart *et al.*, 1995). While there is considerable diversity among triggering molecules, integrins are the dominant adhesion systems in each of the adhesion cascades and their capacity to mediate adhesion is linked to their activation by engaging another surface molecule (Schweighoffer and Shaw, 1992).

The conditions that control the developmental expression of adhesion receptors and ligands in the uterus are regulated precisely through the action of ovarian steroids, growth factors and cytokines. During the pre-receptive phase, the blastocyst is free to move within the uterus, a condition that appears to be supported in a number of species by the non-adhesive nature of the large glycoprotein Muc-1. Muc-1 may extend outward from the apical cell surface further than receptors or triggering molecules and thereby have the capacity to sterically inhibit the interaction of the embryo with adhesion molecules at the maternal–fetal interface (Aplin *et al.*, 1996).

Entry of the uterus into the receptive phase in synchrony with conceptus development and the expression or unmasking of adhesion molecules permits the primary interactions involving low-affinity bonds with triggering molecules. Several candidates for this role have been described in this review including the pig IαIH4 (see Geisert and Yellich, this supplement), and membrane-associated CSF-1 (Tuo *et al.*, 1995).

Although not discussed in this review, cytokines, growth factors and paracrine signals from trophectoderm may also act as triggering mechanisms or signals involved in the implantation cascade. Potential participants in this process include, but are not limited to, transforming growth

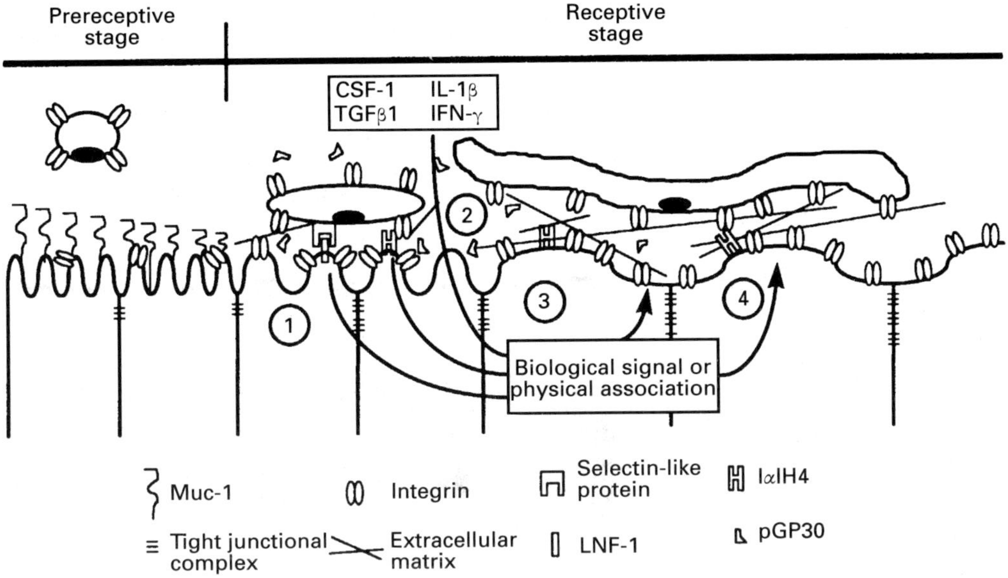

Fig. 6. Implantation as an adhesion cascade. Transition of uterine luminal epithelium from pre-receptive to receptive requires the downregulation of the anti-adhesive mucin, Muc-1, thought to inhibit the interaction of the embryo with adhesion molecules at the interface of the blastocyst and uterine epithelium. The sequential events of implantation (1–4) resemble events of other adhesion cascades. (1) The progesterone-induced structural changes resulting in reduction of epithelial polarity is indicated by loss of apical microvilli, alteration of tight junction organization, and close apposition of conceptus and apical uterine epithelium. (2) Initial interaction between the two genetically distinct cell types occurs with triggering molecules that are either induced or constitutively expressed but unavailable until loss of Muc-1. The triggering molecules that initiate the implantation cascade are involved in low affinity and weak molecular interactions. (3) Transduction of signals from the triggering molecules can include a diverse array of biochemical and physical interactions, including Ca^{2+} fluxes, protein phosphorylation, or cytoskeletal alterations. (4) The final adhesion event is mediated by the high-affinity integrin–ligand interactions. Potential integrin heterodimers that are candidates for the implantation cascade are all members of the fibronectin and vitronectin family of receptors. LNF-1: lacto-*N* fucopentaose; IαIH4: inter–trypsin inhibitor heavy chain-like protein; GM–CSF: granulocyte–macrophage colony-stimulating factor; TGFβ1: transforming growth factor β1; IL-1β: interleukin 1β; CSF-1: colony stimulating factor 1.

factor β1 (TGF-β_1), interferon gamma (IFN-γ), granulocyte–macrophage colony-stimulating factor (GM-CSF), and interleukin 1β (IL-1β) (Robertson *et al.*, 1994), all of which are secreted by pig trophectoderm during the peri-implantation period. These and other cytokines could initiate integrin activation during signalling events.

It should also be noted that changes in the polarized functions of uterine epithelium that occur during the receptive phase result in cytoskeletal alterations and changes in the distribution of plasma membrane proteins and lipids partitioned by tight junctions. These changes may contribute to conformational alterations in the integrins which could lead to their activation (Conforti *et al.*, 1990).

As discussed previously, the involvement of integrins in adhesion, migration, and invasion, and their numerous effects on the organization of the cytoskeleton and bi-directional signalling are consistent with these adhesion molecules as dominant participants in the implantation cascade. The upregulation of several integrin subunits and exposure of others upon loss of Muc-1 during the receptive period provide a number of heterodimer combinations that are members of the fibronectin and vitronectin family of integrin receptors. The redundancy in the potential integrin receptors present and the variety of potential triggering molecules and ligands highlight the physiological complexity of the implantation cascade in mammals.

Supported by USDA grants 92-37203-7952 (RCB), 95-37203-2185 (FWB) and NIH grants HD32534 (FWB) 2S06 GM08094 (GRN). The authors thank S. R. Glasser, L. A. Jaeger and R. D. Geisert for helpful discussions.

References

Aplin JD, Hey NA and Li TC (1996) MUC1 as a cell surface and secretory component of endometrial epithelium: reduced levels in recurrent miscarriage *American Journal of Reproductive Immunology* **35** 261–266

Bischof P, Haenggeli L and Campana A (1995) Gelatinase and oncofetal fibronectin secretion is dependent on wintergreen expression on human cytotrophoblasts *Human Reproduction* **10** 734–742

Bjorkman N, Dantzer V, Hasselager E, Holm H and Kjaersgaard P (1981) Perfusion *in vivo* of the porcine placenta. Fixation for EM *Placenta* **2** 287–302

Blair RM, Geisert RD, Zavy MT, Yellin T, Fulton RW and Short EC (1991) Endometrial surface and secretory alterations associated with embryonic mortality in gilts administered estradiol valerate on days 9 and 10 of gestation *Biology of Reproduction* **44** 1063–1079

Bowen JA, Newton GR, Weiss DW, Bazer FW and Burghardt RC (1996a) Characterization of a polarized porcine uterine epithelial cell model system *Biology of Reproduction* **55** 613–619

Bowen JA, Bazer FW and Burghardt RC (1996b) Spatial and temporal analyses of integrin and Muc-1 expression in porcine uterine epithelium and trophectoderm *in vivo. Biology of Reproduction* **55** 1098–1106

Bowen JA, Bazer FW and Burghardt RC (1997) Spatial and temporal analyses of integrin and Muc-1 expression in porcine uterine epithelium and trophectoderm *in vitro. Biology of Reproduction* **56** 409–415

Braga VM and Gendler SJ (1993) Modulation of Muc-1 mucin expression in the mouse uterus during the estrus cycle and early pregnancy *Journal of Cell Science* **105** 397–405

Burton GJ (1992) Human and animal models: limitations and comparisons. In *The First Twelve Weeks of Gestation* pp 469–485 Ed. ER Barnea J Hustin and E Janiaux Springer–Verlag, Berlin

Conforti G, Zanetti A, Paxquali-Ronchetti I, Quaglino D, Neyroz P and Dejana E (1990) Modulation of vitronectin receptor binding by membrane lipid composition *Journal of Biological Chemistry* **265** 4011–4019

Cunha GR, Chung LWK, Shannon JM, Tagichi O and Fujii H (1983) Hormone-induced morphogenesis and growth: role of mesenchymal–epithelial interactions *Recent Progress in Hormone Research* **39** 559–598

Damsky C, Sutherland A and Fisher S (1993) Extracellular matrix 5: adhesive interactions in early mammalian embryogenesis, implantation and placentation *FASEB Journal* **7** 1320–1329

Dantzer V (1985) Electron microscopy of the initial stages of placentation in the pig *Anatomy & Embryology* **172** 281–293

Denker H-W (1993) Implantation: a cell biological paradox *Journal of Experimental Zoology* **266** 541–558

Enders AC and Schlafke S (1974) Surface coats of the mouse blastocyst and uterus during the preimplantation period *Anatomical Record* **180** 31–45

Etzioni A (1996) Adhesion molecules – their role in health and disease *Pediatric Research* **39** 191–198

Fazleabas AT, Bell SC, Flemming A, Sun J and Lessey B (1997) Distribution of integrins and the extracellular matrix proteins in the baboon endometrium during the menstrual cycle and early pregnancy *Biology of Reproduction* **56** 348–356

Feinberg RF, Kliman HJ and Wang C-L (1994) Transforming growth factor-β stimulates trophoblast oncofetal fibronectin synthesis *in vitro*: implications for trophoblast implantation *in vivo. Journal of Clinical Endocrinology and Metabolism* **78** 1241–1248

Feinberg RF, Kliman HJ, Bedian V, Monzon-Bordonaba F, Menzin A W and Wang C-L (1995) Monoclonal antibody X18A4 identifies an oncofetal fibronectin epitope distinct from the FDC-6 binding site *American Journal of Obstetrics & Gynecology* **172** 1526–1536

Geisert RD, Renegar RH, Thatcher WW, Roberts RM and Bazer FW (1982a) Establishment of pregnancy in the pig: I. Interrelationships between preimplantation development of the pig blastocyst and uterine endometrial secretions *Biology of Reproduction* **27** 925–940

Geisert RD, Brookbank JW, Roberts RM and Bazer FW (1982b) Establishment of pregnancy in the pig: II. Cellular remodeling of the porcine blastocyst during elongation on day 12 of pregnancy *Biology of Reproduction* **27** 941–955

Geisert RD, Brenner RM, Moffatt RJ, Harney JP, Yellin T and Bazer FW (1993) Changes in oestrogen receptor protein, mRNA expression and localization in the endometrium of cyclic and pregnant gilts *Reproduction, Fertility and Development* **5** 247–260

Geisert RD, Pratt TN, Bazer FW, Mayes JS and Watson GH (1994) Immunocytochemical localization and changes in endometrial progestin receptor protein during the porcine oestrous cycle and early pregnancy *Reproduction, Fertility and Development* **6** 749–760

Geisert RD, Mendan JD, Pratt T, Schmitt RAM, Lessey BA and McCann JP (1995) Isolation and characterization of a 30 kDa endometrial glycoprotein synthesized during the estrous cycle and early pregnancy of the pig *Biology of Reproduction* **53** 1038–1050

Geisert RD, Yelich JV, Pratt T and Pomp D (1996) Identification of porcine GP30 as inter-trypsin inhibitor family heavy chain-related protein *Biology of Reproduction* **54** (Supplement 1) 78a

Giancotti FG and Mainiero F (1994) Integrin-mediated adhesion and signaling in tumorigenesis *Biochimica et Biophysica Acta* **1198** 47–64

Glasser SR and Mullholland J (1993) Receptivity is a polarity dependent special function of hormonally regulated uterine epithelial cells *Microscopy Research and Techniques* **25** 106–120

Glasser SR, Julian J, Decker GL, Tang J-P and Carson DD (1988) Development of morphological and functional polarity in primary cultures of immature rat uterine epithelial cells *Journal of Cell Biology* **107** 2409–2423

Guidice LC, Dsupin BA, Jin IH, Vu TH and Hoffman AR (1993) Differential expression of messenger ribonucleic acids encoding insulin-like growth factors and their receptors in human uterine endometrium and decidua *Journal of Clinical Endocrinology and Metabolism* **76** 1115–1122

Gupta A, Bazer FW and Jaeger LA (1996) Differential expression of beta transforming growth factors (TGF-β1,TGF-β2, and

TGF-β3) and their receptors (type I and II) in peri-implantation porcine conceptuses *Biology of Reproduction* **55** 796-802

Hana V and Murphy LJ (1994) Interdependence of epidermal growth factor and insulin-like growth factor-I expression in the mouse uterus *Endocrinology* **135** 107–112

Hakomori S (1992) Le^x and related structures as adhesion molecules *Histochemical Journal* **24** 771–776

Humphries MJ (1993) Mechanisms of ligand binding by integrins *Biochemical Society Transactions* **22** 275–282

Hynes RO (1992) Integrins: versatility, modulation, and signaling in cell adhesion *Cell* **69** 11–25

Kao LC, Caltabiano S, Wu S, Strauus JF and Kilman H (1988) The human villous cytotrophoblast: interactions with extracellular matrix proteins, endocrine function, and cytoplasmic differentiation in the absence of syncytium formation *Developmental Biology* **130** 639–702

Keys JL and King GJ (1990) Microscopic examination of porcine conceptus–maternal interface between days 10 and 19 of pregnancy *American Journal of Anatomy* **188** 221–238

Kimber SJ and Lindenberg S (1990) Hormonal control of a carbohydrate epitope involved in implantation in mice *Journal of Reproduction and Fertility* **89** 13–21

Kimber SJ, Lindenberg S and Lundblad A (1988) Distribution of some Gal β1-24 GlcNAc related carbohydrate antigens on the mouse uterine epithelium in relation to the peri-implantation period *Journal of Reproductive Immunology* **12** 297–313

Kimber SJ, White S, Cook A and Illingworth I (1994) The initiation of implantation: parallels between attachment of the embryo and neutrophil–endothelial interaction. In *Gamete and Embryo Quality* pp 171–178 Ed. L Mastroianni, Jr , HJT Coelingh Bennink, S Suzuki, and HM Vemer. Parthenon, London

Kimber SJ, Illingworth IM and Glasser SR (1995) Expression of carbohydrate antigens in the rat uterus during early pregnancy and after ovariectomy and steroid replacement *Journal of Reproduction and Fertility* **130** 75–87

Kliman H and Feinberg RF (1990) Human trophoblast–extracellular matrix (ECM) interactions *in vitro*: ECM thickness modulates morphology and proteolytic activity *Proceedings of the National Academy of Sciences USA* **87** 3057–3061

Koji T, Chedid M, Rubin JS, Slayden Ov D, Csaky KG, Aaronson SA and Brenner RM (1994) Progesterone-dependent expression of keratinocyte growth factor mRNA in stromal cells of the primate endometrium: keratinocyte growth factor as a progestomedin *Journal of Cell Biology* **125** 393–401

Lafrenie RM and Yamada KM (1996) Integrin-dependent signal transduction *Journal of Biological Chemistry* **61** 543–553

Lessey BA, Damjanovich L, Coutifaris C, Castelbaum A, Albelda SM and Buck CA (1992) Integrin adhesion molecules in the human endometrium. Correlation with the normal and abnormal menstrual cycle *Journal of Clinical Investigation* **90** 188–195

Lessey BA, Ilesanmi AO, Lessey MA, Riben M, Harris JE and Chwalisz K (1996) Luminal and glandular endometrial epithelium express integrins differentially throughout the menstrual cycle: implications for implantation, contraception, and infertility *American Journal of Reproductive Immunology* **35** 195–204

Ligtenberg MJ, Buijs F, Vos HL and Hilkens J (1992) Suppression of cellular aggregation by high levels of episialin *Cancer Research* **52** 2318–2324

Lindenberg S, Sundberg K, Kimber SJ and Lundblad A (1988) The milk oligosaccharide, lacto-*N*-fucopentaose I, inhibits attachment of mouse blastocysts on endometrial monolayers *Journal of Reproduction and Fertility* **83** 149–158

Liu SW, Tuo W and Bazer FW (1996) Expression of keratinocyte growth factor receptor and mRNA in porcine endometrium *Biology of Reproduction* **54** (Supplement 1) 75a

Luscinskas FW and Lawler J (1994) Integrins as dynamic regulators of vascular function *FASEB Journal* **8** 929–938

Meredith JE, Jr, Winitz S, McArthur Lewis J, Hess S, Ren X-D, Renshaw MW and Schwartz MA (1996) The regulation of growth of intracellular signaling by integrins *Endocrine Reviews* **17** 207–220

Miyamoto S, Teramoto H, Coso OA, Gutkind S, Burbelo PD, Akiyama SK and Yamada KM (1995) Integrin function: molecular hierarchies of cytoskeletal and signaling molecules *Journal of Cell Biology* **131** 791–805

Pemberton L, Taylor-Papadimitrou J and Gendler SJ (1992) Antibodies to the cytoplasmic domain of Muc-1 mucin show conservation throughout mammals *Biochemical and Biophysical Research Communications* **185** 167–175

Raab G, Kover K, Paria B, Dey SK and Ezzell RM (1996) Mouse preimplantation blastocysts adhere to cells expressing the transmembrane form of heparin-binding EGF-like growth factor *Development* **122** 637–645

Roberts RM, Trout WE, Mathialagan N, Stallings-Mann M and Ling P (1993) Uterine secretory activity and embryo development *Preimplantation Embryo Development*, pp 229–243 Ed. BD Bavister. Springer-Verlag, New York

Robertson SA, Seamark RF, Guilbert LJ and Wegmann TG (1994) The role of cytokines in gestation *Critical Reviews in Immunology* **14** 239–292

Samuel CA and Perry JS (1972) The ultrastructure of pig trophoblast transplanted to an ectopic site in the uterine wall *Journal of Anatomy* **113** 139–149

Schwartz MA (1993) Spreading of human endothelial cells on fibronectin or vitronectin triggers elevation of intracellular free calcium *Journal of Cell Biology* **120** 1003–1010

Schwartz MA, Ingber DE, Lawrence M, Springer TA and Lechene C (1991) Multiple integrins share the ability to induce elevation of intracellular pH *Experimental Cell Research* **195** 533–535

Schwartz MA and Denninghoff K (1994) α_v Integrins mediate the rise in intracellular calcium in endothelial cells on fibronectin even though they play a minor role in adhesion *Journal of Biological Chemistry* **269** 11133–11137

Schwartz MA, Schaller MD and Ginsberg MH (1995) Integrins: emerging paradigms of signal transduction *Annual Review of Cell and Devopmental Biology* **11** 549–599

Schweighoffer T and Shaw S (1992) Adhesion cascades: diversity through combinatorial strategies *Current Opinion in Cell Biology* **4** 824–829

Stewart M, Thiel M and Hogg N (1995) Leukocyte integrins *Current Opinion in Cell Biology* **7** 690–696

Stroband HWJ and Van der Lende T (1990) Embryonic and uterine development during early pregnancy *Journal of Reproduction and Fertility Supplement* **40** 261–277

Surveyor GA, Gendler AJ, Pemberton L, Das SK, Chakraborty I, Julian J, Pimental RA, Wegner CC, Dey SK and Carson DD (1995) Expression and steroid hormonal control of Muc-1 in the mouse uterus *Endocrinology* **136** 3639–3647

Sutherland AE, Calarco PG and Damsky CH (1993) Developmental regulation of integrin expression at the time of implantation in the mouse embryo *Development* **119** 1175–1186

Svalander PC, Odin P, Nilsson BO and Obrink B (1990) Expression of cell CAM-105 in the apical surface of rat uterine epithelium is controlled by ovarian steroid hormones *Journal of Reproduction and Fertility* **88** 213–221

Tabibzadeh S (1990) Immunoreactivity of human endometrium: correlation with endometrial dating *Fertility and Sterility* **54** 624–631

Tedder TF, Steeber DA, Chen A and Engel P (1995) The selectins: vascular adhesion molecules *FASEB Journal* **9** 866–873

Tuo W and Bazer FW (1996) Expression of oncofetal fibronectin by porcine conceptuses and uterus throughout gestation *Reproduction, Fertility and Development* **8** 1207–1214

Tuo W, Harney JP and Bazer FW (1995) Colony-stimulating factor-1 in conceptus and uterine tissues in pigs *Biology of Reproduction* **53** 133–142

Valdizan MC, Julian J and Carson DD (1992) WGA-binding, mucin glycoproteins protect the apical cell surface of mouse uterine epithelium *Journal of Cellular Physiology* **151** 451–465

Wegner CC and Carson DD (1994) Cell adhesion processes in implantation. In *Oxford Reviews of Reproductive Biology* pp 87–138 Ed. HM Charlton. Oxford University Press, New York

Wesseling J, van der Valk SW, Vos HL, Sonnenberg A and Hilkens J (1995) Episialin (MUC1) overexpression inhibits integrin-mediated cell adhesion to extracellular matrix components *Journal of Cell Biology* **129** 255–265

Whyte A and Robson T (1984) Saccharides localized by fluorescent lectins on trophectoderm and endometrium prior to implantation in pigs, sheep, and equids *Placenta* **5** 533–540

Whyte A and Binns RM (1994) Adhesion molecule expression and infiltrating maternal leucocyte phenotypes during blastocyst implantation in the pig *Cell Biology International* **18** 759–766

Journal of Reproduction and Fertility Supplement **52**, 165–176

Embryonic and fetal development in different genotypes in pigs

S. P. Ford

Department of Animal Science, Iowa State University, Ames, IA 50011–3150, USA

It is widely accepted that uterine capacity, not ovulation rate, is the greatest restraint on litter size in pigs. Recently, the reproductive strategy(s) of the Chinese Meishan pig, a breed which farrows three to five more piglets per litter than US or European pig breeds, has come under intense scrutiny. It was initially determined that the Meishan female could farrow more viable piglets per litter than US or European pig breeds, with a uterine size and ovulation rate equivalent to those of less prolific breeds. It has become apparent that the Meishan conceptus exhibits a reduced trophectoderm mitotic rate during the preimplantation period, elongates from fewer cells and remains smaller throughout gestation compared with conceptuses from less prolific US or European pig breeds. This strategy by the Meishan conceptus for a lower growth rate results in a marked reduction in conceptus loss through day 18 of gestation compared with less prolific breeds. An additional strategy is required in the Meishan to allow the larger number of viable fetuses to survive after day 30 of gestation when uterine capacity becomes limiting. Our research has demonstrated that the rapid growth of the fetus in US pig breeds appears to require continued placental growth to increase the surface area for nutrient exchange. In contrast, the increased numbers of smaller Meishan fetuses achieve the same increase in placental efficiency by markedly increasing the density of placental blood vessels at the fetal–maternal interface. This proliferation of placental blood vessels obviates the need for marked increases in placental size.

Introduction

Evidence has rapidly accumulated suggesting that from the time a conceptus enters the uterus, it is subject to modulations of growth, differentiation or both processes through its exposure to a bewildering array of uterine environmental factors (Paria and Dey, 1990; Pollard, 1990; Simmen and Simmen, 1991; Simmen *et al.*, 1995). Furthermore, far from being passive in this process, all evidence points to the fact that both the conceptus and the uterus actively contribute to placental and fetal growth and development throughout gestation in a variety of mammalian species (Venge, 1950; Hunter, 1958; Ferrell, 1991a,b; Youngs *et al.*, 1994).

In this review, I will emphasize the two periods during gestation in the pig that I consider are the most important in determining the number of viable piglets at farrowing. These periods include the peri-implantation period (i.e. days 12–18), and the period from midgestation to term. It is during the peri-implantation period that the greatest numbers of viable conceptuses in a litter are lost (about 75% of the total loss; reviewed by Ford and Youngs, 1993). Fetal weight increases most markedly during the latter half of gestation in pigs, as in other mammalian species, in association with very limited placental growth (Warwick, 1928; Reynolds and Redmer, 1995). Furthermore, on the basis of a number of studies, increased blood flow at the fetal–maternal interface appears to be a primary determinant of increased transplacental exchange during this period of late gestation in livestock species (Ford, 1995; Reynolds and Redmer, 1995). Reynolds *et al.* (1985) demonstrated that in pigs as the number of fetuses in a uterine horn increased (range from two to eight fetuses), the blood flow allotted to each during late gestation progressively declined. Furthermore, as uterine blood flow per

Table 1. Comparisons between breeds in ovulation status and early conceptus development 48–54 h after the onset of oestrus

Parameter	Meishan (n = 66)	Yorkshire (n = 55)
Ovulation status (% of gilts)		
All follicles ovulated	89.4%	92.7%
Some but not all follicles ovulated	7.6%	5.5%
No follicles ovulated	3.0%	1.8%
Conceptus development		
Minimum number of cells[a]	1.5±0.9	1.9±0.3
Maximum number of cells[b]	3.4±0.4	3.3±0.4
Range of cell numbers across all conceptuses	1–8	1–8

[a]Average cell number of the least developed conceptus in a litter.
[b]Average cell number of the most developed conceptus in a litter.
n = number of litters.

fetus declined with increasing numbers of fetuses, umbilical blood flow to these fetuses also declined (r = 0.70; P < 0.01). The decrease in placental surface area of each conceptus with increasing numbers of fetuses per horn (Knight *et al.*, 1977) may reduce the size of each placental vascular bed, with concomitant reductions in fetal growth.

However, these periods are by no means independent, since the number of conceptuses surviving the peri-implantation period will dictate the uterine space that can be allotted to each for the remainder of gestation. Thus, it is assumed that prolific pig breeds such as the Chinese Meishan, which exhibits a markedly reduced preimplantation conceptus loss when compared with commercial US and European pig breeds (Galvin *et al.*, 1993; Haley and Lee, 1993), must have a strategy for successfully maintaining these larger litters to term. This is difficult to explain as the size of the uterus in Meishan and US and European pig breeds is similar when adjusted for the slaughter weight of the female (Bazer *et al.*, 1988; Lee *et al.*, 1995).

For the past 20 years, the litter size of commercial pig breeds in the US has been relatively constant, despite repeated attempts to define selection criteria for an increased number of piglets (Pork Facts, 1993). This failure to change litter size significantly was the major impetus for the importation of the prolific Meishan pig into the US from China in 1989. The Meishan is known to farrow three to five more viable piglets per litter than US or European pig breeds (Jin *et al.*, 1992). This review will emphasize the most recent information obtained by our group and others from studies in Meishan pigs and provide insight into its prolificacy.

The Peri-implantation Period

An essential period for the establishment of pregnancy in the pig occurs between day 12 and day 18 after mating. During this period, conceptuses expand to 8–10 mm spheres, space themselves uniformly throughout the uterus (Dhindsa *et al.*, 1967; Dziuk, 1968, 1985), elongate to long filamentous forms (Geisert *et al.*, 1982a), and attach to the uterine wall through interdigitation of trophoblastic and endometrial microvilli (Dantzer, 1985).

Coincident with expansion and elongation, the day 11–12 conceptuses acquire the ability to synthesize oestradiol (Gadsby *et al.*, 1980; Pusateri *et al.*, 1990). Oestradiol is measurable in large spherical blastocysts, and then increases rapidly during tubular expansion into filamentous forms, before declining precipitously after elongation (Pusateri *et al.*, 1990). On days 11–12 after mating, the progesterone-primed endometrium is very sensitive to conceptus oestradiol, which will alter stromal and epithelial development (Geisert *et al.*, 1991) and secretory activity (Geisert *et al.*, 1982b; Morgan *et al.*, 1987). During the late preimplantation period, blastocysts are especially dependent

Table 2. Comparisons between breeds of ovulation rate, conceptus recovery on days 11 to 12 of gestation and litter size

Parameter	Meishan	(n)	Yorkshire	(n)
Ovulation rate[a]	16.5±0.5	(70)	16.3±0.4	(93)
Number of conceptuses collected[a]	12.5±0.5	(69)	12.3±0.5	(82)
Litter size[b]	12.3±0.4	(42)	8.9±0.4	(66)

[a]Means ± SEM are not different between breeds, $P > 0.30$.
[b]Calculated from another group of gilts of similar reproductive age. Means ± SEM are significantly different, $P < 0.01$.
n = number of litters.

upon endometrial secretion of specific hormones, growth factors and nutrients for normal growth and development (Roberts and Bazer, 1988). Owing to the marked variation among littermates in conceptus size at this time of gestation, the most advanced conceptuses in the litter begin the synthesis of oestradiol before other conceptuses (Stroband and Van der Lende, 1990). The increase in uterine luminal oestradiol results in increased uterine blood flow and vascular permeability, and alters the amount and composition of endometrial secretions to provide a suitable micro-environment for the increasingly complex requirements of the conceptus (Flint, 1981; Ford *et al.*, 1982; Van der Meulen *et al.*, 1989; Stroband and Van der Lende, 1990). The less well-developed conceptuses in the litter are thus exposed to this rapidly changing uterine environment, resulting in their eventual demise (Dziuk, 1987; Pope *et al.*, 1990; Pope, 1994).

The high incidence of preimplantation conceptus loss in US and European pig breeds was first noted by Hammond (1914), and has since been confirmed to average approximately 30% (Dziuk, 1987). The period surrounding implantation is considered especially critical for conceptus survival, and most loss occurs between days 12 and 18 after mating (Scofield, 1972).

Uterine capacity has been defined as the maximum number of fetuses that can be successfully carried to term when the number of potentially viable conceptuses is not limiting (Christenson *et al.*, 1987). Treatments imposed to exceed uterine capacity in pigs (for example superovulation, superinduction) will produce large numbers of viable conceptuses, but only to approximately day 30 of gestation (Dziuk, 1968; Pope *et al.*, 1972). Thereafter, the effect(s) of the limiting uterine capacity is evident, in that these females will give birth to numbers of piglets equivalent to those found in untreated pigs.

Meishan versus Yorkshire conceptuses during the preimplantation period

If a large number of Meishan and Yorkshire gilts in our herds were compared at the same reproductive age (i.e. two to five post-pubertal oestrous cycles), no obvious differences among breeds were observed in the interval from the initiation of oestrus to ovulation or the conceptus stages recovered on day 2 (Table 1). Furthermore, ovulation rate and the number of viable conceptuses recovered on day 12 are almost identical, even though these Meishan gilts farrow approximately three more piglets than do the Yorkshire gilts (Table 2). We have used Meishan and Yorkshire gilts of these equivalent reproductive ages for our studies on the control of litter size. This is critical, since ovulation rate increases progressively to reach 30 to 40 ova shed in older multiparous Meishan sows, while remaining at 14 to 16 in US and European pig breeds (Christenson, 1993).

In our herds, Meishan and Yorkshire conceptuses elongate synchronously on days 11–12 after mating (Fig. 1), a time similar to that reported for other US and European pig breeds (Youngs *et al.*, 1994). A similar range of conceptus development occurs in both breeds throughout this period (Fig. 1). The apparent greater absolute diversity within the Yorkshire litters is accounted for by a similar within-litter variation (standard deviation) in litters with an increased mean conceptus diameter. However, if the within-litter coefficient of variation, which takes into consideration the difference in

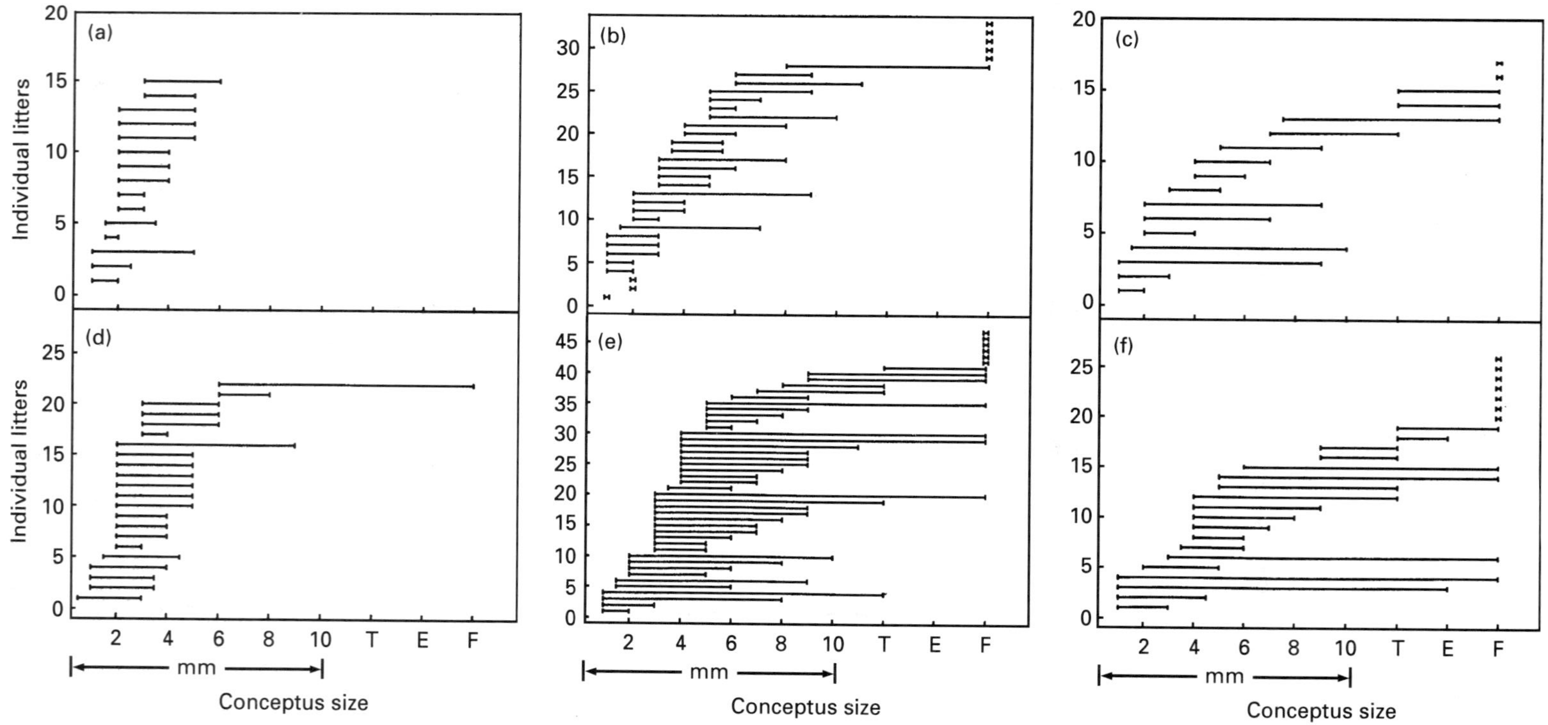

Fig. 1. Range of conceptus development of individual Meishan (a, b, c) and Yorkshire (d, e, f) female pigs from days 11 (a, d), 11.5 (b, e) and 12 (c, f) of gestation (first day of oestrus and mating = day 0). T, tubular; E, elongating; F, filamentous.

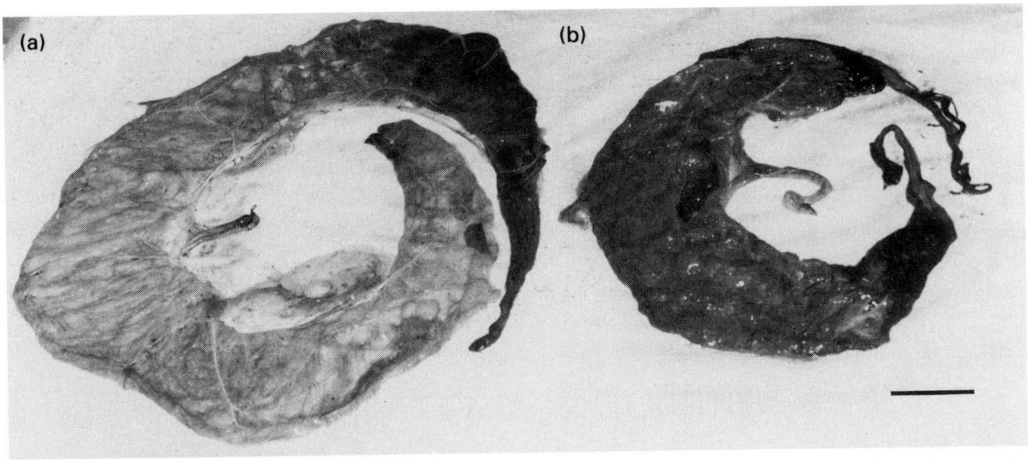

Fig. 2. A large relatively avascular Yorkshire placenta (a) and a small highly vascular Meishan placenta (b) recovered from a single Yorkshire recipient female pig. These placentae were matched to fetuses of very similar birth weight. Scale bar represents 10 cm.

mean conceptus diameter, is calculated, there are breed differences (i.e. differences in Yorkshire litters are greater than Meishan litters). Thus, although the average within-litter diversity is relatively constant across breeds, the absolute diversity in a litter is associated with the degree of conceptus development in that litter. These data are consistent with the published reports that Meishan conceptuses at days 11–12 are smaller when they initiate steroidogenesis (Wilson and Ford, 1997) and elongate (Youngs *et al.*, 1994) than Yorkshire conceptuses. Furthermore, Wilson *et al.* (1995) reported that Meishan filamentous conceptuses were shorter and contained fewer cells than Yorkshire conceptuses on day 14.

The smaller size of the Meishan versus the Yorkshire preimplantation conceptus results from its markedly small numbers of trophectoderm (TE) cells from day 5 (Rivera *et al.*, 1996) to day 12 (Wilson and Ford, 1997) in similar stage conceptuses. Furthermore, the reduction in numbers of TE cells in Meishan versus Yorkshire conceptuses has been shown to result from a lower mitotic rate of these cells in the Meishan conceptus (Wilson and Ford, 1997). Conley *et al.* (1992) have shown that P450 17α-hydroxylase/17–20 lyase is the enzyme most closely associated with pig conceptus oestradiol production, and have localized this enzyme exclusively in the TE (Conley *et al.*, 1994). In association with its reduced population of TE cells on days 11–12, Meishan conceptuses produce and secrete less oestradiol into uterine luminal fluid than do similar stage Yorkshire conceptuses (Wilson and Ford, 1997). Furthermore, these researchers demonstrated that the decreased oestradiol concentrations in uterine luminal fluid of Meishan versus Yorkshire females is associated ($r = 0.57$; $P < 0.001$) with a lower concentration of insulin-like growth factor-I (IGF-I) in the same flushings.

The potential interrelationship between uterine luminal oestradiol and IGF-I may be crucial for the further rapid growth and development of the conceptus as it prepares to undergo elongation. Evidence supporting this hypothesis includes: (1) the presence of receptors for IGF-I on the preimplantation pig conceptus (Corps *et al.*, 1990), (2) the finding that insulin, a hormone known to bind the IGF-I receptor, increases protein synthesis by, and the growth of, the preimplantation pig conceptus (Lewis *et al.*, 1992) and (3) the expression of steroidogenic enzymes important for oestradiol biosynthesis has been correlated with the concentration of IGF-I in uterine luminal fluid (Ko *et al.*, 1994). These and other findings have led Simmen *et al.* (1995) to propose a feed forward mechanism in which the oestradiol produced by the rapidly expanding pig blastocyst stimulates endometrial IGF-I production which, in turn, stimulates conceptus steroidogenic enzymes and further oestradiol synthesis. However, this story is far from complete, as Chastant *et al.* (1994) failed to detect IGF-I receptors on pig conceptuses from day 4 to day 10 of gestation, or on TE cells from the placenta on day 20. These data suggest that the reduced size and oestradiol secretory activity of the

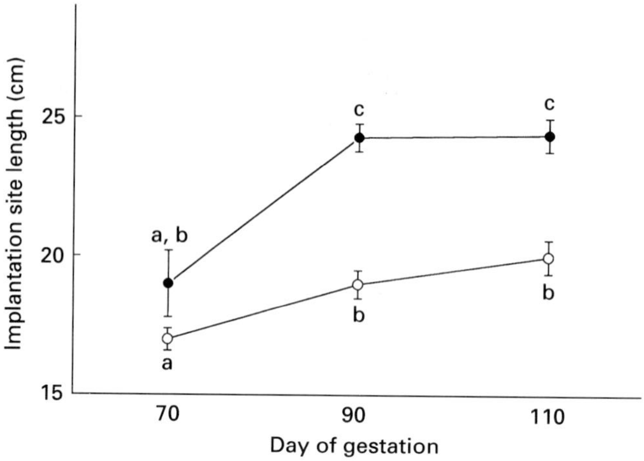

Fig. 3. Average length of the uterine horn occupied by each conceptus on days 70, 90 and 110 of gestation in (○) Meishan and (●) Yorkshire female pigs. [a,b,c] Means ± SEM with different superscripts are significantly different ($P < 0.01$).

Meishan conceptus, when compared with Yorkshire conceptuses throughout blastocyst expansion and elongation, may allow for greater conceptus survival and potential litter size.

Finally, a reciprocal conceptus transfer experiment between Meishan and Yorkshire females was conducted to evaluate the specific influences of uterine environment versus conceptus genotype on preimplantation conceptus growth rate (Youngs *et al.*, 1994). In this study, conceptuses were transferred from Meishan or Yorkshire donors to synchronous recipients of either breed on day 2 and removed on day 12 of gestation. Meishan conceptuses were always smaller than Yorkshire conceptuses at a similar stage on day 12, regardless of uterine environment, but conceptuses of both breeds contained fewer cells and were smaller when recovered from a Meishan uterus than when recovered from a Yorkshire uterus. Furthermore, conceptuses of both genotypes produced markedly less oestradiol when contained in a Meishan uterus than when contained in a Yorkshire uterus. These data suggest that both uterine environmental and conceptus genotype affect preimplantation conceptus growth, and that only uterine environmental differences affect oestradiol secretion. These data are consistent with the concept that the uterine environment (Meishan versus Yorkshire) is the sole determinant of conceptus oestradiol secretory activity.

Latter Half of Gestation

In pigs, it is generally accepted that placentae that are smaller than average size limit fetal growth (Knight *et al.*, 1977). As the demands of the fetus for nutrients and oxygen increase markedly in late gestation, placental insufficiency should become most evident during this time. The pig placenta is diffuse according to the classification of Grosser (1909) and of the epitheliochorial type. Over the mouths of the uterine glands the allantochorionic membrane is not attached to the endometrial epithelium, but forms regular or irregular areolae which first appear on day 30 of gestation (Brambell, 1933). The interareolar placenta is composed of chorionic ridges and their corresponding endometrial folds. Placental efficiency is determined by its surface area of contact with the uterine wall (chorionic villus surface area) and the thickness of the placental membranes. Baur (1977) has shown that the chorionic villus surface area increases progressively throughout gestation in pigs, but lags behind fetal weight. However, there is clear histological evidence of a progressive thinning of the effective placental barrier separating the fetal and maternal bloodstream in pigs (Friess *et al.*, 1980). This thinning, which accelerates during the last third of gestation, results from a marked

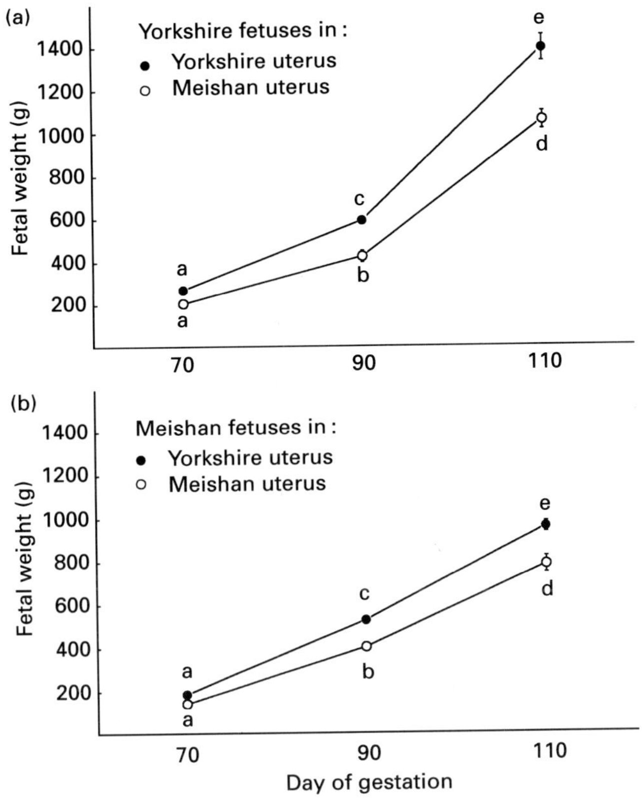

Fig. 4. Weight of (a) Yorkshire and (b) Meishan fetuses in Yorkshire and Meishan uteri on days 70, 90 and 110 of gestation. [a,b,c,d,e] Within a fetal breed, means ± SEM with different superscripts are significantly different ($P < 0.05$).

indentation of fetal vessels into the trophoblast, and the indentation of maternal capillaries into the uterine epithelial cells at the fetal–maternal interface. At the end of pregnancy, the distance between fetal and maternal capillaries is often less than 2 µm.

Meishan versus Yorkshire conceptuses during late gestation

We next investigated the consequences of the observed differences in uterine environment and conceptus genotype on subsequent placental and fetal growth between Meishan and Yorkshire pigs. Rivera *et al.* (1994) transferred 20 conceptuses (ten Meishan plus ten Yorkshire) into the oviducts of synchronous Meishan or Yorkshire recipient females on day 2 after oestrus, and recovered the conceptuses on day 90 of gestation. More ($P < 0.05$) fetuses were recovered from Meishan recipients on day 90 than from Yorkshire recipients (14.8±0.4 versus 10.0±1.9 fetuses, respectively). Placental weight was markedly reduced for conceptuses recovered from Meishan versus Yorkshire recipient females, whereas fetal weight was affected both by recipient breed and by fetal genotype. Meishan and Yorkshire fetuses gestated in Meishan uteri to day 90 were similar in weight, averaging 381±23 g, but were markedly smaller than either Meishan (498±20 g) or Yorkshire (613±16 g) fetuses gestated in Yorkshire uteri. Interestingly, day 90 Yorkshire fetuses were larger than Meishan fetuses when both were gestated in Yorkshire recipient females. These data suggest that the Meishan uterine environment restricts placental growth of both Meishan and Yorkshire conceptuses, ultimately limiting fetal growth by day 90 of gestation. It also appeared that the larger placental sizes of

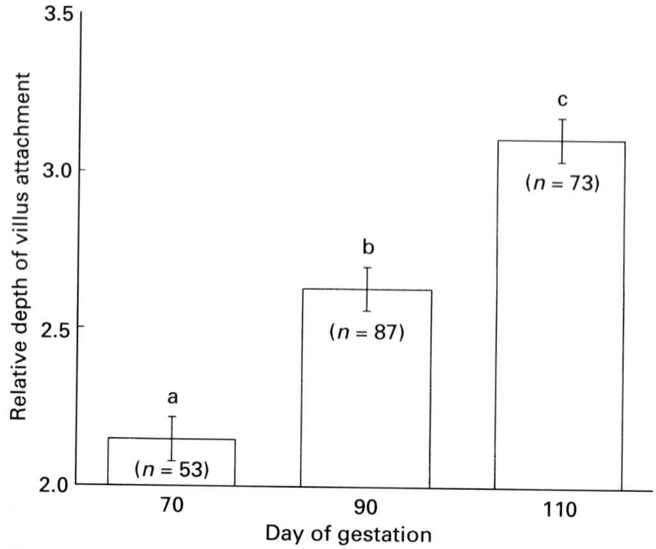

Fig. 5. The relative depth of villus attachment of placentae of Meishan versus Yorkshire pigs on days 70, 90 and 110 of gestation. [a,b,c]Means ± SEM with different superscripts are significantly different ($P < 0.05$).

Meishan and Yorkshire conceptuses in Yorkshire uteri allowed the expression of breed-specific differences in fetal growth.

Owing to this breed-specific (Meishan versus Yorkshire) difference in fetal weight of littermates gestated in Yorkshire recipients, a subsequent experiment was conducted with Yorkshire recipients only. As previously described, both Meishan ($n = 10$) and Yorkshire ($n = 10$) conceptuses were transferred to Yorkshire recipients on day 2, but instead of recovering conceptuses on day 90, the females were allowed to farrow. At farrowing, each piglet was caught as it appeared at the vaginal opening and the exposed umbilical cord was tagged to match the piglet's number, cut and allowed to retract. In this way, each piglet could be matched with the appropriate placenta at expulsion. Unexpectedly, two visually distinct placental types were observed during the farrowing of each recipient Yorkshire female. Relatively large and blanched placentae were exclusively matched to Yorkshire fetuses, while smaller, very reddened (i.e. more vascular) placentae were matched to Meishan fetuses. Two placentae expelled from a single Yorkshire recipient female and matched to a Yorkshire and a Meishan piglet of equal weight are shown in Fig. 2. In these mixed litters of Meishan and Yorkshire fetuses, the piglet weight (g): placental weight (g) ratio was calculated as an estimate of placental efficiency for each conceptus. The piglet weight: placental weight ratio differed markedly ($P < 0.05$) for Meishan conceptuses, which averaged 8.7±0.4 when compared with littermate Yorkshire conceptuses which averaged 3.4±0.8. Interestingly, these breed differences in fetal weight: placental weight ratios are due predominately to differences in placental weight (approximately 66%) rather than to any marked differences in piglet weight (7%).

Meishan and Yorkshire recipient females carrying either Meishan or Yorkshire fetuses were slaughtered on days 70, 90 and 110 of gestation to characterize more fully the specific differences in placental and fetal growth and development between the Meishan and Yorkshire breeds. At slaughter, the length of each uterine horn was determined and fetal measurements were taken (weight and crown–rump length) and a section of each individual placenta and attached endometrium was collected for histological evaluation of the following: (1) number of villi per unit length of placenta (villar number), (2) length of villus attachment per unit length of placenta (villar depth), and (3) percentage of placenta occupied by blood vessels (placental vascular density). Each placenta was then weighed and its length determined as well as the length of its implantation site in

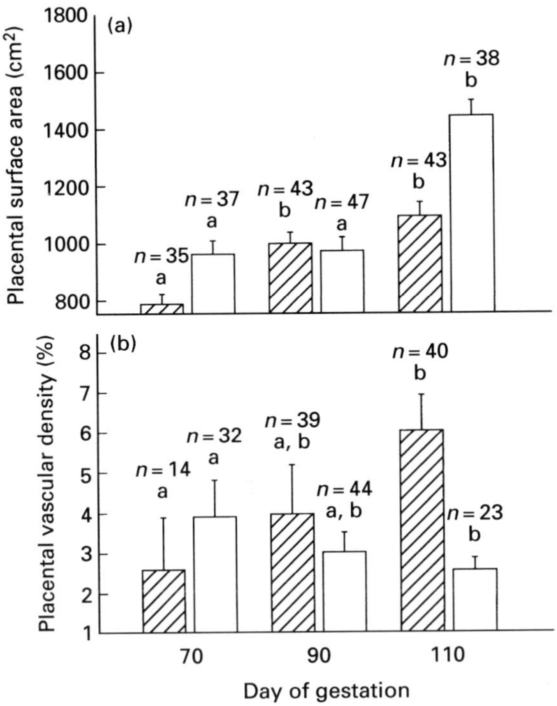

Fig. 6. Surface area (a) and associated blood vessel density (b) of placentae of Meishan (▨) and (□) Yorkshire pigs on days 70, 90 and 110 of gestation. [a,b] Within a breed type, means ± SEM with different superscripts are significantly different ($P < 0.05$).

the uterus. In addition, each placenta was cut along the anti-mesometrial border so that it could be spread out on a flat surface and traced with a planimeter to determine its surface area.

There was no effect of uterine position in which a conceptus resided (i.e. tip, middle or base) or the number of conceptuses in a horn on any fetal or placental measurement. In addition, no effect of breed or day was observed on total uterine length; however, conceptuses in Meishan uteri occupied a smaller ($P < 0.01$) length of uterine horn than conceptuses in Yorkshire uteri (Fig. 3). Furthermore, implantation site length of conceptuses in both Meishan and Yorkshire uteri increased by about 20% from day 70 to day 110 of gestation. The decreased implantation site length of conceptuses in Meishan versus Yorkshire uteri is reflected by the fact that both Yorkshire and Meishan fetuses grew more slowly in a Meishan than in a Yorkshire uterus from day 70 to day 110 (Fig. 4a,b). The number of chorionic villi per unit length of placenta was similar for Meishan and Yorkshire placentae. Thus, the length of villus attachment per unit length placental–endometrial interface reflects the relative depth to which chorionic villi extended into the endometrial folds (Fig. 5). As depicted, villi length increased progressively from day 70 to day 110 of gestation and was similar across uterine and fetal breeds.

However, the pattern of placental growth differed markedly between Meishan and Yorkshire conceptuses. Surface area of Yorkshire placentae remained relatively constant from day 70 to day 90, before increasing ($P < 0.001$) markedly by day 110 (Fig. 6a), while exhibiting a progressive decline in placental vascular density (Fig. 6b). In contrast, the surface area of Meishan placentae increased only moderately from day 70 to day 90, and failed to exhibit any further increase from day 90 to day 110 (Fig. 6a). Throughout this period, and in contrast to the Yorkshire placenta, the Meishan placenta showed a marked increase ($P < 0.01$) in vascular density (Fig. 6b). The fetal weight : placental weight ratio of these conceptuses on day 110 was markedly higher ($P < 0.01$) for Meishan than for Yorkshire conceptuses gestated in Meishan (6.07±0.35 and 4.65±0.20) or Yorkshire (5.08±0.17 and 4.35±0.15) uteri, due predominantly to differences in placental size.

Conclusion

The numbers of blood vessels per unit placental area at the uterine interface will determine in large part the capacity for nutrient exchange between the fetus and dam. As a result of this progressively increasing efficiency of nutrient and oxygen extraction per unit area of placenta, the Meishan conceptus can support the growth and development of a viable fetus with a much smaller placenta than is found in US pig breeds. Since the real limitation to litter size is the available space in the uterus, conceptuses that occupy less uterine space would result in larger litters.

The Meishan pig, therefore, has two separate yet complimentary strategies for increasing litter size. Firstly, the Meishan conceptus develops from fewer cells in the early stages of development than do similar stage conceptuses of US or European pig breeds, resulting in reduced oestradiol production. This reduced oestradiol production by the preimplantation Meishan conceptus results in less marked changes in the composition of uterine histotroph, allowing more conceptuses to survive beyond day 18 of gestation than for US or European pig breeds. Secondly, the Meishan conceptus develops a smaller, more vascular placenta than that found in US or European pig breeds. Furthermore, owing to their smaller size, more Meishan conceptuses are accommodated by the uterus after day 30 of gestation, when uterine size becomes limiting. Therefore, each Meishan placenta requires less endometrial space for nutrient exchange throughout late pregnancy when fetal demands are increasing, leading to the potential for larger litters.

References

Baur R (1977) Morphometry of the placenta exchange area *Advances in Anatomy, Embryology and Cell Biology* **53** 1–65

Bazer FW, Thatcher WW, Martinal-Botte F and Terqui M (1988) Sexual maturation and morphological development of the reproductive tract in Large White and prolific Chinese Meishan pigs *Journal of Reproduction and Fertility* **83** 723–728

Brambell CE (1933) Allantochorionic differentiation of pig *American Journal of Anatomy* **52** 397–459

Chastant S, Monget P and Terqui M (1994) Localization and quantitation of insulin-like growth factor-I (IGF-I) and IGF-II/Mannase-6-Phosphate (IGF-II/M6P) receptors in pig embryos during early pregnancy *Biology of Reproduction* **51** 588–596

Christenson RK (1993) Ovulation rate and embryonic survival in Chinese Meishan and white crossbred pigs *Journal of Animal Science* **71** 3060–3066

Christenson RK, Leymaster KA and Young LD (1987) Justification of unilateral hysterectomy–ovariectomy as a model to evaluate uterine capacity in swine *Journal of Animal Science* **65** 738–744

Conley AJ, Christenson RK, Ford SP, Geisert RD and Mason JI (1992) Steroidogenic enzyme expression in porcine conceptuses during and after elongation *Endocrinology* **131** 896–902

Conley AJ, Christenson LK, Ford SP and Christenson RK (1994) Immunocytochemical localization of cytochrome P450 17α-hydroxylase and aromatase in embryonic cell layers of elongating porcine blastocysts *Endocrinology* **135** 2248–2254

Corps AN, Brigstock DR, Littlewood CJ and Brown KD (1990) Receptors for epidermal growth factor and insulin-like growth factor-I on preimplantation trophectoderm of the pig *Development* **110** 221–227

Dantzer V (1985) Electron microscopy of the initial stages of placentation in the pig *Anatomy and Embryology* **172** 281–293

Dhindsa DS, Dziuk PJ and Norton HW (1967) Time of transuterine migration and distribution of embryos in the pig *Anatomical Record* **159** 325–330

Dziuk PJ (1968) Effect of number of embryos and uterine space on embryo survival in the pig *Journal of Animal Science* **27** 673–676

Dziuk PJ (1985) Effects of migration, distribution and spacing of pig embryos on pregnancy and fetal survival *Journal of Reproduction and Fertility Supplement* **33** 57–63

Dziuk PJ (1987) Embryonic loss in the pig: an enigma. In *Australasia Pig Science Association Committee* pp 28–39 Manipulating Pig Production, Australasia Pig Science Association, Melbourne, Australia

Ferrell CL (1991a) Maternal and fetal influences on uterine and conceptus development in the cow: I. Growth of tissues of the gravid uterus *Journal of Animal Science* **69** 1945–1953

Ferrell CL (1991b) Maternal and fetal influences on uterine and conceptus development in the cow: II. Blood flow and nutrient flux *Journal of Animal Science* **69** 1954–1965

Flint APF (1981) A unifying hypothesis for the control of blastocyst growth based on observations of the pig *Journal of Reproduction and Fertility Supplement* **29** 215–227

Ford SP (1995) Control of blood flow to the gravid uterus of domestic livestock species *Journal of Animal Science* **73** 1852–1860

Ford SP and Youngs CR (1993) Early embryonic development in prolific Meishan pigs *Journal of Reproduction and Fertility Supplement* **48** 271–278

Ford SP, Christenson RK and Ford JJ (1982) Uterine blood flow and uterine arterial, venous and luminal concentrations of oestrogens on Days 11, 13 and 15 after oestrus in pregnant and non-pregnant sows *Journal of Reproduction and Fertility* **64** 185–190

Friess AC, Sinowatz F, Skolek-Winnisch R and Träutner W (1980) The placenta of the pig *Anatomy and Embryology* **158** 179–191

Gadsby JE, Heap RB and Burton RD (1980) Oestrogen production by blastocysts and early embryonic tissue of various species *Journal of Reproduction and Fertility* **60** 409–417

Galvin JM, Wilmut I, Day BN, Ritchie M, Thomsen M and Haley CS (1993) Reproductive performance in relation to uterine and embryonic traits during early gestation in Meishan, Large White and crossbred sows *Journal of Reproduction and Fertility* **98** 277–384

Geisert RD, Brookbank JW, Roberts RM and Bazer FW (1982a) Establishment of pregnancy in the pig: II. Cellular remodeling of the porcine blastocyst during elongation on day 12 of pregnancy *Biology of Reproduction* **27** 941–955

Geisert RD, Renegar RH, Thatcher WW, Roberts RM and Bazer FW (1982b) Establishment of pregnancy in the pig: I. Interrelationship between preimplantation development of the pig blastocyst and uterine endometrial secretions *Biology of Reproduction* **27** 925–939

Geisert RD, Morgan GL, Zavy MT, Blair RM, Gries LK, Cox A and Yellin T (1991) Effect of asynchronous transfer and oestrogen administration on survival and development of porcine embryos *Journal of Reproduction and Fertility* **93** 475–481

Grosser O (1909) *Vergleichende Anatomie und Entwicklungsgeschichte der Eihäute und der Placenta* Braumüller, Leipzig

Haley CS and Lee GJ (1993) Genetic basis of prolificacy in Meishan pigs *Journal of Reproduction and Fertility Supplement* **48** 247–259

Hammond J (1914) On some factors controlling fertility in domestic animals *Journal of Agricultural Science* **6** 263–274

Hunter GL (1958) The maternal influence on size in sheep *Journal of Agricultural Science* (Cambridge) **48** 36–60

Jin RB, Cui HM and Mao JD (1992) Reproductive characteristics of Taihu pigs *Pig News and Information* **13** 99N–102N

Knight JW, Bazer FW, Thatcher WW, Franke DE and Wallace HD (1977) Conceptus development in intact and unilaterally hysterectomized ovariectomized gilts: interrelations among hormonal states, placental development, fetal fluids and fetal growth *Journal of Animal Science* **44** 620–637

Ko Y, Choi I, Green ML, Simmen FA and Simmen RCM (1994) Transient expression of the cytochrome P450 aromatase gene in elongating porcine blastocysts is correlated with uterine insulin-like growth factor levels during peri-implantation development *Molecular Reproduction and Development* **37** 1–11

Lee GJ, Ritchie M, Thomson M, Macdonald AA, Blasco A, Santacreu MA, Argente MJ and Haley CS (1995) Uterine capacity and prenatal survival in Meishan and Large White pigs *Animal Science* **60** 471–479

Lewis AM, Kaye PL, Lising R and Cameron RDA (1992) Stimulation of protein synthesis and expansion of pig blastocysts by insulin *in vitro*. *Reproduction Fertility and Development* **4** 119–123

Morgan GL, Geisert RD, Zavy MT, Shawley RV and Fazleabas AT (1987) Development of pig blastocysts in a uterine environment advanced by exogenous oestrogen *Journal of Reproduction and Fertility* **80** 125–131

Paria BC and Dey SK (1990) Preimplantation embryo development *in vitro*: cooperative interactions among embryos and role of growth factors *Proceedings of the National Academy of Sciences USA* **87** 4756–4760

Pollard JW (1990) Regulation of polypeptide growth factor synthesis and growth-related gene expression in the rat and mouse uterus before and after implantation *Journal of Reproduction and Fertility* **88** 721–731

Pope WF (1994) Embryonic mortality in swine. In *Embryonic Mortality in Domestic Animals* pp 53–77 Eds MT Zavy and RD Geisert. CRC Press Inc, Boca Raton

Pope CE, Christenson RK, Zimmerman-Pope VA and Day BN (1972) Effect of number of embryos on embryonic survival in recipient gilts *Journal of Animal Science* **35** 805–808

Pope WF, Xie S, Broermann DM and Nephew KP (1990) Causes and consequences of early embryonic diversity in pigs *Journal of Reproduction and Fertility Supplement* **40** 251–260

Pork Facts (1993) *National Pork Producers Council* p 6 Ed. C Miller. National Pork Producers Association Des Moines, IA

Pusateri AE, Rothschild MF, Warner CM and Ford SP (1990) Changes in morphology, cell number, cell size and cellular estrogen content of individual littermate pig conceptuses on days 9 to 13 of gestation *Journal of Animal Science* **68** 3727–3735

Reynolds LP and Redmer DA (1995) Utero–placental development and placental function *Journal of Animal Science* **73** 1839–1851

Reynolds LP, Ford SP and Ferrell CL (1985) Blood flow and steroid and nutrient uptake of the gravid uterus and fetus of sows *Journal of Animal Science* **61** 968–974

Rivera RM, Christenson LK, Youngs CR and Ford SP (1994) Competitive survival growth rate and estrogen secretory activity of Meishan and Yorkshire fetuses *Journal of Animal Science* **72** (Supplement 1) 78 (Abstract)

Rivera RM, Youngs CR and Ford SP (1996) A comparison of the number of inner cell mass and trophectoderm cells of preimplantation Meishan and Yorkshire pig embryos at similar developmental stages *Journal of Reproduction and Fertility* **106** 111–116

Roberts RM and Bazer FW (1988) The functions of uterine secretions *Journal of Reproduction and Fertility* **82** 875–892

Scofield AM (1972) Embryonic mortality. In *Pig Production* pp 367–383 Ed. CJA Cole. Butterworths, London

Simmen FA and Simmen RCM (1991) Peptide growth factors and proto-oncogenes in mammalian conceptus development *Biology of Reproduction* **44** 1–5

Simmen RCM, Green ML and Simmen FA (1995) IGF system in peri-implantation uterus and embryonic development. In *Molecular and Cellular Aspects of Peri-implantation Processes* pp 185–204 Ed. SK Dey. Serono Symposia USA, Springer-Verlag, New York

Stroband HWJ and Van der Lende T (1990) Embryonic and uterine development during early pregnancy in pigs *Journal of Reproduction and Fertility Supplement* **40** 261–277

Van der Meulen J, te Kronnie G, van Deursen R and Greelan J (1989) Aromatase activity in individual day-11 pig blastocysts *Journal of Reproduction and Fertility* **87** 783–788

Venge O (1950) The maternal influence on birth weight in rabbits *Acta Zoologica* **31** 1–147

Warwick BL (1928) Prenatal growth of swine *Journal of Morphology and Physiology* **46** 59–84

Wilson ME and Ford SP (1997) Differences in trophectoderm mitotic rate and P450 17α-hydroxylase expression between late preimplantation Meishan and Yorkshire conceptuses *Biology of Reproduction* **56** 380–385

Wilson ME, Kaminski MA, Conley AJ and Ford SP (1995) Development differences between Meishan and Yorkshire preimplantation conceptuses *Biology of Reproduction* **52** (Supplement 1) 178 (Abstract)

Youngs CR, Christenson LK and Ford SP (1994) Investigations into the control of litter size in swine: III. A reciprocal embryo transfer study of early conceptus development *Journal of Animal Science* **72** 725–731

Poster titles

Laforest JP, Arsenault G and GP Martineau
Dead semen affects the *in vitro* production of prostaglandin E_2 by the uterine leucocytes of the sow

Hazeleger W, Maathuis C, Woudstra A and B Kemp
Effect of insemination of recipients with seminal antigens on pregnancy rate after non-surgical embryo transfer in sows

Flechon JE, Degrouard J, Delasalle S and B Flechon
The embryonic disc of the pig blastocyst

Boerjan ML, Tieman M and JJ Taverne-Thiele
Hypoblast differentiation in blastocysts of the pig

Wianny F, Perreau C and MT Hochereau-de Reviers
Proliferation of inner cell mass or epiblast after *in vitro* culture

Persson E, Sahlin L, Tasende C, Masironi B, Eriksson H and H Rodriguez-Martinez
IgF-I, oestrogen receptors and glandular morphology in the uterus of early pregnant gilts

Vallet JL
Purification and characterization of porcine folate binding proteins from uterine flushings and allantoic fluid

Christenson RK, Trout WE, Vallet JL and HG Klemcke
The influence of breed on uterine protein secretion in white crossbred and Meishan gilts

Iyeghe-Erakpotobor GT, Ashworth CJ, Beattie L, Burke MD and AR Pickard
Hepatic progesterone metabolism in Meishan and Large White × Landrace gilts during early pregnancy

Rens BTTM van, Lende T van der, Groot PN de and HAM van der Steen
The estrogen receptor gene and components of litter size in Meishan synthetics

Laforest JP, Narcisse W, Fortin M, Matte JJ and R Lambert
Immunosuppressive activity of the porcine allantoic fluid in early gestation

Laforest JP, Giguère A, Girard CL, Lambert R and JJ Matte
Prostaglandin E_2 might be involved in the development and survival of the porcine embryo in early gestation

Père MC, Dourmad JY and M Etienne
Variation of uterine blood flow in the sow during pregnancy

Yang L-G, Page KR and CJ Ashworth
Nutrient transport across the porcine placenta *in vitro* and its relation to fetal growth

Sterle JA, Cantley TC, Carroll JA, Matteri RL, Lucy MC and WR Lamberson
Effect of recombinant porcine somatotropin (rpST) on fetal and placental growth in gilts with reduced uterine capacity

Leenhouwers, JI, Lende T van der and EF Knol
Stillbirth and preweaning survival in different genetic lines of pig

Thorup F
Stillborn piglets

BOAR FERTILITY AND SEMEN PRESERVATION

Chair
L. A. Johnson

Journal of Reproduction and Fertility Supplement **52**, 181–193

Physiology of the Meishan boar

D. D. Lunstra[1], J. J. Ford[1], J. Klindt[2] and T. H. Wise[1]

USDA, ARS, [1]Reproduction Research Unit and [2]Nutrition Research Unit, R. L. Hruska US Meat Animal Research Center, Clay Center, Nebraska 68933-0166, USA

Onset of puberty (sperm production) occurs at a much younger age (56–84 days) in Meishans than in conventional boars (120–180 days). Throughout postnatal development, Meishans exhibit markedly higher (two- to ten-fold) concentrations of serum FSH, LH and androgens compared with conventional boars, and these high hormone concentrations are maintained at maturity. Increased gonadotrophin concentrations occur only in Meishan males, since concentrations in female Meishans and conventional females do not differ. In the Meishan boar, FSH increases steadily during establishment of sperm production and LH increases markedly after tubule diameter stabilizes, while FSH and LH concentrations are low and change little during this period of rapid testicular growth in conventional boars. The proportion of the testis occupied by Leydig cells is two-fold greater and average Leydig cell size is two- to four-fold larger in Meishans than in conventional boars during the onset of spermatogenesis and into maturity. Testis size in Meishans is only half that of conventional boars at maturity, and the number of Sertoli cells is markedly reduced in Meishans. However, Meishan testes exhibit reduced degeneration of germ cells and spermatids during spermatogenesis, and spermatogenic efficiency per Sertoli cell in Meishans is double that in conventional boars. At maturity, Meishan pituitaries have greater mass, larger gonadotrophs, higher content of FSH, and greater expression of genes for FSH and LH subunits than found in conventional boars. The high concentrations of serum gonadotrophins in the mature Meishan exhibit a typical postcastration rise, respond to GnRH stimulation and are suppressed by gonadal steroids, indicating that the hypothalamo–pituitary–gonadal axis functions normally but with much higher set points in Meishan boars. These unusual endocrine and physiological attributes in boars of the highly prolific Meishan breed offer a unique model for further investigation of factors influencing early onset of puberty and efficiency of male reproductive function.

Introduction

There are over 75 breeds of pig in China today; some of these have been selected for over 7000 years (Li and Enfield, 1989). Hierarchical classification of Chinese breeds based on geographic distribution, history, body size, physical appearance and reproductive characteristics identified one group, known as Taihu pigs, which were characterized by high rates of reproduction and average size (Li and Enfield, 1989). Taihu pigs originate from the Lake Taihu region west of Shanghai in southern China. Meishans are typical of this group; they exhibit early sexual maturity and high prolificacy (Cheng, 1983; Wang, 1990). The importation of Meishan pigs into France (1979), the United Kingdom (1987), United States (1989) and other countries allowed research that documented greater prolificacy in Meishans than in conventional pigs of these countries, averaging three to five more pigs per litter (Haley *et al.*, 1995). Female Meishans reach puberty at very young ages (85–118 days of age), at least 3–4 months earlier than pubertal age (200–235 days of age) in conventional pigs (Wang, 1990; Christenson, 1993; Hunter *et al.*, 1993; White *et al.*, 1993). In this review, conventional breeds refers to typical pig breeds of US and European origin. Meishan boars, compared with conventional breeds of pig, offer a unique model for investigation of factors influencing male reproductive function that relate to high prolificacy, but only limited data on pubertal development and mature reproductive function have been reported for these boars. In addition, no study has characterized relationships between testicular structure and endocrine function in Meishan boars. The purpose of this review is to summarize results on pubertal development, endocrinology, and mature reproductive function in Meishan boars and to compare these with observations in conventional breeds of boar to provide insights concerning the primary mechanisms controlling reproductive function in male pigs.

Testicular Development and Onset of Sperm Production in Meishan Boars

During pubertal development in Meishan boars, body weight (BWT), paired testis weight (PTW) and epididymal weights increase continuously from birth to 16 weeks of age (Fig. 1; Harayama *et al.*, 1991; Lunstra *et al.*, 1993). The testis exhibits a typical rapid increase in mass associated with pubertal development beyond 6 weeks of age (Fig. 1). Elongated spermatids first appear within seminiferous tubules at 6 weeks of age, and homogenization-resistant spermatid nuclei consistently occur in the testes by 8 weeks of age (BWT = 10.3 kg, PTW = 30.9 g), which is slightly earlier than the appearance of elongated spermatids at 60–75 days of age reported for Meishan boars in Japan (Harayama *et al.*, 1991). Percentage of tubules exhibiting elongated spermatids increases steadily and > 80% have elongated spermatids by 12 weeks of age (BWT = 19.2 kg, PTW = 63.3 g). Lumina are first observed in seminiferous tubules at 4 weeks of age; the majority of tubules exhibit lumina by 6 weeks; and spermatozoa are present in the lumina of seminiferous tubules at 8 weeks. Thus, onset of spermatogenesis (spermatozoa formed and released into the lumen of the seminiferous tubules) in the Meishan occurs at 8 weeks of age (Fig. 1). Spermatozoa first appear in the cauda epididymis at 8 weeks of age; first motile spermatozoa (about 20% motile) are present by 10 weeks of age, and appreciable sperm motility (> 70% motile) is attained by 14 weeks and beyond, which is in agreement with results of other workers (Cheng, 1983; Wang, 1990; Harayama and Kato, 1992).

Onset of sperm production (puberty) in Meishan boars occurs much earlier (56–84 days) and at lower testis weight (30–60 g PTW) compared with the 120–180 days of age and 160–240 g PTW at puberty reported for various breeds of conventional boar (FlorCruz and Lapwood, 1978; Allrich *et al.*, 1982, 1983; Colenbrander *et al.*, 1982; Ford, 1983; Young *et al.*, 1986; Lee *et al.*, 1987; Kosco *et al.*, 1987; Trudeau *et al.*, 1992). Data from Meishans extend the range of differences among breeds within a species for body weight and testis weight at given ages during pubertal development (Young *et al.*, 1986). In addition, puberty occurs earlier in males than in females, even within the Meishan, since puberty in Meishan boars occurs 30–60 days earlier than in Meishan females.

Changes in Testicular Structure During Pubertal Development in Meishan Boars

During postnatal testicular development in Meishan boars, the patterns of change in diameter of seminiferous tubules and in volume percentages of the testis occupied by tubules, interstitial tissue and Leydig cells (Fig. 1) are very similar to patterns exhibited by conventional boars but are more compressed in time. The diameter of seminiferous tubules in the Meishan is low at birth (about 70 μm), increases rapidly and linearly through first sperm production (about 200 μm; 8 weeks), and reaches a plateau by 12 weeks of age and thereafter at 250–290 μm (Fig. 1 and Table 1; Harayama *et al.*, 1991). Tubule diameter in conventional boars is also small at birth (about 60 μm), increases peripubertally and reaches a plateau at approximately 24 weeks of age and beyond (FlorCruz and Lapwood, 1978; van Straaten and Wensing, 1977; Allrich *et al.*, 1983). Maximal tubule diameter tends to be larger in Meishan (250–290 μm) than in conventional boars (190–257 μm; Table 1; FlorCruz and Lapwood, 1978; van Straaten and Wensing, 1977; Allrich *et al.*, 1983; Lee *et al.*, 1987). Appearance of lumina in seminiferous tubules indicates establishment of the blood–testis barrier; thus this barrier is established at a markedly younger age in Meishan (about 40 days of age) than in conventional boars

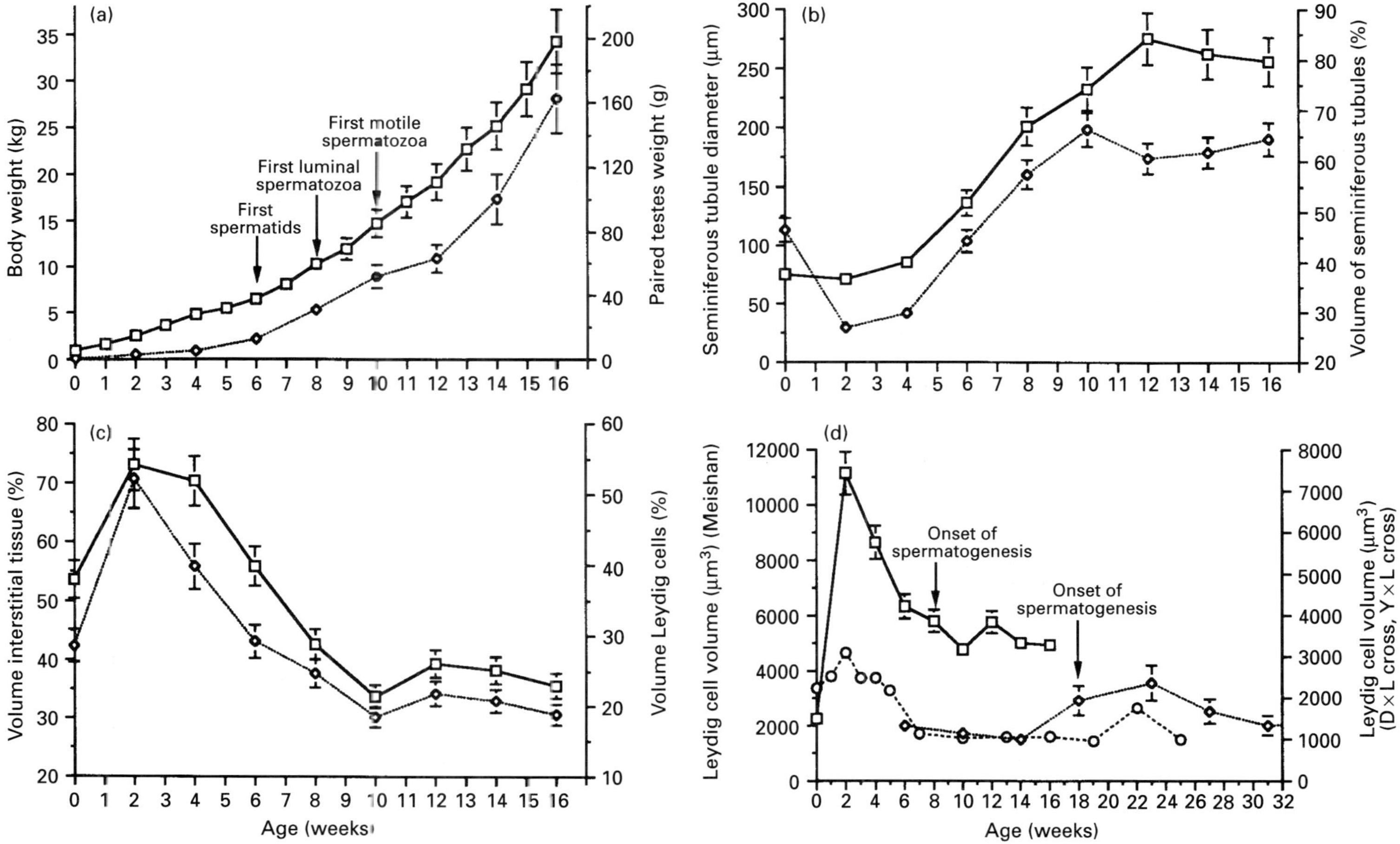

Fig. 1. Patterns of change in (a) body weight (□) and paired testes weight (◊); (b) seminiferous tubule diameter (□) and volume (%) seminiferous tubules (◊); (c) volume (%) interstitial tissue (□) and volume (%) Leydig cells (◊) in Meishan boars from birth through puberty. An average of seven Meishan boars per age point were castrated and evaluated. (d) A comparison of changes in Leydig cell size during postnatal development in Meishan (□) and conventional boars (◊: Allrich *et al.*, 1983; ○: van Straaten and Wensing, 1978) is shown.

Table 1. Differences between mature Meishan and conventional boars in size of testes, sperm production and various aspects of testicular structure[a]

	Meishan (n=5)	Composite[b] (n=5)
Age (days)	225 ± 3	346 ± 2
Body weight (kg)	78 ± 2	133 ± 6
Paired testes weight (g)	269 ± 19	541 ± 35
Volume (%) seminiferous tubules	48.6 ± 1.7	68.7 ± 1.6
Volume (%) interstitial tissue	51.4 ± 1.9	31.3 ± 0.8
Volume (%) Leydig cells	18.5 ± 0.6	7.1 ± 0.2
Volume (%) vascular	10.0 ± 0.9	6.5 ± 0.8
Seminiferous tubules, diameter (μm)	287 ± 15	257 ± 8
Daily sperm production/g^{-1} testis[c]	24.47 ± 1.26	24.77 ± 1.43
Total daily sperm production[d]	6.04 ± 0.52	12.46 ± 1.43
Total no. Sertoli cells[d]	3.87 ± 0.59	14.45 ± 1.33
DSP per Sertoli cell	1.56 ± 0.32	0.86 ± 0.20
Total no. Leydig cells[d]	15.27 ± 0.66	34.79 ± 1.76
Leydig cell size ($μm^3$)	2855 ± 124	960 ± 49

[a]Calculated from data of Okwun *et al.* (1996a,b). All data shown are presented as means ± SEM. Volume (%): volume percentage of testicular parenchyma occupied by respective structure or cell type.
[b]Composite boars: crossbred boars (1/4 Chester White, 1/4 Landrace, 1/4 Large White, 1/4 Yorkshire). Each mean ± SEM for Meishan differs ($P < 0.10$ to $P < 0.01$) from that for Composite boars, except ($P > 0.20$) for average tubule diameter and daily sperm production g^{-1} testis.
[c]Daily sperm production (DSP) × 10^6 per gram of testicular parenchyma.
[d]Total daily sperm production (TDSP) × 10^9 per paired testes. Total number of Sertoli cells × 10^9 per paired testes. Total number of Leydig cells × 10^9 per paired testes.

(90–120 days; FlorCruz and Lapwood, 1978; Allrich *et al.*, 1983; Colenbrander *et al.*, 1982). Rapid proliferation of Sertoli cells occurs primarily during the first month after birth in conventional boars (Tran *et al.*, 1981; Kosco *et al.*, 1987), and establishment of the blood–testis barrier indicates cessation of proliferation of Sertoli cells thereby defining the total population of Sertoli cells and maximum mature testis size (Russell *et al.*, 1990; Chubb, 1992). The markedly earlier onset of puberty in the Meishan requires curtailment of Sertoli cell proliferation sooner in the early postnatal period. Indeed, at sexual maturity, Meishan boars have substantially smaller paired testis weight and fewer Sertoli cells than do conventional boars (Table 1).

In Meishan (Fig. 1) and conventional boars, the percentage of the testis occupied by total interstitial tissue and by Leydig cells exhibits a major early postnatal peak at 1–4 weeks of age followed by a rapid decline through onset of puberty and maintenance at reduced values after puberty (van Straaten and Wensing, 1977, 1978; FlorCruz and Lapwood, 1978; Colenbrander *et al.*, 1982; Allrich *et al.*, 1983). While similar in pattern of change, the absolute values for volume percentage of the testis occupied by interstitial tissue (35–50%) and by Leydig cells (15–25%) in Meishan boars beyond 10 weeks of age and into maturity remains substantially greater than the values reported for conventional boars beyond 20 weeks of age (15–30% interstitial, 7–10% Leydig, respectively). Thus, during the latter part of pubertal development and into maturity (Table 1), Meishan boars consistently exhibit a 1.5- to 2.5-fold higher proportion of the testes occupied by interstitium and Leydig cells than do conventional boars.

Average volume per Leydig cell, in both Meishan (Fig. 1 and Table 1) and conventional boars, exhibits a major early postnatal peak at 1–4 weeks of age, followed by a reduction in size, a transient increase during onset of spermatogenesis, and a slow decline in size that is maintained into maturity (van Straaten and Wensing, 1977, 1978; FlorCruz and Lapwood, 1978; Colenbrander *et al.*, 1982; Allrich *et al.*, 1983). Despite the similarity in pattern, absolute values for Leydig cell size in the

Meishan are markedly larger than in other breeds (Fig. 1 and Table 1), and average two- to four-fold larger at every age examined than reported values for Leydig cell size in conventional boars (van Straaten and Wensing, 1978; Peyrat *et al.*, 1981; Allrich *et al.*, 1983; Lunstra *et al.*, 1986). Concomitant with the increase in cell size, increases in average volume of the nucleus, number of mitochondria and amount of endoplasmic reticulum occur within Leydig cells of Meishan testis (Lunstra, unpublished data), and increases in these intracellular components generally reflect increased steroid-synthesizing capacity (Zirkin *et al.*, 1980; Lunstra *et al.*, 1986). Leydig cell size in conventional boars declines and remains small (smaller than at birth) between 6 and 16 weeks of age before increasing transiently coincident with increasing spermatogenesis (18–26 weeks of age). Meishans also exhibit a less noticeable increase coinciding with increasing spermatogenesis (12 weeks of age) and both the percentage interstitial tissue and Leydig cells exhibit small but consistent increases at 12 and 14 weeks of age (Fig. 1). Thus, the early postnatal and pubertal peaks occur so close together in the Meishan, and overall Leydig cell size is so much larger, that a pronounced regression in prepubertal Leydig cell volume does not occur, and the rebound in Leydig cell size at puberty is relatively small (Fig. 1).

Endocrinology During Pubertal Development in Meishan Boars

In Meishan boars, serum FSH, LH, testosterone and dehydroepiandrosterone sulfate (DHEAS) concentrations all change markedly and exhibit distinct patterns during pubertal development (Fig. 2). One striking characteristic in Meishans is that serum concentrations of all four hormones (FSH, LH, testosterone and DHEAS) are each substantially higher (two- to ten-fold) throughout postnatal development, both before and after puberty, than in conventional boars, as assayed in our laboratory (Table 2, and unpublished data) or reported by others for various breeds of conventional boar (Tan and Raeside, 1980; Allrich *et al.*, 1982, 1983; Colenbrander *et al.*, 1982; Ford, 1983; Schinckel *et al.*, 1984; Kosco *et al.*, 1987; Trudeau *et al.*, 1992).

A major peak in serum concentrations of FSH (1–2 weeks of age), LH (3–4 weeks of age) and androgens (3–4 weeks of age) occurs early in the postnatal period in Meishan boars (Fig. 2). Although the early postnatal peak in testosterone appears to be quite subtle (Fig. 2), the increase in serum testosterone concentrations at 3–4 weeks of age is significantly higher than that during the preceding (2 week) and following (5 week) time frames. Early postnatal peaks in serum hormones also occur at comparable ages or even slightly earlier in conventional boars (Colenbrander *et al.*, 1982; Kosco *et al.*, 1987; Trudeau *et al.*, 1992; Ford *et al.*, 1997), although absolute values are lower (Fig. 2; note smaller scale of *y*-axis for conventional boars). After the early postnatal peak, serum concentrations of gonadotrophins and androgens decline at least until the onset of puberty in both Meishan (8 weeks) and conventional boars (16 weeks). However, with onset of spermatogenesis in Meishans, serum FSH, testosterone and DHEAS all increase steadily and markedly during the period of increasing testis mass (Figs 1 and 2), and very high concentrations of these hormones are maintained in adulthood (Table 2). Serum concentrations of LH in Meishans also increase markedly, but the increase is delayed until maximum tubule diameter is achieved (Figs 1 and 2). In contrast, after the early postnatal peak in conventional boars, serum FSH remains relatively low through puberty and on through sexual maturity, although a small increase in FSH immediately preceding or coinciding with the onset of sperm production has been reported in conventional boars (Colenbrander *et al.*, 1982; Schinckel *et al.*, 1984; Kosco *et al.*, 1987; Trudeau *et al.*, 1992; Mariscal *et al.*, 1996), and serum concentrations of LH remain low into adulthood with no significant peripubertal changes (Allrich *et al.*, 1982; Colenbrander *et al.*, 1982; Schinckel *et al.*, 1984; Kosco *et al.*, 1987). Despite the large differences among breed in serum concentrations of hormones, it is surprising that serum LH declines before and changes relatively little during the onset of spermatogenesis in both Meishan and conventional boars (Fig. 2).

The highest peripubertal concentrations of FSH, LH, testosterone and DHEAS all occur postpubertally at 15–16 weeks of age in Meishan boars. Although serum LH concentrations decline somewhat between 16 weeks of age and maturity in Meishan boars (compare Fig. 2 and Table 2), the

Table 2. Differences between mature Meishan and conventional boars in body weight, testes size and circulating plasma hormone concentrations

Characteristic	Meishan (Borg *et al.*, 1993)	Duroc[a] (Borg *et al.*, 1993)	Meishan (Borg *et al.*, 1993)	Duroc[a] (Borg *et al.*, 1993)	Meishan (Wise *et al.*, 1996)	Composite[b] (Wise *et al.*, 1996)	Meishan (Ford *et al.*, 1997)	Composite[b] (Ford *et al.*, 1997)
Number of boars	9	8	9	8	8	7	22	22
Age (days)	533 ± 5	453 ± 11	706 ± 5	627 ± 11	682 ± 55	426 ± 5	509 ± 44	440 ± 41
Body weight (kg)[c]	140 ± 5	184 ± 4	164 ± 7	197 ± 6	147 ± 4	156 ± 3	115 ± 5	147 ± 6
Paired testes weight (g)[c]	417 ± 37[d]	701 ± 64[d]	408 ± 31[d]	969 ± 31[d]	394 ± 53	509 ± 31	325 ± 24	547 ± 25
Plasma FSH (ng ml^{-1})[c]	634 ± 79	65 ± 11	656 ± 73	91 ± 11	554 ± 50	102 ± 8	704 ± 48	136 ± 48
Plasma LH (ng ml^{-1})[c]	nd	nd	nd	nd	1.3 ± 0.1	0.2 ± 0.1	nd	nd
Plasma testosterone (ng ml^{-1})[c]	15.1 ± 5.0	6.9 ± 1.7	20.8 ± 6.9	7.5 ± 1.7	14.5 ± 0.6	2.6 ± 0.1	16.8 ± 2.6	3.8 ± 1.3
Plasma inhibin (fmol ml^{-1})[c]	97 ± 9	147 ± 18	117 ± 17	269 ± 16	131 ± 13	156 ± 13	nd	nd

[a]Purebred Duroc boars.

[b]Composite boars: crossbred boars (1/4 Chester White, 1/4 Landrace, 1/4 Large White, 1/4 Yorkshire).

[c]Within each study, mean for Meishan boars differs significantly ($P < 0.05$ to $P < 0.01$) from that for conventional boars, except ($P > 0.10$) for body weight (kg) and plasma concentrations of inhibin reported by Wise *et al.* (1996).

[d]Paired testis volume (cm^3) calculated from caliper measurements (equivalent to weight, assuming 1.0 cm^3 = 1.0 g).

nd: not determined. All data shown are presented as means ± SEM. Data from Borg *et al.* (1993); Wise *et al.* (1996); and Ford *et al.* (1997) .

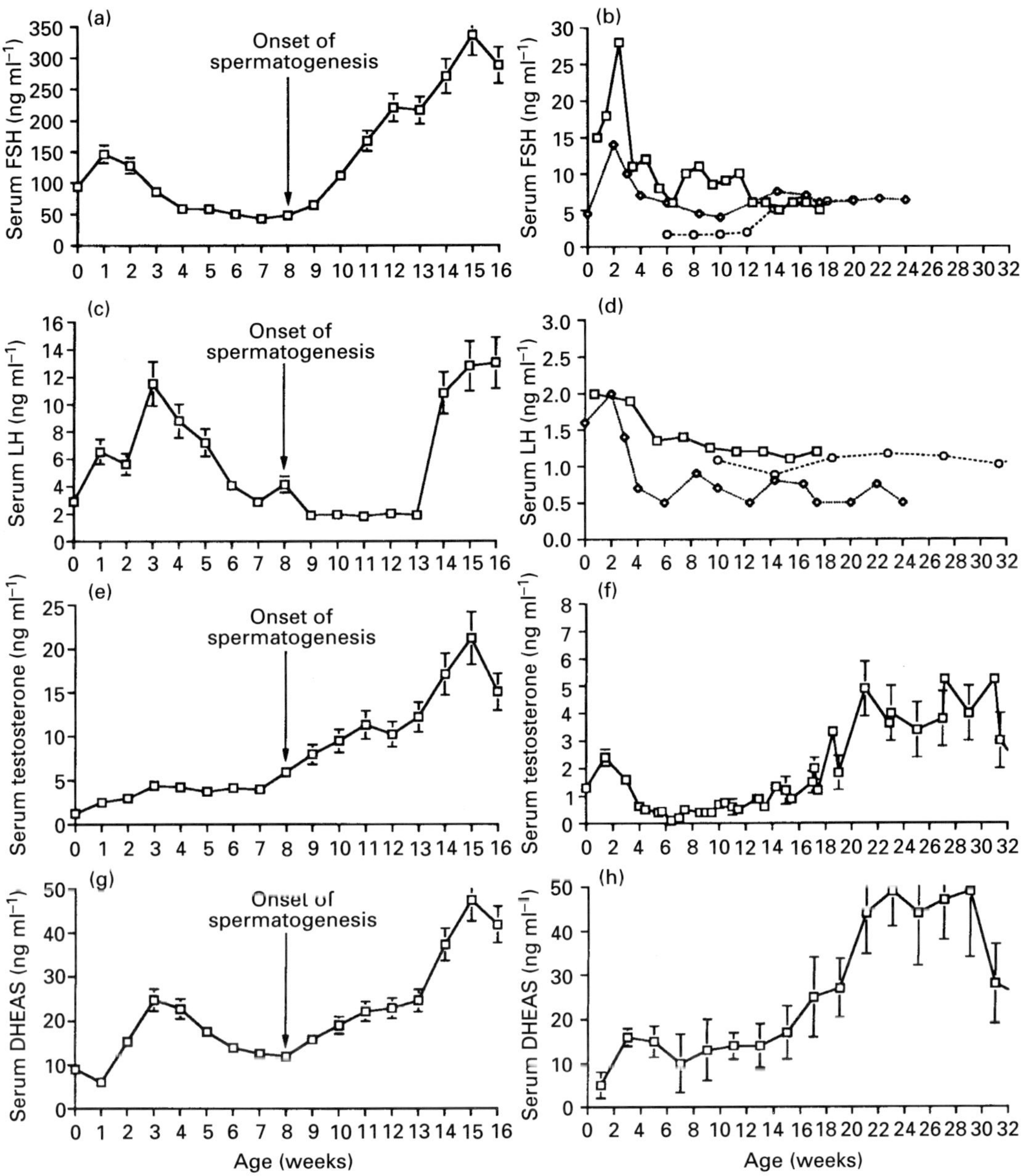

Fig. 2. Patterns of change in (a,b) FSH, (c,d) LH, (e,f) testosterone and (g,h) dehydroepiandrosterone sulfate (DHEAS) concentrations in (a,c,e,g,) Meishan and (b,d,f,h,) conventional boars from birth through puberty. (b) Data from (□) Kosco *et al.*, 1987; (◊) Colenbrander *et al.*, 1982; and (○) Schinckel *et al.*, 1984; (d) data from (□) Kosco *et al.*, 1987; (◊) Colenbrander *et al.*, 1982 and (○) Allrich *et al.*, 1982; (f) data combined from Tan and Raeside (1980), Allrich *et al.* (1982), Ford (1983) and Kosco *et al.* (1987); (h) data from Tan and Raeside (1980).

higher serum concentrations of FSH, LH, and testosterone present at 16 weeks of age are maintained at sexual maturity (Table 2). One possible explanation is that the sensitivity of the hypothalamus to steroid feedback in the Meishan boar may be greatly reduced during this time frame (14–16 weeks of age), but the reduction must be temporary, since feedback sensitivity seems to be fully functional at sexual maturity (see discussion on sexual maturity below). Regardless of the mechanism, Meishan boars continue to exhibit greater concentrations of serum FSH, LH, and androgens than do conventional boars at sexual maturity (Table 2).

In Meishan and conventional boars, the major early postnatal peaks in serum concentrations of FSH, LH and androgens coincide well with the timing of early postnatal peaks observed in volume percentages of the testis occupied by interstitial tissue and Leydig cells, and with the marked early peak in Leydig cell size (compare Fig. 1 and Fig. 2). Although Meishans achieve puberty at a much younger age than do conventional boars, the coincidence between structural and hormonal peaks indicate that there is a strong interdependence and cause-and-effect relationship amongst these characteristics, even in these two very different types of pig. Presumably, the high early postnatal concentrations of gonadotrophins stimulate the marked increase in Leydig cell size, and the increased steroidogenic machinery in the enlarged Leydig cells synthesize and secrete enough steroids to produce the early postnatal peak in serum androgens.

In conventional boars, the general decreases in serum LH and FSH through puberty and into sexual maturity, with perhaps a small transient increase in FSH preceding or coinciding with the onset of sperm production, are not due to decreases in pituitary gonadotrophin content. Indeed, hypothalamic content of GnRH and pituitary content of FSH and LH increases progressively from birth through puberty in conventional male pigs (McNamara *et al.*, 1985; Lee *et al.*, 1987). It is more likely that the rapid increase in testis size, increasing total Leydig cell mass, and increasing numbers of receptors for both LH and FSH (presumably occurring in Meishans, as has been demonstrated in conventional boars) collectively contribute to rising serum concentrations of androgens during this time frame, i.e., an increase in testicular responsiveness. The increasing serum androgens presumably allow steroid feedback to maintain low serum concentrations of LH during this peripubertal period. Assuming that this occurs, the feedback suppression on gonadotrophins appears to be expressed differently in the Meishan, since LH is suppressed (at least until 13 weeks of age) while FSH increases steadily and markedly at 8 weeks of age and beyond (Fig. 2).

General Reproductive Characteristics in Mature Meishan Boars

In general, mature Meishan boars exhibit lower body weight and much smaller testicular size than do conventional breeds of boar at comparable ages (Table 2). The number of spermatozoa per ejaculate from Meishan boars is approximately half that of conventional boars (Borg *et al.*, 1993; Gerfen *et al.*, 1994; Huang and Johnson, 1996), but percentage progressively motile spermatozoa, percentage spermatozoa with normal morphology, and other semen quality characteristics are similar, despite major breed differences in mature testis size. In mammals, the number of Sertoli cells determines mature testicular size (Russell *et al.*, 1990; Chubb, 1992), and findings in boars concur with this. Testes of mature Meishans that are approximately half the size of those of conventional boars (Table 1) have substantially fewer Sertoli cells (Okwun *et al.*, 1996a, 1996b). Of significance, daily sperm production per gram of testis (DSP g^{-1}) does not differ between Meishan and conventional boars (Table 1); therefore, major differences in total DSP primarily reflect breed differences in paired testis weight. However, the percentage of the testis occupied by seminiferous tubules is 1.4-fold lower in Meishan than in conventional boars (Table 1). Meishan boars produce approximately twice as many spermatozoa per Sertoli cell (Table 1) and almost twice as many elongated spermatids per type A spermatogonium compared with conventional boars to accommodate this change in testis composition (Okwun *et al.*, 1996a, 1996b). The yield of spermatids per type A spermatogonium is a reliable measure of efficiency of spermatogenesis (Berndtson and Igboeli, 1989). Thus, mature Meishan boars produce spermatozoa much more efficiently (per gram of seminiferous tubule, per Sertoli cell, and per spermatogonium) than do conventional boars and appear to achieve this via reduced degeneration of germ cells and spermatids during the spermatogenic process. Although the mechanism that allows Meishan boars to achieve this greater efficiency is unknown, it probably relates to constant exposure of the Sertoli cells to greater concentrations of FSH and androgens.

Sensitivity to GnRH Stimulation in Mature Meishan and Conventional Boars

When mature Meishan and conventional boars (Composite = 1/4 Chester White, 1/4 Landrace, 1/4 Large White, 1/4 Yorkshire) of approximately the same body weight (BWT) are challenged with a high (1500 ng kg^{-1} BWT) or a low (150 ng kg^{-1} BWT) dose of GnRH, both breeds respond with significant increases in serum FSH, LH and testosterone concentrations (Wise *et al.*, 1996), and greater increases occur in response to the high dose of GnRH in both types of boar. Despite major pretreatment differences in circulating hormone concentrations, the FSH response to GnRH is comparable (i.e., approximately 125% at low dose and 160% at high dose above pretreatment values) in Meishan and Composite boars. In Meishan boars, the substantially increased serum concentrations of FSH and LH take longer to return to pretreatment values. Although paired testis size is significantly lower in Meishan boars, serum inhibin concentrations differ minimally at pretreatment and changed little during GnRH challenge (Wise *et al.*, 1996). Accepting the limitations of the radioimmunoassay for the α-subunit of inhibin, serum inhibin does not appear to be a primary determinant of high circulating concentrations of FSH in Meishan boars. Thus, under normal physiological conditions, pituitaries of Meishan boars are not functioning under maximal GnRH stimulation and are not secreting maximal amounts of gonadotrophins. High gonadotrophin concentrations in Meishan boars apparently are under control of a functional and responsive GnRH system.

Interestingly, differences in gonadotrophin secretion during the oestrous cycle are not apparent between females of Meishan and European breeds (Hunter *et al.*, 1996). In contrast to the situation in males, ovariectomized European gilts release greater concentrations of FSH and LH in response to GnRH stimulation than do Meishan gilts and are more responsive in their reduction of FSH concentrations after treatment with pig follicular fluid (inhibin effect; Tilton *et al.*, 1994). A mechanism to explain these sex differences in Meishans is not obvious, but in mice that possess a transgene for human FSHβ, expression is greater in males, but not in females, relative to normal mice (Kumar and Low, 1993).

Negative Feedback Regulation of FSH in Mature Meishan and Conventional Boars

When mature Meishan and Composite boars that differed 4- to 9-fold in circulating FSH concentrations were castrated, FSH exhibited a postcastration rise, and the rate of this increase (0–48 h) in both breeds did not differ (Wise *et al.*, 1996). In addition, differences in postcastration FSH concentrations were maintained between breeds through at least 16 days. The pattern of decline in serum testosterone did not differ between breeds after castration, although Meishan boars had greater serum testosterone before castration. These results demonstrate that gonadal hormones exert negative feedback on gonadotrophin concentrations in the mature intact Meishan boar, apparently in a manner similar to that observed in intact conventional boars.

Injection of castrate Meishan and Composite boars with the same dosages of gonadal steroids (testosterone + oestradiol, i.m. on kg^{-1} BWT basis) was equally effective, in both breeds, at reducing the increased concentrations of serum FSH present in castrates (Wise *et al.*, 1996). The rate of decline in FSH after injection did not differ among the breeds, and the relative degree of suppression by gonadal steroids appeared to be similar. The relative difference in serum FSH before steroid injection continued after steroid treatment. Furthermore, serum FSH began to escape exogenous steroid suppression earlier in Meishan than in Composite castrates. Thus, feedback sensitivity to testicular steroids in these two divergent breeds of boar seems to be fully functional and coordinate within their relative concentrations of circulating hormones. Collectively, these data indicate that the high concentrations of gonadotrophins and androgens in Meishans occur under the control of a hypothalamo–pituitary–gonadal axis that functions responsively but with much higher set points in the Meishan boar.

Meishan Pituitary Mass and Biological Activity of FSH

Pituitary size is greater in mature Meishan than in mature Composite boars, and both content and concentration of FSH are substantially higher in anterior pituitaries of Meishan boars (Zanella *et al.*, 1996). The greater expression of the genes for α-subunit and for FSHβ and LHβ subunits accounts, in part, for the observed differences in gonadotrophin concentrations within pituitaries of Meishan boars. In addition, nucleated pituitary cells that are positive immunocytochemically for FSH or LH (gonadotrophs) have larger cell volumes in Meishan than in Composite boars (Macdonald *et al.*, 1995).

Furthermore, in an *in vitro* bioassay for FSH activity that utilized a cell line possessing recombinant FSH receptors coupled with a cAMP-responsive system, the activity of pituitary extracts from Meishan and Composite boars is highly correlated with FSH content determined by radioimmunoassay ($r = 0.92$; Macdonald *et al.*, 1995). In association with this difference in pituitary FSH concentration, gene expression for activin, a stimulator of FSH secretion, is higher in pituitaries of Meishan than in Composite boars (Li *et al.*, 1997). In total, the difference in FSH concentrations between Meishan and Composite boars may be due more to differences in activin regulated synthesis rather than to differences in GnRH regulation of secretion or differences in sensitivity to inhibitors of secretion.

Negative Relationship Between FSH and Mature Testis Size

When the variation in mature testis size and serum FSH concentration was examined, a negative correlation became apparent in mature Meishan boars as well as in mature Composite and Meishan × Composite crossbred boars ($r = -0.74$, Fig. 3; Lunstra *et al.*, 1996). Owing to the positive correlation between mature testis weight and total daily sperm production (TDSP), there is a negative correlation between FSH and TDSP in these boars. Thus, a three times greater serum FSH concentration was associated with 32% smaller mature testis weight and 33% lower TDSP. In Meishan × Composite boars destined to exhibit major differences in serum FSH concentration at maturity, we detected divergence in serum FSH concentrations at 2–8 weeks of age, and this divergence remained during pubertal development (4–6 months of age) and at 1 year of age (Ford *et al.*, 1997). In addition, we have found that serum FSH concentration has a moderate heritability ($h^2 = 0.42$) in Meishan × Composite boars (Rohrer *et al.*, 1997), comparable in magnitude to the heritability of testis size in conventional boars (Young *et al.*, 1986).

The reason for this negative relationship between mature testis size and serum FSH concentration in boars is not clear. In contrast to our findings, FSH secretion (Schinckel *et al.*, 1984; Mariscal *et al.*, 1996) was not influenced by generations of selection for increased testis size in conventional boars (Huang and Johnson, 1996). This lack of an effect on FSH secretion may reflect selection that was not based on testis size at maturity; instead, boars were selected for larger testis size at 150 days of age which would select primarily for earlier pubertal development and secondarily for mature testis size. In contrast to our findings, most other evidence supports a concept that high serum FSH, especially during the early postnatal period, would contribute to greater proliferation of Sertoli cells and result in a larger mature testis size. Treatment of rodents with FSH during early postnatal life causes testicular hypertrophy and increases in numbers of Sertoli cells that persist into adulthood (Meachem *et al.*, 1996). In addition, treatment of conventional boars with FSH from 8 to 40 days of age caused no testicular hypertrophy by 100 days of age, but appeared to delay puberty, to increase the length of seminiferous tubules and the total number of Sertoli cells, and, theoretically, would have increased mature testis size (Swanlund *et al.*, 1995). The latter study contradicts the association between testis size and high serum FSH in Meishan boars (i.e., earlier puberty and reduced mature testis size) and in conventional breeds of boar (i.e., reduced mature testis size). Moreover, other researchers have begun to question the necessity of FSH for regulation of spermatogenesis (Zirkin *et al.*, 1994). Thus, further studies are needed to clarify relationships between extremes in FSH concentrations and extremes in age at puberty and mature testis size.

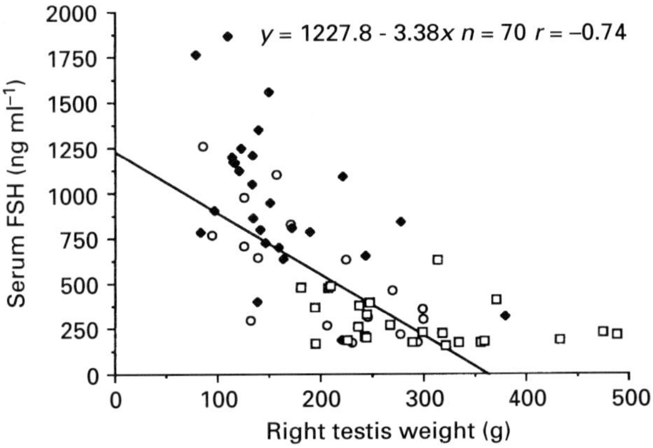

Fig. 3. Relationship between testis size and serum FSH concentration in mature (◆) Meishan, (□) Composite and (○) Meishan × Composite boars. All boars were > 1 year of age. Composite boars: 1/4 Chester White, 1/4 Landrace, 1/4 Large White, 1/4 Yorkshire.

Of particular relevance may be that serum TSH concentration is greater and expression of the β-subunit gene for thyroid-stimulating hormone (TSH-β) in pituitaries is higher in mature Meishan than in conventional boars (Li *et al.*, 1996). Sertoli cells have receptors for TSH, and TSH has potency equivalent to FSH for stimulation of androgen binding protein secretion by Sertoli cells (Hutson and Stocco, 1981). Increased serum TSH concentrations (Ford *et al.*, unpublished data) in combination with high FSH concentrations are associated with reduced testis size exhibited by mature boars with high FSH. Early prepubertal treatment of male rodents with propylthiouracil, a compound that blocks T3 and T4 production (hypothyroidism), causes TSH and FSH to remain at low concentrations into adulthood and results in increased numbers of Sertoli cells, increased mature testis size, and increased sperm production (Hess *et al.*, 1993; Simorangkir *et al.*, 1995). The implication from these propylthiouracil studies is that the opposite effect (high concentrations of TSH and FSH) prepubertally would cause reduced numbers of Sertoli cells and smaller adult testis size. Contrary to prediction, mature testis size of conventional boars was not increased by propylthiouracil treatment from 1 to 9 weeks of age (Tarn *et al.*, 1996), but this treatment failed to decrease serum FSH. At present the prepubertal manipulation of FSH and TSH in conventional boars has not duplicated effects on testicular development comparable to those observed in Meishan boars.

Conclusion

Major endocrine differences and unique differences in the structural and functional efficiency of the testis are exhibited in Meishan boars during pubertal development and at maturity in comparison with boars of conventional breeds. Many fundamental questions remain concerning the factors that control these aspects of reproduction in boars. Why is high FSH secretion throughout development associated with increased spermatogenic efficiency but an unexpected reduction in testis size, while high LH secretion is associated with predictable increases in Leydig cell size and testicular steroidogenesis? These unusual endocrine and physiological attributes in boars of the highly prolific Meishan breed offer a unique model for further investigation of factors influencing male reproductive function. In particular, can the greater efficiency of spermatogenesis in Meishan boars be introgressed into boars of conventional breeds? Hopefully, future results using this model will provide new insights and contributions to the expansion of our understanding of the physiological basis for the reproductive superiority of hyperprolific pigs.

The authors thank A. Kruger, P. Sherman and S. Hassler for technical assistance, personnel of USMARC Swine Operations for care of animals, and D. Griess for secretarial assistance. Mention of trade names is necessary to report factually on available data; however, the USDA neither guarantees nor warrants the standard of the product, and the use of the same by USDA implies no approval of the product to the exclusion of others that may also be suitable.

References

Allrich RD, Christenson RK, Ford JJ and Zimmerman DR (1982) Pubertal development of the boar: testosterone, estradiol-17β, cortisol and LH concentrations before and after castration at various ages *Journal of Animal Science* **55** 1139–1146

Allrich RD, Christenson RK, Ford JJ and Zimmerman DR (1983) Pubertal development of the boar: age-related changes in testicular morphology and *in vitro* production of testosterone and estradiol-17β *Biology of Reproduction* **28** 902–909

Berndtson WE and Igboeli G (1989) Numbers of Sertoli cells, quantitative rates of sperm production, and the efficiency of spermatogenesis in relation to the daily sperm output and seminal quality of young beef bulls *American Journal of Veterinary Research* **50** 1193–1197

Borg KE, Lunstra DD and Christenson RK (1993) Semen characteristics, testicular size, and reproductive hormone concentrations in mature Duroc, Meishan, Fengjing, and Minzhu boars *Biology of Reproduction* **49** 515–521

Cheng PL (1983) A highly prolific pig breed of China – the Taihu breed (Parts I and II) *Pig News and Information* **4** 407–425

Christenson RK (1993) Ovulation rate and embryonic survival in Chinese Meishan and white crossbred pigs *Journal of Animal Science* **71** 3060–3066

Chubb C (1992) Genes regulating testis size *Biology of Reproduction* **47** 29–36

Colenbrander B, Frankenhuis M and Wensing CJG (1982) Male sexual development In *Control of Pig Reproduction* pp 3– 24; Eds D Cole and G Foxcroft. Butterworth Scientific, London

FlorCruz SV and Lapwood KR (1978) A longitudinal study of pubertal development in boars *International Journal of Andrology* **1** 317– 330

Ford JJ (1983) Serum estrogen concentrations during postnatal development in male pigs *Proceedings of the Society for Experimental Biology and Medicine* **174** 160–164

Ford JJ, Wise TH and Lunstra DD (1997) Negative relationship between blood concentrations of follicle stimulating hormone (FSH) and testicular size in mature boars *Journal of Animal Science* **75** 790–795

Gerfen RW, White BR, Cotta MA and Wheeler MB (1994) Comparison of semen characteristics of Fengjing, Meishan and Yorkshire boars *Theriogenology* **41** 461–469

Haley CS, Lee GJ and Ritchie M (1995) Comparative reproductive performance in Meishan and Large White pigs and their crosses *Animal Sciences* **60** 259–267

Harayama H and Kato S (1992) Changes in motility and morphology of spermatozoa during their transit through the epididymis in Meishan boars at various ages *Animal Science Technology* **63** 462–467

Harayama H, Nanjo I, Kanda S and Kato S (1991) Testicular development in Chinese Meishan boars *Theriogenology* **36** 637–643

Hess RA, Cooke PS, Bunick D and Kirby JD (1993) Adult testicular enlargement induced by neonatal hypothyroidism is accompanied by increased Sertoli and germ cell numbers *Endocrinology* **132** 2607–2613

Huang Y-T and Johnson RK (1996) Effect of selection for size of testes in boars on semen and testis traits *Journal of Animal Science* **74** 750–760

Hunter MG, Biggs C, Foxcroft GR, McNeilly AS and Tilton JE (1993) Comparisons of endocrinology and behavioural events during the periovulatory period in Meishan and Large-White hybrid gilts *Journal of Reproduction and Fertility* **97** 475–480

Hunter MG, Picton HM, Biggs C, Mann GE, McNeilly AS and Foxcroft GR (1996) Periovulatory endocrinology in high ovulating Meishan sows *Journal of Endocrinology* **150** 141–147

Hutson JC and Stocco DM (1981) Regulation of Sertoli cell function by thyrotropin *Biology of Reproduction* **25** 303–306

Kosco MS, Bolt DJ, Wheaton JE, Loseth KJ and Crabo BG (1987) Endocrine responses in relation to compensatory testicular growth after neonatal hemicastration in boars *Biology of Reproduction* **36** 1177–1185

Kumar TR and Low MJ (1993) Gonadal steroid hormone regulation of human and mouse follicle stimulating hormone β-subunit gene expression *in vivo. Molecular Endocrinology* **7** 898–906

Lee K-H, Diekman MA, Moss GE and Allrich RD (1987) Pituitary gonadotropins, hypothalamic gonadotropin-releasing hormone, and testicular traits of boars exposed to natural or supplemental lighting during pubertal development *Biology of Reproduction* **36** 1164–1169

Li MD and Enfield FD (1989) A characterization of Chinese breeds of swine using cluster analysis *Journal of Animal Breeding Genetics* **106** 379–388

Li MD, Matteri RL, Macdonald GJ, Wise TH and Ford JJ (1996) Overexpression of β-subunit of thyroid-stimulating hormone in Meishan swine identified by differential display *Journal of Animal Science* **74** 2104–2111

Li MD, Macdonald GJ and Ford JJ (1997) Breed differences in expression of inhibin/activin subunits in porcine anterior pituitary glands *Endocrinology* **138** 712–718

Lunstra DD, Ford JJ, Christenson RK and Allrich RD (1986) Changes in Leydig cell ultrastructure and function during pubertal development in the boar *Biology of Reproduction* **34** 145–158

Lunstra DD, Borg KE and Klindt J (1993) Changes in porcine testicular structure during development in Meishan boars *Biology of Reproduction* **48** (Supplement 1) 126 (Abstract)

Lunstra DD, Ford JJ and Wise TH (1996) Selection for extremes in serum FSH concentrations results in reduced testis size and fertility in Meishan and White composite boars. In *Advances in Swine in Biomedical Research* Vol. 2 pp. 523–531 Eds ME Tumbleson and LB Schook. Plenum Press, New York

Macdonald GJ, Wise T, Sluss PM, Lunstra DD and Ford JJ (1995) Differences in pituitary gonadotropins from male pigs representing divergent breeds *Biology of Reproduction* **52** (Supplement 1) 77 (Abstract)

Mariscal DV, Wolfe PL, Bergfeld EG, Cupp AS, Kojima FN, Fike KE, Sanchez T, Wehrman ME, Johnson RK, Kittok RJ, Ford JJ and Kinder JE (1996) Comparison of circulating concentrations of reproductive hormones in boars of lines selected for size of testes or number of ovulations and embryonal survival to concentrations in respective control lines *Journal of Animal Science* **74** 1905–1914

McNamara M, Vandalem JL, Hennen G, Colenbrander B, MacDonald AA, Van De Wiel D and Meijer JC (1985) The ontogeny of hypothalamic pituitary function in the pig. I Pituitary LH and FSH in the fetus and neonate *Molecular and Cellular Endocrinology* **39** 119–124

Meachem SJ, McLachlan RI, de Kretser DM, Robertson DM and Wreford NG (1996) Neonatal exposure of rats to recombinant follicle stimulating hormone increases adult Sertoli and spermatogenic cell numbers *Biology of Reproduction* **54** 36–44

Okwun OE, Igboeli G, Ford JJ, Lunstra DD and Johnson L (1996a) Sertoli cell number and function, spermatogonial number and yield, and daily sperm production in three breeds of boars *Journal of Reproduction and Fertility* **107** 137–149

Okwun OE, Igboeli G, Lunstra DD, Ford JJ and Johnson L (1996b) Testicular composition, number of A spermatogonia, germ cell ratios, and number of spermatids in three different breeds of boars *Journal of Andrology* **17** 301–309

Peyrat J-P, Meusy-Dessolle N and Garnier J (1981) Changes in Leydig cells and luteinizing hormone receptors in porcine testis during postnatal development *Endocrinology* **108** 625–631

Rohrer GA, Ford JJ, Wise TH and Young LD (1997) Estimates of genetic correlations between serum FSH concentrations in males and measures of female reproduction in a crossbred Meishan population of pigs *Journal of Animal Science* **75** (Supplement 1) 31 (Abstract)

Russell LD, Ping Ren H, Sinha Hikim I, Schultze W and Sinha Hikim AR (1990) A comparative study in twelve mammalian species of volume density, volumes, and numerical densities of selected testis components, emphasizing those related to the Sertoli cell *American Journal of Anatomy* **188** 21–30

Schinckel AP, Johnson RK and Kittok RJ (1984) Testicular development and endocrine characteristics of boars selected for either high or low testis size *Journal of Animal Science* **58** 675–685

Simorangkir DR, de Kretser DM and Wreford NG (1995) Increased numbers of Sertoli and germ cells in adult rat testes induced by synergistic action of transient neonatal hypothyroidism and neonatal hemicastration *Journal of Reproduction and Fertility* **104** 207–213

Swanlund DJ, N'Diaye MR, Loseth KJ, Pryor JL and Crabo BG (1995) Diverse testicular responses to exogenous growth hormone and follicle-stimulating hormone in prepubertal boars *Biology of Reproduction* **53** 749–757

Tan HS and Raeside JI (1980) Developmental patterns of plasma dehydroepiandrosterone sulfate and testosterone in male pigs *Animal Reproduction Science* **3** 73–81

Tarn CY, Rosenkrans CF, Jr and Kirby JD (1996) Effects of 6-*N*-propyl-2-thiouracil on boars. II Growth and reproductive traits *Journal of Animal Science* **74** (Supplement 1) 155 (Abstract)

Tilton JE, Biggs C, Hunter MG and Foxcroft GR (1994) Gonadotropin secretion in ovariectomized Chinese Meishan and hybrid large white gilts: responses to challenges with estradiol benzoate, gonadotropin-releasing hormone, or porcine follicular fluid *Biology of Reproduction* **51** 963–970

Tran D, Meusy-Dessolle N and Josso N (1981) Waning of anti-Mullerian activity: an early sign of Sertoli cell maturation in the developing pig *Biology of Reproduction* **24** 923–931

Trudeau VL, Meijer JC, Erkens JHF, Van De Wiel DFM and Wensing CJG (1992) Pubertal development in the male pig: effects of treatment with a long-acting gonadotropin-releasing hormone agonist on plasma luteinizing hormone, follicle stimulating hormone and testosterone *Canadian Journal of Veterinary Research* **56** 102–109

van Straaten HWM and Wensing CJG (1977) Histomorphometric aspects of testicular morphogenesis in the pig *Biology of Reproduction* **17** 467–472

van Straaten HWM and Wensing CJG (1978) Leydig cell development in the testis of the pig *Biology of Reproduction* **18** 86–93

Wang RX (1990) Reproductive characteristics of the Fengjing pigs. In *Proceedings of Chinese Pig Symposium* Toulouse pp. 6–16 Eds M Molenat and C Legault, , INRA (Institut National de la Recherche Agronomique), Jouy-en-Josas Cedex, France

White BR, McLaren DG and Dzuik PJ (1993) Age at puberty, ovulation rate, uterine length, prenatal survival and litter size in Chinese Meishan and Yorkshire females *Theriogenology* **40** 85–97

Wise T, Lunstra DD and Ford JJ (1996) Differential pituitary and gonadal function of Chinese Meishan and European white composite boars: effects of GnRH stimulation, castration, and steroidal feedback *Biology of Reproduction* **54** 146–153

Young LD, Leymaster KA and Lunstra DD (1986) Genetic variation in testicular development and its relationship to female reproductive traits in swine *Journal of Animal Science* **63** 17–26

Zanella EL, Ford JJ, Wise T and Hamernik DL (1996) Pituitary relationship of gonadotropins and messenger ribonucleic acid for gonadotropin subunits in white composite and Meishan boars *Biology of Reproduction* **54** 154–159

Zirkin BR, Ewing LL, Kromann N and Cochran RC (1980) Testosterone secretion by rat, rabbit, guinea pig, dog, and hamster testes perfused *in vitro*: correlation with Leydig cell ultrastructure *Endocrinology* **107** 1867–1874

Zirkin, BR, Awoniyi C, Griswold MD, Russell LD and Sharpe R (1994) Is FSH required for adult spermatogenesis? *Journal of Andrology* **15** 273–276

Journal of Reproduction and Fertility Supplement **52**, 195–211

Sperm plasma membrane characteristics and boar semen fertility

R. A. P. Harrison

Department of Signalling, The Babraham Institute, Babraham, Cambridge CB2 4AT, UK

Much effort is being made to establish relationships between the molecular events that take place in spermatozoa under fertilizing conditions and actual sperm function during fertilization. During capacitation, the process that 'primes' spermatozoa for interaction with the egg, components of the sperm's environment, notably bicarbonate, provoke various specific changes in the architecture and functioning of the sperm plasma membrane in a large number of cells. The individual changes have been found to proceed on different time scales, and may therefore represent sequential stages in the capacitation process. However, each change takes place at different rates in individual cells, revealing considerable functional heterogeneity within the sperm population. Recent work on membrane changes provoked by cooling has indicated similarities with capacitational changes. The effect of cooling may therefore be to induce premature capacitation (and destabilization). Such an effect would greatly compromise sperm fertilizing potential. A pig sperm–egg interaction model was used to examine quantitative details of zona binding and zona penetrating abilities within capacitated sperm populations, and sperm behaviour was found not to accord with generally held beliefs. In particular, individual spermatozoa that have bound to the zona pellucida show great variation in the delay before penetrating: no evidence has been found for a specially competent subgroup. Even in sperm samples incubated to undergo maximal capacitational membrane changes, cells with actual penetrating potential represent less than 15% of the total number that attach initially to the zona pellucida. Thus detection of capacitational membrane changes appears greatly to overestimate zona penetrating capability. Future studies linking sperm membrane characteristics with semen fertility in the field will need to consider differences between *in vitro* and *in vivo* conditions. The need for survival in the female tract may require much slower sperm responses than are considered optimal for *in vitro* fertilization.

Introduction

In seeking to preserve semen, the essential goal is obviously to preserve fertilizing ability. However, if definition of the goal is sought in terms of cell biology, i.e. which particular sperm functions need special protection, an answer is much less obvious – indeed, clear answers cannot yet be given. We still lack detailed understanding of the various reproductive processes that affect the fertilizing success or otherwise of individual semen inseminates. Only when we have identified the crucial processes and the way in which they impinge on sperm function are we likely to be able to discover better ways of preserving sperm fertilizing ability.

Inevitably, research on sperm function and semen preservation procedures must rely heavily on *in vitro* investigations and assessment methods. However, linking any observed biochemical or cellular changes to biological action, even to *in vitro* fertilization (IVF) success, is very problematical. Linking *in vitro* findings to the *in vivo* situation is even more challenging. Many articles have reviewed semen fertility and semen assessment in recent years; of those, Amann (1989), Amann and Hammerstedt (1993), Amann *et al.* (1993), Saacke *et al.* (1994), Berger *et al.* (1996) and Dzuik (1996) are especially to be recommended. Rather than reiterate the vast body of relevant work in this area, I have preferred to present in this paper a more personal and selective view of the current position, using results from my laboratory's recent work to illustrate my remarks. First, I shall discuss studies of sperm membrane changes that have been found to occur under IVF conditions. Then I shall present some preliminary quantitative studies of pig sperm–egg interaction *in vitro*, the findings of which conflict somewhat with current concepts. I shall also discuss some boar sperm characteristics that appear to hinder quantitative comparisons between sperm samples. Finally, I shall propose some overall aspects of sperm behaviour that may play an important role *in vivo* and that should be considered as we plan our future approaches for assessing sperm fertilizing capability. Limitations of space have resulted in arbitrary choices when referring to others' published findings.

A Brief Overview of Fertilization

The events and processes involved in the actual act of fertilization have been very well studied in mammals (see Yanagimachi, 1994). The current view is that the fertilizing spermatozoon initiates fertilization by binding in a specific fashion to the egg's zona pellucida envelope. In a process activated by this binding, it exocytoses its acrosome ('acrosome reaction'). Then, with the aid of the action of enzymes exposed as a result of the exocytosis and of the enhanced motive power generated through expression of additional flagellar flexure (hyperactivated motility), the sperm forces its way through the zona pellucida to the perivitelline space beneath. At this point, the sperm attaches to the egg's plasma membrane via specific receptor–ligand interactions in its head equatorial region, the plasma membrane of the sperm begins to fuse with that of the egg, whence the sperm body is adsorbed into the egg and syngamy ensues. From the spermatozoon's point of view, fertilization involves two vital membrane fusion processes – fusion of the sperm head plasma membrane with the underlying outer acrosomal membrane (as the crucial step in the acrosome reaction), and fusion of the sperm plasma membrane with the oolemma during syngamy itself. Biological membranes are intrinsically stable, and require prior architectural modulation and destabilization in order to fuse. In the case of the spermatozoon, the head plasma membrane must remain stable during maturation and throughout subsequent storage in the epididymis. It must also remain stable as the sperm proceeds on its journey up the female tract. Destabilization should only occur in the vicinity of the egg. It is well known that spermatozoa require a priming process known as capacitation in order for them to acquire the ability of immediate interaction with the egg (see Harrison, 1996). While binding to zona components is believed to initiate in the fertilizing spermatozoa the cascade of molecular events that bring about actual membrane fusion during the acrosome reaction (for example Bailey and Storey, 1994; Roldan *et al.*, 1995), current thinking ascribes to capacitation the destabilizing action needed to render the sperm head plasma membrane potentially fusible. Attention is now being focused on capacitation because, clearly, without this priming procedure, the spermatozoa are incapable of responding to any egg stimuli. Moreover, considerable evidence has accrued that capacitation does indeed bring about alterations in the sperm plasma membrane (see Cohen-Dayag and Eisenbach, 1994). It is also worth pointing out that a thorough understanding of capacitation is of practical importance, because capacitation must be achieved during any *in vitro* testing of sperm fertilizing ability.

Capacitation and Sperm Membrane Changes

Although the sperm's requirement for capacitation was discovered some 46 years ago, knowledge of the processes involved is still sketchy despite much research effort having been expended on the topic; the many alterations in sperm characteristics that have been described remain essentially phenomenological (see Cohen-Dayag and Eisenbach, 1994). Our lack of success in gaining understanding may in part be ascribed to having taken 'a wrong view' of the process (see Harrison, 1996).

Insofar as sperm fertility and semen preservation are concerned, there are two important conceptual questions to be posed. First, do spermatozoa respond to capacitating conditions simply as passive targets of an external action exerted on their surfaces, or are positive alterations induced in the innate cellular processes of the spermatozoa? Second, do all spermatozoa undergo capacitation changes in parallel or in synchrony, or are some cells affected more or faster than others? If the latter is the case, obvious supplementary questions concern differences between boars or even ejaculates in their response to capacitating conditions, and to the modulating effects of preservation treatments.

The answers to these questions are now emerging. They are leading to the contention that capacitation is a positive process that involves alterations in dynamic sperm plasma membrane functions. Within the sperm population, individuals respond to capacitating conditions at widely different rates; there appear to be considerable differences between ejaculates, and preservation treatments modulate considerably sperm sample responses.

Some of the evidence for these views comes from recent studies, in which we have been able to profit from the application of flow cytometry to living sperm suspensions to detect changes in individual cells. From such measurements, we have been able to observe trends within the overall population in response to experimental 'capacitating' treatments. Using three different sorts of fluorescent probe, fluo-3 to detect changes in intracellular calcium (Harrison *et al.*, 1993), fluorescein-conjugated lectins to detect changes in surface glycoproteins (Ashworth *et al.*, 1995), and merocyanine to detect changes in plasma membrane lipid architecture (Harrison *et al.*, 1996a), we have found that bicarbonate specifically induces major alterations in these parameters within the sperm population. We have also demonstrated that bicarbonate enhances considerably the development of ability to bind solubilized zona proteins to the sperm head plasma membrane (Harkema *et al.*, 1998). As several studies have demonstrated that bicarbonate is a key effector of capacitation and fertilization *in vitro* (see Suzuki *et al.*, 1994; Harrison, 1996), the potential relevance of such membrane changes is clear.

Capacitation of freshly ejaculated spermatozoa appears to take a minimum of several hours in most mammalian species studied. While such estimates are based on 'average population response' (i.e. capacitation is detected as having occurred because a sufficient proportion of capacitated spermatozoa have accumulated in the sample to achieve rapid fertilization), the slowness of the overall process implies that it results from a series of sequential events. The bicarbonate-induced changes described above take place on different time scales, hence we believe that they represent different stages in the capacitation sequence. The merocyanine-detectable change (Harrison *et al.*, 1996a) is of particular interest because it takes place within 5 min of exposing the sperm cells to bicarbonate. We therefore suspect that it represents one of the earliest steps. Merocyanine is a fluorescent lipophilic molecule that is membrane-impermeable but that binds to the outer leaflet of the plasma membrane of intact cells in relation to the degree of lipid disorder in the leaflet. The action of bicarbonate in enhancing the ability of live spermatozoa to bind merocyanine can thus be interpreted as inducing lipid disorder in the external leaflet of the sperm plasma membrane.

Details of the mechanism of this action remain to be elucidated. However, preliminary evidence indicates that the bicarbonate activates adenylyl cyclase (c.f. Okamura *et al.*, 1985; Visconti *et al.*, 1995) to produce increases in intracellular cyclic AMP concentrations (Fig. 1), which in turn stimulate a protein kinase A to phosphorylate an unknown protein component. This protein controls directly or indirectly plasma membrane lipid architecture. The control mechanism appears to be a dynamic one, involving a continual cycle of phosphorylation and dephosphorylation. Protein kinase inhibitors block the increase in merocyanine binding without blocking cAMP increases; a protein phosphatase inhibitor enhances merocyanine binding (Table 1). It is also noteworthy that the bicarbonate induction of merocyanine binding is temperature dependent and reversible, occurring at 38°C but not at 25°C (see Fig. 1), and declining as the spermatozoa are transferred from 38° to 25°C (Harrison *et al.*, 1996a). Such reversibility is, of course, in accord with the concept of control being mediated through a dynamic phosphorylation– dephosphorylation cycle. Bicarbonate thus exerts a positive effect on intrinsic sperm metabolic processes, whence we deduce that capacitation is a specific, initiated and controllable process rather than a natural continuation of maturation (c.f.

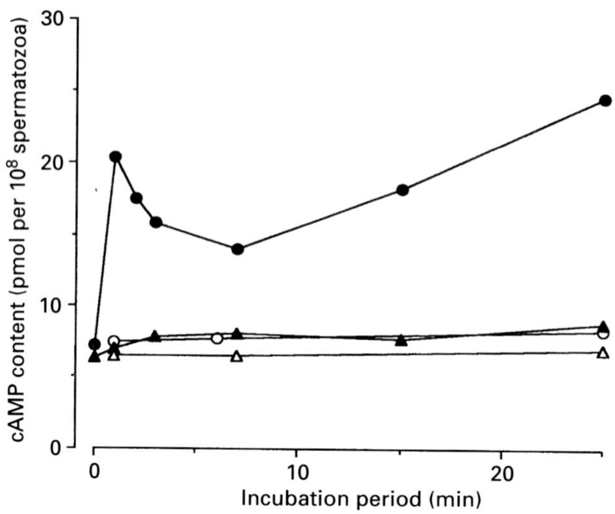

Fig. 1. Cyclic AMP concentrations in boar sperm suspensions exposed to bicarbonate/CO_2. Samples of washed boar spermatozoa were added to variants of a Ca^{2+}-free Hepes-buffered Tyrode's-based medium containing 3 mg BSA ml^{-1}, either at 38°C or at 25°C (Harrison *et al.*, 1996a). One variant (Tyr+bic) contained 15 mmol bicarbonate l^{-1} in equilibrium with 5% CO_2; the other (Tyr) contained 15 mmol NaCl l^{-1} in place of the bicarbonate and was not exposed to CO_2. At intervals after initiating incubation, perchloric acid was added to stop cellular activity (0.3 mol l^{-1} final concentration), and samples were placed on ice for 30 min. Perchlorate was removed from the protein-free supernatants with tri-*n*-octylamine and freon according to Sharps and McCarl (1982). Cyclic AMP was assayed in the final supernatants using an enzyme-linked immunosorbent assay (Amersham Life Science, Little Chalfont, Bucks). (●) Tyr+bic at 38°C; (○) Tyr at 38°C; (▲) Tyr+bic at 25°C; (△) Tyr at 25°C. Note that only Tyr+bic at 38°C provokes a rise in cAMP. The effect of these incubation conditions on merocyanine binding (Harrison *et al.*, 1996a) was tested in parallel. Only in Tyr+bic at 38°C were significant numbers of high merocyanine-binding cells induced (62% of the total population after 25 min); less than 5% of high merocyanine-binding spermatozoa were found under other conditions.

These results are from a single experiment; similar cAMP profiles during incubation in Tyr+bic and Tyr at 38°C have been found in four other experiments.

Bedford, 1974; Mann and Lutwak-Mann, 1981). It is noteworthy (see Harrison, 1996) that bicarbonate levels are maintained low in the epididymis (where the spermatozoa need to be kept stable), but are high in the female reproductive tract (where capacitation is brought about under natural conditions). Our findings also indicate that capacitation should not be considered as being a release from stabilization conditions within the epididymis (c.f. Bedford, 1991): washed spermatozoa do not readily undergo capacitational changes unless bicarbonate is specifically included in their environment (Harrison *et al.*, 1993; Ashworth *et al.*, 1995; Harrison *et al.*, 1996a).

In linking capacitational changes to fertility, the key point of the bicarbonate-inducible increase in merocyanine binding is that semen samples vary considerably in the quantitative nature of their

Table 1. Effect of protein kinase and protein phosphatase inhibitors on merocyanine-detectable response to bicarbonate in boar spermatozoa[a]

	% High merocyanine-binding live cells[b] after	
	10 min	25 min
Expt A: inhibition of protein kinases		
5 mmol bicarbonate l^{-1}/ 1.7% CO_2	43.4	39.3
Bicarbonate/CO_2+1 µmol staurosporine l^{-1}	31.6	30.4
Bicarbonate/CO_2+10 µmol staurosporine l^{-1}	6.2	4.6
No addition (control)	1.9	3.0
Expt B: inhibition of protein kinases		
6 mmol bicarbonate l^{-1}/ 2% CO_2	59.3	53.9
Bicarbonate/CO_2+10 µmol staurosporine l^{-1}	7.3	9.0
No addition (control)	5.0	8.4
Expt C: inhibition of protein phosphatases		
6 µmol papaverine[c] l^{-1}	2.6	17.3
Papaverine+1 µmol okadaic acid l^{-1}	20.7	57.9
No addition (control)	2.9	6.7
Expt D: inhibition of phosphatases and kinases		
10 µmol papaverine[c] l^{-1}	12.5	28.5
Papaverine+1 µmol okadaic acid l^{-1}	23.1	35.5
Papaverine+okadaic acid+10 µmol staurosporine l^{-1}	6.4	5.5
No addition (control)	1.6	2.5

[a]Washed boar spermatozoa were added to variants of a Hepes-buffered Tyrode's-based medium at 38°C; all suspensions contained 3 mg BSA ml^{-1}, and also Yo-Pro-1 as viability probe. After 10 and 25 min, samples of the suspensions were removed, mixed with merocyanine, and subjected to flow cytometry to detect cell-associated merocyanine and Yo-Pro fluorescence (Harrison *et al.*, 1996a). Because of variation between semen samples in terms of overall response, data from four different experiments are given.
[b]Computer-generated thresholds, set empirically in each experiment, used the boundary of the major sperm population in the control medium to categorize the limits of low merocyanine fluorescence and low Yo-Pro fluorescence. Cells found beyond these limits were classified as showing high merocyanine-binding or as dead, respectively. Results are expressed as percentages of the total sperm sample.
[c]Low concentrations of papaverine, a powerful phosphodiesterase inhibitor, were used for sub-maximal stimulation of the cAMP-dependent process (see Harrison *et al.*, 1996a)

response. In some samples, about 80% of the live cells undergo the change within 10 min of bicarbonate addition; in others, the same treatment results in only a small percentage showing any increase in binding (Harrison *et al.*, 1996b). There is some evidence that the merocyanine-detectable response to bicarbonate may be related to *in vivo* fertility, although this is still very preliminary. What is intriguing is that storage of spermatozoa in preservation diluents gradually reduces the response, although again individual samples show considerable differences in the degree to which their response may decline (Harrison *et al.*, 1996b).

Another effect of bicarbonate treatment is to increase cell death within the sperm population, apparently by extreme destabilization of the plasma membrane (Harrison *et al.*, 1993). While this destabilization appears to be closely associated with the merocyanine-detectable change (indeed, increased lipid disorder implies increased instability), the cell death observed is not directly related to the degree of merocyanine-detectable response (Fig. 2). When measurements of cell death are combined with measurements of membrane changes in the live sperm population as a response to bicarbonate, a profile with time is obtained that suggests that sperm samples consist of two broad groups, cells that respond rapidly and cells that respond slowly (Fig. 3). It is not yet clear as to which of these two groups may contain the spermatozoa that fertilize *in vivo*. (This point is made in more detail in a later section.)

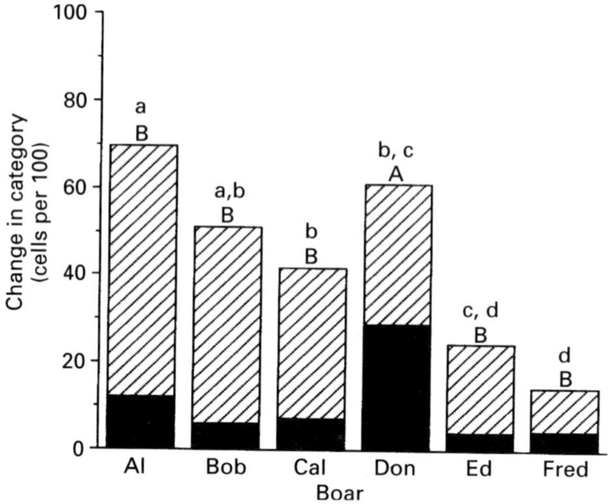

Fig. 2. Variation among boars in sperm response to bicarbonate. Sperm-rich fractions from fresh ejaculates of six different boars were diluted in Beltsville Thawing Solution (Johnson *et al.*, 1988) and kept overnight at ambient temperatures. Spermatozoa were isolated from the extended samples by washing through Percoll and put to incubate at 38°C in bicarbonate-containing or bicarbonate-free Hepes-buffered Tyrode's-based medium; Yo-Pro-1 was included as viability probe. After 10 min, samples were removed from the sperm suspensions, mixed with merocyanine, and subjected to flow cytometry. The data were analysed to provide estimates of the relative proportions of high merocyanine-binding live cells and dead cells (see Harrison *et al.*, 1996a for all experimental details). The results are given as the bicarbonate-induced change in either high merocyanine-binding live cells (▨) or dead cells (■), and are the means of four assessments (carried out at weekly intervals); standard error of difference = 6.3 for changes in merocyanine binding, 5.0 for changes in dead cells. Columns that do not share a common letter are significantly different ($P < 0.05$); lower case letters refer to changes in merocyanine binding, upper case letters to changes in dead cells.

In general terms, however, since capacitation is considered to be a destabilizing process (see Harrison, 1996), any artefactual acceleration of capacitation must be regarded as undesirable. Evidence is now accumulating that treatments associated with cryopreservation may induce capacitation-like changes, at the same time shortening the lifespan of the spermatozoa. The physiological implications of this concept are discussed in more detail in a later section, but of particular interest are the recent demonstrations that cooling can induce changes in the pattern of chlortetracycline (CTC) staining over the sperm head (Johnson *et al.*, 1996; Watson, 1996; see also Maxwell and Watson, 1996). These changes closely resemble those seen in capacitating sperm samples, a transition from uniform overall staining of the head (F pattern) to a weak staining of the post-acrosomal region (B pattern), which has been applied to several species as a means of identifying individual capacitated cells (in boar by Wang *et al.*, 1995; Mattioli *et al.*, 1996). Although the precise molecular interpretation of CTC staining has not been elucidated, the fact that both capacitation and cooling induce similar changes in staining pattern strongly suggests that similar destabilizing changes in membrane architecture may have occurred (see Maxwell and Watson, 1996; Watson, 1996). A molecular explanation of the CTC staining changes is clearly urgently needed.

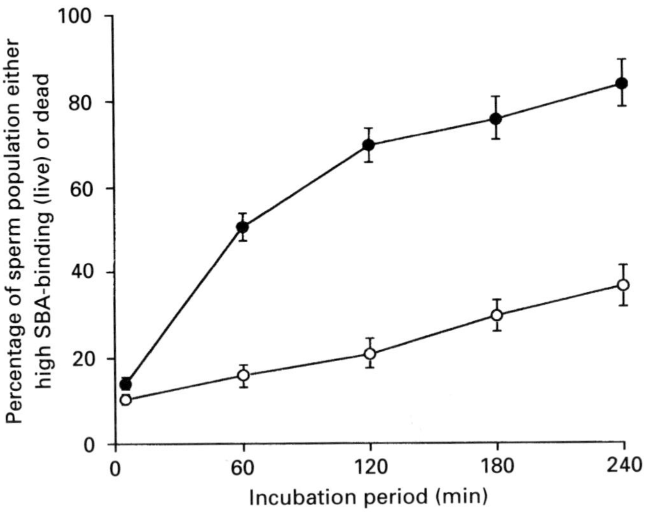

Fig. 3. Effect of bicarbonate on changes in boar sperm samples during lengthy incubation. Washed boar spermatozoa were incubated at 38°C in a bicarbonate-containing (●) or a bicarbonate-free (○) Hepes-buffered Tyrode's-based medium with 2 mmol Ca^{2+} l^{-1} and 3 mg BSA ml^{-1} (Ashworth *et al.*, 1995); propidium iodide was included as viability probe. At intervals, samples were removed from the suspensions, mixed with fluorescein–conjugated soybean agglutinin (F-SBA), and subjected to flow cytometry. The data were analysed to provide estimates of the proportions of high F-SBA-binding live cells and dead cells (see Ashworth *et al.*, 1995). The results are expressed as the percentage of the total population that were either high F-SBA-binding live or dead (mean ± SD from nine independent sperm samples). At each time point, the difference between the two media was highly significant ($P < 0.001$).

Membrane Changes and Sperm–Egg Interaction

Relating specific sperm parameters to fertilizing capability poses great practical problems, particularly in pigs. Laboratory tests compare differences between individual sperm samples, but the fertility of such samples *in vivo* cannot be assessed in field trials because of the numbers of spermatozoa needed to achieve reasonable fertilization rates (quite apart from such trials being very time-consuming, complex and expensive). However, a satisfactory but simple *in vitro* test system for assessing sperm–egg interaction remains to be developed. The use of conventional IVF systems (e.g. Matas *et al.*, 1996; Xu *et al.*, 1996) is impractical: lengthy incubation protocols are involved, assessment of categories of spermatozoa other than those that have actually penetrated the egg vitellus is difficult and laborious, and the provision of the matured live eggs needed to obtain the sperm head decondensation within the egg cytoplasm that is the criterion of fertilization is very demanding. There is now a great body of evidence that the zona pellucida provides the major barrier to sperm entry into the egg; the fertilizing ability of a spermatozoon hinges on its ability to cross this barrier (Yanagimachi, 1994; Johnson *et al.*, 1995). Penetration of the zona is thought to require concomitant expression of several sperm functions, including motility (Yanagimachi, 1994; Olds-Clarke, 1996). Since our ability to determine multiple functions in individual living cells is currently strictly limited, direct measurement of a sperm population's ability to penetrate the zona pellucida would seem to offer at present the best means of assessing fertilizing ability *in vitro*. Whole eggs are required as test material, since zona penetrating ability cannot realistically be estimated using zona

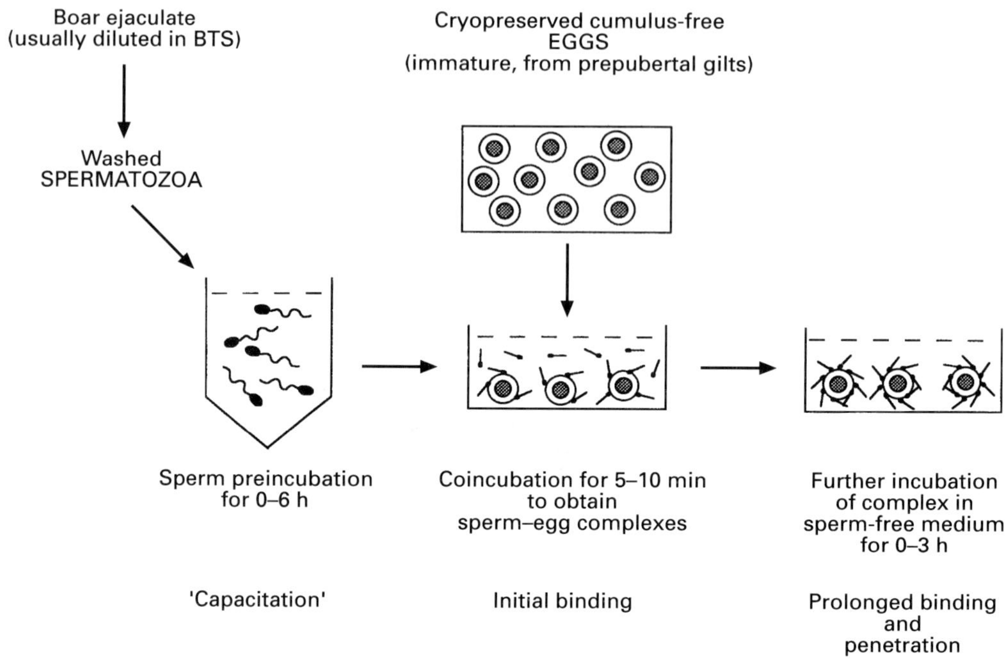

Fig. 4. Multi-step procedure for testing boar sperm interaction with cryopreserved pig eggs. For details of the procedure, see Lynham and Harrison (1998). Sperm–egg complexes resulting from the possible incubation sequences were subjected to one of three different physical treatments to assess the degree of sperm–egg association. The total number of spermatozoa attached to the egg ('total binding') was assessed by transferring complexes directly into 1% (w/v) paraformaldehyde fixative. The number of spermatozoa tightly bound was assessed by removing loosely bound spermatozoa by passaging groups of complexes six times in and out of a specially enlarged 'yellow' Gilson pipette tip before transferring them to fixative. The number of penetrated spermatozoa was assessed by stripping off spermatozoa bound to the zona surface by passaging groups of complexes 12 times in and out of a narrow-bore Gilson Round Gel Loading Tip before transferring them to fixative. After 15 min in fixative, complexes were washed through a glycine buffer (to quench unreacted formaldehyde) and wet-mounted in a solution containing the nuclear stain Yo-Pro-1. The sperm heads associated with each egg were then counted using fluorescence microscopy.

fragments (i.e. hemi-zonae: Fazeli *et al.*, 1995); for consistency of comparison as well as for convenience, the whole eggs should be available as stored 'pools'.

A number of groups have already used killed eggs successfully to detect zona penetration in several species including pig (Ivanova and Mollova, 1993). On the other hand, doubt has been expressed on the ability of zona penetration tests to inform on fertility *in vivo* (see Codde and Berger, 1995). However, the systems used so far have not allowed detailed investigation of the different stages of sperm–egg interaction, and the number of eggs have been limited, usually because they have been matured *in vitro* before storage. In a new approach, therefore, we have used cryopreserved immature eggs isolated in large quantities from the ovaries of prepubertal gilts, to examine separately the stages of sperm function that eventually result in zona penetration (Lynham and Harrison, 1998). In a multi-step model (Fig. 4), the sperm sample is first preincubated in a suitable medium to induce capacitation, then co-incubated briefly with a group of 10–20 eggs to allow a representative subsample of the spermatozoa to bind to the zona surfaces, after which the resultant sperm–egg complexes are transferred to fresh sperm-free medium and incubated further so that the penetrating ability of the already bound spermatozoa can be monitored. Our findings have indicated features of sperm–egg interaction that are not altogether in agreement with current dogma.

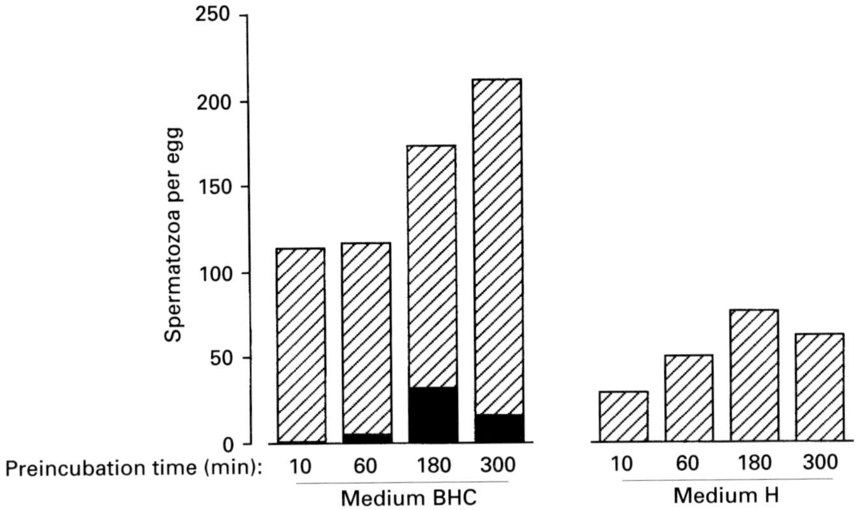

Fig. 5. Development of zona-binding and zona-penetrating abilities during incubation of boar spermatozoa before mixing with eggs. Washed boar spermatozoa (2×10^6 ml^{-1}) were preincubated for various periods, either in medium BHC (known to support fertilization) or in medium H (known not to support fertilization) (for details of these media, see Suzuki *et al.*, 1994). Samples of the sperm suspensions were then coincubated for 10 min with groups of cumulus-free pig eggs in the same medium at a concentration of 1.6×10^6 sperm ml^{-1}. The groups of complexes were then transferred to fresh sperm-free medium, after which some of each group were taken and fixed immediately for assessment of total binding (▨), while the others were incubated for a further 60 min before being repeatedly passaged through a narrow-bore pipette and fixed for assessment of penetration (■). (See Fig. 4 for general explanation of procedures.)

Three replicate experiments were performed with independent semen samples of known high fertility. Each group of eggs yielded a 'mean' data point (spermatozoa per egg) for a particular treatment × replicate. The data were subjected to analysis of variance after log transformation, and back-transformed means are illustrated. Effects of media and time of preincubation on total binding were both highly significant ($P < 0.005$); effects of media and also of time of preincubation in BHC on penetration were very highly significant ($P < 0.001$).

First, boar spermatozoa bind to the zona pellucida in considerable numbers under non-capacitating conditions. Binding is generally no more than 2–3 times greater under capacitating conditions. While binding ability develops rapidly during preincubation in either capacitating or non-capacitating medium, zona penetrating ability develops slowly and only in capacitating medium (Fig. 5). Thus, although it has been widely monitored in several species as a parameter of fertility (for example in the hemi-zona assay: see Fazeli *et al.*, 1995), zona binding ability does not appear to be a reliable indicator of zona penetrating ability in pigs. Peterson *et al.* (1984) and Berger *et al.* (1989) noted the ability of non-capacitated boar spermatozoa to bind to the zona pellucida, but did not at the same time compare zona binding with zona penetrating ability.

A second point of interest is that while we were able to distinguish spermatozoa that were tightly bound to the zona from those that were loosely bound, we were unable to detect any change in the relative proportions of the two types (Fig. 6), either following different periods of preincubation ('capacitation') or during incubation of pre-formed sperm–egg complexes (when numbers of attached spermatozoa remained constant – i.e. attachment was not reversible). Early studies of sperm–zona interaction in hamsters and mice noted transitions in strength of association that could be detected by passaging sperm–egg complexes through wide-bore micropipettes: spermatozoa removed by this

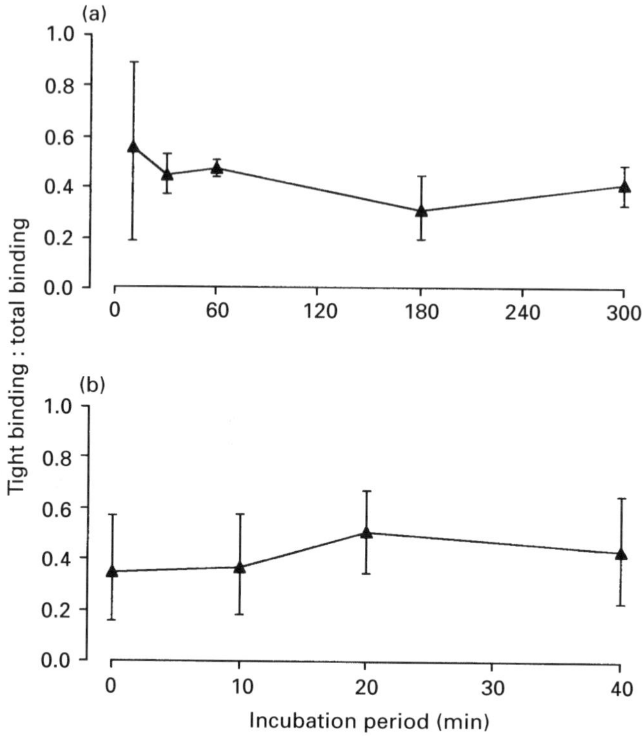

Fig. 6. Strength of boar sperm binding to the zona pellucida during incubation under fertilizing conditions.

(a) Effect of preincubation. Washed spermatozoa were preincubated for various periods in the medium BHC known to support fertilization (see legend to Fig. 5), and then coincubated for 10 min with groups of eggs in the same medium. Some of the complexes in each group were then fixed immediately for assessment of total binding while others were passaged through a wide-bore pipette prior to fixation for assessment of tight binding (see legend to Fig. 4).

(b) Effect of incubation of sperm–egg complexes. Washed spermatozoa were preincubated for 3 h in medium BHC, co-incubated with groups of eggs for 10 min, and then the resultant sperm–egg complexes were transferred to fresh sperm-free medium and incubated further. Complexes were removed at intervals after transfer and treated for assessment as before. Each group of eggs yielded a 'mean' data point (spermatozoa per egg). For each time point, the number of spermatozoa tightly bound was divided by the total number of bound sperm to obtain a ratio of tight binding: total binding. Three replicate experiments of each type were performed using independent semen samples of known high fertility. The results were analysed after arcsin transformation, and are presented as back-transformed means ± SD; differences in binding strength were not significant in either experiment type.

procedure were termed 'attached' while spermatozoa that remained on the zona were termed 'bound' (see Wassarman, 1987). Such transition has been interpreted as the establishment of specific ligand–receptor interactions that lead to the induction of the acrosome reaction (see Wassarman, 1987; Yanagimachi, 1994; Thaler and Cardullo, 1996), and evidence from several species has been presented to indicate that such zona receptor activity increases during capacitation (Leyton and Saling, 1989; Burks *et al.*, 1995; Harkema *et al.*, 1998). Our sperm–egg model system does not behave according to this concept. Moreover, when we used fluorescein-conjugated peanut agglutinin to stain (specifically) the acrosomes of zona-bound spermatozoa, we found that sperm–egg complexes from which loosely bound spermatozoa had been removed bore many stained amorphous fragments on their zona surface (Table 2). We concluded that these fragments were acrosomal remnants and deduced that they had been derived from the loosely bound spermatozoa. Since the zona surface is not smooth in relation to a sperm head (e.g. Dietl and Czuppon, 1984; Phillips *et al.*, 1985; Familiari *et al.*, 1992), we believe that spermatozoa often attach by 'embedding' themselves in the rugosities. The strength of attachment of a particular sperm cell more likely reflects the surface area involved in binding rather than the association constant of the particular form of ligand–receptor interaction involved (c.f. Thaler and Cardullo, 1996).

A third finding at odds with dogma concerns the temporal relationship of zona penetration to zona binding. We formed sperm–egg complexes from sperm suspensions that had been preincubated for 3 h in capacitating medium to develop fully their penetrating ability. After coincubation of spermatozoa and eggs for only 10 min, we transferred the complexes to sperm-free medium and assessed numbers of zona-penetrated spermatozoa after increasing intervals of incubation. Very few penetrated spermatozoa were found in the freshly formed complexes. Numbers of penetrated spermatozoa slowly increased with time in a curvilinear fashion, appearing to reaching a maximum after about 3 h (Fig. 7). There was no evidence of a rapidly penetrating sperm subgroup. Even after these 3 h of 'post'incubation, the number of penetrated spermatozoa represented less than 15% of the total number bound. (Our assay counted all spermatozoa that had entered the zona as 'penetrated'; many fewer would have actually passed through the zona into the perivitelline space.)

These findings need to be viewed against recent studies identifying membrane changes in sperm cells that are specifically engendered by capacitating conditions. The changes, which affect large proportions of the sperm population, have been seen as maximizing in boar sperm samples after some 2 h of incubation (Ashworth *et al.*, 1995; Harkema *et al.*, 1998). This time course correlates well with other studies based on IVF success which have indicated that optimal capacitation *in vitro* takes 2–3 h in pigs (e.g. Yoshida *et al.*, 1993); moreover, in our sperm–egg model also, judged by the number of spermatozoa that were able to penetrate the zona after 1 h's postincubation of sperm–egg complexes, zona penetrating ability peaked after 3 h sperm preincubation. Current thinking (e.g. Wassarman, 1987; Yanagimachi, 1994) considers that the binding of capacitated spermatozoa to the zona pellucida initiates the signalling sequence leading to the acrosome reaction, and that penetration of the zona matrix begins once lytic enzymes have been exposed by the acrosome reaction. In our sperm–egg model, the proportion of zona-bound spermatozoa that actually showed zona penetrating ability was much less than the proportions of spermatozoa reported in previous studies to undergo capacitational changes (e.g. Ashworth *et al.*, 1995; Mattioli *et al.*, 1996; Harkema *et al.*, 1998). Moreover, the time course of the build-up of penetrated spermatozoa indicated varying delays between binding and initiation of penetration, rather than the presence of a subgroup that initiated penetration soon after binding. In other words, the membrane changes induced in spermatozoa by capacitating conditions do not seem to relate directly to the ability to initiate zona penetration on zona contact. Since the latter might arguably be considered as the indicator of true capacitation (see Harrison, 1996), the value of the membrane parameters may need to be reconsidered. Zona penetration appears to demand both the ability to undergo the acrosome reaction in response to zona binding and the expression of vigorous (or even hyperactive) motility (Yanagimachi, 1994; Olds-Clarke, 1996). As the two functions stem from different and essentially independent cellular processes, one might propose that no more than half the cells with one attribute would show the other at the same time; indeed, one could argue that very few at any one time would possess both attributes simultaneously.

Boar Sperm Membrane Properties That May Interfere With Sperm–Egg Interaction *In Vitro*

Soon after dispersion in fertilization-supporting medium, boar spermatozoa show a strong tendency to agglutinate head-to-head. This characteristic has been observed previously (Harayama *et al.*, 1994; Suzuki *et al.*, 1994), but its potential effect on the quantitation of sperm–egg interaction *in vitro* has

Table 2. Acrosomal fragments on pig sperm–egg complexes before and after rinsing[a]

Sperm preincubation	Unrinsed		Rinsed	
	Number of sperm-associated acrosomes per egg[b]	Acrosomal fragments[c] (per 100 sperm-associated acrosomes)	Number of sperm-associated acrosomes per egg[b]	Acrosomal fragments[c] (per 100 sperm-associated acrosomes)
Time in capacitating medium (min)				
10	18.8	3	13.3	5
60	32.0	16	18.4	21
180	83.4	12	35.4	36
300	77.3	16	48.7	46
Time in non-capacitating medium (min)				
10	8.6	1	0.56	(60)
60	14.9	4	6.1	15
180	44.7	5	19.1	24
300	12.7	27	13.7	31

[a]Washed boar spermatozoa were preincubated for various periods either in a medium (BHC) that supported fertilization ('capacitating') or in a medium (H) that did not support fertilization ('non-capacitating') (Suzuki *et al.*, 1994). Samples of spermatozoa were then coincubated for 10 min with groups of cumulus-free pig eggs in the same medium. Some of the sperm–egg complexes in each group were then fixed immediately with 1% paraformaldehyde ('unrinsed')), while the others were passaged six times through a wide-bore pipette ('rinsed') before fixation. After washing, the complexes were stained with fluorescein-conjugated peanut agglutinin (F-PNA: Vector Laboratories, Peterborough, Cambs; 20 µg ml^{-1} final concentration) in Hepes-buffered saline containing 150 µg saponin ml^{-1}; F-PNA binds specifically to acrosomal components (Ashworth *et al.*, 1995), but does not stain the zona pellucida (Yoshida *et al.*, 1993). After further washing, complexes were wet-mounted in a solution containing propidium iodide (to counterstain the sperm heads), and examined under fluorescence microscopy. The combined results of two replicate experiments are presented. Analysis of variance of the incidence of acrosomal remnants (after log transformation) showed that differences between rinsed and non-rinsed eggs were significant ($P = 0.012$).

[b]Morphologically recognizable acrosomes associated with sperm heads (numbers represent means from 9–11 eggs).

[c]Fragmentary material on zona surface, staining with F-PNA but not associated with sperm heads.

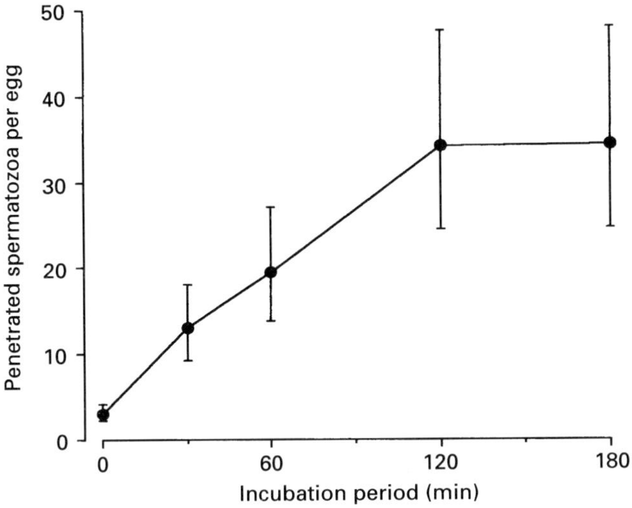

Fig. 7. Zona penetration by boar spermatozoa during incubation of pre-formed sperm–egg complexes. Washed spermatozoa were preincubated for 3 h in medium BHC, and then coincubated for 10 min with groups of eggs. Sperm–egg complexes were transferred to fresh sperm-free medium, and incubation continued for various periods. Before fixation, some of each group of complexes were passaged through a wide-bore pipette (for assessment of tight binding), while the others were repeatedly passaged through a narrow-bore pipette for assessment of penetration (see legend to Fig. 4). Six replicate experiments were performed using semen samples of known high fertility. Each group of eggs yielded a 'mean' data point (spermatozoa per egg). Because of variation between samples with respect to innate binding capabilities, penetration was calculated as a proportion of the number bound. The data were then subjected to analysis of variance after log transformation. The results are shown as back-transformed means that have been multiplied by the overall mean of numbers bound (to indicate the actual number of spermatozoa involved). Standard deviation limits for these means are also shown. The differences between consecutive time points from 0 to 120 min were each individually significant ($P < 0.05$).

not been considered. On the one hand, such agglutination causes sedimentation of the spermatozoa, whence concentrations in the vicinity of the eggs (at the bottom of the container) will be enhanced artificially unless the cells are kept in suspension throughout co-incubation. However, agglutination will restrict access of spermatozoa to the eggs, both because agglutinated spermatozoa are effectively sequestered and because the resultant cell groups block mobility of the remaining free cells. The degree of head-to-head agglutination shown by boar spermatozoa under fertilizing conditions can vary greatly from boar to boar.

Another property of boar spermatozoa that could seriously affect quantitative studies of zona penetration is the innate tendency to adhere to the zona (see above). Not only do large numbers of spermatozoa bind under non-capacitating incubation conditions, but the ability to bind varies considerably between samples. Such variation could have disparate effects. On the one hand, more spermatozoa might be expected to penetrate if the number of associated spermatozoa in the sperm–egg complexes were higher. On the other hand, a strong tendency of the overall sperm population to bind might impede any enhanced binding by spermatozoa with zona penetrating ability.

Why should boar spermatozoa show a general tendency to bind to the zona pellucida, and why should there be much variation between animals or ejaculates? One possibility is that spermadhesins are responsible (see Berger, 1996). Spermadhesins are members of a family of relatively low molecular weight seminal plasma proteins which coat the spermatozoa and also show strong affinity for the zona pellucida (Töpfer-Petersen and Calvete, 1996). Although some spermadhesins are present in epididymal fluid, most are derived from the seminal vesicles. Spermadhesin coating of spermatozoa increases greatly following ejaculation, when the epididymal cells come into contact with seminal vesicle secretions; it then declines again during capacitation (Dostàlovà *et al.*, 1994). If spermadhesins are indeed responsible for the 'non-specific' zona binding we have observed, they pose a serious problem, because the spermadhesin content of a given sperm sample will depend not only upon the intrinsic spermadhesin levels in that boar but also on the amount of seminal vesicle secretion present in the actual semen sample. This latter will vary, particularly within the sperm-rich fractions that are usually taken for research or for commercial purposes: fractionation of the ejaculate is almost always subjective, and the degree to which the sperm-rich (epididymal) fractions may be mixed with adjoining seminal vesicle contributions is very unpredictable. Differences between boars in terms of ejaculate profile have been presented by Lavon and Boursnell (1975) and Xu *et al.* (1996); the latter authors also demonstrated differences in fertilizing performance between spermatozoa in the different ejaculate fractions.

Head-to-head agglutination and 'non-specific' zona binding are both manifestations of boar sperm membrane characteristics. They may relate to specific physiological functions within the overall *in vivo* fertilization process in which an inseminate participates. Immediately, however, either may interfere considerably with quantitative *in vitro* assessments of sperm ability to interact with the egg. The fact that individual semen samples show great variation in each characteristic is of particular concern. Detailed studies of the two phenomena are required.

Tests of Fertilizing Capability: Future Considerations

In vitro tests of sperm–egg interaction are needed as a key step in relating specific sperm functional characteristics with fertilizing ability. However, as pointed out in the Introduction, the eventual goal must be to identify those characteristics that are important for fertilizing success *in vivo*: it is those that must be maintained if semen preservation is to be successful. The conditions under which spermatozoa are induced to fertilize *in vitro* clearly differ very greatly from the situation *in vivo* (see discussion in Harrison, 1996). As our knowledge of the overall fertilization process improves, so these differences can be highlighted, to provide a better insight as to what sperm attributes might be advantageous *in vivo*.

Recent studies have suggested two aspects that merit particular consideration. The first relates to semen longevity in the female reproductive tract. Under natural conditions, sows may be mated at the beginning of oestrus, some 40 h before ovulation occurs (see Hunter, 1994). Even under conditions of competent husbandry, sows will probably be inseminated 12–24 h before ovulation (see Waberski *et al.*, 1994). The inseminated spermatozoa must therefore remain viable within the female reproductive tract for a considerable period before the eggs become available to them. Investigations in cows, sheep and pigs (see Hunter, 1995) have revealed that, regardless of the time of insemination, the fertilizing spermatozoa do not ascend the upper reaches of the oviduct until shortly before ovulation. Instead, following insemination, the spermatozoa rapidly colonize a specific region of the oviduct, the isthmus, close to the uterotubal junction, from which they are later released to encounter the eggs. The role of the oviductal isthmus as a specific sperm reservoir or storage site is now established for laboratory as well as farm animal species (see Smith and Nothnick, 1997). Hunter (e.g. Hunter, 1995) has proposed that periovulatory signals from the ovary are responsible for modifying the local isthmic environment so as to induce sperm release at this time with concomitant completion of capacitation. Inherent in the concept is protection of spermatozoa by the isthmus prior to the engagement of such signals, and indeed oviductal epithelial cells have been shown to exert a strong protective effect on spermatozoa *in vitro* (Smith and Nothnick, 1997, and references therein). Waberski *et al.* (1994) have shown that the well-established decline in fertilization rates following long-term storage of semen under liquid conditions is increasingly pronounced the earlier after onset of oestrus that insemination is performed, but that high fertilization rates continue to be achievable provided insemination is closely synchronized with ovulation. The decline in fertilization rate is accompanied by an even greater decline in the number of accessory spermatozoa found embedded within egg zonae. Linking these observations with the concepts of sperm storage and protection in the oviductal isthmus, it appears that prolonged preservation in a liquid state renders the spermatozoa less able to survive in the female tract. Cryopreservation treatments also appear to compromise sperm survival in the female tract (over and above losses during freezing and thawing). Normal intrauterine insemination of cryopreserved boar semen results in poor fertilization rates, whereas acceptable rates can be achieved by oviductal insemination around the time of ovulation (see Hunter, 1994); similarly, vaginal insemination of cryopreserved ram semen results in very low fertility, whereas uterine insemination achieves satisfactory results (Hunter, 1994; Maxwell and Watson, 1996). A clue to what may be happening during cryopreservation is provided by IVF studies in several species: spermatozoa from frozen semen are able to effect fertilization more rapidly than control spermatozoa from fresh ejaculates, in other words cryopreserved spermatozoa appear prematurely capacitated (see Watson, 1996). Since current beliefs equate capacitation with destabilization (see Harrison, 1996), a shortened lifespan can be expected. At the present time, understanding of the cellular mechanisms that control sperm survival is essentially entirely lacking. There is little doubt that the key organelle is the sperm plasma membrane. Lacking transcriptional or translational activity, the spermatozoon responds to its environment only via the transducing abilities of its plasma membrane, the function of which also controls entirely the internal compartmentation necessary for normal cellular activity. In the long term, therefore, development of improved semen preservation technology will depend on better knowledge of the dynamic functioning of the sperm plasma membrane and the molecular processes that affect its stability. More immediately, however, it is possible that fertility *in vivo* should not be equated with conventional fertility *in vitro*. In effect, IVF seeks fertilization as fast as possible (both for practical reasons and because of the detrimental effects of post-ovulatory ageing of the eggs: see Hunter, 1994). However, this will tend to ascribe quantitative 'success' to those sperm samples in which many individual cells undergo capacitation rapidly. As we have seen, such a response may be in direct conflict with the need to survive for long periods in the female tract. It may be, therefore, that our *in vitro* assessments should include tests of the ability of spermatozoa to survive for long periods under physiological conditions.

At the same time, Hunter's concept of sperm release from the isthmus in response to ovarian signalling (Hunter, 1995) must be coupled with the need for a sufficient supply of capacitated spermatozoa throughout the period of egg release at ovulation. In the case of pigs, the normal complement of 10–20 eggs will be shed over some 2–3 h. Thus, sperm population response to full capacitating conditions should be diverse rather than rapid. In a recent study of the exposure of specific zona-binding sites on the sperm surface under capacitating conditions (Harkema *et al.*, 1998), attention has been drawn to the presence of two subpopulations differing in their speed of response: it was pointed out that the slowly responding subpopulation might represent the spermatozoa that would be more likely to fertilize *in vivo*. It is noteworthy that in a new study Holt *et al.* (1997) observed that higher litter size in pigs after artificial insemination may be correlated with lower frequency of both spontaneous and zona pellucida-induced acrosome reactions during *in vitro* incubation.

Conclusion

Improvements in semen preservation must focus on preserving the sperm functions that are of paramount importance to the fertilization process. To do so, we need to identify these functions, and we need to be able to monitor them by means of simple laboratory tests. The sperm plasma membrane clearly plays a pivotal role in controlling sperm fertilizing ability; not only does it directly mediate the contact interactions between the spermatozoa and the epithelial cells of the female tract and subsequently between the spermatozoa and the egg itself, but it also acts as the receiver of the environmental signals that induce the positive response the spermatozoon must make in order to achieve fertilization. We are still rather far from understanding the relevant aspects of sperm membrane function. However, new knowledge, derived from a wide variety of multidisciplinary approaches, is beginning to reveal key details of the complex processes that enable an inseminate to achieve fertilization. This knowledge should offer the opportunity to derive by logical inference more focused investigations of sperm membrane characteristics as well as more relevant whole-cell approaches to assessing semen fertility. Best progress overall will undoubtedly be achieved by co-ordinating as closely as possible these two areas of study with further investigations of sperm physiology in the female reproductive tract.

The author would like to record the essential contributions of his colleagues, Nigel Miller, Peter Ashworth, and Jim Lynham with respect to the research results presented in this paper. Semen samples and financial support from the member companies of the UK Association of Pig AI Operators (in conjunction with the BBSRC: Grant 202/A06447), which enabled the studies on stored eggs, are also most gratefully acknowledged.

References

Amann RP (1989) Can the fertility potential of a seminal sample be predicted accurately? *Journal of Andrology* **10** 89–98

Amann RP and Hammerstedt RH (1993) *In vitro* evaluation of sperm quality: an opinion *Journal of Andrology* **14** 397–406

Amann RP, Hammerstedt RH and Veeramachaneni DNR (1993) The epididymis and sperm maturation: a perspective *Reproduction, Fertility and Development* **5** 361–381

Ashworth PJC, Harrison RAP, Miller NGA, Plummer JM and Watson PF (1995) Flow cytometric detection of bicarbonate-induced changes in lectin binding in boar and ram sperm populations *Molecular Reproduction and Development* **40** 164–176

Bailey JL and Storey BT (1994) Calcium influx into mouse spermatozoa activated by solubilized mouse zona pellucida, monitored with the calcium fluorescent indicator, fluo-3. Inhibition of the influx by three inhibitors of the zona pellucida induced acrosome reaction: tyrphostin A48, pertussis toxin, and 3-quinuclidinyl benzilate *Molecular Reproduction and Development* **39** 297–308

Bedford JM (1974) Report of a workshop. Maturation of the fertilizing ability of mammalian spermatozoa in the male and female reproductive tract *Biology of Reproduction* **11** 346–362

Bedford JM (1991) The coevolution of mammalian gametes. In *A Comparative Overview of Mammalian Fertilization* pp 3–35 Eds BS Dunbar and MG O'Rand. Plenum Press, New York

Berger T (1996) Fertilization in ungulates *Animal Reproduction Science* **42** 351–360

Berger T, Davis A, Wardrip N and Hedrick JL (1989) Sperm binding to the pig zona pellucida and inhibition by solubilized components of the zona pellucida *Journal of Reproduction and Fertility* **86** 559–586

Berger T, Anderson D and Penedo MCT (1996) Porcine sperm fertilizing potential in relationship to sperm functional capacities *Animal Reproduction Science* **44** 231–239

Burks DJ, Carballada R, Moore HDM and Saling PM (1995) Interaction of a tyrosine kinase from human sperm with the zona pellucida at fertilization *Science* **269** 83–86

Codde JM and Berger T (1995) *In vivo* fertility of rams in relation to sperm–zona pellucida binding and sperm–zona pellucida penetration of ovine oocytes *Theriogenology* **44** 901–906

Cohen-Dayag A and Eisenbach M (1994) Potential assays for sperm capacitation in mammals *American Journal of Physiology* **267** C1167–C1176

Dietl J and Czuppon AB (1984) Ultrastructural studies of the porcine zona pellucida during the solubilization process by Li-3,5-diiodosalicylate *Gamete Research* **9** 45–54

Dostàlovà Z, Calvete JJ, Sanz L and Töpfer-Petersen E (1994) Quantitation of boar spermadhesins in accessory sex gland fluids and on the surface of epididymal, ejaculated and capacitated spermatozoa *Biochimica et Biophysica Acta* **1200** 48–54

Dzuik PJ (1996) Factors that influence the proportion of offspring sired by a male following heterospermic insemination *Animal Reproduction Science* **43** 63–88

Familiari G, Nottola SA, Macchiarelli G, Micara G, Aragona C and Motta PM (1992) Human zona pellucida during *in vitro* fertilization: an ultrastructural study using saponin, ruthenium red, and osmium thiocarbohydrazide *Molecular Reproduction and Development* **32** 51–61

Fazeli AR, Holt C, Steenweg W, Bevers MM, Holt WV and Colenbrander B (1995) Development of a sperm hemizona binding assay for boar semen *Theriogenology* **44** 17–27

Harayama H, Miyano T, Miyake M, Kusunoki H and Kato S (1994) Identification of anti-agglutinin for spermatozoa in epididymal boar plasma *Molecular Reproduction and Development* **37** 436–445

Harkema W, Harrison RAP, Miller NGA, Topper EK and Woelders H (1998) Enhanced binding of zona pellucida proteins to the acrosomal region of intact boar spermatozoa in response to fertilizing conditions: a flow cytometric study *Biology of Reproduction* **58** 421–430

Harrison RAP (1996) Capacitation mechanisms, and the role of capacitation as seen in eutherian mammals *Reproduction, Fertility and Development* **8** 581–594

Harrison RAP, Mairet B and Miller NGA (1993) Flow cytometric studies of bicarbonate-mediated Ca^{2+} influx in boar sperm populations *Molecular Reproduction and Development* **35** 197–208

Harrison RAP, Ashworth PJC and Miller NGA (1996a) Bicarbonate/CO_2, an effector of capacitation, induces a rapid and reversible change in the lipid architecture of boar sperm plasma membranes *Molecular Reproduction and Development* **45** 378–391

Harrison RAP, Ashworth PJC and Miller NGA (1996b) Assessment of sperm function under fertilizing conditions *Reproduction in Domestic Animals* **31** 25–30

Holt C, Holt WV, Moore HDM, Reed HCB and Curnock RM (1997) Objectively measured boar sperm motility parameters correlate with the outcomes of on-farm inseminations: results of two fertility trials *Journal of Andrology* **18** 312–323

Hunter RHF (1994) Causes for failure of fertilization in domestic species. In *Embryonic Mortality in Domestic Species* pp 1–22 Eds MT Zavy and RD Geisert. CRC Press, Boca Raton, Florida

Hunter RHF (1995) Ovarian endocrine control of sperm progression in the Fallopian tubes *Oxford Reviews of Reproductive Biology* **17** 85–124

Ivanova M and Mollova M (1993) Zona-penetration *in vitro* test for evaluating boar sperm fertility *Theriogenology* **40** 397–410

Johnson LA, Aalbers JG and Grooten HJG (1988) Artificial insemination of swine: fecundity of boar semen stored in Beltsville TS (BTS), Modified Modena (MM), or MR-A and inseminated on one, three and four days after collection *Zuchthygiene* **23** 49–55

Johnson LA, Maxwell WMC, Dobrinsky JR and Welch GR (1996) Staining sperm for viability assessment *Reproduction in Domestic Animals* **31** 37–47

Johnson LR, Pilder SH, Bailey JL and Olds-Clarke P (1995) Sperm from mice carrying one or two *t* haplotypes are deficient in investment and oocyte penetration *Developmental Biology* **168** 138–149

Lavon U and Boursnell JC (1975) The split ejaculate of the boar: contributions of the epididymides and seminal vesicles *Journal of Reproduction and Fertility* **42** 541–552

Leyton L and Saling P (1989) 95 kd sperm proteins bind ZP3 and serve as tyrosine kinase substrates in response to zona binding *Cell* **57** 1123–1130

Lynham JA and Harrison RAP (1998) The use of stored pig eggs to assess boar sperm fertilizing functions *in vitro Biology of Reproduction* **58** (in press)

Mann T and Lutwak-Mann C (1981) *Male Reproductive Function and Semen* pp 26 and 141. Springer-Verlag, Berlin

Matas C, Martinez E, Vazquez JM, Roca J and Gadea J (1996) *In vitro* penetration assay of boar sperm fertility: effect of various factors on the penetrability of immature pig oocytes *Theriogenology* **46** 503–513

Mattioli M, Barboni B, Lucidi P and Seren E (1996) Identification of capacitation in boar spermatozoa by chlortetracycline staining *Theriogenology* **45** 373–381

Maxwell WMC and Watson PF (1996) Recent progress in the preservation of ram semen *Animal Reproduction Science* **42** 55–65

Okamura N, Tajima Y, Soejima A, Masuda H and Sugita Y (1985) Sodium bicarbonate in seminal plasma stimulates the motility of mammalian spermatozoa through direct activation of adenylate cyclase *Journal of Biological Chemistry* **260** 9699–9705

Olds-Clarke P (1996) How does poor motility affect sperm fertilizing ability? *Journal of Andrology* **17** 183–186

Peterson RN, Russell LD and Hunt WP (1984) Evidence for specific binding of uncapacitated boar spermatozoa to porcine zonae pellucidae *in vitro*. *Journal of Experimental Zoology* **231** 137–147

Phillips DM, Shalgi R and Dekel N (1985) Mammalian fertilization as seen with the scanning electron microscope *American Journal of Anatomy* **174** 357–372

Roldan ERS, Murase T and Shi Q-X (1994) Exocytosis in spermatozoa in response to progesterone and zona pellucida *Science* **266** 1578–1581

Saacke RG, Nadir S and Nebel RL (1994) Relationship of semen quality to sperm transport, fertilization, and embryo quality in ruminants *Theriogenology* **41** 45–50

Sharps ES and McCarl RL (1982) A high performance liquid chromatographic method to measure ^{32}P incorporation into phosphorylated metabolites in cultured cells *Analytical Biochemistry* **124** 421–424

Smith TT and Nothnick WB (1997) Role of direct contact between spermatozoa and oviductal epithelial cells in maintaining rabbit sperm viability *Biology of Reproduction* **56** 83–89

Suzuki K, Ebihara M, Nagai T, Clarke NGE and Harrison RAP (1994) Importance of bicarbonate/CO_2 for fertilization of pig oocytes *in vitro*, and synergism with caffeine *Reproduction, Fertility and Development* **6** 221–227

Thaler CD and Cardullo RA (1996) The initial molecular interaction between mouse sperm and the zona pellucida is a complex binding event *Journal of Biological Chemistry* **271** 23289–23297

Töpfer-Petersen E and Calvete JJ (1996) Sperm-associated protein candidates for primary zona pellucida-binding molecules: structure–function correlations of boar spermadhesins *Journal of Reproduction and Fertility Supplement* **50** 55–61

Visconti PE, Moore GD, Bailey JL, Leclerc P, Connors SA, Pan D, Oldsclarke P and Kopf GS (1995) Capacitation of mouse spermatozoa. II. Protein tyrosine phosphorylation and capacitation are regulated by a cAMP-dependent pathway *Development* **121** 1139–1150

Waberski D, Weitze KF, Lietmann C, Lübbert zur Lage W, Bortolozzo FP, Willmen T and Petzoldt R (1994) The initial fertilizing capacity of longterm-stored liquid boar semen following pre- and postovulatory insemination *Theriogenology* **41** 1367–1377

Wang WH, Abeydeera LR, Fraser LR and Niwa K (1995) Functional analysis using chlortetracycline fluorescence and *in vitro* fertilization of frozen–thawed ejaculated boar spermatozoa incubated in a protein-free chemically defined medium *Journal of Reproduction and Fertility* **104** 305–313

Wassarman PM (1987) Early events in mammalian fertilization *Annual Review of Cell Biology* **3** 109–142

Watson PF (1996) Cooling of spermatozoa and fertilizing capacity *Reproduction in Domestic Animals* **31** 135–140

Xu X, Ding J, Seth PC, Harbison DS and Foxcroft GR (1996) *In vitro* fertilization of *in vitro* matured pig oocytes: effects of boar and ejaculate fractions *Theriogenology* **45** 745–755

Yanagimachi R (1994) Mammalian fertilization. In *The Physiology of Reproduction* (2nd Edn) 19 Vol. 1, pp 189–317 Eds E Knobil and JD Neill. Raven Press, New York

Yoshida M, Cran DG and Pursel VG (1993) Confocal and fluorescence microscopic study using lectins of the distribution of cortical granules during the maturation and fertilization of pig oocytes *Molecular Reproduction and Development* **36** 462–468

Journal of Reproduction and Fertility Supplement **52**, 213–222

Assessment of boar sperm function in relation to freezing and storage

W. V. Holt and A. Medrano

Institute of Zoology, Zoological Society of London, Regent's Park, London NW1 4RY, UK

The functions necessary for normal fertilization to occur *in vivo* or *in vitro* are examined and a rational approach to identifying the main features of a fertilizing spermatozoon are developed. It is concluded that methods for testing the quality of spermatozoa must probe the dynamic changes experienced by the spermatozoa during capacitation or under stressful incubation conditions. Recent developments in the multivariate analysis of sperm motility data are used to illustrate the success that can be achieved by this approach. Ideally, changes in sperm motility characteristics should be correlated with an assessment of capacitation status. However, until the capacitation status of any individual cell can be clearly defined this will remain problematic.

Introduction

All sperm function tests share the practical constraint that they inevitably operate by measuring parameters in ejaculated or epididymal spermatozoa, but the results are extrapolated forwards in time to predict the outcome of fertilization *in vivo* or *in vitro*. Viewing fertilization in its broadest sense, sperm function tests are required to predict the outcomes of numerous processes, many of which are poorly understood. These include efficiency of sperm transport, speed of capacitation, ability to bind to the egg surface, undergo the acrosome reaction, penetrate the zona pellucida and fuse with the oolemma. The developmental competence of the male pronucleus formed from the spermatozoon can be included in this sequence. To compound the difficulty, these attributes can themselves be reduced to a series of integrated molecular interactions, the efficacy of each being dependent upon developmental processes that took place during spermatogenesis and sperm maturation. This could have been several weeks before the time of sperm assessment.

Assessment of stored semen is performed for reasons that include (1) comparison of storage methods under experimentally controlled conditions, (2) identification of males who, if not subfertile under natural conditions, may provide semen that responds poorly to liquid or frozen storage, and (3) routine quality assurance. As these purposes aim mainly to compare animals, ejaculates or individual samples of semen, exact fertility levels are rarely required. In these situations sperm function tests therefore have a very useful role, and provide valid information that can be interpreted reasonably well on a comparative basis.

What Are the Tests and What Do They Measure?

During storage, whether by cryopreservation or dilution in an extender, a proportion of the spermatozoa lose their capacity to fertilize. This can be through sustaining a major lesion, such as loss of plasma membrane integrity, acrosomal degeneration, or through more subtle processes. The gross damage is relatively easy to distinguish, and many tests have been developed for the purpose. Some depend upon the exclusion of impermeant dyes by intact cells (e.g. propidium iodide or eosin) by intact cells, or the uptake and subsequent differential loss or retention of fluorescent dyes such as carboxyfluorescein or calcein homodimer (Harrison and Vickers, 1990; Johnson *et al.*, 1996). Use of the fluorescent dyes allows sperm viability testing to be performed by both microscopy and flow

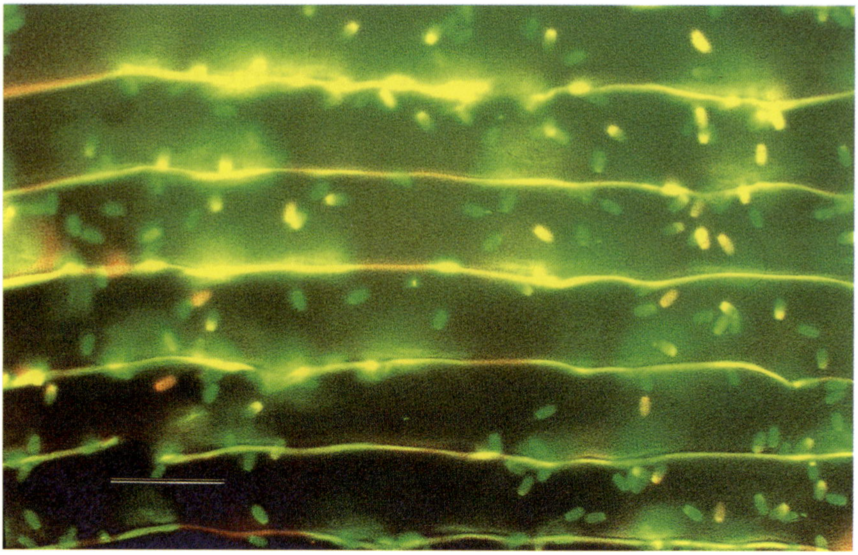

Fig. 1. Photomicrograph showing boar spermatozoa frozen in cryopreservation medium (BF5) on a cryomicroscope stage at –20°C. SYBR-14 and propidium iodide were included in the BF5 diluent as probes of membrane integrity. Live cells (SYBR-14 positive) fluoresce green, and dead, or membrane damaged, cells (propidium iodide positive) fluoresce red. Scale bar represents 20 μm.

cytometry; with the latter technique, integrity of the cell membrane can be monitored quickly on large numbers of cells. One of the fluorescence-based tests, which uses SYBR-14 and propidium iodide to distinguish intact and damaged spermatozoa (Garner and Johnson, 1995), has proved useful for direct observations of spermatozoa in ice during cryomicroscopy (Medrano and Holt, 1996) (Fig. 1). It is also suitable as a long-acting detector of sperm viability within egg yolk-containing media and thus useful for assessing cryopreservation protocols. A different approach to the recognition of membrane-intact spermatozoa involves the hypo-osmotic swelling test (HOS). This is based on the principle that spermatozoa swell in hypotonic media, their flagellae becoming coiled, but only if the plasma membrane remains intact. The HOS test has been combined with the use of fluorescent probes to evaluate acrosomal and plasma membrane integrity simultaneously. Richardson and Aitken (1993) discussed the value of the HOS test in predicting human *in vitro* fertilization results, and concluded that it does not offer satisfactory predictability relative to the zona-free hamster egg penetration test. While osmotic fragility is likely to vary among individuals within a species, there may be more important species-dependent differences in membrane elasticity. It is therefore inadvisable to extrapolate HOS test conditions between species. Although phase contrast microscopy of fixed spermatozoa has been used extensively to evaluate acrosomal status, generating data about the proportions of normal acrosomal ridges (% NAR) seen in many publications, this approach has also been largely superseded by the use of fluorescent probes. Many lectins, proteins that interact with the glycoconjugates of the acrosomal membranes or matrix, have been tested for use in acrosomal evaluation (for review, see Cross, 1995). These include peanut agglutinin, Concanavalin A, *Pisum sativum* and *Ricinus communis* agglutinins. Monoclonal antibodies are also used for acrosomal evaluation, but in practical terms they are more time consuming (see Cross, 1995). Acrosomal integrity is not necessarily indicative of an intact plasma membrane; therefore, combining two tests for these attributes results in a more discriminatory test than either on its own.

Tests that aim to probe the ability of the cell to respond appropriately to agents that promote the acrosome reaction are now used widely. Methods that test ability to bind and penetrate the zona pellucida also fall into this category. Whitfield and Parkinson (1992) used heparin to induce the acrosome reaction in bull spermatozoa and, in a fertility trial using frozen semen, observed a high correlation between 90 day non-return rates and the extent of acrosome reaction induction. This study was remarkable in that the correlation was found even though the acrosome reactions were detected using fresh, not cryopreserved, spermatozoa. One interpretation of this is that freezing itself did not change the heparin responsiveness of the structurally intact cells. Furthermore, the correlation between fresh and frozen spermatozoa also implies that the acrosome reacting ability may be related to capacity to reach the site of fertilization, a function governed mainly by cell motility. Graham and Foote (1987), who used liposomes instead of heparin, found broadly similar results. In human clinical investigations, the principle of examining acrosomal responses to various substances, for example ionophore A23187 and more recently progesterone has been developed, although the validity of this approach has yet to be confirmed. A slightly different strategy for testing the functional integrity of the plasma membrane–acrosome complex has been evaluation of the sperm interaction with isolated zonae pellucidae. As acrosome reactions are induced physiologically by zona pellucida proteins, incubation of spermatozoa with fresh, frozen or salt-stored zonae has been developed as a test. The end-points can be either the number of spermatozoa bound to the zona surface, or the number that undergo the acrosome reaction and commence penetration (see for example, Fazeli *et al.*, 1993). This type of test was developed further with the introduction of the hemi-zona assay (Fazeli *et al.*, 1995), in which one half of a zona is reacted with a reference semen sample while the other is incubated with the sample of interest. Sperm–zona interactions are discussed in more detail by Harrison (this supplement).

If a large proportion of spermatozoa in an inseminate sustains any degree of gross damage, the 'effective concentration' of spermatozoa is reduced, and the number of spermatozoa potentially capable of participating in fertilization is thereby diminished. In some previous reviews (for example Amann, 1989; Den Daas, 1992), the relationship between fertility and number of spermatozoa inseminated has been presented graphically as a dose–response curve with an asymptote, where continued increases in sperm dose do not further increase fertility. Given that for practical purposes the inseminated sperm dose would normally be set within the asymptotic region of the response, reducing the effective concentration of spermatozoa in an inseminate is equivalent to shifting the total sperm concentration back into the sensitive region of the dose–response curve; this could potentially cause a serious reduction in fertility. Assessment of the proportion of grossly intact cells, including those that display normal motility, is thus clearly useful for comparing samples with one another and detecting this type of shift.

Arguments about sperm viability or motility can in principle be extended to include ever more subtle levels of sperm dysfunction; with the application of each new test a subset of spermatozoa can be regarded as non-functional, thereby affecting the dose–response relationship. However, although gross lesions are easily understood, it is more difficult to categorize other differences between individual cells as rendering them unambiguously 'fertile' or 'infertile'. Morphological assessment of spermatozoa provides some clear examples where infertile spermatozoa can be recognized, but this is more applicable to the diagnosis of infertility than to semen assessment from normal donors. For example, occasional cases of human male infertility are attributable to absence of the acrosome or to genetic lesions leading to complete immotility (Kartagener's syndrome). The introduction of highly systematized methods for assessing sperm morphology, the so-called 'strict criteria' (Kruger *et al.*, 1987), may help to consolidate this approach for animal as well as human spermatozoa.

Having recognized that only a proportion of cells survive storage processes with their fertilizing ability intact, the problem of sperm function testing becomes more complex. Two cells may both be intact and responsive, but if one undergoes capacitation and the acrosome reaction before reaching the oocyte, it has rendered itself infertile. On this basis, the definition of fertility becomes very elusive; assuming that an oocyte is already present, the short-lived spermatozoon will be the one that fertilizes. The only conclusion to be drawn from this is that fertility is dependent upon circumstances. Watson (1995) has argued that after freezing and thawing, spermatozoa experience accelerated capacitation which renders them inadequately fertile if they are expected to behave like fresh spermatozoa. If, however, allowed to fertilize within a new shortened time frame, they perform very well. This introduces the concept that sperm assessment should include aspects that test the dynamics of capacitation and fertilization, as well as simply testing the ability of the cell to undergo these processes. Viewed in this way, it could be argued that the greatest chances of conception will be associated with semen samples that contain subpopulations of spermatozoa whose dynamics differ; thus the rapidly responding cells will fertilize oocytes if ovulation is early with respect to insemination, while slowly responding cells will be able to remain quiescent and then respond if the interval to ovulation is prolonged.

Regarding sperm function assessment in this way accords well with current concepts of sperm behaviour in the female reproductive tract. Although it has been established that spermatozoa appear within the oviduct within about 15 min of insemination, these are often a selected population of dead cells carried there by contractions of the female tract. A combination of the intrinsic vigour of the sperm cell and uterine muscular contractions allows only a selected subset of highly motile cells to colonize the uterus and oviduct. To achieve this goal they have to traverse the cervix in species such as cattle, sheep and humans, and/or the uterotubal junction in other species. These processes are highly selective, probably involving aspects of plasma membrane surface biochemistry to allow avoidance of phagocytosis by leucocytes as well as the expression of motility (Harper, 1994). In rodent species it has been shown that traversing the uterotubal junction requires activated, rather than hyperactivated, motility (Olds-Clarke, 1996) emphasizing once more the need for temporal control of capacitation processes.

Once the spermatozoa have reached the oviductal isthmus, it seems that at least a proportion bind to the epithelium and may become quiescent until ovulation (Overstreet and Cooper, 1975; Hunter and Nichol, 1983). The dynamics of this process are poorly understood, but sperm survival is thought to be crucially dependent upon sperm–oviduct interactions. Ovulation itself may provide the trigger for sperm reactivation; the spermatozoa then pass from the isthmus to the ampulla of the oviduct, perhaps using hyperactivated motility to help extricate themselves from oviductal crypts if they become trapped (Suarez, 1987). On reaching the ampulla, the spermatozoa are in principle ready to respond to contact with the oocyte by binding to its surface and undergoing the acrosome reaction. This is another dynamically unstable process, since the spermatozoa can undergo spontaneous acrosome reactions in the absence of an egg. Those that remain intact until receiving a physiological stimulus could, therefore, be considered to constitute the most highly fertile population. Data from recent pig artificial insemination trials lend support to that view; when boar ejaculates were classified as producing large or small litters, large litters were consistently associated with lower frequency of *in vitro* induced acrosome reactions (Holt *et al.*, 1997).

The above discussion repeatedly identifies the dynamic aspects of sperm function as being important for functional assessment, implying that tests are most informative when they attempt to mimic *in vitro* various physiological aspects of sperm survival or capacitation. This principle has previously been recognized in some of the older literature that used thermoresistance tests as a crude way of determining whether spermatozoa could survive prolonged heating (Larsson, 1985). The ability to assess capacitation status provides a useful insight into the effects of preservation methods on subsequent sperm dynamics, but there is a need to develop a wider range of techniques. As the topic of sperm function assessment has been reviewed on several occasions recently, the remainder of this review will be confined to considering some of the most recent conceptual advances in sperm motility and capacitation assessment.

Recent Developments in Sperm Motility Analysis

The value of sperm motion measurements for the diagnosis of human subfertility was demonstrated in the early 1980s (Aitken *et al.*, 1982a,b). However, the manual techniques needed at the time were extremely laborious and consequently were used by few investigators. Computer-assisted semen analysis (CASA) systems have undergone considerable development over the past decade, and now have the capacity to measure multiple motion parameters on hundreds or thousands of individual spermatozoa in any given sample. The value of these systems lies in their ability to perform the motion analyses, and this discussion will concentrate on this aspect of their worth. Mainly in response to clinical demand, they also calculate sperm concentrations and the proportion of motile cells; however, they are not ideally suited for this purpose owing to problems of cell versus debris discrimination. Furthermore, although CASA systems are now widely used in clinical laboratories investigating cases of infertility, they are mostly used to make snapshot measurements of sperm behaviour. In view of the arguments presented above which demonstrate the need to measure the dynamics of sperm behaviour, this seems somewhat ill considered.

Studies of human sperm survival *in vitro* provided a clue to the possibility that examining changes in sperm behaviour is one means of obtaining information relevant to fertility. Sperm swimming speeds in frozen human donor semen samples measured shortly after thawing, and again after incubation for 3.5 h at 37°C, revealed that the decline in velocity over that period was associated with differences in fertility when the frozen semen was used in a donor insemination programme (Holt *et al.*, 1989). Higher cumulative pregnancy rates were associated with a greater decline in sperm velocity *in vitro*. It is important to note that these differences were all derived from measurements of the live, motile, sperm population, indicating that the results cannot be attributed simply to sperm death. This study has since been repeated on a larger scale with similar results (MacLeod and Irvine, 1995).

These studies led us to test the hypothesis that changes in sperm velocity during incubation are correlated with fertility *in vivo*, this time using the pig as the model system (Holt *et al.*, 1997). Various motion parameters of boar spermatozoa were measured in aliquots of semen prepared for artificial insemination. The semen was transferred from its extender, Beltsville Thawing Solution (BTS), into a Tris-buffered medium containing 40 mmol calcium l^{-1}, incubated at 39°C for up to 6 h, and motion parameters measured by CASA. Two separate fertility trials were performed, in which the insemination dose was set to a minimum compatible with commercial AI usage (1.5×10^9 spermatozoa per dose). The high calcium medium was chosen following the demonstration that this medium supported the capacitation of boar spermatozoa, as judged by the zona-free hamster oocyte penetration test (Berger and Horton, 1988). Both fertility trials resulted in significant relationships between sperm velocity during incubation and on-farm fertility, estimated as mean litter size or conception rate. Up to approximately 20% of variance could be explained by the sperm velocity measurements. In one trial, the correlations with fertility were apparent in the mean velocity parameters over the first 2 h of incubation, while in the other they were found in the differences in velocity detected over the first 2 h of incubation. Larger litters and conception rates were associated with increased straight line velocity and track linearity during the first 2 h *in vitro* (Fig. 2). Although 20% of variation seems insignificant in a practical context, it should be remembered that this is against a background where variation is also due to factors such as the skill of the AI operators and differences in female fertility. As these sources of variation become more controlled, the explained variability due solely to semen effects will be relatively increased.

This type of analysis was taken a step further by Moore and Akhondi (1996), who examined rat sperm motility data in relation to *in vitro* fertilization results. Their studies used an automated CASA system, in which epididymal spermatozoa were incubated in 24-well culture plates and regularly revisited for sampling every 30 min for 5 h, with the aid of a computer-controlled automatic stage. The fertilizing capacity of spermatozoa correlated exceptionally well with the decline in mean straight line velocity (VSL) over the 5 h ($r = -0.933$; $P < 0.001$); conversely, there was no relationship between the initial VSL value and fertilizing capacity. The objective of this work was to evaluate an approach that might be useful in reproductive toxicology, where the aim is mainly to find an index of sperm function that reflects events happening during spermatogenesis and sperm maturation; however, this technical approach should also find general application under any circumstances involving inter-sample comparisons.

Under *in vitro* fertilization conditions there is obviously no sperm selection due to passage through the female tract, so it could be argued that measures obtained with this system are mainly a reflection of the capacity to penetrate the zona pellucida. Although this argument is difficult to refute without further evidence, it is likely that the main attribute being measured is one aspect of a related set of flagellar functions that could include anything from membrane ion permeability and calcium regulation to protein phosphorylation. It would, therefore, be more surprising if *in vivo* and *in vitro* studies were found to be completely independent.

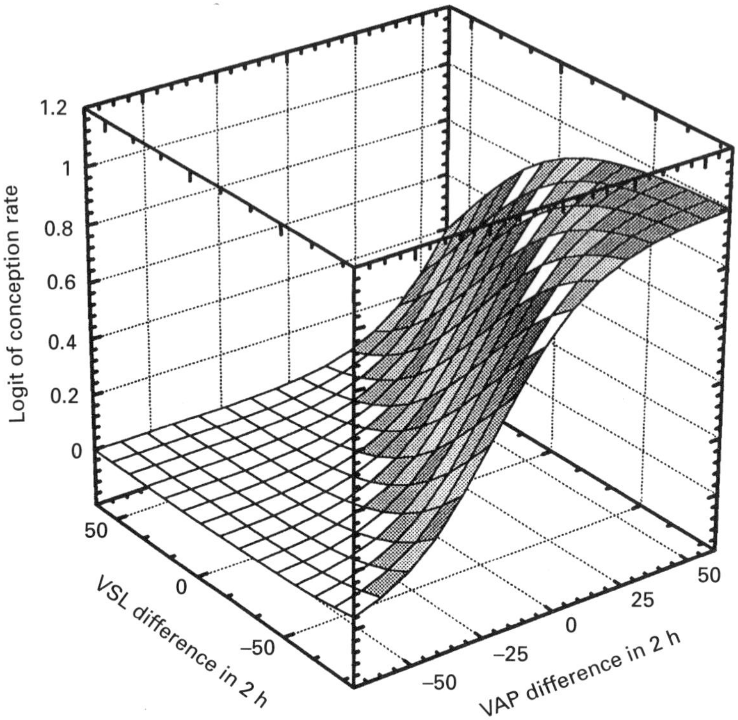

Fig. 2. Three-dimensional plot representing the relationships between conception rate and boar sperm velocity changes during incubation for 2 h at 39°C in Tris-buffered medium, containing 40 mmol calcium l^{-1}. Increases in VSL (straight line velocity) over 2 h, represented as negative differences, and decreases in VAP (average path velocity), represented as positive differences, were associated with higher conception rates (logit transform) in a fertility trial. These data were from a study undertaken in collaboration with the Meat & Livestock Commission and JSR Healthbred Ltd. For further details see Holt *et al.* (1997).

In the studies cited above the sperm motion data were always analysed in terms of population mean values; this method of analysis is almost universal in studies using CASA. The main assumption with this approach is that samples of semen contain a homogeneous population of spermatozoa, which tend to behave synchronously throughout their lifetime. However, several lines of evidence would contradict this view; and it is evident that there are a number of sperm subpopulations that behave asynchronously. Perhaps one of the clearest examples of this principle is that of transmission distortion in mice heterozygous for the t^{w32} gene, where almost all eggs are fertilized *in vivo* by the t^{w32}-carrying spermatozoa, even though the 50% of spermatozoa that do not carry this gene are normally fertile when tested *in vitro*. The explanation for this effect probably includes molecular differences in flagellar membrane function between the two sperm subpopulations, which lead to variations in motility and thus differential ability to gain access to the oviduct (Olds-Clarke and Sego, 1992). While this example probably represents an extreme form of subpopulation structure within a semen sample, other examples also occur for which there is currently no known explanation. Sex ratio distortion has been related to the timing of insemination with respect to ovulation (Clutton-Brock *et al.*, 1986), an effect that could be attributable to differences in capacitation rate between X- and Y-chromosome bearing spermatozoa.

At present it is not possible to correlate unequivocally the motion parameters of *individual* spermatozoa with their genetic content; this will require the development of innocuous DNA probes that can be applied to spermatozoa without killing them. However, the wealth of multivariate motion data available from CASA systems has stimulated interest in the problem of characterizing sperm subpopulations on the basis of their motility. Recognition of hyperactivated human spermatozoa as a discrete subset has been explored by several research groups (see Irvine, 1995), some of which have then used their criteria of hyperactivation as the basis for correlating the proportion of hyperactivated cells in an ejaculate with various fertility outcomes. This approach seems very promising, and in some cases the incidence of hyperactivation has proved highly predictive of *in vitro* fertilization rate. Sukcharoen *et al.* (1996) could correctly predict over 70% of good or poor IVF results using hyperactivation criteria alone. This approach to the identification of sperm subsets in semen was also used by Davis *et al.* (1995) to see whether characteristics of the sperm population substructure in fresh samples could be used to predict how well the samples would survive cryopreservation. This study is especially interesting as it provides statistical methodology for the objective identification of subsets by cluster analysis, but also because it was concluded that a small set of kinematic variables measured in fresh samples could be used to predict the proportion of motile cells present after freezing and thawing. The interpretation of this somewhat surprising result is presumably that the motion parameters of the live cells measured before freezing are identifying subpopulations that are particularly resistant or vulnerable to the freeze–thawing process. The authors commented that their best overall regression model predicted motility from the difference in straight line velocity (VSL) between the cluster with the highest VSL and the cluster with the lowest VSL.

These novel approaches to the use of CASA have been explored to a much lesser extent with animal spermatozoa. Holt (1996) outlined cluster analysis approaches to the identification of sperm subsets within boar semen samples. It was shown that when bicarbonate and caffeine were used to induce different types of sperm motion, the changing population substructures could also be detected numerically. Earlier it had also been demonstrated that if artificial data sets were constructed by combining measurements from three sperm populations showing different types of movement, defined as circular, progressive or erratic, cluster analysis algorithms could disentangle the data with reasonable efficiency (Holt *et al.*, 1992).

All fertility trials conducted in animal species use either homospermic or heterospermic insemination protocols. In the latter case, semen samples from two or more individuals are mixed prior to insemination. Correlations of CASA data with fertility results have been attempted within both paradigms; in some cases marked differences in the outcomes of these correlations have been observed. When Budworth *et al.* (1988) studied frozen–thawed bull semen, no significant relationships between CASA parameters and homospermic inseminations were detected (all r values <0.1), but the motion parameters curvilinear velocity (VCL) and straightline velocity (VSL) accounted for approximately 70% of variance in competitive fertility index under the heterospermic insemination protocol. Such contrasting relationships must point to fundamental differences between the biology of the two systems, and suggest that the two types of test are addressing different questions. With homospermic inseminations the question is 'will conception occur, and if so how many eggs will be fertilized?', while for heterospermic inseminations there is an implicit assumption that conception will definitely occur and the question is then 'which subpopulation of sperm is more successful?'. This is similar to the analysis presented above where there are also subpopulations within a single sample of semen. As the questions are different, the answers can hardly be the same. The arguments presented here strongly indicate that although the technology for measuring sperm motion parameters is now well developed (see, Boyers *et al.*, 1989; Davis and Katz, 1992, 1993), much is yet to be done on methods of data analysis. Exploration of the multivariate analysis approach combined with dynamic aspects of sperm motility is likely to be the best option for extracting useful information from the use of CASA.

Integrating such approaches into the routine working environment of a commercial pig AI centre, where hundreds of semen samples have to be collected and diluted for efficient distribution to farmers, may require some compromises to be made. The time delay introduced by preparing semen samples for CASA, and the need for repeat assessments over 1 or 2 h, will obviate use of the technique as a routine screening method for all ejaculates. However, the method is likely to be useful in planned quality assurance programmes or for testing the effects of modified practices upon the quality of stored semen.

Developments in the Detection of Capacitation

Ideally, it would be useful to combine the measurement of sperm motion with assessment of capacitation status. This approach would allow us to answer the type of questions recently raised by Watson (1995), who proposed that cryopreservation causes accelerated capacitation and consequently shortens the longevity of spermatozoa *in vivo* and *in vitro*. Watson (1995) reviewed evidence from *in vitro* studies of human and ram semen, both of which supported the principle that frozen–thawed spermatozoa were able to fertilize oocytes almost immediately, whereas fresh spermatozoa required an interval of preincubation before acquiring the same readiness to fertilize. Similar arguments can be adduced from *in vivo* studies, where the lifespan of frozen–thawed spermatozoa in the female tract has been demonstrated to be considerably shorter than that of fresh semen. Whether this is an artificial form of destabilization, rather than a true manifestation of capacitation, requires further research. For example, it is known that capacitation is accompanied by membrane hyperpolarization (Zeng *et al.*, 1995), which possibly permits the spermatozoa to switch rapidly between hyperactivated and activated motility, and allows them to respond appropriately to the zona pellucida stimulus.

Current concepts of capacitation were reviewed recently by Harrison (1996), who confirmed that individual spermatozoa within a population become capacitated at different rates. For studies of sperm dynamics, precise knowledge of capacitation in individual cells is crucial, as it would permit characterization of the sperm population behaviour in great detail. To date there are no tests that discriminate microscopically between capacitated and non-capacitated spermatozoa; this is partly because recognizing that an individual spermatozoon has moved from the uncapacitated to the capacitated state requires foreknowledge of its ability to fertilize the oocyte. Most data relevant to this problem have come from flow cytometric studies, observation of responses to the zona pellucida or calcium ionophore stimulus, or through the application of a fluorescent antibiotic, chlortetracycline (CTC). However, all of this is circumstantial, and based on the behaviour of sperm populations rather than of individuals. The interesting aspect of CTC is that individual cells can be distinguished on the basis of sperm head fluorescence patterns. These have been considered as (1) uncapacitated (F pattern), (2) acrosome reacted (AR) or (3) B pattern, where the fluorescence is diminished over part of the sperm head; the B pattern has been widely interpreted as indicating capacitation (Wang *et al.*, 1995). Watson (1996) showed that cooling boar spermatozoa induced an increased incidence of B patterns, thus supporting the contention that cryopreservation accelerates capacitation. The exact mechanism of CTC staining remains controversial; one explanation is that it may interact with calcium at the sperm surface, and thereby be indicative of calcium ATPase activity (Adeoya-Osiguwa and Fraser, 1996). So far, experiments involving CTC have involved fixation of the spermatozoa before their evaluation; this necessarily precludes the simultaneous assessment of motility and capacitation status, although such an approach would be highly informative.

It seems self evident that any spermatozoon that fertilizes an egg must have been through the capacitation process; however, recognizing the process unequivocally may be impossible as any test has, by definition, to be predictive and therefore subject to uncertainty. The problem is hampered by the likelihood that at any given moment, only a small proportion of spermatozoa have the ability to fertilize. As hyperactivation is at least a functional property required for zona penetration, the efficient recognition of this attribute may ultimately prove more effective than the search for visible changes in plasma membrane organization.

Conclusions

In this review we take the position that rather than listing a series of sperm function tests and their technical merits, it is more important and interesting to analyse the meaning of such tests and from that discussion derive arguments that guide future research. The most important conclusions from this exercise seem to be that sperm function testing is not always aimed at estimating fertility on a quantitative basis, and that having realized this it is possible to derive a considerable amount of comparative information of experimental or practical value. The overriding aspect of the discussion is that as more becomes known about the dynamic changes undergone by spermatozoa during physiological processes such as capacitation, it will be possible to derive ever better objectively based function tests. The disadvantage of the logical approach to sperm function tests is that they may become too complex for implementation in routine laboratories, where the aim is quality assurance rather than diagnosis or research. It will be a challenge to simplify the available tests without losing sight of the requisite functional assessments.

References

Adeoya-Osiguwa SA and Fraser LR (1996) Evidence for Ca^{2+}-dependent ATPase activity, stimulated by decapacitation factor and calmodulin, in mouse sperm *Molecular Reproduction and Development* **44** 111–120

Aitken RJ, Best FSM, Richardson DW, Djahanbakhch O and Lees MM (1982a) The correlates of fertilizing capacity in normal fertile men *Fertility and Sterility* **38** 68–76

Aitken RJ, Best FSM, Richardson DW, Djahanbakhch, O and Templeton A (1982b) An analysis of semen quality and sperm function in cases of oligospermia *Fertility and Sterility* **38** 705–711

Amann RP (1989) Can the fertility potential of a seminal sample be predicted accurately? *Journal of Andrology* **10** 89–98

Berger T and Horton MB (1988) Evaluation of assay conditions for the zona-free hamster bioassay of boar sperm fertility *Gamete Research* **19** 101–111

Boyers SP Davis R and Katz DF (1989) Automated semen analysis *Current Problems in Obstetrics and Gynaecology* **12** 172–200

Budworth PR, Amann RP and Chapman PL (1988) Relationships between computerized measurements of motion of frozen–thawed bull spermatozoa and fertility *Journal of Andrology* **9** 41–54

Clutton-Brock TH, Albon SD and Guinness FE (1986) Great expectations – dominance, breeding success and offspring sex-ratios in red deer *Animal Behaviour* **34** 460–471

Cross N (1995) Methods for evaluating the acrosomal status of human sperm. In *Human sperm acrosome reaction* pp 277–285 Eds P Fenichel and J Parinaud. John Libby Eurotext Ltd, Montrouge

Davis RO and Katz DF (1992) Standardization and comparability of CASA instruments *Journal of Andrology* **13** 81–86

Davis RO and Katz DF (1993) Computer-aided sperm analysis: technology at a crossroads *Fertility and Sterility* **59** 953–955

Davis RO, Drobnis EZ and Overstreet JW (1995) Application of multivariate cluster, discriminate function, and stepwise regression analyses to variable selection and predictive modeling of sperm cryosurvival *Fertility and Sterility* **63** 1051–1057

Den Daas N (1992) Laboratory assessment of semen characteristics *Animal Reproduction Science* **28** 87–94

Fazeli AR, Steenweg W, Bevers, MM, De Loos FAM, Van den Brock J and Colenbrander B (1993) Development of a sperm zona pellucida binding assay for bull sperm *Veterinary Record* **132** 14–16

Fazeli AR, Holt C, Steenweg W, Bevers MM, Holt WV and Colenbrander B (1995) Development of a sperm hemizona binding assay for boar semen *Theriogenology* **44** 17–27

Garner DL and Johnson LA (1995) Viability assessment of mammalian sperm using SYBR-14 and propidium iodide *Biology of Reproduction* **53** 276–284

Graham JK and Foote RH (1987) Dilaurylphosphatidyl-choline liposome effects on the acrosome reaction and *in vitro* penetration of zona-free hamster eggs by bull spermatozoa: A fertility assay for frozen-thawed semen *Gamete Research* **16** 147–158

Harper MJK (1994) Gamete and zygote transport. In *The Physiology of Reproduction* pp 123–187 Eds E Knobil and JD Neill. Raven Press, New York

Harrison RAP (1996) Capacitation mechanisms, and the role of capacitation as seen in eutherian mammals *Reproduction, Fertility and Development* **8** 581–594

Harrison RAP and Vickers SE (1990) Use of fluorescent probes to assess membrane integrity in mammalian spermatozoa *Journal of Reproduction and Fertility* **88** 343–352

Holt C, Moore HDM, Reed HCB and Holt WV (1992) Can sperm subpopulations be recognized within boar semen samples? An approach using cluster analysis *Journal of Reproduction and Fertility **Abstract** Series* **10** Abstract 81

Holt C, Holt WV, Moore HDM, Reed HCB and Curnock RM (1997) Objectively measured boar sperm motility parameters correlate with the outcomes of on-farm inseminations. Results of two fertility trials *Journal of Andrology* **18** 20–31

Holt WV (1996) Can we predict fertility rates? Making sense of sperm motility *Reproduction in Domestic Animals* **31**, 17–24

Holt WV, Shenfield F, Leonard T, Hartman TD, North RD and Moore HDM (1989) The value of sperm swimming speed measurements in assessing the fertility of human frozen semen *Human Reproduction* **4** 294–297

Hunter RHF and Nichol R (1983) Transport of spermatozoa in the sheep oviduct: preovulatory sequestering of cells in the caudal isthmus *Journal of Experimental Zoology* **228** 121–128

Irvine DS (1995) Computer-assisted semen analysis systems: sperm motility assessment *Human Reproduction* **10** 53–59

Johnson LA, Maxwell WMC, Dobrinsky JR and Welch GR (1996) Staining sperm for viability assessment *Reproduction in Domestic Animals* **31** 37–47

Kruger TF, Acosta AA, Simmons KF, Swanson BJ, Matta JF, Veeck LL, Morshedi M and Brugo S (1987) New method of evaluating sperm morphology with predictive value for human *in vitro* fertilization *Urology* **30** 248–592

Larsson K (1985) Boar sperm viability after freezing and thawing. In *Deep Freezing of Boar Semen* pp 177–187 Eds E Knobil and JD Neill. Raven Press, New York

MacLeod IC and Irvine DS (1995) The predictive value of computer assisted semen analysis in the context of a donor insemination programme *Human Reproduction* **10** 580–586

Medrano A and Holt WV (1996) Protective effects of glycerol during cold shock in boar spermatozoa: a cryomicroscope study using propidium iodide and Sybr-14 *Reproduction in Domestic Animals* **31** 281–282

Moore HDM and Akhondi MA (1996) Fertilizing capacity of rat spermatozoa is correlated with decline in straight-line velocity measured by continuous computer-aided sperm analysis – epididymal rat spermatozoa from the proximal cauda have a greater fertilizing-capacity *in-vitro* than those from the distal cauda or vas deferens *Journal of Andrology* **17** 50–60

Olds-Clarke P (1996) How does poor motility alter sperm fertilizing ability *Journal of Andrology* **17** 183–186

Olds-Clarke P and Sego R (1992) Calcium alters capacitation and progressive motility of uterine spermatozoa from +/+ and congenic $t^{w32/+}$ mice *Biology of Reproduction* **47**, 629–635

Overstreet JW and Cooper, GW (1975) Reduced sperm motility in the isthmus of the rabbit oviduct *Nature* **258** 718–719

Richardson DW and Aitken RJ (1993) Techniques for examining the fertilizing capacity of spermatozoa. In *Donor Insemination* pp 50–70 Eds CLR Barratt and ID Cooke. Cambridge University Press, Cambridge

Suarez SS (1987) Sperm transport and motility in the mouse oviduct – observations *in situ. Biology of Reproduction* **36**, 203–210

Sukcharoen N, Keith J, Irvine DS and Aitken RJ (1996) Prediction of the *in vitro* fertilization (IVF) potential of human spermatozoa using sperm function tests: the effect of the delay between testing and IVF *Human Reproduction* **11** 1030–1034

Wang WH, Abeydeera LR, Fraser LR and Niwa K (1995) Functional-analysis using chlortetracycline fluorescence and *in vitro* fertilization of frozen–thawed ejaculated boar spermatozoa incubated in a protein-free chemically defined medium *Journal of Reproduction and Fertility* **104**, 305–313

Watson PF (1995) Recent developments and concepts in the cryopreservation of spermatozoa and the assessment of their post-thawing function *Reproduction, Fertility and Development* **7** 871–891

Watson PF (1996) Cooling of spermatozoa and fertilizing capacity *Reproduction in Domestic Animals* **31** 135–140

Whitfield CH and Parkinson TJ (1992) Relationship between fertility of bovine semen and *in vitro* induction of acrosome reactions by heparin *Theriogenology* **38** 11–20

Zeng Y, Clark EN and Florman HM (1995) Sperm membrane potential-hyperpolarization during capacitation regulates zona pellucida-dependent acrosomal secretion *Developmental Biology* **171** 554–563

Poster titles

Hochereau-de Reviers MT, Perreau C, Rahal-Perola V and D Royère
Onset of spermatogenesis in Meishan and Large White male piglets

Ford JJ, Lunstra DD, Matteri RL and T Wise
Endocrine traits associated with differences in plasma FSH in boars

Kommisrud E, Paulenz H, Sehested E and IS Grevle
Routine assessment of boar semen motility

Palomo MJ and RAP Harrison
A sperm subpopulation in boar semen resistant to hypo-osmotic shock, revealed by assessment of plasma membrane integrity

Lucidi P, Barboni M and M Mattioli
Expanded cumuli induce AR in capacitated boar semen

Harkema W, Visser I, Kemp B, Soede NM and H Woelders
The relation between the capacity of boar sperm to bind zona pellucida proteins and fertilization rates *in vivo*

Petrounkina A, Petzoldt R, Töpfer-Petersen E and KF Weitze
Influence of fresh semen preservation on osmotic activity of boar spermatozoa during a posterior capacitation treatment

Eriksson B, Lundeheim N and H Rodriguez-Martinez
Viability of cryopreserved boar spermatozoa frozen in flat plastic containers

Mileham AJ, Haven D, Rohl J and HAM van der Steen
Porcine semen cryopreservation in a commercial setting

Ferreira FM, Fazeli AR, Colenbrander B, Waberski D, Borchardt Neto G and KF Weitze
Porcine sperm binding to the hemizona in relation to other sperm functional parameters *in vitro*

Calvete JJ, Ensslin M, Mburu J, Iborra A, Martinez P, Adermann K, Waberski D, Sanz L, Töpfer-Petersen E, Weitze K-F, Einarsson S and H Rodriguez-Martinez
Mapping of a continuous antigenic determinant in boar sperm zona pellucida-binding protein awn-1 and immunolocalization of awn epitopes in inseminated sows

Lende T van der, Willemsen MHA, Arendonk JAM van and EBPG van Haandel
Genetic analysis of the service sire effect on litter size in swine

ADVANCES IN BIOTECHNOLOGY IN PIG REPRODUCTION

Chair
H. Niemann

Journal of Reproduction and Fertility Supplement **52**, 227–236

Molecular approaches to improved pig fertility*

M. F. Rothschild, L. A. Messer and A. Vincent

225 Kildee Hall, Department of Animal Science, Iowa State University, Ames, IA 50011–3150, USA

Geneticists have made limited progress in improving reproductive traits in pigs through traditional selection and crossbreeding systems. Recent advances in molecular genetics and the human genome project have allowed progress in gene identification and gene mapping in pigs. The pig genetic linkage map now has over 1700 genetic markers, and about 200 of these are genes. Furthermore, comparative genomic maps are improving and are becoming useful tools. Molecular approaches such as candidate gene identification and genomic scans permit new genes and chromosomal regions to be discovered which may influence reproduction. One significant finding is that the oestrogen receptor locus (ESR) is associated with increased litter size. Differences between female Chinese pig ESR BB and AA homozygotes have averaged 2.3 pigs born alive for first parity, and in commercial Large White crosses the differences are about 0.9 pigs per litter. Additional new genes have been discovered which may significantly affect litter size. Initial genome scans have revealed that there may be a gene or genes influencing ovulation rate and litter size on chromosome 8. Approaches such as positional comparative candidate gene analysis and eventually positional cloning will probably identify the genes controlling reproduction in pigs.

Introduction

By their very nature, reproductive processes must be under a high level of genetic control for the maintenance of the species. Despite this level of control, there is considerable variability in heritable traits such as total number of pigs born. Genetic improvement of reproductive traits has traditionally been restricted to use of quantitative genetic methods including selection, inbreeding and crossbreeding but gains have been limited using these methods (Rothschild, 1996).

Molecular genomic analysis has revolutionized how geneticists examine the genetic differences within domesticated pigs. In the past few years, efforts have been directed toward the development of a genomic map consisting of anonymous genetic markers. This initial advance is now being followed by the discovery and mapping of genes of biological and economic interest. Present and future genetic improvements will result from the growing understanding of the function and structure of the individual genes and gene families that are responsible for the reproductive traits in livestock species. In this review the recent discoveries of individual genes involved in reproduction are discussed and molecular genetic approaches that have and will be useful in the discovery of new genes controlling increased fertility in pigs are described.

Gene Mapping Overview

Genome mapping in livestock is an on-going process that has far exceeded early expectations. Our progress has in part been due to the success of the well funded US Human Genome Project (Collins and Galas, 1993). Livestock genome mapping is a parallel process for producing comprehensive maps of the genomes of domestic animal species. In general, there are two basic types of gene maps.

* Journal Paper No. J-17315 of the Iowa Agriculture and Home Economics Experiment Station, Ames, Iowa. Project No. 3043.

The genetic linkage map is a group of genes and markers that are linked together on a chromosome. These loci are separated by distances called centimorgans (cM), which represent the frequency of crossovers or recombination frequency. The smaller the distance between two loci, the more likely it is that the loci and their effects will be inherited together. In humans, the average map distance between loci is less than 1 cM apart. In pigs, the genetic linkage maps are at an average of 5 cM spacing of loci. The second type of genetic map is a physical or cytogenetic map which represents the physical location of a gene or marker on a chromosome based on the G-banding pattern of metaphase chromosomes. Physical maps are useful for establishing large regions of chromosomes that are syntenic between species, or in finding the location of an individual gene.

A more recent emphasis is to develop extensive comparative genome maps (O'Brien, 1993; O'Brien *et al.*, 1993). A comparative map only includes genes, so that the location and the linkage and physical relationships of genes can be compared across species. 'Map rich' species like humans and mice are the foundation for comparative maps, as the information from these species is then used to infer the location of genes in 'map poor' species such as pigs. This is increasingly true since the human and mouse genomes have provided a growing number of expressed sequence tags (ESTs). ESTs are representative of genes and provide an additional resource for the identification of genes that could be mapped in livestock species. Combining the information from comparative, genetic linkage and the physical maps allows the geneticist to form a useful picture of the genetic composition of pigs.

Status of the Pig Genome Maps

For the significant effort required to generate genetic maps, scientists and administrators realized that a cooperative effort would be the most productive approach. The PiGMaP project (Archibald *et al.*, 1995) was initiated in Europe and was funded by the European Economic Community. PiGMaP now involves 18 European laboratories and a total of seven other laboratories in the USA, Japan and Australia. In the USA, the USDA launched two efforts. The USDA-ARS directs a sizeable gene mapping project at the Meat Animal Research Center in Clay Center, Nebraska. Secondly, the USDA-CSRS sponsored National Animal Genome Research Program was developed in 1993 to provide a structure that would stimulate coordination and collaboration of gene mapping in livestock species, including pigs. Scientists from state and private universities and federal laboratories cooperatively created a Swine Genome Technical Committee with a coordinator. The US Pig Genome Coordinator activities, in concert with activities of the USDA-ARS and international gene mapping projects have allowed the pig gene maps to be produced quickly. In 1989, only 50 genes and markers were mapped in pigs, and many of these were isozyme or blood group markers. At present, nearly 1700 markers and genes have been mapped in pigs (Ellegren *et al.*, 1994; Archibald *et al.*, 1995; Rohrer *et al.*, 1996), although not all have been published. The most extensive pig map has over 1100 loci (Rohrer *et al.*, 1996), and consists mostly of anonymous markers. The new PiGMaP genetic linkage map will have over 650 loci on the map, including nearly 200 genes. Coverage for the different linkage maps varies, and while the average distance between markers is approximately 5 cM, some large gaps still remain. The physical genetic map (see Fig. 1 for example) in pigs currently consists of approximately 700 genes and markers in part resulting from some excellent resources such as somatic cell hybrids for physical mapping (Yerle *et al.*, 1996). Despite these concerted efforts, the pig genome map still pales in comparison with the human and mouse maps and indicates the need for additional funding and effort. Over 20 000 genes and markers have been mapped in humans and over 15 000 in mice. As more genes are placed on the livestock comparative map, the human genome map will have increased use in the search for individual genes responsible for traits of economic interest.

Early Approaches to Finding Genes Controlling ReproductiveTraits

Quantitative trait loci, or QTL, are single genes that have a 'large' effect on a given trait and are also often called Economic Trait Loci (ETL). A QTL is likely to have been identified when there is a mean

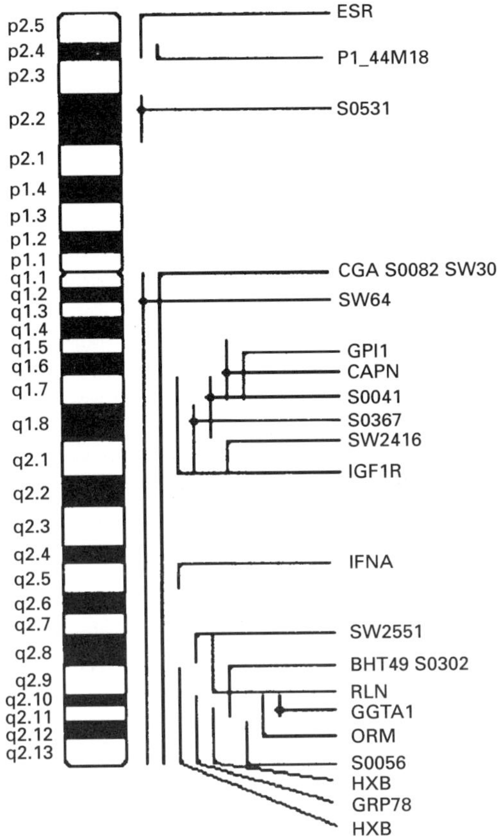

Fig. 1. The cytogenetic map for chromosome 1 in the pig. The map contains several genes including the oestrogen receptor (ESR) and several microsatellite markers. See http://www.iastate.edu/ pigmap.

difference in trait measurements between alleles of a gene or marker. The researcher must be able to follow the pattern of segregation of the alleles of the gene with the performance differences of the traits of interest. This requires the use of resource populations or families that usually comprise two or three generations. There are several resource families using diverse breed crosses in the world, as well as additional resource families from crosses of specialized lines. Resource family material is the foundation of genetic QTL identification and mapping. While there are several resource families for growth and production traits, reproduction trait family resources remain limited.

Two primary strategies used to detect QTL or major genes include the use of anonymous markers such as microsatellites and the use of candidate genes. The first approach uses microsatellites that are well spaced across the genome to genotype all the animals in the resource families. Then sophisticated statistical methods are used to determine the closest markers to a QTL affecting a given trait. This is called a genomic scan and can be an effective way to identify chromosomal regions that may contain a QTL. However, the size of such experiments must be considerable to detect QTL of moderate or small effects. Large-scale genotyping is becoming increasingly cost-efficient, but the cost of gathering phenotypic data on many individuals is significant.

An alternate strategy is to investigate a specific candidate gene as a potential QTL based upon its chromosomal location or its direct or indirect involvement in reproductive trait physiology. Once a candidate gene is identified and mapped in pigs, genotypes are obtained for individuals within a resource population. These genotypes are analysed with phenotypic data to identify possible associations between the gene and the trait. This approach relies considerably on available gene sequence information and genetic comparative maps from other species.

Early Results

Early genetic investigations into reproductive traits centered on the identification of blood groups and protein polymorphisms and primarily their association with litter size. The first evidence of such associations was obtained by Jensen *et al.* (1968) and Rasmusen and Hagen (1973), who reported an association between the H locus on pig chromosome 6 and litter size with an unfavourable effect of the Ha allele. Kristjansson (1964) and Imlah (1970) reported an apparent effect of alleles at the transferrin locus (pig chromosome 13) on pig fertility and prolificacy but this was not confirmed by other investigators (Jensen *et al.*, 1968; Huang and Rasmusen, 1982). Several other associations between blood group loci and reproductive traits have led to contradictory results (for example Kennedy *et al.*, 1973; Rasmusen 1975; Renard *et al.*, 1988). The most widely studied gene has been the ryanodyne receptor or halothane sensitivity (Hal) locus. Although best known for its major effects on carcass and meat quality traits in pigs, various and inconsistent effects of the Hal locus on male and female reproductive performance have been reported (Carden *et al.*, 1985; Lampo *et al.*, 1985).

During the 1980s, there was considerable investigation of the role of the major histocompatibility complex (MHC) called the swine leukocyte antigen (SLA) complex on male and female reproductive traits (reviewed in Warner and Rothschild, 1991). The swine MHC is a large group of genes on pig chromosome 7. Swine MHC effects on male testicular traits, with 1.5–5% of the phenotypic variation in male testicular size and hormone differences explained, have been reported (Rothschild *et al.*, 1986a). In females, the MHC also appears to be associated with several traits including ovulation rate (Rothschild *et al.*, 1984; Conley *et al.*, 1988), litter size, number of pigs born alive and number weaned (reviewed in Vaiman *et al.*, 1988; Warner and Rothschild, 1993). There is evidence to suggest that embryo homozygosity within the MHC may be a disadvantage (Vaiman *et al.*, 1988). Results in NIH miniature pigs have also indicated the presence of an MHC gene affecting embryo development and its relationship to litter size (Ford *et al.*, 1988). Other reports (Rothschild *et al.*, 1986b; Vaiman *et al.*, 1988; Warner and Rothschild, 1991) have demonstrated an association of MHC with birth and weaning weight. Whether these effects are the direct result of genes within the MHC or are due to linkage with other genes outside the complex is unknown. This question can only be resolved by further cloning and identification of individual genes on pig chromosome 7.

Recent discoveries, like the Booroola fecundity gene (FEC^B) locus in sheep, have encouraged researchers to search for individual genes influencing reproductive traits in pigs. Initial interest included the investigation of the prolificacy of some breeds of pig, like the Chinese Meishan. In 1991, Rothschild and colleagues began a candidate gene investigation of the role of the oestrogen receptor (ESR) gene in controlling litter size. Given that oestrogen is central to all reproductive processes, it was assumed that ESR would be an important candidate gene. Early results demonstrated that one ESR allele, found first in the Meishan and later in the Large White breed, was associated with improved litter size in a Meishan × Large White composite line (Rothschild *et al.*, 1994). An example of the ESR test is given (Fig. 2). More recent results (Rothschild *et al.*, 1995; Rothschild *et al.*, 1996; Short *et al.*, 1997) in Chinese crosses show first parity differences between ESR homozygotes to be nearly 2.3 pigs per litter or an additive effect of an increase of 1.15 pigs per litter for each copy of the B allele. In subsequent litters, the effect of the B allele is about 0.5 pigs per litter and appears to act in a dominant manner (Table 1). The B allele is also segregating in several Large White populations (Rothschild *et al.*, 1995; Legault *et al.*, 1996; Rothschild *et al.*, 1996; Short *et al.*, 1997b), but the effect of the B allele seems to differ between populations. In the largest study involving four commercial lines (Short *et al.*, 1997b) differences were additive for the B allele of about 0.42 pigs per litter or a difference of 0.84 pigs per litter between the AA and BB homozygotes (Table 1). The effect is only slightly smaller in later parities when the B allele shows more of a dominant effect than an additive

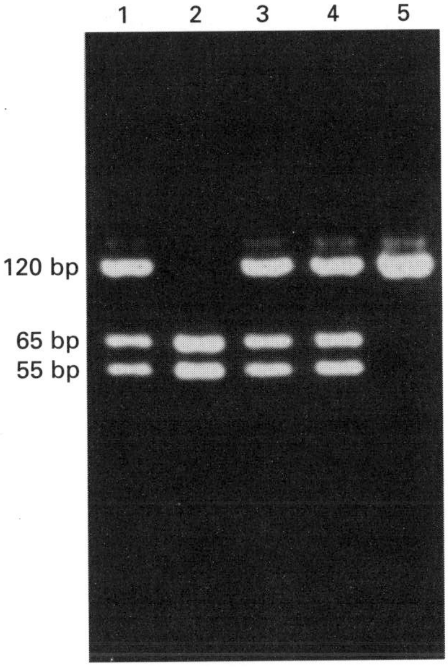

Fig. 2. The PCR-RFLP test for the oestrogen receptor polymorphism. Lane 2 contains a sample from a homozygous animal for the favourable B allele. Lane 5 contains a sample from an animal homozygous for the A allele and lanes 1, 3 and 4 contain samples from heterozygous animals. Reproduced with permission from the Journal of Animal Science.

effect. Line differences may either indicate that the ESR locus is only a linked marker or that differences in the genetic background have an impact on the expression of the ESR gene. Investigations are continuing to uncover the underlying mechanism of the favourable ESR allele.

It has been suggested that a statistical approach such as segregation analysis should precede molecular approaches to search for a major gene. This approach can be used to identify animals that are expected to carry different alleles for a gene, making a good starting point for a genomic scan. Janss *et al.* (1997) investigated the presence of major genes for litter size in an F_1 and F_2 Meishan and Western pig cross-population. A dominant gene was identified which resulted in a difference of five to six piglets between homozygotes. No significant effect was found in second parity litters. In the experiment, young sows were mated at a fixed age such that variation in the onset of puberty could affect first parity litter size. Therefore, it cannot be excluded that the major gene identified is partly related to the onset of puberty.

Genomic scans for reproductive traits have been limited. Using a cross between Meishan and Large White breeds, Wilkie *et al.* (1996) reported marker gene associations with number of pigs born per litter on a region of chromosome 6. Regions of chromosomes 4 and 7 were associated with the number of stillbirths. An association between the microsatellite marker SW444 region on chromosome 8 and ovulation rate and uterine length was also reported (Wilkie *et al.*, 1996). A second group has also identified an association between microsatellite markers (S0017, S0086) in this region of chromosome 8 in a similar synthetic population (G. Plastow, personal communication). In another study, Rathje *et al.* (1996) found a QTL for ovulation rate on pig chromosome 8 in a cross between a line selected for ovulation rate and a control line. A large study involving commercial pigs also suggests that this second chromosome 8 may contain a QTL with a major effect for litter size (Short

Table 1. Effects of oestrogen receptor (ESR) genotype on total number born (TNB) and number born alive (NBA) of Meishan synthetic and Large White lines

	Meishan synthetics					
	First parities			All parities		
ESR genotype	TNB	NBA	N	TNB	NBA	N
AA	10.1[a]	9.1[a]	62	11.0[a]	10.2[a]	98
AB	11.4[b]	10.5[b]	73	12.4[b]	11.5[b]	125
BB	12.4[b]	11.4[b]	26	12.5[b]	11.7[b]	53

	Large White lines					
	First parities			Later parities		
ESR genotype	TNB	NBA	N	TNB	NBA	N
AA	10.14[a]	9.42[a]		11.36	10.03[a]	
AB	10.59[b]	9.87[b]		11.86[b]	10.54[b]	
BB	10.97[c]	10.22[c]		12.04[b]	10.71[b]	

[a, b, c]Means in the same column with different superscripts are significantly different ($P < 0.01$).
Data from Rothschild *et al.* (1996) and Short *et al.* (1997).

et al., 1997a). This part of chromosome 8 is syntenic with sheep chromosome 6 where FEC^B is located. Taken together, the results of these studies strongly suggest that a region or regions on chromosome 8 influence litter size, possibly through ovulation rate.

Recently, studies in our laboratory have begun to investigate several candidate genes for litter size. These genes are of interest because of their role in reproductive trait physiology and include retinoic acid receptor gamma (RARG), retinol binding protein 4 (RBP4), melatonin receptor 1A (MTNR1A), prolactin (PRL), prolactin receptor (PRLR) and leptin receptor (LEPR).

Two of these genes, retinoic acid receptor-gamma (RARG) and retinol-binding protein 4 (RBP4), are expressed during rapid trophoblast elongation, a critical time of pregnancy in pigs. These genes were mapped by linkage analysis to pig chromosomes 5 and 14, respectively. The allele frequencies of the RBP4 gene (Messer *et al.*, 1996a) and RARG gene (Messer *et al.*, 1996b), although based upon a small representative sample of several domestic breeds of pig, may indicate a higher allele frequency in the maternal lines which may be due in part to selection for litter size in these breeds. Two French Large White lines were genotyped at these loci to investigate further possible effects of these genes. The first line consisted of French Hyperprolific (LWH) pigs (32 sows with 216 litter records). The second line consisted of French control (LW) pigs (27 sows with 242 litter records). The average additive effects of the RARG gene, estimated as the linear regression of litter size on genotype, was 0.21 ± 0.30 pigs per litter increase in LWH and 0.14 ± 0.38 pigs per litter increase in LW. The RBP4 gene had an additive effect for litter size of 0.52 ± 0.30 pigs per litter in LWH and 0.45 ± 0.43 in LW. The allelic substitution effect for litter size ranges from 5 to 17% of the phenotypic standard deviation (Messer *et al.*, 1996c). These initial data suggest that alleles of RARG and RBP4 may account for increased litter size in pigs.

The French lines were also genotyped at the MTNR1A locus. Melatonin is a pineal gland hormone that is involved in circadian rhythm entrainment and seasonal breeding in some livestock, and exerts its effects by binding to receptors located in the brain. The frequencies of the 4.3 kb and 3.9 kb alleles were 0.23 and 0.77 in the hyperprolific line, and 0.04 and 0.96 in the control line (Messer, 1996). The differences in gene frequency at this locus between lines was strikingly different from those observed with the RBP4 and RARG genes. The 3.9 kb allele was also observed at high frequency in the Meishan, Wild Boar, Yorkshire and Large White breeds of pig (Messer *et al.*, 1997) and suggests that an allele of MTNR1A may play a role in increased litter size. The differences in frequency might also be explained by factors such as breed differences, founder effects or genetic drift. Further investigation in large populations with several litter records is necessary to estimate better the possible effect of MTNR1A alleles on increased litter size.

Prolactin is an anterior pituitary peptide hormone involved in many different endocrine activities and is essential for reproductive success. The mapping of prolactin on pig chromosome 7 has been recently refined by linkage and physical mapping (Vincent *et al.*, 1996). The polymorphism used for the genetic linkage mapping was also used to screen unrelated animals of seven different breeds. Maternal breeds such as Meishan, Large White and Chester White contained genotypes not found in the paternal breeds tested. This marker is currently being investigated for associations with litter size in a crossbred population.

The prolactin receptor (PRLR) has been identified in various tissues including mammary gland, corpus luteum and uterus (Kelly *et al.*, 1991) and is considered a litter size candidate gene because of its integral role in the prolactin pathway. A polymorphism was identified and used to linkage map PRLR to pig chromosome 16. Initial screening of a commercial line indicated that the A allele had a favourable effect on litter size; AA females have litters of 0.66 pigs more per litter than do AB or BB females (Vincent *et al.*, in press). This marker is being used to screen various other commercial lines to confirm the early findings.

Another gene which may influence reproductive traits is the leptin receptor (LEPR). Leptin is a protein secreted by adipose tissue and is thought to signal satiety by binding to leptin receptors in the brain. Mice with mutations in LEP (ob/ob) and LEPR (db/db) genes are infertile as well as obese. Treatment with leptin has been shown to stimulate the reproductive endocrine system in female and male ob/ob mice (Barash *et al.*, 1996). A different isoform of the receptor than that found in the brain has been shown to be expressed in human ovary and prostate (Cioffi *et al.*, 1996). LEPR has been physically (Ernst *et al.*, 1997) and linkage mapped to pig chromosome 6 (Vincent and Rothschild, personal communication). Again, seven different breeds were genotyped using the LEPR polymorphism. There were no obvious indications of a maternal allele, but future work will consider possible links between obesity and reproduction.

Very large populations with several litter records are necessary in these studies for a more representative estimate across breeds and to allow for a better understanding of the size of the effects of reproduction genes. Considerable investment in resources is needed to accomplish the collection of both genotypes and phenotypic information. It will be interesting to determine how many of these new candidate effects are confirmed when the studies are completed.

Beyond Candidate Genes and Initial Genome Scans

As indicated in the previous section, both candidate gene investigations and genomic scans have found genes and regions of chromosomes that are worthy of further investigation. Finding the gene causing the desired effect in such regions is a complicated and difficult process. In many cases genome scans will find regions that are from 10–25 cM in size. This region of $10–25 \times 10^6$ nucleotides could contain hundreds of genes and it is very difficult by linkage mapping approaches to determine which single gene is responsible for the effect. Efforts are needed to narrow the QTL to a 1–2 cM region, making a positional cloning approach more feasible.

When the QTL region has been narrowed to 20 cM or less, the researcher may try to identify a positional candidate gene. The comparative gene map is a valuable resource to indicate candidate genes in the QTL region. If some of these genes have been mapped in other species or are known to cause similar phenotypes, these comparative positional candidates genes can be evaluated further. Candidate genes identified in other species must be genetically and physically mapped in pigs to ensure that their map position is closely linked to the QTL. Some positional candidate genes will be eliminated because they do not map to the region of interest in pigs. If the new candidate gene does map to the area of interest, further analysis can be attempted.

If, however, there are no obvious positional candidate genes, or those chosen do not prove to be the QTL, then additional genetic markers are needed to saturate the QTL region. There are several techniques for finding new genetic markers. Chromosomal flow sorting (Yerle *et al.*, 1993) allows libraries of specific chromosomes to be constructed. The region can be narrowed further by microdissection of chromosomes which permits libraries consisting of unique chromosomal regions to be generated (Schmutz *et al.*, 1994). Libraries can then be screened to find region-specific microsatellites. Another approach is to use amplified fragment length polymorphisms (AFLPs) to develop new anonymous polymorphic markers (Plastow, 1997). New markers must be mapped both genetically and physically. The development of radiation hybrid panels will allow high-resolution physical mapping of newly identified markers (Rexroad and Womack, 1997).

New markers that flank the QTL may then be used to identify yeast artificial chromosomes (YACs) which might contain the gene of interest or additional anonymous markers. YAC libraries consist of large genomic insert clones, which contain fragments of up to several hundred kilobases. These YACs could be used to screen smaller insert libraries to converge upon a more manageable fragment. The desired gene or region of the gene can then be identified by several techniques including exon trapping, cDNA selection and CpG island rescue (Valdes *et al.*, 1994; Gardiner *et al.*, 1995; Krizman, 1996).

All new genes that are identified in the QTL region by these approaches must be evaluated further to determine an association of the gene with trait differences. This might be accomplished by sequence analysis, the identification of restriction fragment length polymorphisms (RFLPs), or expression level variation. Finally, the gene effect should be validated in a variety of populations. All of these processes are laborious and expensive. Clearly excluding a gene is much easier than proving it is the QTL.

Marker-Assisted Selection

Once QTL or major genes are identified, their use in commercial situations is accomplished through marker assisted selection (MAS). MAS is best applied to traits of low heritability or traits measured in one sex. For these reasons, reproductive traits are often suggested as prime targets for MAS. Selection is the best approach when the gene or QTL is in well-established lines and breeds and the beneficial allele is at an intermediate gene frequency. Animals are simply selected to increase the gene frequency of the desired QTL or decrease the frequency of an undesirable gene. Another application is the introgression (repeated back crossing) of a QTL or major gene discovered in an exotic or uncommon breed into a preferred line. For example, both of these techniques have been used by the pig breeding company, PIC, to increase the frequency of the favourable ESR allele in their commercial pig stocks. The use of MAS should only be done in concert with selection based on standard performance measures and care should be taken that antagonistic pleiotropic effects do not exist (see Rothschild *et al.*, 1996; Short *et al.*, 1997b). Another method for using QTL or a major gene would be the introduction of the gene into a new population using gene transfer (transgenics). However, this approach has not been undertaken for reproductive traits in pigs. Further technology will probably be developed before this approach can be used. Consumer concerns about transgenic pigs may prevent a widespread acceptance of such an approach in commercial pigs.

Conclusion

It is now possible to localize QTL and genes controlling complex traits, but additional developments are needed before the potential can be realised and the genetics of complex traits in pigs are more effectively dissected. Improved reproduction records for large resource families are necessary to map genes with moderate to small effects as are efficient genotyping methods to make these projects more economically feasible. Current maps, with markers at 3–5 cM spacing will need to be saturated with markers at intervals of 1–2 cM. These saturated maps are crucial for the eventual cloning of the genes affecting reproductive traits. Advanced comparative genetic maps will certainly aid in the process of finding positional candidate genes. An improved understanding of complex physiological traits is also imperative for identifying candidate genes. The discovery of useful major genes and QTL should be valuable for pig producers. It is likely that several gene tests will be available in the future, aiding in the development of specialized lines and new genetic packages for the producer. The ultimate beneficiary of these new developments will be the consumer who can purchase improved livestock products at a reasonable price.

The authors wish to express their appreciation to G. Plastow for his comments and suggestions. Contributions of their colleagues at Iowa State University, PiGMaP and NC-210 collaborators are appreciated. Financial support and technical assistance from PIC, Dalgety FTC, and the Biotechnology Research Development Corporation, for several research projects and the USDA-CSREES for Pig Genome Coordination is also appreciated.

References

Archibald A, Brown J, Couperwhite S, McQueen H, Nicholson D, Haley C, Couppieters W, van de Weghe A, Stratil A, Wintero A, Fredholm M, Larson N, Nielsen V, Milan D, Woloszyn N, Robic A, Dalens M, Riquet J, Gellin J, Caritez J-C, Hue D, Burgaud G, Ollivier L, Bidanel J-P, Vaiman M, Renard C, Geldermann H, Davoli R, Ruyter D, Verstege E, Groenen M, Davies W, Hoyheim B, Keiserud A, Andersson L, Ellegren H, Johansson M, Marklund L, Miller R, Dear D, Signer E, Jeffreys A, Moran C, Le Tissier P, Muladno, Rothschild M, Tuggle C, Vaske D, Helm J, Liu H-C, Rahman A, Yu T-P, Larson RG and Schmitz C (1995) The PiGMaP consortium linkage map of the pig (*Sus scrofa*). *Mammalian Genome* **6** 157–175

Barash I, Clement C, Weigle D, Hongping R, Kabigting E, Kuijper J, Clifton D and Steiner R (1996) Leptin is a metabolic signal to the reproductive system *Endocrinology* **137** 3144–3147

Carden AE, Hill WG and Webb AJ (1985) The effects of halothane susceptibility on some economically important traits in pigs *Animal Production* **40** 351–358

Cioffi J, Shafer A, Zupancic T, Smith-Gbur J, Mikhail A, Platika D and Snodgrass H (1996) Novel B219/OB receptor isoforms: possible role of leptin in hematopoiesis and reproduction *Nature Medicine* **2** 585–589

Collins F and Galas D (1993) A new five-year plan for the U.S. human genome project *Science* **262** 43–46

Conley A J, Jung YC, Schwartz NK, Warner CM, Rothschild MF and Ford SP (1988) Influence of SLA haplotype on ovulation rate and litter size in miniature pigs *Journal of Reproduction and Fertility* **22** 595–601

Ellegren H, Chowdhary BP, Johansson M, Marklund L, Fredholm M, Gustavsson I and Andersson L (1994) A primary linkage map of the porcine genome reveals a low rate of genetic recombination *Genetics* **137** 1089–1100

Ernst C, Kapke P, and Rothschild MF (1997) The leptin receptor gene (LEPR) maps to porcine chromosome 6 *Mammalian Genome* **8** 226

Ford, SP, Schwartz NK, Rothschild MF, Conley AJ and Warner CM (1988) Influence of SLA haplotype on preimplantation embryonic growth rate in miniature swine *Journal of Reproduction and Fertility* **84** 99–104

Gardiner K, Ichikawa H, Ohki M, Patterson D and Cheng JF (1995) Localization of cDNAs to a region poorly represented in the CEPH chromosome 21 YAC contig: candidate genes for genetic diseases mapped to 21q22.3 *Genomics* **30** 562–570

Huang MY and Rasmusen BA (1982) Parental transferrin types and litter size in pigs *Journal of Animal Science* **54** 757–762

Imlah P (1970) Evidence for the Tf locus being associated with an early lethal factor in a strain of pig *Animal Blood Groups and Biochemical Genetics* **1** 5–13

Janss LLG, van Arendonk JAM and Brascamp EW Segregation analysis for presence of major genes to affect growth, back fat and litter size in Dutch Meishan crossbreds *Journal of Animal Science* (in press)

Jensen EL, Smith C, Baker LN and Cox DF (1968) Quantitative studies on blood group and serum protein systems in pigs II – Effects on production and reproduction *Journal of Animal Science* **27** 856–862

Kelly P, Djiane J, Postel-Vinay M, and Edery M (1991) The prolactin/growth hormone receptor family *Endocrine Reviews* **12** 235–251

Kennedy BW, Moxley JE and Saison R (1973) Some relationships between blood group factors and economic traits in swine *Canadian Journal of Animal Science* **53** 389–397

Kristjansson FK (1964) Transferrin types and reproductive performance in the pig *Journal of Reproduction and Fertility* **8** 311–317

Krizman DB (1996) Gene identification by 3′ terminal exon trapping *Genetic Engineering (NY)* **18** 49–56

Lampo P, Nauwynck W, Bouquet Y and vand Zeveren A (1985) Effect of stress susceptibility on some reproductive traits in Belgian Landrace pigs *Livestock Production Science* **13** 279–287

Legault C, Gruand J, Lebost J, Garreau H, Ollivier L, Messer LA and Rothschild MF (1996) Frequence et effect sur la prolificite du gene ESR dams deux lignees Large White en France *Journees Recherche Porcine en France* **28** 9–14

Messer LA (1996) *Mapping of candidate genes for litter size in the pig* p77 Masters Thesis. Iowa State University, Iowa

Messer L, Wang L, Yelich J, Pomp, D, Geisert RD and Rothschild MF (1996a) Linkage mapping of the retinol-binding protein 4 (RBP4) gene to porcine chromosome 14 *Mammalian Genome* **7** 396

Messer L, Wang L, Yelich J, Pomp D, Geisert RD and Rothschild MF (1996b) Linkage mapping of the retinoic acid receptor-gamma (RARG) gene to porcine chromosome 5 *Animal Genetics* **27** 175–177

Messer L, Wang L, Legault C and Rothschild MF (1996c) Mapping and investigation of candidate genes for litter size in French Large White pigs *Animal Genetics* **27** (Supplement 2) 114

Messer L, Wang L, Tuggle CK, Yerle M, Chardon P, Pomp D, Womack JE, Barendse W, Crawford AM, Notter DR and Rothschild MF (1997) Mapping of the melatonin receptor 1A (MTNR1A) gene in the pigs, sheep and cattle *Mammalian Genome* **8** 368–370

O'Brien SJ (1993) *Genetic Maps* Sixth Edition, Book 4 Nonhuman Vertebrates Cold Spring Harbor Laboratory Press, New York

O'Brien SJ, Womack, JE, Lyons LA, Moore KJ and Jenkins NA (1993) Anchor reference loci for comparative genome mapping in mammals *Nature Genetics* **3** 103–112

Plastow G (1997) Demonstration of AFLP and its application to animal genome mapping and farm animal selection (molecular breeding with AFLP) *In Examples of Demonstration Projects in the Life Sciences Programmes* Eds A Herrero, M Jarvenpaa and I Norstedt European Commission DGXII Science Research and Development, Brussels **EUR 16981** (in press)

Rasmusen BA (1975) A and O blood types and reproduction in pigs *Journal of Animal Science* **41** 256 (Abstract)

Rasmusen BA and Hagen KL (1973) The H blood-group system and reproduction in pigs *Journal of Animal Science* **37** 568–573

Rathje TA, Rohrer GA and Johnson RK (1996) Quantitative trait loci affecting reproductive traits in swine *Journal of Animal Science* **74** (Supplement 1) 122 (Abstract)

Renard C, Bidanel JP, Palovics A, Vaiman M, Guérin G and Runavot JP (1988) Relations entre des marqueurs génétiques et les caractères de production *Journées Recherche Porcine en France* **20** 315–320

Rexroad CE and Womack JE (1997) Use of a whole genome radiation hybrid cell panel in constructing a high density map of bovine chromosome 1 *Proceedings of the Plant and Animal Genome V*, (Abstract) p 297

Rohrer GA, Leeson JA, Hu Z, Smith TP, Keele JW and Beattie CW (1996) A comprehensive map of the porcine genome *Genome Research* **6** 371–391

Rothschild MF (1996) Genetics and reproduction in the pig *Animal Reproduction Science* **42** 143–151

Rothschild MF, Zimmerman DR, Johnson RK, Venier L and Warner CM (1984) SLA haplotype differences in lines of pigs which differ in ovulation rate *Animal Blood Groups and Biochemical Genetics* **15** 155–158

Rothschild MF, Renard Ch, Sellier P, Bonneau M and Vaiman M, (1986a) Swine lymphocyte antigen (SLA) haplotype effects on male genital trait development and androsterone level. In *Proceedings of the 3rd World Congress on Genetics Applied to Livestock Production* **11** 197–202. Eds GE Dickerson and RK Johnson, University of Nebraska, Lincoln

Rothschild MF, Renard Ch, Bolet G, Dando P and Vaiman M (1986b) Effect of swine lymphocyte antigen haplotypes on birth and weaning weights in pigs *Animal Genetics* **17** 267–272

Rothschild MF, Jacobson C, Vaske DA, Tuggle CK, Short T, Sasaki S, Eckardt GR and McLaren DG (1994) A major gene for litter size in pigs In *Proceeding of the 5th World Congress on Genetics Applied to Livestock Production* **21** 225–228 Eds C Smith, JS Gavora, B Benkel, J Chesnais, W Fairfull, JP Gibson BW Kennedy and EB Burnside. University of Guelph, Guelph

Rothschild MF, Vaske DA, Tuggle CK, Messer LA, McLaren DG, Short TH, Eckardt GR, Mileham AJ and Plastow GS, (1995) Estrogen receptor locus is a major gene for litter size in the pig. In *Proceedings of the European Association of Animal Production* Ed. JAM van Arendonk p 53. Wageningen Pers, Wageningen

Rothschild MF, Jacobson C, Vaske DA, Tuggle CK, Wang L, Short T, Eckardt GR, Sasaki S, Vincent A, McLaren DG, Southwood O, van der Steen H, Mileham A and Plastow G (1996) The estrogen receptor locus is associated with a major gene influencing litter size in pigs *Proceedings of the National Academy of Sciences USA* **93** 201–205

Schmutz SM, Berryere TG, Moker JS, Thue TD and Winkelman DC (1994) Gene mapping from a bovine 1:29 DNA library prepared with chromosome microdissection *Mammalian Genome* **5** 138–141

Short TH, Southwood OI, DeVries AG, McLaren DG, Evans GJ, Mileham AJ, and Plastow G (1997) Evidence of a new genetic marker for litter size in pigs *Journal of Animal Science* (in press)

Short TH, Rothschild MF, McLaren DG, Southwood OI, DeVries A, van der Steen H, Tuggle CK, Helm J, Vaske DA, Mileham AJ and Plastow GS (1997) The effect of the estrogen receptor locus on reproduction and production traits in four commercial lines of pigs *Journal Animal Science* (in press)

Vaiman M, Renard Ch and Bourgeux N (1988) SLA, the major histocompatibility complex in swine: its influence on physiological and pathological traits. In *The Molecular Biology of the Major Histocompatibility Complex of Domestic Animal Species* pp 23–38 Eds C. Warner, M Rothschild and S. Lamont Iowa State Press, Ames IA

Valdes JM, Tagle DA and Collins FS (1994) Island rescue PCR: a rapid and efficient method for isolating transcribed sequences from yeast artificial chromosomes and cosmids *Proceedings of the National Academy of Sciences USA* **91** 5377–5381

Vincent A, Wang L and Rothschild M (1996) Placement of prolactin on pig chromosome 7 by linkage and physical mapping *ISU 1996 Swine Research Reports* Ed. P. Holden pp 60–61 Iowa State University Press, Ames IA

Vincent AL, Short TH, Southwood OI, Plastow GS, Tuggle CK and Rothschild MF The prolactin receptor gene is associated with increased litter size in pigs *Proceedings of the Sixth World Congress on Genetics Applied to Livestock Production* (in press)

Warner CM and Rothschild MF (1991) The swine major histocompatibility complex (SLA) In *Immunogenetics of the MHC* Ed. R Srivasava, pp 368–397 Verlag Chemie International Publications, New York

Wilkie PJ, Paszek AA, Flickinger GH, Rohrer GA, Alexander LJ, Beattie CW and Shook LB (1996) Scan of 8 porcine chromosomes for growth, carcass and reproductive traits two likely quantitative trait loci *Animal Genetics* 27 (Supplement 2) 117–118

Yerle M, Schmitz A, Milan D, Chaput B, Monteagudo L,Vaiman M, Grelat G, and Gellin J (1993) Accurate characterization of porcine bivariate flow karyotype by PCR and fluorescence in situ hybridization *Genomics* **16** 97–103

Yerle M, Echard G, Robic A, Mairal A, Dubut-Fontana C, Riquet J, Pinton P, Milan D, Lahbib-Mansais Y and Gellin J (1996) A somatic cell hybrid panel for pig regional gene mapping characterized by molecular cytogenetics *Cytogenetic and Cell Genetics* **73** 194–202

Journal of Reproduction and Fertility Supplement **52**, 237–244

Developments in transgenic techniques in pigs

M. B. Nottle, H. Nagashima, P. J. Verma, Z. T. Du,
C. G. Grupen, R. J. Ashman and S. MacIlfatrick

BresaGen Limited, PO Box 259, Rundle Mall Adelaide, South Australia 5000, Australia

Manipulation of the pig genome is currently restricted to the random insertion of new DNA using pronuclear microinjection. This method suffers from a number of inherent limitations, the majority of which result from the inability to control the site at which the transgene becomes integrated. These drawbacks, together with the need to be able to target existing genes, will result in the replacement of pronuclear injection by new methods that have the capability to direct insertion to a particular genomic site that does not influence expression. Currently, it is possible to control the site of insertion in mice using embryonic stem (ES) cell and homologous recombination technologies. However, pluripotent ES cells have yet to be isolated in pigs. The possibility of using nuclear transfer to reprogramme early differentiated embryonic cells as well as somatic cells from adult animals may provide an alternative method for generating precise genetic modifications. Methods that allow these changes to be carried out *in situ* are also likely to be developed in the future.

Introduction

The ability to manipulate the genome of the pig either by inserting new genetic material or deleting existing genetic information will greatly improve pig production in the coming decades and has a number of important biomedical applications such as xenotransplantation (reviewed by Pursel and Rexroad, 1993; Wall, 1996; Muller *et al.*, in press). In the first part of this review, the production of transgenic pigs using pronuclear injection is reviewed and areas where there is potential to improve existing methodology is highlighted. The second half of the review takes a more speculative approach and describes several approaches currently being developed, which if successful will greatly improve the efficiency of current transgenesis programmes.

Pronuclear Microinjection

Use of prepubertal gilts as embryo donors

Currently, transgenic pigs are produced by injecting DNA directly into the pronucleus of a recently fertilized egg or zygote. Several methods for obtaining pronuclear zygotes from prepubertal and sexually mature animals have been described (Brem *et al.*, 1990; Williams *et al.*, 1992a; Bolamba and Sirad, 1996). Embryos obtained from prepubertal gilts appear to be less competent in their developmental ability than those obtained from mature animals (Pinkert *et al.*, 1989). However, when more embryos are transferred back to recipients to take account for this, there would appear to be little difference in terms of pregnancy rates (Brem *et al.*, 1990). There are considerable cost advantages in using prepubertal gilts compared with mature animals as well as the potential for more embryos. We have compared the number of transgenic pigs that can be produced using both types of donors (M. Nottle, H. Nagashima, P. Verma, Z. Du, C. Grupen and S. MacIlfatrick, unpublished; Tables 1 and 2). The results from this study demonstrated that prepubertal gilts produce similar numbers of ova but produce fewer fertilized ova compared with adult animals. In

Table 1. Number of corpora lutea, ova recovered, fertilized and injected from superovulated prepubertal or sexually mature donor pigs

Donor	Corpora lutea	Ova recovered	Ova fertilized	One-cell injected
Mature* (n = 63)	19.3±0.6[b]	15.1±0.7[b]	13.6±0.8[b]	10.7±1.0[b]
Prepubertal (n = 63)	20.5±1.5[b]	13.6±1.2[b]	9.3±1.3[c]	7.5±1.0[c]

*Sexually mature animals were treated as described by Nottle *et al.* (1995). Prepubertal gilts (23 week old) were treated as per sexually mature gilts except that 1500 iu of PMSG was used.
Values are means ± SEM and are expressed on a per donor basis. Within columns means with different superscripts are significantly different ($P < 0.05$).

this study pregnancy and integration rates were not different but litter size tended to be lower, although the difference was not significant ($P < 0.1$), when prepubertal donors were used. The efficiency with which transgenic pigs were produced was similar for prepubertal gilts compared with sexually mature donors (3.4% versus 4.0%).

Improving the response to superovulation

Other groups have reported that larger numbers of injectable embryos can be obtained from prepubertal gilts than we obtained in our original comparison (Pinkert *et al.*, 1989; Brem *et al.*, 1990; Williams *et al.*, 1992a). Individual prepubertal gilts appear to respond particularly well to superovulation (Fig. 1), suggesting that there is considerable potential for reducing the number of donors if the overall response could be improved. A number of methods have been examined for improving the response to superovulation in cattle and sheep (reviewed by Armstrong, 1993), some of which may be worthy of consideration in pigs. Several groups have shown that treatment with bovine growth hormone can improve the response to superovulation in cattle (Rieger *et al.*, 1991; Gong *et al.*, 1992; Kuehner *et al.*, 1993). Growth hormone (GH) has direct as well as indirect effects on follicular growth and oocyte maturation (reviewed by Bevers *et al.*, 1997). In a preliminary study, we found that treatment with recombinant pig GH tended to increase the number of small and medium follicles on the surface of the ovary of non-superovulated gilts treated before their second oestrus but the difference was not significant ($P > 0.05$; M. Nottle, C. Grupen and R. Campbell, unpublished; Table 3). Future work will determine whether GH can improve the response to superovulation in prepubertal gilts.

In pigs, PMSG is used exclusively for superovulation. In cattle and sheep, FSH has largely replaced PMSG as the gonadotrophin of choice because it overcomes a number of deficiencies associated with the relatively long half-life of PMSG (reviewed by Armstrong, 1993). The possibility of using FSH in pigs warrants further examination as work by us and others (M. Nottle and R. Ashman, unpublished; see also Bolamba *et al.*, 1996) suggests that larger numbers of injectable embryos may be obtained when FSH rather than PMSG is used to superovulate pigs. The use of GnRH as an alternative to hCG may also be worthy of investigation. Support for this suggestion comes from relatively large AI programmes which have demonstrated that a GnRH analogue increased pregnancy rates in gilts that were not superovulated compared with hCG (Brussow *et al.*, 1996).

The potential for IVM/IVF systems (reviewed by Nagai, 1996) to replace the need for embryo donors, together with further developments in the non-surgical collection and transfer of pig embryos (Li *et al.*, 1996) will result in further efficiencies in microinjection programmes in the future.

Increasing integration rate

Transgenes normally integrate in the form of multiple copies arranged in head to tail arrays at a single site in the genome (Hammer *et al.*, 1985; Palmiter and Brinster, 1986; Burdon and Wall, 1992).

Table 2. Pregnancy rate, litter size and integration rates using embryos obtained from sexually mature or prepubertal pigs

Donor	Transfers	Recipients farrowing (%)	Litter size	Transgenics/ piglets liveborn (%)
Mature	21[a]	17 (81)	8.1±0.5	26/137 (18.9)
Prepubertal	15	12 (80)	6.6±0.6	16/79 (20.3)

[a]Between 30 and 35 injected embryos were transferred per recipient for both groups.
Values are means ± SEM. Means are not significantly different ($P > 0.05$).

Between 0.3% and 4.0% of injected embryos result in the production of a transgenic pig (Pursel and Rexroad, 1993). Increasing the rate at which injected DNA becomes incorporated into the genome is one method whereby the efficiency with which transgenic pigs are generated could be improved. In mice, the concentration at which the DNA is microinjected (or more strictly the number of copies injected) does not appear to influence integration rate (Brinster *et al.*, 1985). However, in pigs, injecting DNA at concentrations greater than 5 ng μl^{-1} appears to result in relatively high integration rates (M. Nottle, H. Nagashima, P. Verma, R. Ashman, Z. Du, C. Grupen, S. MacIlfatrick, M. Harding, C. Cheah, D. Harrison, B. Luxford, R. Campbell, R. Crawford and A. Robins, unpublished; Table 4). Further work is needed to confirm these preliminary results.

Maximizing pregnancy rates

It is well established that microinjection reduces embryo survival (Williams *et al.*, 1992b; Hajdu *et al.*, 1994; Martin *et al.*, 1996). Several groups have examined the number of injected embryos that need to be transferred to maximize pregnancy rates in pigs (Wei *et al.*, 1993; Lancaster *et al.*, 1996; Pursel and Wall, 1996). On the basis of these results, transferring between 30 and 35 injected embryos would appear to be optimal. Similar numbers of embryos would appear to be appropriate for prepubertal donors (Brem *et al.*, 1990; M. Nottle, this supplement). Where sufficient numbers of embryos are not available, the co-transfer of two or three non-injected one-cell (i.e. those with attached spermatozoa but no visible pronuclei) or two-cell embryos may improve pregnancy rates (Hammer *et al.*, 1985; Pursel and Wall 1996; M. Nottle, H. Nagashima, P. Verma, Z. Du, C. Grupen and S. MacIlfatrick, unpublished).

Donor animals can also be used as recipients, reducing the number of animals required for microinjection programmes (Brem *et al.*, 1990; Pursel and Wall, 1996). Pursel and Wall (1996) reported that donors whose ovulation rate was 21 or more had fewer pregnancies than those whose ovulation rates were 20 or less. An increase in the production of ovarian steroids may have been responsible for this difference (reviewed by Armstrong, 1993). Acceptable pregnancy rates have also been reported when prepubertal gilts have been used as embryo recipients (Brem *et al.*, 1990). The method described by Bolamba and Sirad (1996) using prepubertal gilts induced to ovulate and used as donors at their subsequent oestrus may also be useful for producing suitably synchronized recipients.

Methods for Controlling Transgene Insertion

Pronuclear microinjection suffers from a number of inherent limitations, the majority of which arise from the inability to control the site at which the transgene integrates. This results in variable expression depending on the proximity of regulatory elements (so called 'position effects') as well as the potential to disrupt existing genes (reviewed by Bishop, 1997). Transgenes normally integrate as

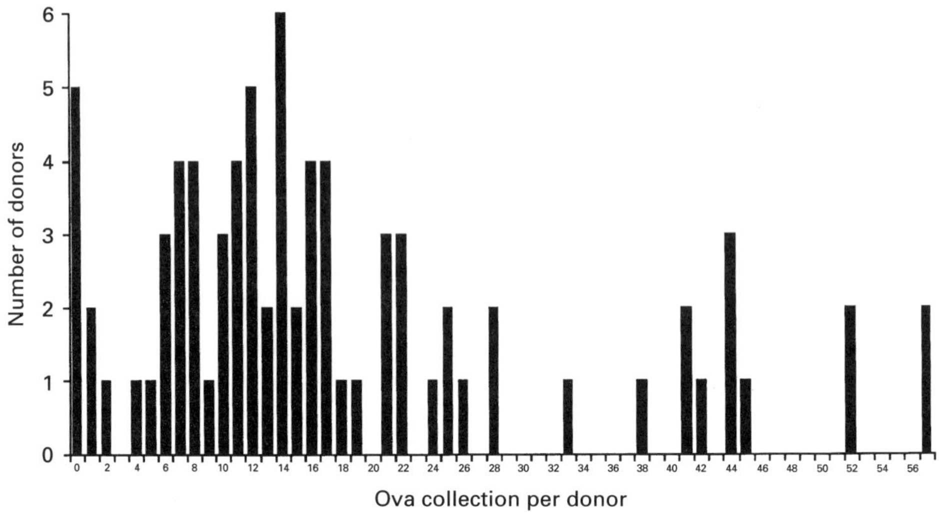

Fig. 1. The number of ova obtained from prepubertal gilts superovulated with 1500 iu PMSG and induced to ovulate 72 h later with 500 iu hCG. Animals were mated at 28–30 h, followed by AI at 30–32 h after hCG. Ova were recovered 50–56 h after hCG by mid-ventral laparotomy under general anaesthesia.

multiple copies, making them prone to methylation (Mehtali *et al.*, 1990) and heterochromatin formation which can also affect expression (Martin and Whitelaw, 1996). Furthermore, the injected DNA does not always integrate at the one-cell stage, resulting in the production of mosaics (Burdon and Wall, 1992; Whitelaw *et al.*, 1993). Mosaicism would appear to be a particular problem in pigs: more than half of transgenic founders transmit the transgene to less than half of their progeny (Brem *et al.*, 1990; Nottle *et al.*, 1996).

A number of methods have been developed in mice for improving transgene expression, including the use of locus control regions (Grosveld *et al.*, 1987) or sequences that insulate the transgene from surrounding elements (Stief *et al.*, 1989). Simpler methods for inserting DNA have also been developed, such as the retroviral infection of early embryos and the use of adenoviruses (Rubenstein *et al.*, 1986; Tsukui *et al.*, 1996). However, none of these has the ability to control where the injected DNA becomes inserted in the genome. The use of yeast artificial chromosomes (YACS) in providing a method for delivering transgenes with the necessary regulatory elements to guarantee expression may prove useful in this regard (reviewed by Montoliu *et al.*, 1994; Brem *et al.*, in press).

The costs associated with producing large numbers of founder animals in order to screen for animals in which the transgene is appropriately expressed is particularly prohibitive for domestic species. The ability to insert a single copy of a transgene at a site in the genome that does not interfere with its expression will be essential for the pig as well as other livestock species.

Embryonic stem cells

The use of embryonic stem (ES) cells (Evans and Kaufman 1981; Martin, 1981) and primordial germ (PGC) cells (Matsui *et al.*, 1992; Resnick *et al.*, 1992) can overcome many of the limitations associated with pronuclear injection and is the subject of a separate review in these proceedings (Piedrahita *et al.*, this supplement). Together with homologous recombination (Capecchi, 1989), it is possible to transform and select for the appropriate insertion *in vitro* before injecting these cells into the developing embryo to produce a chimaera. This is then bred to produce progeny that are heterozygous for the genomic modification. This approach can be used to disrupt the function of existing genes (gene knockout), allowing the effect of a particular gene to be studied (reviewed by

Table 3. Effect of porcine growth hormone on ovulation and follicular growth in pigs

	Ovulation rate	Small follicles (1.0–2.9 mm)	Medium follicles (3.0–5.9 mm)	Large follicles (> 6.0 mm)
Control (*n* = 11)	13.1±0.9	30.7±6.5	25.1±4.8	1.1±.04
pST[a] (*n* = 9)	15.2±0.7	43.4±6.0	36.3±6.0	3.0±1.3

[a]50 μg kg^{-1} porcine growth hormone (Reporcin, Southern Cross Biotech, Melbourne) was administered for 14 days from 7 days after the start of the first oestrus.
Values are means ± SEM. Means are not significantly different ($P > 0.05$).

Joyner, 1993). However, pluripotent ES cells are yet to be isolated in pigs (Piedrahita *et al.*, 1990; Notariana *et al.*, 1991; Wheeler, 1994). The ability of nuclear transfer to reprogramme early differentiated embryonic cells (Campbell *et al.*, 1996) as well as somatic cells from adult animals (Wilmut *et al.*, 1997) may circumvent the need for pluripotent cells. The use of nuclear transfer would also overcome the need to generate chimaeras and screen for germline transmission. Nuclear transfer is still in its infancy in pigs (Prather *et al.*, 1989; Nagashima *et al.*, 1992, 1997). However, we have recently demonstrated that it is possible to generate blastocysts from blastomeres obtained as late as the morula stage (Nagashima *et al.*, 1997; Table 5).

Alternative methods for gene insertion

In mice, the time required to transform and select cells that have undergone homologous recombination and reintroduce these into the developing embryo to produce a chimaera can be 6–12 months by the time germline transmission is demonstrated. In pigs the same process would take longer and cost considerably more because of the longer generation interval involved. Furthermore, a number of generations may be required to breed the mutation on to the desired genetic background. Given these limitations, it is likely that alternative methods that allow transgenes to be inserted *in situ* without the need to transform cells *in vitro* may be developed in the future.

Pronuclear injection. The most direct way of inserting genes at a predetermined site would be to use existing pronuclear injection technology and simply inject targeting constructs that contained sufficient homology to guarantee the insertion of a transgene at a predetermined site. However, homologous recombination appears to be a rare event in pronuclear embryos (Brinster *et al.*, 1989a). Methods are being designed to improve this frequency, such as the possibility of using exonuclease-resistant hairpin bends to allow injected DNA to persist to embryonic stages capable of supporting homologous recombination (Horie and Shimada, 1994).

Gene insertion using the gametes. Spermatogonia and oogonia undergo homologous recombination during meiosis. In oogonia this occurs before birth and as such is not readily accessible. Several workers have suggested that spermatozoa can bind DNA *in vitro* and can be used to carry transgenes into the oocyte, which then become incorporated in the genome (Brackett *et al.*, 1971; Lavitrano *et al.*, 1989; Hochi *et al.*, 1990). Considerable controversy surrounds this approach and the consensus appears to be emerging that, although foreign DNA can be detected in the developing embryo or fetus, it is rearranged (Brinster *et al.*, 1989b; Wall, 1996; Kim *et al.*, 1997).

Several groups have demonstrated that injection of DNA–liposome complexes into the testes can transform sperm cells *in situ* (Sato *et al.*, 1994; Ogawa *et al.*, 1995; Kim *et al.*, 1997). Of particular interest is the suggestion (Kim *et al.*, 1997) that this method may transform the male stem cell population. In this study, treatment with busulfan destroyed the majority of developing

Table 4. Summary of BresaGen Pig Transgenesis Programs

DNA concentration (ng μl^{-1})[a]	Transgenics liveborn/ total liveborn (%)	Transgenics liveborn/ ova injected (%)
10	43/148 (29)	43/1327 (3.2)
10	45/141 (32)	45/1835 (2.4)
7.5	76/302 (25)	76/1810 (4.1)
5.0	18/192 (9)	18/1012 (1.8)
1.0	8/185 (4)	8/1137 (0.7)

[a]Constructs ranged in size from 1.8 to 7.1 kb.

Table 5. Development of pig nucleus transfer embryos reconstructed with blastomeres from cryopreserved delipated four–eight-cell or morula stage embryos

Donor blastomeres	Fusion rates (%)	Number of embryos cultured	Number developed to (%)			
			Two-cell	Four-cell	Morula	Blastocyst
4–8 cell	104/113 (92)	100	53 (53)	34 (34)	17 (17)	6 (6)[a]
Morula	56/58 (97)	53	29 (55)	20 (38)	16 (30)	9 (17)[b]

[a, b]Significantly different ($P < 0.05$).
Reproduced with permission from Nagashima *et al.* (1997).

spermatocytes, causing the stem cells to undergo meiosis and proliferate. Whether homologous recombination would occur at frequencies high enough simply to allow transgenes containing sufficient homology to be injected, animals to be mated and progeny to be screened for the appropriate insertion (or deletion) remains to be determined. Alternatively, it may be possible to screen spermatozoa or embryos for homologous recombination before their transfer. It may also prove possible in the future to isolate spermatogonia and use them in the same way as ES cells are used presently in mice to insert transgenes (Brinster and Zimmermann, 1994).

In situ insertion. A number of possible approaches involving the use of DNA or DNA–RNA oligonucleotides to generate genetic modifications in mammalian cells have been reported. Perhaps the most promising of these is the recent report that DNA–RNA duplexes can be used to correct single base mutations at relatively high frequencies in tissue culture (Cole-Strauss *et al.*, 1996). It may be possible in the future to use a similar approach in fertilized eggs using pronuclear injection as an alternative method for gene knockout applications. Whether similar technology can be used to direct transgene insertion remains to be determined.

Conclusion

Technologies aimed at improving the efficiency with which transgenic pigs can be produced using pronuclear injection continue to be developed. In the future, methods will be developed for pigs as well as other domestic species and these methods will allow transgenes to be inserted at genomic sites that do not influence their expression. It is possible to direct insertion in the mouse using ES cells and homologous recombination technologies. The use of nuclear transfer to reprogramme differentiated cells as well as the development of other methods that allow genetic modifications to be carried out *in situ* may circumvent the need for pluripotent cells in the future.

The authors are grateful to P. Wigley and P. Tolstoshev for critically reading the manuscript. The assistance of Bunge Meat Industries Limited, Corowa, NSW, Australia in many of the studies reported is also gratefully acknowledged.

References

Armstrong DT (1993) Recent advances in superovulation of cattle *Theriogenology* **39** 7–24

Bevers MM, Dielemann SJ, van den Hurk R and Izadyar F (1997) Regulation and modulation of oocyte maturation in the bovine *Theriogenology* **47** 13–22

Bishop JO (1997) Chromosomal insertion of foreign DNA *Reproduction, Nutrition and Development* **36** 607–618

Bolamba D and Sirad MA (1996) *In vitro* development of embryos from superovulated gilts treated with the progesterone agonist, altrenogest (Regu-mate) or the prostaglandin analogue, cloprostenol (Planate) *Theriogenology* **46** 1045–1052

Bolamba D, Dubuc A, Dufour JJ and Sirard MA (1996) Effects of gonadotropin treatment on ovarian follicle growth, oocyte quality and *in vitro* fertilization of oocytes in prepubertal gilts *Theriogenology* **46** 717–726

Brackett BG, Baranksa W, Sawichi W and Kaprowski H (1971) Uptake of heterologous genome by mammalian spermatozoa and its transfer to ova through fertilization *Proceedings of the National Academy Sciences USA* **68** 353–357

Brem G, Springmann K, Meier E, Krausslich H, Brenig B, Muller M and Winnacker EL (1990) Factors in the success of transgenic pig programs. In *Transgenic Models in Medicine and Agriculture* pp 61–72 Ed. RB Chruch. Wiley-Liss, New York

Brem G, Besenfelder U, Aigner B, Muller M, Liebl I, Schutz G and Montoliu L (1996) YAC transgenesis in farm animals: rescue of albinism in rabbits *Molecular Reproduction and Development* (in press)

Brinster RL and Zimmermann JW (1994) Spermatogenesis following male germ cell transplantation *Proceedings of the National Academy of Sciences USA* **91** 11 298–11 302

Brinster RL, Chen HY, Trumbauer ME, Yagle ML and Palmiter RD (1985) Factors affecting the efficiency of introducing foreign DNA into mice by microinjecting eggs *Proceedings of the National Academy of Sciences USA* **82** 4438–4442

Brinster RL, Braun RE, Lo D, Avarbock MR, Oram F and Palmiter RD (1989a) Targeted correction of a major histocompatability class II *E*α gene by DNA microinjected into mouse eggs *Proceedings of the National Academy of Sciences USA* **86** 7087–7091

Brinster RL, Sandgren EP, Behringer RR and Plamiter RD (1989b) No simple solution for making transgenic mice *Cell* **59** 239–241

Brussow KP, Jochle W and Huhn U (1996) Control of ovulation with a GnRH analog in gilts and sows *Theriogenology* **46** 925–934

Burdon TG and Wall RJ (1992) Fate of microinjected genes in preimplantation mouse embryos *Molecular Reproduction and Development* **33** 436–442

Campbell KHS, McWhir J, Ritchie WA and Wilmut I (1996) Sheep cloned by nuclear transfer from a cultured cell line *Nature* **380** 64–66

Capecchi MR (1989) Altering the genome by homologous recombination *Science* **244** 1288–1292

Cole-Strauss A, Yoon K, Xiang Y, Byrne BC, Rice MC, Gryn J, Holloman WK and Kmiec EB (1996) Correction of the mutation responsible for sickle cell anemia by an RNA–DNA oligonucleotide *Science* **273** 1386–1389

Evans MJ and Kaufman MH (1981) Establishment in culture of pluripotent cells from mouse embryos *Nature* **292** 154–156

Gong JG, Wilmut I, Bramely TA and Webb R (1992) Effects of recombinant bovine somatotropin (BST) on superovulatory response in heifers *Proceedings of the 12th International Congress on Animal Reproduction* **1** 214–216

Grosveld F, van Assendelft GB, Greaves DR and Kollias G (1987) Position independent, high level expression of the human beta-globin gene in transgenic mice *Cell* **51** 975–985

Hajdu MA, Knight JW, Krisher RL, Canseco RS, Velander WH, Pearson RE and Gwazdauskas FC (1994) Effect of culture conditions, donor age, and injection site on *in vitro* development of DNA microinjected porcine zygotes *Journal of Animal Science* **72** 1299–1305

Hammer RE, Pursel VG, Rexroad CE, Jr, Wall RJ, Bolt DJ, Ebert JM, Palmiter RD and Brinster RL (1985) Production of transgenic rabbits, sheep and pigs by microinjection *Nature* **315** 680–683

Hochi S, Ninomiya T, Mizuno A, Honma M and Yuki A (1990) Fate of exogenous DNA carried into mouse eggs by spermatozoa *Animal Biotechnology* **1** 21–31

Horie K and Shimada K (1994) Gene targeting by a vector with hairpin-shaped oligonucleotide caps *Biochemistry and Molecular Biology International* **32** 1041–1048

Joyner AL (1993) *Gene Targeting: a Practical Approach* IRL Press, Oxford

Kuehner LF, Rieger D, Walton JS, Zhao X and Johnson WH (1993) The effect of a depot injection of recombinant bovine somatotropin on follicular development and embryo yield in superovulated holstein heifers *Theriogenology* **40** 1003–1013

Lancaster RT, Elsome KT, Yannoutsos N, Langford GA, Richards A, Cozzi E, Tolan M, Wallwork J and White DJG (1996) Production of pigs transgenic for human regulators of complement activation; a source of donor organs for transplantation *Proceedings of the 13th International Congress on Animal Reproduction* **3** P26–3 (Abstract)

Lavitrano M, Camaioni A, Fazio VM, Dolci S, Farace MG and Spadafora C (1989) Sperm cells as vectors for introducing foreign DNA into eggs: genetic transformation of mice *Cell* **57** 717–723

Li J, Rieke A, Day BN and Prather RS (1996) Technical note: porcine non-surgical embryo transfer *Journal of Animal Science* **74** 2263–2268

Kim JH, Jung-Ha HS, Lee HT and Chung KS (1997) Development of a positive method for male stem cell mediated gene transfer in mouse and pig *Molecular Reproduction and Development* **46** 515–526

Martin DIK and Whitelaw E (1996) The vagaries of variegating transgenes *BioEssays* **18** 919–923

Martin GR (1981) Isolation of a pluripotent cell line from early mouse embryos cultured in medium conditioned by teratocarcinoma stem cells *Proceedings of the National Academy Sciences USA* **78** 7634–7638

Martin MJ, Houtz J, Adams C, Thomas D, Freeman B, Keirns J and Cottrill F (1996) Effect of pronuclear DNA microinjection on the development of porcine ova *in utero. Theriogenology* **46** 695–701

Matsui Y, Zsebo K and Hogan BLM (1992) Derivation of pluripotential embryonic stem cells from murine primordial germ cells in culture *Cell* **70** 841–847

Mehtali M, LeMeur M and Lathe R (1990) The methylation-free status of a housekeeping transgene is lost at high copy number *Gene* **91** 179–184

Montoliu L, Schedl A, Kelsey G, Zentgraf H, Lichter P and Schultz G (1994) Germ line transmission of yeast artificial chromosomes in transgenic mice *Reproduction Fertility and Development* **6** 577–584

Muller M, Besenfelder U and Brem G (1996) Transgenics and modern reproduction technologies. In *Genetics Of The Pig* Eds M Rothschild and A Ruvinsk. CAB International, Wallingford (in press)

Nagai T (1996) *In vitro* maturation and fertilization of pig oocytes *Animal Reproduction Science* **42** 153–162

Nagashima H, Yamakawa H and Saito S (1992) Transplantation of porcine blastomere nuclei into oocytes collected from prepubertal gilts *Journal of Reproduction and Development* **38** 73–78

Nagashima H, Ashman RJ and Nottle MB (1997) Nucleus transfer of porcine embryos using cryopreserved delipated blastomeres as donor nuclei *Theriogenology* **47** 234 (Abstract)

Notarianni E, Laurie S, Moor RM and Evans MJ (1991) Maintenance and differentiation in culture of pluripotent embryonic cell lines from pig blastocysts *Journal of Reproduction and Fertility Supplement* **41** 51–56

Nottle MB, Nagashima H, Verma PJ, Ashman RJ, Du Z, Grupen CG, MacIlfatrick SM, Harding MP, Cheah C, Crawford RJ and Robins AJ (1995) Production of pigs containing a metallothionein porcine growth hormone gene construct *Proceedings of the 26th Annual Conference of The Australian Society for Reproductive Biology* 33 (Abstract)

Nottle MB, Nagashima H, Verma PJ, Ashman RJ, Du Z, Grupen CG, MacIlfatrick SM, Harding MP, Cheah C, Harrison DT, Luxford BG, Campbell RG, Crawford RJ and Robins AJ (1996) Inheritance of a metallothionein porcine growth hormone transgene in pigs *Proceedings of the 13th International Congress on Animal Reproduction* **3** P26–2 (Abstract)

Ogawa S, Hayashi K, Tada N, Sato, Kurihara T and Iwaya M (1995) Gene expression in blastocyst following direct injection of DNA into testis *Journal of Reproduction and Development* **41** 379–382

Palmiter RD and Brinster RL (1986) Germ-line transformation of mice *Annual Reviews in Genetics* **20** 465–499

Piedrahita JA, Anderson GB and BonDurant RH (1990) On the isolation of embryonic stem cells: comparative behaviour of murine, porcine, and ovine embryos *Theriogenology* **34** 879–901

Pinkert C A, Kooyman D L, Baumgartner A and Keisler D H (1989) *In-vitro* development of zygotes from superovulated prepubertal and mature gilts *Journal of Reproduction and Fertility* **87** 63–66

Prather RS, Sims MM and First NL (1989) Nuclear transplantation in early pig embryos *Biology of Reproduction* **41** 414–418

Pursel VG and Rexroad CE, Jr (1993) Recent progress in the transgenic modification of swine and sheep *Molecular Reproduction and Development* **36** 251–254

Pursel VG and Wall RJ (1996) Effects of transferred ova per recipient and dual use of donors as recipients on production of transgenic swine *Theriogenology* **46** 201–209

Resnick JL, Bixler LS, Cheng L and Donovan PJ (1992) Long-term proliferation of mouse primordial germ cells in culture *Nature* **359** 550–551

Rieger D, Walton JS, Goodwin ML and Johnson WH (1991) The effect of co-treatment with recombinant bovine somatotropin on plasma progesterone concentration and number of embryos collected from superovulated holstein heifers *Theriogenology* **35** 863–868

Rubenstein JL, Nicolas JF and Jacob F (1986) Introduction of genes into preimplantation mouse embryos by use of a defective recombinant retrovirus *Proceedings of the National Academy Sciences USA* **83** 366–368

Sato M, Iwase R, Kasai K and Tada N (1994) Direct injection of foreign DNA into mouse testes as a possible alternative of sperm-mediated gene transfer. *Animal Biotechnology* **5** 19–31

Stief A, Winter DM, Stratling WH and Sippel AE (1989) A nuclear attachment element mediates elevated and position-independent gene activity *Nature* **341** 343–345

Tsukui T, Kanegae Y, Saito I and Toyoda Y (1996) Transgenesis by adenovirus-mediated gene transfer into mouse zona-free eggs (1996) *Nature Biotechnology* **14** 982–985

Wall RJ (1996) Transgenic livestock: progress and prospects for the future *Theriogenology* **45** 57–68

Wei Q, Fan J and Chen D (1993) Effect of number of transferred microinjected embryos on pregnancy rate and litter size of pigs *Theriogenology* **39** 338 (Abstract)

Wheeler MB (1994) Development and validation of swine embryonic stem cells: a review *Reproduction, Fertility and Development* **6** 563–568

Whitelaw CBA, Springbett AJ, Webster J and Clark J (1993) The majority of G_0 transgenic mice are derived from mosaic embryos *Transgenic Research* **2** 29–32

Williams BL, Sparks AET, Canseco RS, Knight JW, Johnson JL, Velander WH, Page RL, Drohan WN, Kornegay ET, Pearson RE, Wilkins TD and Gwarzdauskas FC (1992a) Evaluation of systems for collection of porcine zygotes for DNA microinjection and transfer *Theriogenology* **38** 501–511

Williams BL, Sparks AET, Canseco RS, Knight JW, Johnson JL, Velander WH, Page RL, Drohnan WN, Young JM, Pearson R E, Wilkins TD and Gwarzdauskas FC (1992b) *In vitro* development of zygotes from prepubertal gilts after microinjection of DNA *Journal of Animal Science* **70** 2207–2211

Wilmut I, Schnieke AE, McWhir J, Kind AJ and Campbell KHS (1997) Viable offspring derived from fetal and adult mammalian cells *Nature* **385** 810–813

Journal of Reproduction and Fertility Supplement **52**, 245–254

Advances in the generation of transgenic pigs via embryo-derived and primordial germ cell-derived cells

J. A. Piedrahita*[1,2,3], K. Moore[1,3], C. Lee[2,3], B. Oetama[1,3], R. Weaks[1,3], J. Ramsoondar[1,3], J. Thomson[1,3] and J. Vasquez[2,3]

[1]*Department of Veterinary Anatomy and Public Health*, [2]*Department of Animal Sciences, and* [3]*Center for Animal Biotechnology, Texas A&M University, College Station, TX 77843–4458, USA*

The development of new technologies that would increase the efficiency for generation of transgenic livestock and would overcome some of the problems associated with random insertion of the transgene will greatly benefit animal agriculture. A potential alternative technology to pronuclear injection for the generation of transgenic pigs involves the isolation, culture and genetic manipulation of cell lines that can be reintroduced into the embryo for participation in the formation of the germ cells. We have isolated and cultured pig primordial germ cells (PGC) while maintaining them in an undifferentiated state as determined by morphology and alkaline phosphatase (AP) activity. More importantly, PGC-derived cells were stably transformed with the green fluorescent protein marker driven by the cytomegalovirus promoter. After visual identification of transgenic colonies, the pluripotential characteristics of the transgenic PGC-derived cells were tested by chimaera formation and to date we have identified, by genomic Southern blots, two chimaeric fetuses that contain tissues with the transgene incorporated into their chromosomes. To our knowledge, this is the first report of a chimaeric transgenic pig fetus obtained via a cultured cell line.

Methods for the Genetic Modification of Pigs

Introduction

Genome modification of selected domestic species has been the focus of intense research effort for the past 15 years. At present, the only techniques available for the generation of transgenic domestic animals are pronuclear injection or viral vectors. Production of transgenic pigs by pronuclear injection has demonstrated the difficulty and inefficiency of the procedure (Wall, 1996). This is due in part to the nature of the pronuclear injection technique itself, which leads to the random integration of the introduced DNA with unpredictable and, in some cases, undesirable results (Pursel *et al.*, 1989), and due in part to the high cost associated with the generation of transgenic pigs. Although viral transformation tends to be more efficient than pronuclear injection, it is also accompanied by problems associated with random insertion, mosaicism due to multiple integrations, and technical difficulties associated with the generation of replication defective recombinant viral vectors. Thus, in domestic animals, the lack of complete information on the regulatory sequences required for proper expression and regulation of the transgene are compounded by the technical disadvantages associated with the generation of transgenic animals.

Some of the drawbacks of pronuclear injection and viral vectors can be overcome by the use of a technique known as homologous recombination (Koller and Smithies, 1992). This technique allows the precise modification of existing genes, overcomes the problems of positional effects and insertional inactivation, and allows the inactivation of specific genes (Piedrahita *et al.*, 1992) as well as the replacement of one gene for another. Unfortunately the efficiency of the procedure is so low that it cannot be used directly on embryos, but must make use of a carrier cell line. The cell line used to date is the embryonic stem (ES) cell as it allows easy manipulation and selection *in vitro* followed by the generation, by ES-blastocyst injection, of a transgenic animal carrying those changes. Unfortunately, homologous recombination can be performed only in mice, as it has not been possible to isolate domestic animal ES cells that permit the same genetic manipulations. Therefore, a system is required that allows the culture and genetic manipulation of domestic animal cell lines with ES-like qualities.

Embryonic stem (ES) cells and primordial germ cell-derived (EG) cells

Embryonic stem cells isolated from the inner cell mass (ICM) of the preimplantation embryo can proliferate indefinitely in an undifferentiated state, are capable of differentiating *in vitro* and *in vivo*, and can contribute to the formation of normal tissues and organs of a chimaeric individual when injected into a host embryo. Embryonic stem cells can differentiate into several tissue types (Doetschman *et al.*, 1985; Koller and Smithies, 1992) and their differentiation can be stimulated *in vitro* by modifying culture conditions, and *in vivo* by injection of ES cells into athymic mice (Doetschman *et al.*, 1985). When allowed to differentiate *in vitro*, ES cells form structures known as embryoid bodies, which closely resemble the embryo-proper portion of the mouse embryo at day 5 (Doetschman *et al.*, 1985).

The ability to colonize the germ line following culture and genetic manipulation has made ES cells a powerful tool for the modification of the genome in mice. Chimaeras produced between genetically modified ES cells and normal embryos have been used to study gene regulation *in vivo* (Stewart *et al.*, 1985) as well as germ-line transmission of introduced genes (Smithies, 1991). In addition, ES cells have been used to study targeted modification of genes by homologous recombination (Smithies, 1991). Although most of the research on ES cells has been done in mice, attempts at developing the ES technology in other mammalian species have been reported by several investigators (see Stice and Strelchenko, 1995 for review). Doetschman *et al.* (1988) showed that ES cells can be isolated from hamster embryos using feeders composed of murine primary embryonic fibroblasts. Several investigators, using STO (a transformed mouse embryonic fibroblast cell line) as feeder layers, have reported the isolation of pig embryo-derived cell lines with ES-like morphology and a limited ability to differentiate *in vitro* and *in vivo* (Evans *et al.*, 1990; Notarianni *et al.*, 1990; Piedrahita *et al.*, 1990; Gerfen and Wheeler, 1995). In pigs, not only has it been demonstrated that injection of ICM into the blastocoel of a developing blastocyst results in chimaeric pigs (Anderson *et al.*, 1994; Onishi *et al.*, 1994) but, in addition, there have been reports of the ability of cultured ICM-derived ES-like cells to contribute to the generation of a chimaeric pig (Wheeler, 1994). However, not only was the degree of reported chimaerism low, but also there has been no report of germ-line transmission of the ES genotype (Wheeler, 1994; Gerfen and Wheeler, 1995). Without germ-line transmission, any genetic changes that have been introduced into the ES cell cannot be passed on to the next generation and as a result the animals have little, if any, practical value.

Recently, it has been reported that mouse cell lines derived from primordial germ cells behave in a similar way to ES cells and are capable of contributing to the germ line (Labosky *et al.*, 1994). These cells, referred to as EG cells or PGC-derived cells (Labosky *et al.*, 1994; Strelchenko, 1996), are similar to ES cells in terms of markers of the undifferentiated state, as well as their ability to colonize the germ line following injection into a host blastocyst (Labosky *et al.*, 1994; Stewart *et al.*, 1994). Thus, even though the starting tissue source or cellular phenotype differs from the ICM-derived cell lines, once established they have similar properties. Shim *et al.* (in press) have reported the ability of PGC-derived cell lines to contribute to the formation of a pig chimaera, demonstrating the pluripotential characteristics of these cell lines. We have extended this observation by demonstrating the ability of genetically transformed PGCs to contribute to chimaera formation, indicating that the cells we have isolated have pluripotential characteristics, and that the genetic transformation and selection procedure do not interfere with the ability of the cells to participate in chimaera formation.

Progress Towards Isolation and Characterization of Pig Embryonic Stem Cells

Our initial experiments in this area were concentrated on identifying homologous or heterologous cytokines that could maintain ICM-derived cells in an undifferentiated state. Several parallel approaches were used. First, we developed a simple *in vitro* assay capable of detecting the effects of added cytokines on the differentiation of pig ICM (Moore and Piedrahita, 1996; 1997). Second, we isolated and characterized several pig cytokines that may assist in the isolation of pluripotential pig embryo-derived cell lines. Third, we examined the potential use of PGCs for the development of pluripotential or totipotential cell lines.

Development and validation of an in vitro *differentiation assay*

An exogenous supply of haematopoietic cytokines is essential for maintaining mouse embryonic stem (ES) cells in a proliferative yet undifferentiated state. Ip *et al.* (1992) and Stahl *et al.* (1993) demonstrated that haematopoietic cytokines use the gp130 signal transduction pathway to maintain this phenotype, yet their involvement towards maintaining pig ES cell pluripotency has not been established. Isolation and maintenance of pig ES cells have been hindered by the inability to inhibit differentiation of the pig inner cell mass (pICM) *in vitro*. Optimizing culture conditions for the pICM is essential. Therefore, the objective of these studies was to determine the effectiveness of several heterologous haematopoietic cytokines and culture medium at maintaining the isolated pig ICM in an undifferentiated state.

We have developed a grading system to detect changes in the differentiation status of *in vitro* cultured pICM (Moore and Piedrahita, 1996, 1997). In the first experiment pig ICMs (day 7) were isolated by immunosurgery and cultured for 4 days in DMEM-based medium (D medium) or DMEM/Hams F-10 (1:1)-based medium (D/H medium) with or without human leukaemia inhibitory factor (hLIF; 1000 u ml^{-1}). For the second experiment pICMs were collected as above and cultured for 4 days in one of six treatments: control medium, human leukaemia inhibitory factor (hLIF; 1000 u ml^{-1}), human interleukin 6 (hIL-6; 100 ng ml^{-1}), hIL-6+hIL-6 soluble receptor (hIL6+sR; 100 ng ml^{-1}+2.5 μg ml^{-1}), human oncostatin M (hOSM; 100 ng ml^{-1}), or rat ciliary neurotrophic factor (rCNTF; 100 ng ml^{-1}). All cytokines were prepared in Dulbecco's Modified Eagles Medium:Hams F-10 (1:1)-based medium. Colonies were photographed daily for morphological analysis. pICMs were categorized into one of two types based on their morphological profile: type A, non epithelial or type B, epithelial-like. Eight investigators evaluated pICM differentiation using standardized differentiation profiles. Each pICM series was graded on a scale of 1 (fully undifferentiated) to 5 (fully differentiated) for each time point. Differentiation was verified by alkaline phosphatase activity, cytokeratin staining and scanning electron microscopy. In Expt 1, neither hLIF nor culture medium delayed differentiation of the developing pICM. In Expt 2, differentiation was significantly lower on day 2 for rCNTF cultured pICMs versus hLIF cultured pICM (2.07 ± 0.15 versus 2.70 ± 0.16; $P < 0.05$). Furthermore, addition of rCNTF gave the lowest overall mean differentiation score (2.53 ± 0.15). However, none of the cytokines significantly delayed differentiation over the control for the 4 day culture period. The grading system used was an effective tool for detecting treatment effects on differentiation of the developing pICM.

Since these heterologous cytokines were unable to inhibit differentiation significantly, it is unlikely that they will be beneficial in isolating pig ES cell lines under current conditions. Future work with homologous cytokines, protein half-lifes, and dose effects may prove more beneficial.

Isolation and characterization of pig cytokines

To determine whether homologous cytokines are more effective in inhibiting the differentiation of pig embryo-derived cells, we have isolated several of the pig haematopoietic cytokine genes. To date, we have isolated the genomic sequences for LIF, LIF receptor, ciliary neurotrophic factor (CNTF) and oncostatin M (OSM).

Leukaemia inhibitory factor, LIF, is another haematopoietic cytokine that has been shown to promote the survival and proliferation of embryonic stem cells in culture (Smith *et al.*, 1988; Williams *et al.*, 1988; Wilson *et al.*, 1992). Like CNTF, the heterologous protein is ineffective in isolating pig ES cell lines (Moore and Piedrahita, 1996; Moore and Piedrahita, 1997). As a result, we are attempting to express the pig LIF protein using a synthetic gene and test the effect of the homologous protein on the isolation of pig ES cells. In previous experiments with expression of the pLIF in mammalian cells (CHOs: Chinese hamster ovarian cells), we were able to detect only very low expression of pLIF by western blot analysis. As a result we decided to develop artificial genes that can be expressed in yeast using the method of Stemmer *et al.* (1995). One clone spanned the entire gene and had a minimal number of mismatches. These mistakes include: six single bp deletions, two of which lie within the linker and need not be repaired and five substitutions which are GCT/GCA: alanine, GCT/ACT: alanine to threonine, AGA/GGA: arginine to glycine, GCT/GGT: alanine to glycine, and AAG/GAG: lysine to glutamic acid. The first substitution does not change the amino acid that is incorporated and thus no repair is needed; however, the remaining mistakes change the amino acid and must be repaired before the protein can be expressed. Repairs will be made by performing *in vitro* site-directed mutagenesis using the Transformer Mutagenesis Kit (Clontech, Palo Alto, CA). Alternatively, PCR conditions can be optimized for a polymerase, such as Pfu, that has a higher fidelity and processivity than Taq for synthesizing the gene from the original twelve oligonucleotides. Once all repairs are complete and verified by sequencing, the synthetic gene will be cloned into pMLD8, the recombinant protein produced in *Pichia pastoris*, and its effectiveness on the isolation of pig ES cell lines tested in culture.

Ciliary neurotrophic factor (CNTF), a haematopoietic cytokine, can maintain mouse embryonic stem (ES) cells in culture (Conover *et al.*, 1993). By interacting with its receptor and forming a hexamer that includes a gp130–LIFR heterodimer (Deserio *et al.*, 1995), CNTF facilitates the activation of the gp130 signal transduction pathway (Ip *et al.*, 1992; Stahl *et al.*, 1993), which ultimately leads to maintenance of the undifferentiated, proliferating phenotype of the ES cell (Yoshida *et al.*, 1994). As we have shown that heterologous CNTF does not positively contribute to the isolation of stable pig embryo-derived cell lines with characteristic ES cell morphology (Moore and Piedrahita, 1996), we have isolated and characterized the pig CNTF to determine the effects of the homologous protein on the inhibition of differentiation of early pig embryonic cells.

The pig CNTF gene, isolated from a phage genomic library, has been cloned, sequenced and assigned to chromosome 2. Briefly, a 369 bp reverse transcription polymerase chain reaction (RT–PCR) probe synthesized from pig activated macrophage mRNA was used to identify positive phage clones. Of seven initial positives, one phage clone was purified through a tertiary screening and a 6 kb ApaI fragment identified by southern blotting as containing the whole gene was cloned into pBluescript (Stratagene Inc., La Jolla, CA) using shotgun cloning. Four subclones were constructed and all clones were sequenced using dye terminator automated cycle sequencing with the M13–20 and reverse primers; in addition, other primers were designed and synthesized as needed to sequence both strands of each clone entirely. The pig CNTF gene includes two exons separated by a 1258 bp intron (GenBank Accession no. U57644). The gene encodes a deduced polypeptide of 200 amino acids which showed 83%, 82%, 82% and 81% similarity to known sequences in the rabbit (Lin *et al.*, 1989), rat (Stockli *et al.*, 1989), human (Negro *et al.*, 1991) and mouse (GenBank Accession no. U05342), respectively. Seven non-conservative amino acid changes were identified in the pig protein compared with other species which, if important in the function of the protein, may explain the inability of heterologous CNTF to activate the pig gp130 pathway. Comparison of the 5′ region of the pig CNTF and other mammalian cytokines indicates the presence of several conserved transcription factor binding motifs, suggesting their importance for controlling the specific expression of these proteins. In addition, CNTF has been localized to pig chromosomes by fluorescence *in situ* hybridization performed essentially as described by Niebergs *et al.* (1993). Pig metaphase chromosomes were prepared from whole blood. All specific hybridization signals were localized on chromosome 2p, band 1.6, and arm length ratios calculated for 40 early metaphase chromosomes confirmed that the pig CNTF gene is located on chromosome 2p1.6. This assignment was further confirmed by aligning R-banded FISH labelled chromosomes to the standard idiogram and R-banded chromosome 2 of the pig (Committee for the Standard Karyotype of the Domestic Pig, 1988).

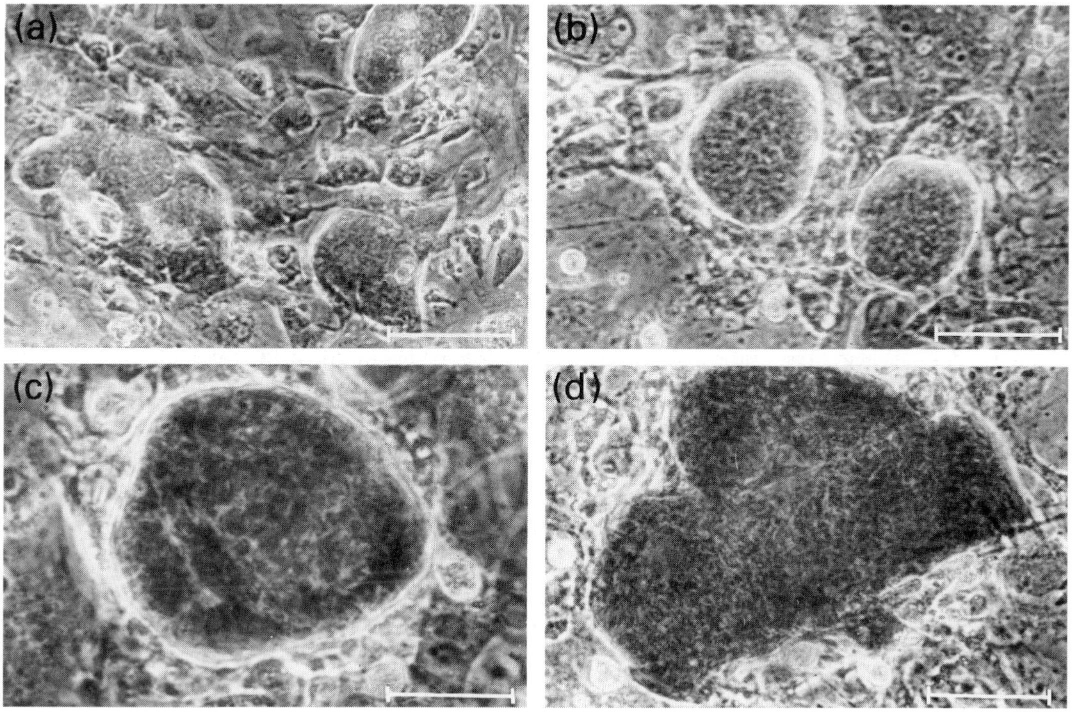

Fig. 1. Morphology and alkaline phosphatase (AP) activity of pig primordial germ cell-derived colonies after several passages. Cultured colonies were passaged at least twice before morphology and AP activity was recorded. (a) and (b) Pig embryonic stem (ES) cell-like cells (Scale bar represents (a) 50 μm and (b) 25 μm); (c) and (d) pig AP-positive colonies (Scale bar represents 50 μm). Note in all cases the typical ES-like morphology with small cytoplasm, large nucleus and prominent nucleoli.

At present, we are preparing to introduce a synthetic version of pig CNTF in *Pichia pastoris* for expression and testing the bioactivity of the homologous protein compared with heterologous CNTF.

Development and characterization of PGC-derived cell lines

Isolation and maintenance of EG-like pig cells. Uteri of pigs at 25–27 days of pregnancy were collected at surgery, the fetuses dissected, and germ cells isolated as previously described by Labosky *et al.* (1994). Essentially, the genital ridge of the developing fetus was removed and incubated in phosphate-buffered saline (PBS) containing 10 mmol EDTA l^{-1} for 20 min to dissociate PGCs from the gonadal ridge. After incubation, the PGCs were gently released in culture media, centrifuged and resuspended in Dulbecco's modified Eagle's media:Ham's F10, supplemented with 0.01 mmol non-essential amino acids l^{-1}, 2 mmol glutamine l^{-1}, 15% fetal bovine sera (selected batches, Summit Biotechnology), 0.1 mmol 2-mercapthoethanol l^{-1}, soluble recombinant human stem cell factor at 40 ng ml^{-1}, human basic fibroblast growth factor at 20 ng ml^{-1}, and LIF at 20 ng ml^{-1} (PEG medium).

The cell suspension at a density of 10 000 PGCs ml^{-1} was plated onto a feeder layer of STO cells prepared as described by Piedrahita *et al.* (1990). After 7–10 days of culture, colonies with ES-like morphology were passaged to fresh feeder layers for establishment of cell lines. Resulting colonies were passaged by trypsinization to fresh feeder layers at 6–9 day intervals. State of differentiation of isolated cell lines was determined by morphology and expression of alkaline phosphatase (AP), a marker of undifferentiated embryonic cells (Talbot *et al.*, 1993). As seen in Fig. 1, colonies with typical

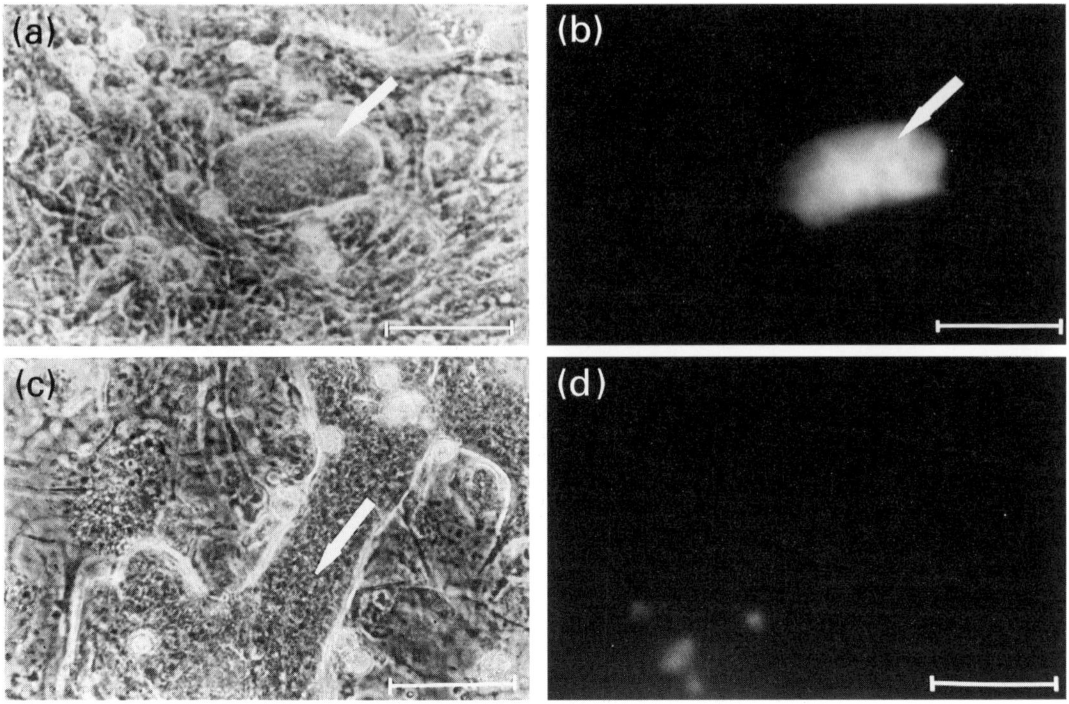

Fig. 2. Genetically transformed pig primordial germ cell (PGC)-derived colonies. PGCs were isolated and electroporated with a plasmid containing the green fluorescent protein (GFP) under the control of the cytomegalovirus promoter. Fourteen days after electroporation, colonies were examined under a fluorescent microscope. (a) Typical PGC-derived colony. The morphology of the colonies shown is similar to that seen in murine embryonic stem cell and embryonic germ cell colonies. (b) Same colony as (a) under fluorescent light. Note all cells are expressing GFP, indicating that the colony probably originated from a single transgenic cell. (c) Large non-transgenic PGC-derived colony in the same plate as (a) and (b). (d) Colony C under fluorescent light. Note the absence of any signal. Arrows indicate the location of the colony. Scale bar represents 50 μm.

ES and EG-like morphology were isolated from the PGCs. Moreover, we have been able to maintain these cell lines for up to 14 passages and 4 months while still retaining the alkaline phosphatase expression and the appropriate morphology.

Genetic transformation of pig PGCs. PGC-derived colonies were electroporated with a plasmid consisting of the 'humanized' green fluorescent protein (GFP) attached to the cytomegalovirus (CMV) promoter. After electroporation, cells were plated on fresh feeder layers and cultured for 8–12 days, and then observed under ultraviolet light (FITC filters) and fluorescent colonies identified (Fig. 2). Fluorescent colonies were picked and passed to fresh feeders for establishment of transgenic EG cell lines.

Ability of transgenic pig cultured PGCs to contribute to formation of chimaera. Transgenic PGCs isolated and transformed as described previously were identified, dissociated by trypsinization, and 10–15 cells injected into the blastocoel of the developing blastocyst or into the inner portion of compacted morulae. Injected embryos were transferred to synchronized recipients and fetuses collected at day 25 of pregnancy. For evaluation, fetuses were observed under fluorescent light for detection of the marker GFP. Unfortunately, owing to the high background fluorescence, the GFP could not be effectively used to identify chimaeric fetuses (Fig. 3). After visual inspection fetuses

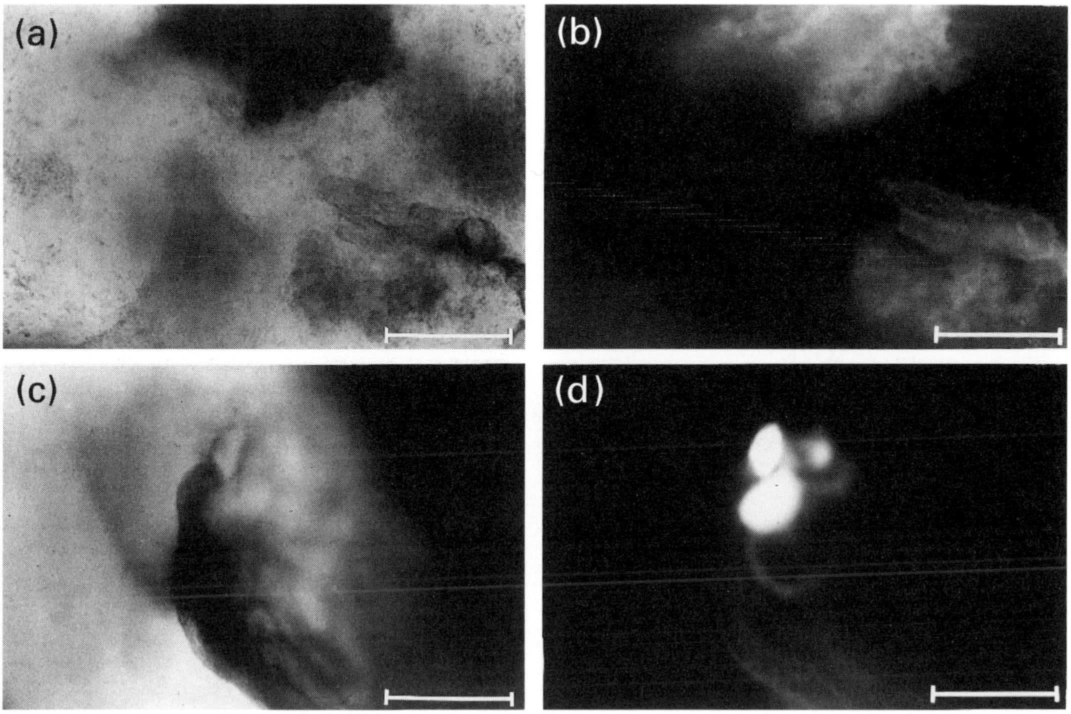

Fig. 3. Autofluorescence of fetal pig tissues. Analysis of non-transgenic 25-day-old pig fetuses under a fluorescent microscope equipped for fluorescein isothiocyanate excitation indicated that fetal tissues give a high autofluorescence signal. The intensity of the signal varied from a low, homogeneous signal (a, transmitted light and b, fluorescent light) to a strong signal associated with particular structures (c, light and d, fluorescent). Unfortunately, both types of autofluorescence makes use of GFP expression for the detection of transgenic chimaeric fetuses difficult. Scale bar represents 200 µm.

were minced and a fraction of the tissue used for DNA analysis, which consisted of detection of the GFP construct by genomic Southern blots. Of seven recipients receiving injected embryos, two were pregnant, and 14 normal fetuses and one resorbed fetus were obtained. Two of these were positive for the transgene by Southern blot analysis. The restriction enzymes selected can distinguish between a plasmid contaminant and DNA that is inserted in the genome (Fig. 4). For instance, the BstXI digest showed a hybridizing band of 8 kb. As the initial construct is only 6.2 kb, it is not possible to obtain such a band unless the construct is embedded in genomic DNA flanked by two BstXI sites. The *Hin*dIII digest cuts once inside the plasmid as well as in the flanking genomic DNA. From the pattern obtained, it is evident that both chimaeric animals had contributions from the same transformation event. As five colonies were isolated, trypsinized, and the cells used to inject into blastocysts, these results indicate that either only one of the five colonies had pluripotential abilities or that by chance, the two resulting chimaeras were derived from cells obtained from the same colony. To our knowledge this is the first report of transgenic cultured pig cells being able to take part in chimaera formation. The data presented above clearly indicate that transgenic pig EG cells can contribute to the development of a fetus when injected into a host blastocyst.

A disadvantage of the present technology in swine is the need for the generation of a germ-line chimaera and the required extra generation before a transgenic founder is produced. Although this is not a problem in mice which have short generation intervals, in pigs and other domestic species this extra generation can increase the costs per transgenic founder quite substantially. We are presently exploring the use of nuclear transfer technology (Liu *et al.*, 1995; Ouhibi *et al.*, 1996) as well

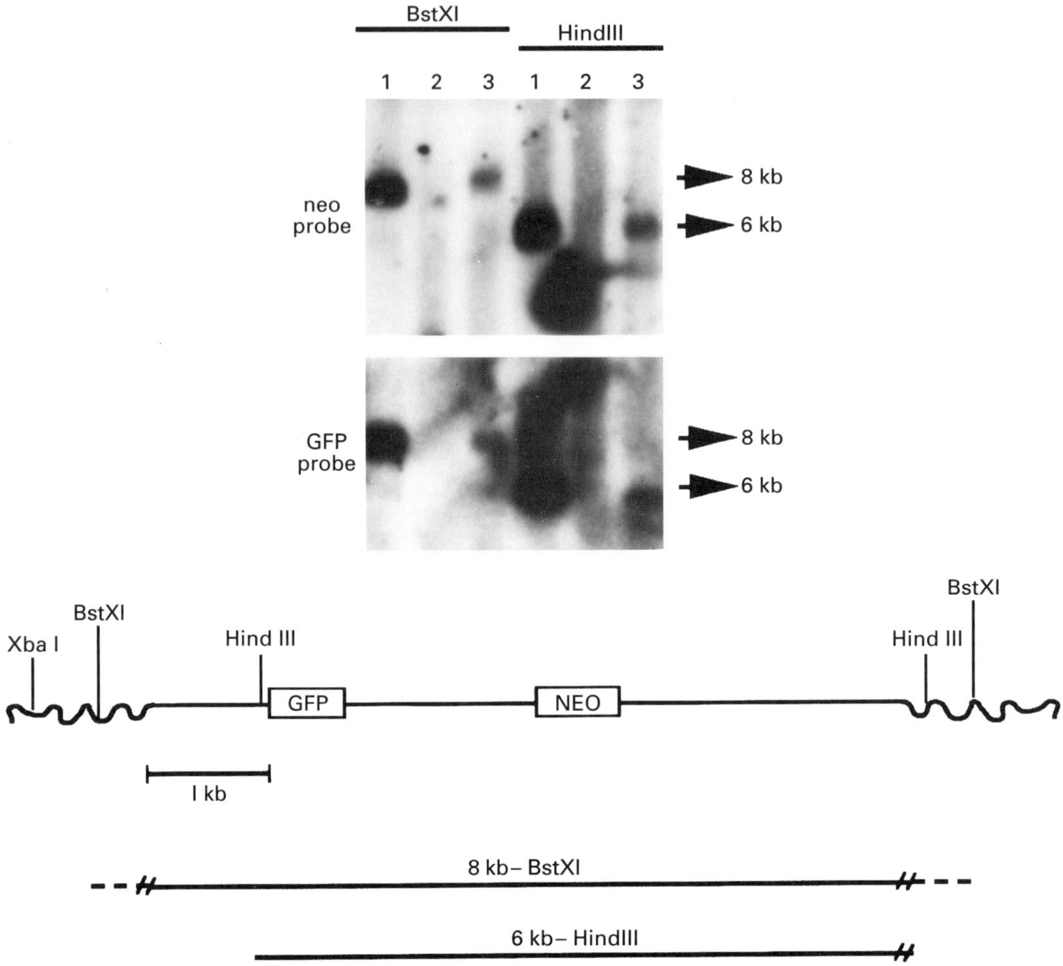

Fig. 4. Southern blot analysis of chimaeric fetal tissues. DNA samples obtained from pig fetuses previously identified as transgenic were digested with either *Bst*XI or *Hin*dIII, separated in 0.7% agarose and sequentially hybridized with a green fluorescent protein (GFP) probe and a neomycin (neo) probe. Lane (1) fetus number 9; lane (2) negative control fetus; lane (3) fetus number 10. The enzymes used for digestion were selected for their ability to differentiate between plasmid contamination and genomic incorporation as well as for determining the number of copies inserted. The 8 kb band obtained with the *Bst*XI digest unequivocally rules out the possibility of contamination as the initial plasmid was only 6.2 kb in length. Moreover, *Bst*XI does not cut within the introduced plasmid indicating that only one copy of the construct has incorporated into the genome. In addition, both digests corroborated the presence of both the GFP and the neogene in the inserted DNA, suggesting that no major DNA rearrangement occurred.

as the use of tetraploid embryos as hosts for chimaera formation (Nagy *et al.*, 1990; Prather *et al.*, 1996; Ueda *et al.*, 1995) as potential methods reducing the time required for generation of a founder animal.

Conclusion

Recent advances in the isolation, culture and genetic manipulation of pig pluripotent stem cells (ES and EG) suggest that within the next few years it will be possible to introduce subtle genetic modification into the pig genome. Although biomedical applications of this technology, such as xenotransplantation and production of biopharmaceuticals in the milk, are obvious, the use of transgenic technologies in agricultural production systems is not as clear. This is due in part to the high costs of generating the transgenic animals but more importantly it is our lack of knowledge of how phenotypes of agricultural importance can be manipulated by modifying single genes. However, progress in the genetic manipulation of pigs combined with progress in the elucidation of the genetic control of phenotypes of agricultural importance will soon lead to the generation of transgenic pigs of use in agriculture.

References

Anderson GB, Choi SJ and BonDurant RH (1994) Survival of porcine inner cell masses in culture and after injection into blastocysts *Theriogenology* **42** 212

Conover JC, Ip NY, Poueymirou WT, Bates B, Goldfarb MP, DeChiara TM and Yancopoulos GD (1993) Ciliary neurotrophic factor maintains the pluripotentiality of embryonic stem cells *Development* **119** 559–565

Deserio A, Graziani R, Laufer R, Ciliberto G and Paonessa G (1995) *In vitro* binding of ciliary neurotrophic factor to its receptors – evidence for the formation of an IL-6 type hexameric complex *Journal of Molecular Biology* **254** 795–800

Doetschman T, Williams P and Maeda N (1988) Establishment of hamster blastocyst-derived embryonic stem ES cells *Developmental Biology* **127** 224–227

Doetschman T, Eistetter CH, Katz M, Schmidt W and Kemler R (1985) The *in vitro* development of blastocyst-derived embryonic stem cell lines: formation of visceral yolk sac, blood islands and myocardium *Journal of Embryology and Experimental Morphology* **87** 27–45

Evans MJ, Notarianni E, Laurie S and Moor RM (1990) Derivation and preliminary characterization of pluripotent cell lines from porcine and bovine blastocysts *Theriogenology* **33** 125–128

Gerfen RW and Wheeler MB (1995) Isolation of embryonic cell-lines from porcine blastocysts *Animal Biotechnology* **6** 1–14

Ip NY, Nye SH, Boulton TG, Davis S, Taga T, Li Y, Birren SJ, Yasukawa K, Kishimoto T, Anderson DJ, Stahl N and Yancopoulos GD (1992) CNTF and LIF act on neuronal cells via shared signaling pathways that involve the IL-6 signal transducing receptor component gp130 *Cell* **69** 1121–1132

Koller B and Smithies O (1992) Altering genes in animals by gene targeting *Annual Reviews of Immunology* **10** 705–730

Labosky PA, Barlow DP and Hogan BLM (1994) Mouse embryonic germ (EG) cell lines: transmission through the germline and differences in the methylation imprint of insulin-like growth factor 2 receptor (*Igf2r*) gene compared with embryonic stem (ES) cell lines *Development* **120** 3197–3204

Lin L-FH, Mismer D, Lile JD, Armes LG, Butler ET, III, Vannice JL and Collins F (1989) Purification, cloning, and expression of ciliary neurotrophic factor (CNTF) *Science* **246** 1023–1025

Liu L, Moor RM, Laurie S and Notarianni E (1995) Nuclear remodelling and early development in cryopreserved, porcine primordial germ cells following nuclear transfer into *in vitro*-matured oocytes *International Journal of Developmental Biology* **39** 871–880

Moore K and Piedrahita JA (1996) Effects of heterologous hematopoietic cytokines on *in vitro* differentiation of cultured porcine inner cell masses *Molecular Reproduction and Development* **45** 139–144

Moore K and Piedrahita JA (1997) The effects of human leukemia inhibitory factor (hLif) and culture medium on *in vitro* differentiation of cultured porcine inner cell mass (pICM) *In Vitro Cellular and Developmental Biology – Animal* **33** 62–71

Nagy A, Gocza E, Diaz EM, Prideaux VR, Ivanyi E, Markkula M and Rossant J (1990) Embryonic stem cells alone are able to support fetal development in the mouse *Development* **110** 815–821

Negro A, Tolosano E, Skaper SD, Martini I, Callegaro L, Silengo L, Fiorini F and Altruda F (1991) Cloning and expression of human ciliary neurotrophic factor *European Journal of Biochemistry* **201** 289–294

Niebergs HL, Gallagher DS, Georges M, Sargeant LS, Dietz AB and Womack JE (1993) Physical mapping of inhibin B-A in domestic cattle *Mammalian Genome* **4** 328–332

Notarianni E, Laurie S, Moor RM and Evans MJ (1990) Maintenance and differentiation in culture of pluripotential embryonic cell lines from pig blastocysts *Journal of Reproduction and Fertility* **41** 51–56

Onishi A, Takeda K, Komatsu M, Akita T and Kojima T (1994) Production of chimeric pigs and the analysis of chimerism using mitochondrial deoxyribonucleic acid as a cell marker *Biology of Reproduction* **51** 1069–1075

Ouhibi N, Fulka J, Jr, Kanka J and Moor RM (1996) Nuclear transplantation of ectodermal cells in pig oocytes: ultrastructure and radiography *Molecular Reproduction and Development* **44** 533–539

Piedrahita JA, Anderson GB and BonDurant RH (1990) On the isolation of embryonic stem (ES) cells: comparative behavior of murine, porcine, and ovine embryos *Theriogenology* **34** 879–901

Piedrahita JA, Zhang SH, Hagaman JR, Clark PM and Maeda N (1992) Generation of mice carrying a mutant apolipoprotein E gene inactivated by gene targeting in mouse embryonic stem cells *Proceedings of the National Academy of Sciences, USA* **89** 4471–4475

Prather RS, Hoffman KE, Schoenbeck RA, Stumpf TT and Li J (1996) Characterization of DNA synthesis during the 2-cell stage and the production of tetraploid chimeric pig embryos *Molecular Reproduction and Development* **45** 38–42

Pursel VG, Pinkert CA, Miller KF, Bolt DJ, Campbell RJ, Palmiter RD, Brinster RL and Hammer RE (1989) Genetic engineering of livestock *Science* **244** 1281–1288

Shim H, Gutiérrez-Adán A, Chen LR, Bon Durant RH, Behboodi E and Anderson GB (1997) Isolation of pluripotent stem cells from cultured porcine primordial germ cells *Biology of Reproduction* **57** 1089–1095

Smith AG, Heath JK, Donaldson DD, Wong GG, Moreau J, Stahl M and Rogers D (1988) Inhibition of pluripotential embryonic stem cell differentiation by purified polypeptides *Nature* **336** 688–690

Smithies O (1991) Altering genes in humans and animals In *Etiology of Human Disease at the DNA Level* pp 221–230 Eds J Lindsten and U Petterson. Raven Press, New York

Stahl N, Davis S, Wong V, Taga T, Kishimoto T, Ip NY and Yancopoulos GD (1993) Cross-linking identifies leukemia inhibitory factor-binding protein as a CNTF receptor component *Journal of Biological Chemistry* **268** 7628–7631

Stemmer WPC, Crameri A, Ha KD, Brennan TM and Heyneker HL (1995) Single-step assembly of a gene and entire plasmid from large numbers of oligodeoxyribonucleotides *Gene* **164** 49–53

Stewart CL, Vanek M and Wagner EG (1985) Expression of foreign genes from retroviral vectors in mouse teratocarcinoma chimeras *EMBO Journal* **4** 3701–3709

Stewart CL, Gadi I and Blatt H (1994) Stem cells from primordial germ cells can re-enter the germline *Developmental Biology* **161** 626–628

Stice SL and Strelchenko NS (1995) Domestic animal embryonic stem cells: progress toward germ-line contribution. In *Biotechnology's Role in the Genetic Improvement of Farm Animals* p. 189 Eds RH Miller, VG Pursel and H Norman. Beltsville Symposia in Agricultural Research, Beltsville, MD

Stockli KA, Lottspeich F, Sendtner M, Masiakowski P, Carroll P, Gotz R, Lindholm D and Thoenen H (1989) Molecular cloning, expression, and regional distribution of rat ciliary neurotrophic factor *Nature* **342** 920–923

Strelchenko N (1996) Bovine pluripotent stem cells *Theriogenology* **45** 131–140

Talbot NC, Rexroad CE, Pursel VG and Powell AM (1993a) Alkaline phosphatase staining of pig and sheep epiblast cells in culture *Molecular Reproduction and Development* **36** 139–147

Ueda O, Jishage K, Kamada N, Uchida S and Suzuki H (1995) Production of mice entirely derived from embryonic stem (ES) cell with many passages by coculture of ES cells with cytochalasin B induced tetraploid embryos *Experimental Animals* **44** 205–210

Wall RJ (1996) Transgenic livestock: progress and perspectives for the future *Theriogenology* **45** 57–68

Wheeler MB (1994) Development and validation of swine embryonic stem cells – a review *Reproduction Fertility and Development* **6** 563–568

Williams RL, Hilton DJ, Pease S, Willson TA, Stewart CL, Gearing DP, Wagner EF, Metcalf D, Nicola NA and Gough NM (1988) Myeloid leukaemia inhibitory factor maintains the developmental potential of embryonic stem cells *Nature* **336** 684–687

Wilson TA, Metcalf D and Gough NM (1992) Cross-species comparison of the sequence of the leukemia inhibitory factor gene and its protein *European Journal of Biochemistry* **204** 21–30

Yoshida K, Chambers I, Nichols J, Smith A, Mikiyoshi S, Yasukawa K, Shoyab M, Taga T and Kishimoto T (1994) Maintenance of the pluripotential phenotype of embryonic stem cells through direct activation of gp130 signalling pathways *Mechanisms of Development* **45** 163–171

Journal of Reproduction and Fertility Supplement **52**, 255–266

Advances in gender preselection in swine

L. A. Johnson

Germplasm and Gamete Physiology Laboratory, Agricultural Research Service, US Department of Agriculture, Beltsville, MD 20705, USA

Gender preselection using isolated populations of X- and Y-chromosome bearing spermatozoa that have been separated on the basis of DNA content is currently possible in swine and other farm animals, as well as in humans. Semen from most livestock species can now be successfully separated into predominantly X or Y sperm populations before their use for intra-tubal insemination, deep-uterine insemination or for *in vitro* fertilization (IVF) to produce sexed offspring. Birth of progeny of the desired sex in cattle, sheep and swine under semi-practical conditions has successfully validated the sexing technology. Spermatozoa are separated on the basis of inherent differences in DNA content in the X- and Y-chromosome bearing sperm population using modified flow cytometry/cell sorting technology. Spermatozoa are stained with Hoechst 33342 which binds to the DNA in an amount proportional to the amount of DNA present in the individual spermatozoa. Over 300 animals from various species at several locations have been born using the USDA-Beltsville Sperm Sexing Technology for separating X and Y spermatozoa. Sex ratios are shifted from the normal 50:50 to 85 to 90% of one sex or the other. In swine, offspring have been born as the result of surgical intratubal insemination of separated spermatozoa and also from IVF and embryo transfer. At the present time, standard swine artificial insemination techniques are not optimized for use with the small numbers of flow cytometrically separated X or Y sperm populations. Cattle, swine and rabbit offspring have been reproduced through the second generation with normal morphology and reproductive function. Numerous improvements have been made in the sexing technology since it was first reported in 1989. Increasing the speed of the sexing process to make the application of the technology available to a larger segment of the livestock industry is paramount, even with insemination technology designed for small numbers of spermatozoa.

Introduction

Use of sexed semen for production of livestock of preselected sex is dependent on economics, effectiveness, efficiency and ease of use. The most cost effective approach for achieving sex preselection involves separating the X- from the Y-chromosome bearing spermatozoa followed by use for artificial insemination (AI) or for *in vitro* fertilization (IVF) and subsequent embryo transfer (ET). Since one is pre-determining the outcome of conception rather than the outcome after fertilization, the use of X- or Y- bearing spermatozoa has a clear advantage over embryo sexing. Livestock management schemes can benefit from sex preselection because of the ability to plan matings for a specific sex. In addition to faster genetic progress from sexed spermatozoa or embryos, there are additional advantages of management efficiency and facility flexibility. The swine industry would have a more efficient and economically advantageous means of producing male and female hybrid lines.

Our report (Johnson *et al.*, 1989) of effective sex preselection in rabbits was the first report detailing the use of DNA as a marker for sex preselection and subsequent birth of offspring of the predicted sex. This was followed by publication of results confirming use of the method with boar semen (Johnson, 1991). Since 1989, more than 300 animals have been born using this procedure

Table 1. Markers that have been used or proposed for separation of X and Y spermatozoa

Proposed marker	Parameter	Reference
DNA	X spermatozoa > Y	Moruzzi, JF., 1979
Size	X spermatozoa > Y	Cui, K., 1997
Motility	Y faster than X	Ericsson *et al.*, 1973
F-Body	Long arm on Y chromosome	Barlow and Vosa 1970
H-Y antigen	Spermatozoa surface	Hendriksen *et al.*, 1993
Protein	Spermatozoa surface	Hendriksen *et al.*, 1996

(Cran *et al.*, 1993,1995; Rath *et al.*, 1997; Seidel *et al.*, 1997; Catt *et al.*, 1996). Many of the animals were produced at other locations where the instrumentation has been duplicated thus confirming the reproducibility of the technology. This review will describe the process involved in using the Beltsville Sperm Sexing Technology with particular emphasis on swine. Current usage as well as future prospects for use in practical situations will be discussed. Since the Beltsville technology is the only sex preselection technology that has been proven effective in livestock to date, it will be the primary emphasis of this review.

Basis for Separation of Spermatozoa

Physical separation methods

Numerous attempts to separate X- from Y-chromosome bearing spermatozoa have been tested over the last 70 years (Table 1); most protocols are based on some physical aspect of the spermatozoa. None of the various protocols that use a supposed physical difference between X and Y spermatozoa have proven effective in obtaining a higher proportion of either sex. Our experience in testing these methods (Johnson, 1988) is that no method developed on the basis of an inherent physical difference within the spermatozoa or on the sperm surface has been effective in separating X- from Y-bearing spermatozoa, except a method based on DNA content of the spermatozoa. Overviews of the various methods that have been tried have been published by Kiddy and Hafs (1971), Amann and Seidel (1982), Gledhill (1988) and Johnson (1992, 1994, 1995).

Difference in DNA content for separation of X and Y spermatozoa

It is well established that the DNA content of X- and Y-chromosome bearing spermatozoa is different in most mammals. The Y chromosome is smaller and carries less DNA than the X chromosome, while the autosomes carried by X- or Y-bearing spermatozoa are identical in DNA content. Taking advantage of this difference to separate X- and Y-bearing spermatozoa has only been possible since the development of flow cytometry. Initial attempts to differentiate X and Y spermatozoa using DNA and flow cytometric analysis did not succeed (Gledhill *et al.*, 1976). However, with the combination of improved staining methods and a realization that aspherical cells such as spermatozoa must be orientated to the excitation source (Dean *et al.*, 1978), the relative difference in DNA could be measured (Pinkel *et al.*, 1982). These initial advances in analysis of sperm DNA formed the basis for routine analysis of sperm DNA using modified commercial instrumentation (Johnson and Pinkel, 1986). Sorting sperm populations on the basis of difference in DNA followed (Johnson *et al.*, 1987a; Johnson and Clarke, 1988). Improved cellular staining and the use of vital fluorochromes to label DNA led to the separation of X- and Y-chromosome bearing spermatozoa for the production of offspring of pre-determined sex (Johnson *et al.*, 1987b, 1989; Johnson, 1991).

Separation of Spermatozoa by Flow Cytometry

The sorting of the spermatozoa is done using a modified version of any full scale flow cytometer/cell sorter. Much of the equipment in use today was designed and built in the early 1980s. Basic flow cytometric system principles have not changed since their introduction in the late 1960s and early 1970s. The adaptation of commercial flow cytometers with modifications for aspherical cells such as spermatozoa (Johnson and Pinkel, 1986) can still be applied to more recent models of flow cytometer/cell sorters (Fig. 1). The older units continue to work as effectively as the newer ones with the exception that the data acquisition system is slower. A flow cytometer/cell sorter system adapted for sperm sorting uses a five watt water cooled argon laser as an excitation source. Although fluorochromes designed to bind to DNA generally give off a bright signal, the five watt water cooled laser remains the laser of choice for most DNA fluorochromes.

Resolving X and Y spermatozoa based on DNA difference

The resolution of X- and Y-bearing spermatozoa using flow cytometric DNA analysis is difficult compared with that of other types of cell. The inordinate compactness of chromatin in the morphologically flat, paddle or ovoid shaped sperm head that is characteristic for domestic animals causes a high index of refraction. The difference in refractive index between the sperm head and the surrounding medium, coupled with the flat shape of the sperm head, results in preferential emission of light in the plane of the cell (from the edge of the sperm head). The orientation of the sperm head with respect to the excitation laser beam and the optical detectors is therefore critical for resolution. The modification of the flow cytometer/cell sorter system consists of a forward fluorescence detector in place of the light scatter detector that is standard to orthogonal configured flow systems. This is necessary so that fluorescent light from both the edge of the spermatozoa (90°) as well as from the flat side (0°) of the spermatozoa can be collected. The second modification is to replace the usual cylindrical sample input needle with a similar but bevelled sample input needle (Johnson and Pinkel, 1986). The modified flow cytometer/cell sorter is essential for attaining separate populations of X- and Y-chromosome bearing spermatozoa based on differences in DNA content and for reanalysing sorted spermatozoa to determine the proportions of X or Y spermatozoa in a given sorted sample (Fig. 1; Johnson *et al.*, 1987a, 1989).

Instrumentation options for sexing spermatozoa by flow sorting

Differentiation of the amount of DNA present in the X- and Y-chromosome bearing spermatozoa for sorting can be done on nearly all commercial flow cytometer/cell sorters that have been manufactured in the past 15–20 years if they have been properly modified. To date flow cytometer/cell sorter systems manufactured by the Coulter Corporation, Hialeah, FL (EPICS V, EPICS 750 series, ELITE) and the Becton Dickinson Co., Mountain View, CA (FACS Vantage; FACStar Plus; FACStar; FACS 440) have been modified for sperm sorting. Recently we have modified a high speed flow cytometer/cell sorter manufactured by Cytomation Inc., Fort Collins, CO (MoFlo). The high speed MoFlo system has the capability of sorting at a rate four to six times that of the older standard systems (L. Johnson and G. Welch, unpublished). Each of the systems mentioned was modified to provide for a bevelled sample injection needle and a second fluorescence detector located 0° to the laser (Fig. 1). This modification allows for orientation of the spermatozoa as it passes from the bevelled needle and past the laser beam and for the selection of the orientated population through electronic gating (Johnson and Pinkel, 1986).

Sperm preparation prior to sorting

Protocols for the preparation of fresh semen for flow sorting have been developed based on the principle that the less insult imposed on the spermatozoa, the greater the likelihood that the ultimate

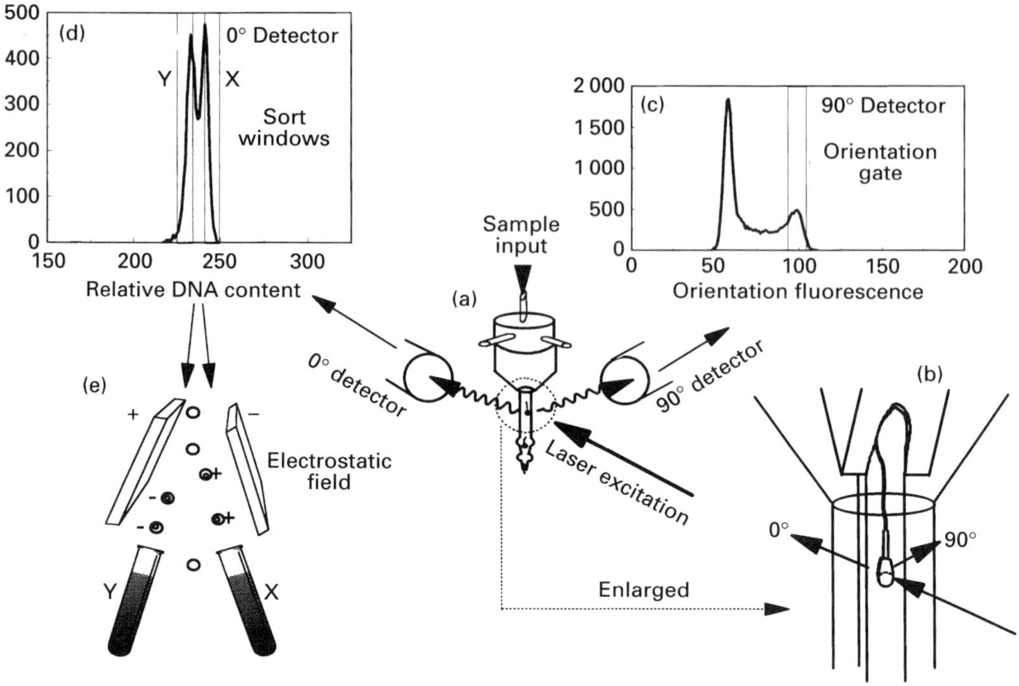

Fig. 1. A diagram of the sample input and data handling capability of a flow cytometer/cell sorter modified for resolution of X- from Y-chromosome bearing spermatozoa based on their difference in DNA content according to the Beltsville Spermatozoa Sexing Technology. Spermatozoa enter through a needle labelled 'sample input' and are passed into the flow cell (a). They leave the needle into the flow nozzle in single file and are oriented according to the position of the needle (b) which has been ground to a bevel on the exit end. Orientation occurs when the opposing flatter surfaces of the spermatozoa head simultaneously face the laser and 0° forward detector, with the edge directed to the 90°C. (c) The fluorescence distribution from the 90° detector. Note the population within the vertical gated area: this portion of the spermatozoa is properly oriented to the laser beam. (d) The separation of the X and Y chromosome populations. Each peak population has a sort window represented by vertical lines on the peaks. The area between the two sort windows shows spermatozoa that are mixed (X and Y) in various percentages and so they are discarded through the waste. The simultaneously collected 0° fluorescence distribution of the properly oriented spermatozoa shows a partially resolved bimodal distribution (d). Spermatozoa containing the smaller Y chromosome and thus less total DNA comprise the dimmer (left) population and similarly the larger X-chromosome bearing spermatozoa comprise the brighter population. (e) A schematic of the sorting process, depicting the electrostatic field through which the charged droplets containing X or Y spermatozoa must fall and be deflected to the waiting tubes.

population of sorted spermatozoa will keep and maintain their fertilizing capacity (Johnson *et al.*, 1989). In particular, the current protocol for swine semen has been refined from our original protocol (Johnson, 1991) so that an aliquot of fresh semen containing 15×10^6 spermatozoa are diluted to 1 ml with BTS extender; Hoechst 33342 is added to that suspension at the concentration of 4 µg per 15×10^6 spermatozoa, giving a final concentration of about 7.1 $\mu mol\ l^{-1}$. This suspension is then incubated at 32°C for approximately 45 min. After the incubation, spermatozoa are uniformly stained so that there is minimal variation in the staining intensity within the X- or Y-bearing sperm population, as the most critical part of the separation process depends on uniform staining to maintain resolution of the X and Y sperm populations. The original procedure (Johnson *et al.*, 1989) required incubation at 35°C and about 5 µg of stain per 10×10^6 spermatozoa and incubation for 1 h. We believe the refined protocol uses the least amount of stain at the lowest temperature to maintain the highest sperm viability after the sorting process. Modifications to the preparation procedures continue as we adapt to new instrumentation. For example, spermatozoa and stain concentration is being increased to provide a more concentrated sample for the high speed MoFlo cell sorter now in use in our laboratory (Mo Flo, Cytomation, Inc.; L. Johnson and G.Welch, unpublished).

Collection of the sorted X and Y spermatozoa populations

A critical aspect to maintaining viability of sperm populations that have been put through a cell sorter is to provide an environment that will minimize the dilution that accompanies sperm sorting. The concentration of the stained boar spermatozoa to be sorted is lower (15×10^6 ml^{-1}) than that of the original ejaculates (about 300×10^6 ml^{-1}) and even the concentrations used for insemination ($25–30 \times 10^6$ ml^{-1}). Once the fluorescently stained spermatozoa pass into the flow cell, the stream is surrounded by a sheath fluid of phosphate-buffered saline (PBS) containing 0.1% (w/v) BSA which contributes to an increasing dilution effect. Dilution is minimized by sorting spermatozoa into a tube containing Test extender containing hens' egg yolk (Johnson *et al.*, 1989). The Test-yolk (20%) has been the most successful for this purpose (Johnson, 1991). Our procedure uses 50 µl of Test-yolk per 0.5 ml microcentrifuge tube and this provides a concentrated environment for spermatozoa to migrate to. Under sorting conditions the fluid volume in the tube increases as sorting progresses. The spermatozoa are falling or being projected into the tube and the motile spermatozoa continue their downward movement into the Test-yolk extender at the bottom of the tube and remain there in a concentrated environment. Once sorting is terminated, spermatozoa may be incubated for 20 min at room temperature. The resultant supernatant containing immotile spermatozoa is drawn off and the concentrated portion is used for insemination. Centrifugation (300 *g* for 5–8 min) can also be used to concentrate the sorted spermatozoa with minimal effect on sperm motility. Centrifugation is necessary if the sorted sample is to be used for IVF.

All of the sorting process takes place at about 22°C. The use of the sort collection process described above is most effective for spermatozoa to be used for insemination. We also found that when the spermatozoa are being used for IVF to produce sexed embryos, the percentage of yolk in the Test-yolk can be reduced to less than 5% so that it will not interfere with fertilization during IVF (Johnson, 1995; Rath *et al.*, 1995). A concentration of 2% yolk is sufficient to provide a concentrated environment for sorted swine spermatozoa (Rath *et al.*, 1997).

Seminal plasma may also be used as a portion of the collection fluid to minimize the dilution effect. In some recent experiments Maxwell *et al.* (1996), and Maxwell and Johnson (1997) showed that if Test-yolk (2%) contained 10% seminal plasma, the percentage of spermatozoa undergoing the acrosome reaction during or after sorting was reduced. However, additional results using the seminal plasma protocol in combination with IVF in pigs showed that fertilization by sorted spermatozoa could be inhibited when the spermatozoa were collected into a Test-yolk (2%) medium containing 10% seminal plasma (W. Maxwell, C. Long and L. Johnson, unpublished). Premature capacitation is clearly a problem with sorted boar spermatozoa, just as it is with frozen boar semen. Much more research on the characterization and composition of the collection fluid remains to be done with or without seminal plasma as a component of the collection fluid.

Rate of production of sexed spermatozoa

Use of cell sorting systems such as the Epics V and 750 (Johnson *et al.*, 1989) and the FacStar equipment (Cran *et al.*, 1995) allows sorting rates of viable X and Y spermatozoa of about 3×10^5 to 4×10^5 spermatozoa h^{-1}. The rate is slow because increased flow rate of the spermatozoa through the system tends to increase the coefficient of variation in the analysis. Spermatozoa must be orientated to the laser beam to differentiate their X and Y spermatozoa DNA content accurately. Only 20–25% of the spermatozoa that go through the system can be resolved into X- and Y-bearing populations. The lower rate is based on the need to achieve a high purity of the individual sperm populations. Although 100% purity cannot be achieved in a livestock sperm population with this flow sorting system under normal sorting conditions, 95% purity can be achieved (Johnson, 1992). In some species in which the difference in DNA is greater, such as in the creeping vole (*Microtus oregoni*; 12.5%; Johnson and Clarke, 1990) and the Chinchilla *Chinchilla laniger* (7.5%; Johnson *et al.*, 1987a) 100% purity can be achieved. Highest purities attained in flow sorting depend on (i) genetically determined difference in DNA content between the X- and Y-chromosome bearing spermatozoa, (ii) proper configuration and sensitivity of the flow system to allow selection of a properly orientated population of spermatozoa coincident with optimized sort windows, and (iii) uniform and consistent binding of the fluorochrome to the DNA in the sperm head.

Purity of sorted spermatozoa

The aforementioned factors are critical for achieving high sort purity. If 20×10^4 spermatozoa are used for sorting, sorting for 2 h will use about 65% of the spermatozoa prepared. This results in the accumulation of approximately 8×10^8 sorted X spermatozoa and the same number of sorted Y spermatozoa with sort windows set to achieve 80–90% purity of either X or Y spermatozoa, i.e. about 8% of all droplets formed are available for sorting. Decreasing the sort window by 15% results in a 10% increase in purity (> 90%). This can result in about a 20% reduction in spermatozoa that can be sorted over 2 h. Viability remains unaffected by manipulations of the sort windows. Sperm motility and acrosomal integrity of sorted spermatozoa are slightly lower than those of the original sample.

Laser and stain effects

Hoechst stained spermatozoa require an excitation source in the ultraviolet (UV) range. Since Hoechst 33342 is the only effective stain for sorting viable intact spermatozoa (Johnson *et al.*, 1989), there is currently no alternative to a UV laser for sorting spermatozoa based on DNA content. The duration of exposure to illumination is determined by the flow velocity of the spermatozoa as it traverses the laser beam (2–3 μs). The intensity of laser illumination is determined by the output power of the laser and is readily adjustable. We have investigated the impact of laser power on both the ability to resolve X from Y spermatozoa and their fertilizing capability.

Since spermatozoa are being subjected to UV irradiation during the sorting process, it is useful to reduce laser power to the minimum necessary to obtain resolution between the X and Y peaks (Fig. 2). Much of our initial work was done at about 175 mW of laser power. More recently we found that resolution could be maintained if the power was reduced to 75–125 mW. At below 75 mW the loss of resolution became more pronounced. Power reductions affected the resulting purities of the Y sperm population to the greatest extent. Even at low laser power (below 125 mW) where the resolution of X and Y spermatozoa appeared diminished, the X sperm population could still be sorted to 90% purity while Y spermatozoa purity fell to 70%. Fertilization rates in rabbits with spermatozoa sorted using various laser powers showed a more pronounced change at 125 mW. The percentage of unfertilized eggs decreased from 60% at 200 mW to 25% at 125 mW. Similarly, the percentage of 8–16-cell and early morula embryos increased from 30% at 200 mW to 70% at 125 mW (Johnson *et al.*, 1996).

Potential for Alternative Method of Sexing Semen Based on Surface Protein

It is well understood that a method for sexing semen on a large scale, at least a scale that could be applied to semen production practice at an AI Centre, would be advantageous. The method most often referred to in this context is that of isolating a protein from the surface of the X- or Y-bearing spermatozoa that is chromosome specific and thus sex specific. The theory goes that if such a marker could be isolated, an antibody could be developed to attach to that X- or Y-bearing population of spermatozoa. The assumption is that the use of affinity chromatography or magnetic beads would provide a large-scale separation process for separating X- from Y-bearing spermatozoa and be readily adaptable for AI Centre use. The first step in this process is to identify a surface protein. Boar spermatozoa were sorted for X or Y populations at purities of 90%. Approximately 1000 proteins

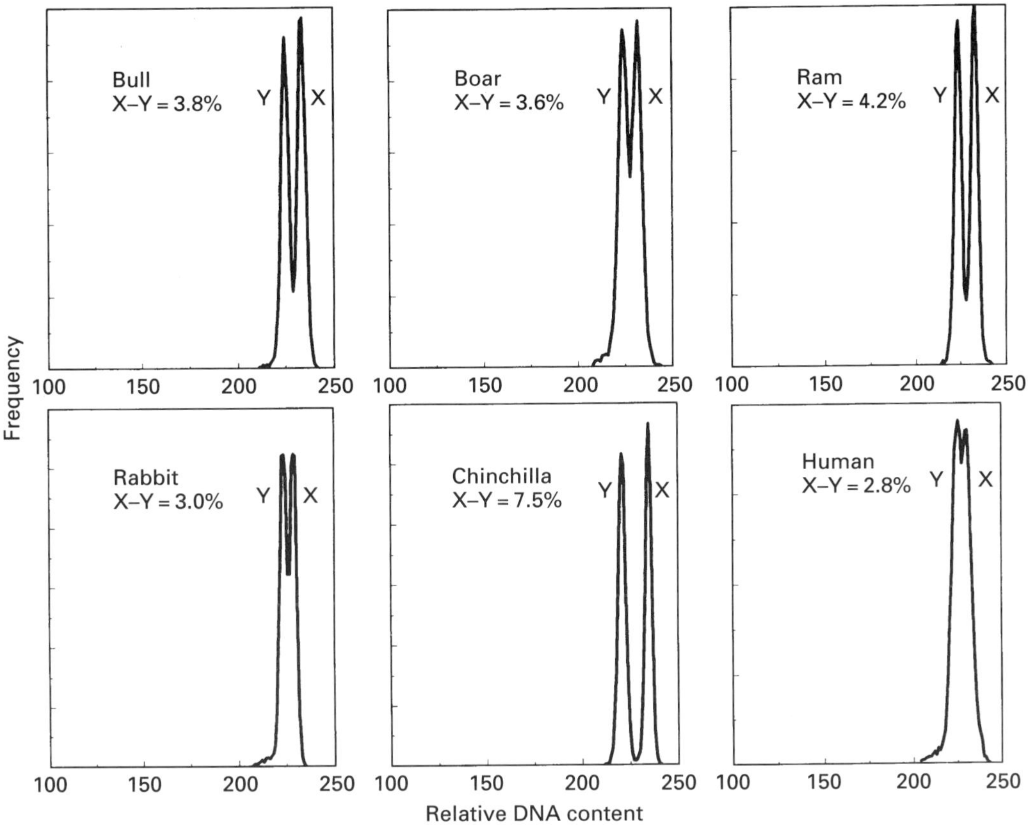

Fig. 2. The relative DNA content of spermatozoa measured by flow cytometry from six different species. DNA content is not comparable between species, as instrument settings dictate histogram position. (These data are also summarized in Johnson *et al.*, 1987a; Johnson 1992; 1994; 1995).

were mapped on the sperm surface. However, no difference could be detected between the proteins isolated from X spermatozoa versus Y spermatozoa. These results would lead to the contention that there are no sex specific proteins on the surface of the spermatozoa (Hendriksen *et al.*, 1996) and that the likelihood of developing a batch process based on an antibody separation is doubtful.

Embryo Development

Fertilization and early development in vivo

The observation that reduced litter size in does and sows inseminated with flow sorted X- or Y-bearing spermatozoa (Johnson *et al.*, 1989; Johnson, 1991) led us to evaluate early embryo development to determine the impact of flow sorted spermatozoa in rabbits. McNutt and Johnson (1996) found an influence of sorting on development of embryos flushed at 42 h after surgical insemination. Embryos had reached the 8–16-cell stage, whereas embryos from the controls (unstained, unsorted spermatozoa) had already progressed to the early morula stage ($P < 0.01$). The impact seen early in development was carried through pregnancy since there were significantly fewer fetuses from inseminations with flow-sorted spermatozoa than from inseminations with unsorted, unstained, control spermatozoa. Results from cattle (Cran *et al.*, 1993) indicated that most embryos produced from IVF using non-stained, non-sorted spermatozoa produced fully expanded blastocysts 7 days after insemination. When stained, sorted spermatozoa were used for IVF, there was a delay of 12 h in blastocyst development. These studies provided some evidence that when flow sorted spermatozoa are used for fertilization there is a definite impact on the early and subsequent fetal development. Questions are frequently raised about the fertilization process and competence of embryos produced by IVF. To test this hypothesis, we added the Hoechst 33342 stain to fresh boar semen and then inseminated gilts. At two different concentrations of Hoechst 33342, no impact was seen on the developmental competence of the resultant embryos (Garner *et al.*, 1996).

Production of embryos through in vitro *fertilization in swine*

The usefulness of the sexing technology in combination with IVF was demonstrated in our laboratory using oocytes matured *in vivo* and IVF (Rath *et al.*, 1997). Cleavage rates after IVF were 56% ($n = 367$) and resultant 2–4-cell embryos were transferred to 24 h asynchronous ($n = 4$) gilts. Two pregnancies resulted from these initial trials and produced litters of six and four pigs. All embryos transferred were produced from IVF using X-sorted spermatozoa and all offspring were female. These were the first pigs born from a combination of the Beltsville Sperm Sexing Technology and IVF in the pig (Table 2). Techniques to improve IVF for swine continue in our laboratories (C. Long, J. Dobrinsky, L. Johnson, unpublished). Once cleavage rates and blastocyst development rates reach the 70% rate with coincident reductions in polyspermy, the use of IVF/sexed spermatozoa in pigs will be advanced. As mentioned earlier, one factor that needs to be controlled carefully in using sexed spermatozoa is that they are used soon after sorting. This need is based on the premature capacitation that is induced by the sorting process. Recent results with small numbers of liquid stored spermatozoa and artificial insemination in cattle suggest that even at the current sexed sperm production rates, there may be applicability for AI of $1–2 \times 10^5$ per insemination (Seidel *et al.*, 1997). It is unlikely that the technology for small numbers of spermatozoa from cattle will be directly applied to swine because of the anatomy of the pig uterus which provides an impediment to small numbers of spermatozoa. This is an area of research that could benefit greatly from more intensive research effort. Indeed, the whole swine artificial insemination industry would benefit from greater research emphasis, since the increasing use of AI will bring about increasing pressure for greater economy of spermatozoa per insemination dose.

Laboratory Validation of Sperm Sorting Results

Reanalysis of sorted spermatozoa for DNA

At some point during a sort or immediately after the completion of a sort to be used for insemination or IVF, a presiliconized BSA-treated tube is placed in position to collect the sorted spermatozoa. No Test-yolk is added to the tube. Approximately 10^5 spermatozoa are sorted into each tube. After the aliquot has been sorted, the spermatozoa are sonicated to remove the tails, and Hoechst 33342 added to maintain staining uniformity and then flow cytometrically reanalysed but not resorted (Johnson *et al.*, 1989). The proportion of X- and Y-bearing spermatozoa are determined based on the DNA difference and histograms are analysed by computer fitting to double gaussian peaks (Fig. 2; Johnson *et al.*, 1987b).

Molecular genetic analysis of single sorted spermatozoa

A qualitative determination of the presence of the X or Y chromosome in an individual spermatozoan can be made by using the allele-specific polymerase chain reaction (PCR). By analysing several hundred individual spermatozoa a quantitative determination of the proportion of X and Y spermatozoa in a given population can be made (Welch *et al.*, 1995). This method also has application for genetic testing using single cell technologies. Single spermatozoa determined to be X or Y by flow cytometric analysis were deposited into each well of a 48 well plate. The wells are

Table 2. Farrowing results from gilts receiving IVM/IVF embryos produced from sexed spermatozoa[a]

Treatment	Litter size	Offspring		Predicted sex[b] (%)
		Male	Female	
Sorted for X	6	0	6	89
Sorted for X	4	0	4	89

[a]Data from Rath *et al.* (1997).
[b]By reanalysis of sorted spermatozoa for DNA.

shallow enough to allow microscopic verification of the contents of the wells after single spermatozoa sorting. The highly conserved zinc finger allele which resides on both the X and Y chromosome (ZFx and Zfy) was amplified using nested allele-specific PCR. The wells also serve as the first round amplification vessel for PCR. Specific X and Y products from a second amplification are separated by agarose gel electrophoresis; product bands are identified and counted; and the resulting purities are determined. Results from 100 amplifications of each X and Y sort showed 94% purity for X and 90% purity for Y (Welch *et al.*, 1995). These purities are consistent with results obtained by flow cytometric reanalysis of sorted spermatozoa described above.

Embryo sexing by PCR

Embryos that have been produced from IVF with sorted spermatozoa are frozen by standard methods and stored for later analysis of gender by PCR. After thawing and incubation with proteinase K, the DNA is probed using primers derived from a multiple repeat bovine Y-chromosome specific sequence (Cran *et al.*, 1995). This is an effective way to ensure almost 100% accuracy in the sex of the embryo.

Fluorescence in-situ hybridization (FISH) for validating proportions of X and Y spermatozoa

Application of FISH to determination of X and Y proportions is fairly straightforward with human spermatozoa because of the availability of commercial microsatellite centromeric DNA probes. Our initial work on humans was validated using FISH (Johnson *et al.*, 1993). Spermatozoa were sorted on to slides, fixed with methanol and acetic acid, exposed to biotinylated DNA probes for X or Y chromosomes and incubated; the signals indicating an X or Y chromosome were then counted using fluorescence microscopy. This technique is particularly helpful when the difference between X and Y DNA (< 3.0%) is small and DNA reanalysis by flow cytometry is inconsistent (Johnson *et al.*, 1993).

Recently we have completed a study using FISH and a Y-specific DNA probe developed by T. Kawarasaki for pig spermatozoa. In a comparative study using sorted boar spermatozoa, there was no significant difference between the purities obtained using FISH and those obtained from reanalysing an aliquot of the sorted spermatozoa for DNA (Kawarasaki *et al.*, 1997).

Birth of Offspring From Sorted X and Y Spermatozoa

Studies to validate the efficacy of the Beltsville Sperm Sexing Technology were initially conducted in rabbits and swine. The sex ratios of the litters produced from females surgically inseminated with sorted X or Y spermatozoa supported the results of reanalysing aliquots of sorted sperm populations (Johnson *et al.*, 1989; Johnson, 1991). A summary of the results obtained in these studies is presented in Tables 2 and 3. In the latter study with pigs, spermatozoa were sorted in the usual way into Test-yolk

Table 3. Progeny produced from intratubal insemination of pig spermatozoa sorted into X- or Y-chromosome bearing populations

Treatment	Number inseminated	Number farrowed	Number born	Offspring: Actual (%) Female	Male	Predicted sex[c] (%) X	Y
Early technology[a]							
Sorted for Y	8	4	37	32	68	23	77
Sorted for X	10	5	34	74	26	80	20
Unsorted/stained	11	5	40	57	43	50	50
Unsorted/unstained	7	5	46	48	52	50	50
Recent technology[b]							
Sorted for Y	1	1	7	14	86	15	85
Sorted for X	1	1	8	88	12	89	11

[a]Johnson (1991).
[b]Rath *et al.* (1997).
[c]By reanalysis of sorted spermatozoa for DNA.

(20%) extender and inseminated via the uterus into the oviduct. In most cases the gilts were synchronized and ovulated (42 h after hCG administration) about 4 h before surgical intervention. The results demonstrate a significant shift in the sex ratio of the litters born. However, the proportion of one sex or the other (68% for males and 74% for females) is lower than that achieved with rabbits. One of the reasons for this is that boar spermatozoa do not orientate as efficiently as rabbit spermatozoa. With current refinements and technology, the shift in the sex ratio for swine has improved to 85–90% using standard conditions. No effort has been made to inseminate gilts with regular AI since the number of spermatozoa required is prohibitive with respect to sorted spermatozoa. As mentioned earlier, sorted spermatozoa have been used for IVF and two litters of pigs have been produced; all were female (Table 2). This work (Rath *et al.*, 1997) demonstrated the feasibility of using IVF and embryo transfer in conjunction with the sexing technology.

Conclusions

Preselection of sex of progeny in livestock has been validated on the basis of live births, on the basis of reanalysis of sorted spermatozoa for DNA content on the basis of PCR and on the basis of determining the sex of the resulting embryos. The cell sorting method to separate X- and Y- bearing spermatozoa can be used in conjunction with IVF in swine and in cattle for the production of sexed embryos to be transferred to eligible recipients for the duration of gestation. This semi-practical sexing method, although impractical for some production systems, could be used to provide a more flexible progeny producing option in many livestock operations. Flow cytometric sorting of spermatozoa for reproductive purposes requires significant commitment in expertise and resources in order to apply the systems commercially or in research situations. Research continues to improve the efficiency of sorting more spermatozoa per unit time. Current research involving the removal of dead spermatozoa from the sorting process through the addition of the dead membrane permeant propidium iodide as well as other dyes in combination with Hoechst 33342 can result in a more efficiently sorted population (Johnson *et al.*, 1994). High speed sorting (MoFlo, Cytomation, Inc.) as is now used in our laboratory will increase the throughput of sorted spermatozoa per unit time. Improvements in the technology will no doubt lead to much greater use of sexed spermatozoa, depending on the species involved. Solving the problems associated with inseminating small numbers of sexed spermatozoa in pigs would be a great benefit to AI and swine production.

References

Amann RP and Seidel GE (Eds) (1982) *Prospects for Sexing Mammalian Spermatozoa* Colorado University Press, Boulder

Barlow P and Vosa CG (1970) The Y chromosome in human spermatozoa *Nature* **226** 961–962

Catt SL, Catt JW, Gomez MC, Maxwell WMC and Evans G (1996) Birth of a male lamb derived from an *in vitro* matured oocyte fertilised by intracytoplasmic injection of a single presumptive male spermatozoa *Veterinary Record* **139** 494–495

Cran DG, Johnson LA , Miller NG, Cochrane D and Polge C (1993) Production of bovine calves following separation of X- and Y-chromosome bearing spermatozoa and *in vitro* fertilisation *Veterinary Record* **132** 40–41

Cran DG, Johnson LA and Polge C (1995) Sex preselection in cattle: a field trial *Veterinary Record* **136** 495–496

Cui K (1997) Size differences between human X and Y spermatozoa and prefertilization diagnosis *Molecular Human Reproduction* **3** 61–87

Dean PN, Pinkel D and Mendelsohn ML (1978) Hydrodynamic orientation of spermatozoa heads for flow cytometry *Biophysics Journal* **23** 7–13

Ericsson RJ, Langerrin CN and Nishirro M (1973) Isolation of fraction-rich in human Y spermatozoa *Nature* **246** 421–424

Garner DL, Dobrinsky JR, Welch GR and Johnson LA (1996) Porcine spermatozoa viability, oocyte fertilization and embryo development after staining spermatozoa with SYBR-14 *Theriogenology* **45** 1103–1113

Gledhill BL (1988) Gender preselection: historical, technical, and ethical perspectives *Seminars in Reproductive Endocrinology* **6** 385–395

Gledhill BL, Lake S, Steinmetz LL, Gray JW, Crawford JR, Dean PN and Van Dilla MA (1976) Flow microfluorometric analysis of spermatozoa DNA content: effect of cell shape on the fluorescence distribution *Journal of Cell Physiology* **87** 367–376

Hendriksen PJM, Tieman M, Van Der Lende T and Johnson LA (1993) Binding of anti-H-Y monoclonal antibodies to separated X and Y chromosome bearing porcine spermatozoa *Molecular Reproduction and Development* **5** 189–196

Hendriksen PJM, Welch GR, Grootegoed JA, Van Der Lende T and Johnson LA (1996) Comparison of detergent-solubilized membrane and soluble proteins from flow cytometrically sorted X- and Y-chromosome bearing porcine spermatozoa by high resolution 2-D electrophoresis *Molecular Reproduction and Development* **45** 342–350

Johnson LA (1988) Flow cytometric determination of spermatozoa sex ratio in semen purportedly enriched for X or Y bearing spermatozoa *Theriogenology* **29** 265

Johnson LA (1991) Sex preselection in swine: altered sex ratios in offspring following surgical insemination of flow sorted X- and Y-bearing spermatozoa *Reproduction in Domestic Animals* **26** 309–314

Johnson LA (1992) Gender preselection in domestic animals using flow cytometrically sorted spermatozoa *Journal of Animal Science* **70** (Supplement 2) 8–18

Johnson LA (1994) Isolation of X- and Y-bearing spermatozoa for sex preselection In *Oxford Reviews of Reproductive Biology* Ed. HH Charlton, pp 303–3226. Oxford University Press, Oxford

Johnson LA (1995) Sex preselection by flow cytometric separation of X and Y chromosome-bearing spermatozoa based on DNA difference: a review *Reproduction Fertility and Developoment* **7** 893–903

Johnson LA and Clarke RN (1988) Flow sorting of X and Y chromosome-bearing mammalian spermatozoa: activation and pronuclear development of sorted bull, boar and ram spermatozoa microinjected into hamster oocytes *Gamete Research* **21** 335–343

Johnson LA and Clarke RN (1990) Spermatozoa DNA and sex chromosome differences between two geographical populations of the creeping vole, *Microtus oregoni. Molecular Reproduction and Development* **27** 159–162

Johnson LA and Pinkel D (1986) Modification of a laser-based flow cytometer for high resolution DNA analysis of mammalian spermatozoa *Cytometry* **7** 268–273

Johnson LA, Flook JP, Look MV and Pinkel D (1987a) Flow sorting of X and Y chromosome-bearing spermatozoa into two populations *Gamete Research* **16** 1–9

Johnson LA, Flook JP and Look MV (1987b) Flow cytometry of X and Y chromosome-bearing spermatozoa for DNA using an improved preparation method and staining with Hoechst 33342 *Gamete Research* **17** 203–212

Johnson LA, Flook J P and Hawk HW (1989) Sex preselection in rabbits: live births from X and Y spermatozoa separated by DNA and cell sorting *Biology of Reproduction* **41** 199–203

Johnson LA, Welch GR, Keyvanfar K, Dorfmann A, Fugger EF and Schulman JD (1993) Gender preselection in humans? Flow cytometric separation of X and Y spermatozoa for the prevention of X-linked diseases *Human Reproduction* **8** 1733–1739

Johnson LA, Welch GR and Garner DL (1994) Improved flow sorting resolution of X and Y chromosomes bearing viable spermatozoa separated using dual staining and dead cell gating *Cytometry* **17** (Supplement 7) 83

Johnson LA, Cran DG, Welch GR and Polge C (1996) Gender preselection in mammals In *Beltsville Symposium XX Biotechnology's Role in the Genetic Improvement of Farm Animals* Ed. RH Miller, VG Pursel and HD Norman. pp 151–164 American Society of Animal Science, Savoy, IL

Kawarasaki T, Welch GR, Long CR, Yoshida M and Johnson LA (1997) Rapid estimation of flow sorted X- and Y-chromosome bearing porcine spermatozoa by fluorescence in situ hybridization *Molecular Reproduction and Development* (in press)

Kiddy CA and Hafs HD (Eds) (1971) Sex ratio at birth – prospects for control *Journal of Animal Science Symposium – Supplement American Society for Animal Science* Champaign, IL

McNutt TL and Johnson LA (1996) Flow cytometric sorting of spermatozoa: influence on fertilization and embryo/fetal development in the rabbit *Molecular Reproduction and Development* **43** 261–267

Maxwell WMC and Johnson LA (1997) Chlortetracycline analysis of boar spermatozoa after incubation, flow cytometric sorting, cooling, or cryopreservation *Molecular Reproduction and Development* **46** 408–418

Maxwell WMC, Welch GR and Johnson LA (1996) Viability and membrane integrity of spermatozoa after dilution and flow cytometric sorting in the presence or absence of seminal plasma *Reproduction Fertility and Development* **8** 1165–1168

Moruzzi JF (1979) Selecting a mammalian species for the separation of X- and Y-chromosome bearing spermatozoa *Journal of Reproduction and Fertility* **57** 319–323

Pinkel D, Gledhill BL, Van Dilla MA, Stephenson D and

Watchmaker G (1982) High resolution DNA measurements of mammalian spermatozoa *Cytometry* **3** 1–9

Rath D, Johnson LA, Welch GR and Niemann H (1995) Successful gamete intrafallopian transfer (GIFT) in the porcine *Theriogenology* **41** 1173–1179

Rath D, Johnson LA, Dobrinsky JR and Welch GR (1997) Production of piglets preselected for sex following *in vitro* fertilization with X and Y chromosome-bearing spermatozoa sorted by flow cytometry *Theriogenology* **47** 795–800

Seidel GE, Allen CH, Johnson LA, Holland MD, Brink Z, Welch GR, Graham JK and Cattell MB (1997) Insemination of heifers with very low numbers of nonfrozen spermatozoa and sexed spermatozoa *Theriogenology* (in press)

Welch GR, Waldbieser GC,Wall RJ and Johnson LA (1995) Flow cytometric spermatozoa sorting and PCR to confirm separation of X- and Y-chromosome bearing bovine spermatozoa *Animal Biotechnology* **6** 131–139

Poster titles

Dixon WT, Jiang L-H, Ransoondar JJ and J Turchinsky
Cloning of porcine CSF-1 from JAG-1 trophoblast cells: a candidate autocrine growth factor for regulation of trophoblast cell proliferation

Hansen AM, Harrison B, O'Brien JK, Catt SL, Reed K and G Evans
Verification of sex-sorting of spermatozoa separated by flow cytometry

Forni M, Fioretti D, Bacci ML, Varzi V, Pucci L, Ceretta S, Di Stefano C, Seren E, Frati L and M Lavitrano
Sperm-mediated gene transfer: production of pigs genetically transformed with a human regulator of complement activation

Ratky J, Brüssow K-P, Solti L and P Szabó
Ovarian response to estrous synchronization of a Hungarian native pig breed (Mangaliza)

Hühn U and K-P Brüssow
Factors affecting the results of estrous cycle synchronization in gilts

Armstrong TA, Flowers WL and JH Britt
The use of hCG and PGF to control rebreeding in early-weaned sows

Wähner M and Huhn U
Control of parturition in sows by using a combined treatment with cloprostenol® plus depotocin®

ASPECTS OF OOCYTE AND EMBRYONIC DEVELOPMENT IN THE PIG

Chair
B. N. Day

Journal of Reproduction and Fertility Supplement **52**, 271–283

Advances in *in vitro* production of pig embryos

H. Funahashi[1] and B. N. Day[2]

[1]*Faculty of Agriculture, Okayama University, Tsushima-naka, Okayama 700 Japan; and*
[2]*Department of Animal Sciences, University of Missouri, Columbia, MO 65211, USA*

A series of integrated, effective techniques is required to produce pig embryos from follicular oocytes *in vitro*. The failure to form a male pronucleus and polyspermy have been serious problems in efforts to produce embryos efficiently *in vitro* from pig oocytes. The former problem is now considered to be due to oxidative stress and the latter has been partially solved by reducing the number of capacitated spermatozoa reaching the oocytes. By the use of new technology for *in vitro* production of embryos, an acceptable rate of blastocyst formation and the birth of live piglets has been achieved. However, even with the use of these improved *in vitro* maturation (IVM) and fertilization (IVF) conditions, the efficiency of production of *in vitro* blastocysts and offspring still remains relatively low. More recently the developmental competence of embryos matured and fertilized *in vitro* has been investigated through modification of culture conditions of oocytes during the germinal vesicle stage. Oocyte competence for early embryonic development appears to be achieved by active communication between the oocyte and follicular cells. Since the ovarian oocytes available for IVM are primarily those present in mid-size antral follicles of prepubertal gilts, more research is needed to gain an improved understanding of the culture conditions required to induce developmental competence in oocytes from both preantral and antral follicles as well as additional modifications in IVF systems to overcome the problem of polyspermic penetration.

Introduction

The pig ovary contains a relatively large number (approximately 210 000) of primordial follicles compared with other mammals (Gosden and Telfer, 1987). An objective of *in vitro* production of pig embryos is to use the large number of follicular oocytes from ovaries in abattoirs to reduce the cost of procurement of embryos for various research purposes. Since successful culture techniques for early development of pig embryos from the one-cell to blastocyst stages have recently been developed (reviewed by Petters and Wells, 1993), efforts to produce embryos *in vitro* have focused on the development of a system for efficient *in vitro* maturation (IVM) and *in vitro* fertilization (IVF) of oocytes, with particular attention on the historical problems of a reduced incidence of male pronuclear formation and a high incidence of polyspermy. Continuous progress towards the solution of traditional problems is being made (see reviews by Day and Funahashi, 1996; Niwa, 1993; Mattioli, 1994; Nagai, 1994). The incidence of blastocyst formation following IVM, IVF and *in vitro* culture has been improved significantly by modification of conditions during IVM (Funahashi *et al.*, 1997a; Funahashi *et al.*, 1997b). However, the simple integration of these improved IVM–IVF techniques and the successful culture system for early embryonic development still does not appear to result in a high incidence of viable blastocyst formation *in vitro*. Piglets from oocytes matured and fertilized *in vitro* have been produced only in a few laboratories (Mattioli *et al.*, 1989; Yoshida *et al.*, 1993c; Funahashi *et al.*, 1996a) Therefore, the present review aims to discuss not only recent technological developments in IVM and IVF but also the establishment of competence of pig oocytes for early embryonic and fetal development.

Historical Problems: Male Pronuclear Formation

Although an abnormally low incidence of male pronuclear formation had frequently been observed in earlier studies when IVM oocytes were fertilized *in vitro* (Nagai *et al.*, 1984; Mattioli *et al.*, 1988), research in several laboratories has solved this problem. As discussed in our previous review (Day and Funahashi, 1996), factors affecting the ability of oocytes to form a male pronucleus are hormonal conditions, follicular secretions, intracellular ionic strength and especially oxidative stress. The incidence of male pronuclear formation is highly correlated with the glutathione content of the oocyte (Yoshida *et al.*, 1993a; Funahashi *et al.*, 1994b) which is associated with a required reduction in the disulfide bond cross-linking of sperm protamine to initiate sperm decondensation (Perreault, 1990). Both the incidence of male pronuclear formation and glutathione content in pig oocytes is increased during IVM when cysteine, which is a thiol, is added to the maturation medium, whereas the content decreases in the absence of cysteine (Yoshida *et al.*, 1993a). The incidence of male pronuclear formation is also increased by supplementation of maturation medium with other thiols such as cysteamine (Grupen *et al.*, 1995) and β-mercaptoethanol (H. Funahashi and B. N. Day, unpublished observation). Therefore, the previous problem of failure in male pronuclear formation appears to have been at least partially due to oxidative stress. Although the presence of cumulus cells surrounding the oocyte is required throughout IVM to maintain a high oocyte glutathione concentration (Funahashi and Day, 1995), the presence of cysteine from 36 h after the start of IVM will maintain both a high oocyte glutathione content and a high incidence of male pronuclear formation, and the absence of cysteine from 36 h after the start of IVM induces a significant reduction in both of these (Sawai *et al.*, 1997). Thus, the oxidative stress may be especially detrimental at relatively late stages of IVM when active intercellular coupling between the oocyte and cumulus cells is significantly reduced (Motlik *et al.*, 1986; Mattioli *et al.*, 1988).

Persistent Problem: Polyspermic Penetration

In mammals, polyspermic penetration results in the developmental failure of the zygote (Hunter, 1996). Since nearly all mammals inherit their centrosomes from their fathers (Schatten, 1994; Navara *et al.*, 1995) and the paternal centrosome is known to organize the microtubules to form a sperm aster in pigs (Kim *et al.*, 1996a), polyspermic zygotes have an abnormal cleavage during early embryonic development (Fig. 1). Many IVF conditions that have been developed for pig oocytes result in a high incidence of polyspermic penetration (Niwa, 1993).

Possible methods to prevent polyspermic penetration in vivo

In normal gilts, polyspermic fertilization appears to be reduced to a very low incidence by two major mechanisms, namely by the female reproductive tract controlling the number of capacitated spermatozoa reaching the site of fertilization and the direct exocytotic reaction of oocytes to prevent further sperm attachment and penetration. The female reproductive tract, especially the ovary, utero–tubal junction, and fallopian tube isthmus, exerts an important role in the control of the number of spermatozoa that reach the site of fertilization (Hunter, 1991; Hunter, 1996). During transport of spermatozoa through the female reproductive tract, spermatozoa are capacitated and the number that finally reach the site of fertilization number less than a hundred cells per oocyte (Hunter, 1982; Mburu *et al.*, 1996). The direct reactions of oocytes to sperm penetration are cortical granule exocytosis and secondarily the reactions to exocytosis in the perivitelline space and zona pellucida (Cran and Esper, 1990; Jones, 1990). During fertilization in mammals, exocytosis of the cortical granule contents containing hydrolytic enzymes and saccharide components to the perivitelline space leads to the formation of a cortical granule envelope (Dandekar and Talbot, 1992), modification of zona pellucida proteins and the inactivation of sperm receptors.

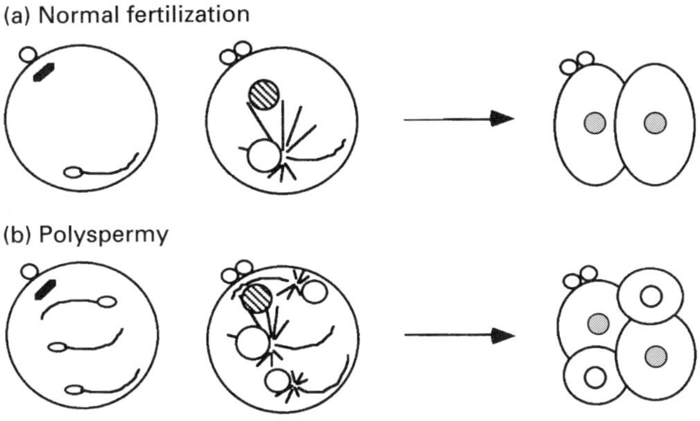

Fig. 1. Schematic diagram of the first cleavage of pig oocytes fertilized normally (a) or polyspermy (b). Sperm aster microtubules are formed around the paternal centrosome. Multiple sperm asters are formed in association with each male chromatin in polyspermy oocytes, resulting in mosaicism.

Is sperm capacitation induced efficiently in vitro?

In most IVF systems, a relatively large number of spermatozoa has been added to the fertilization medium containing mature pig oocytes (Nagai, 1996). Furthermore, even the current IVF systems may be unsuitable for inducing sperm capacitation because there is a large variation among boars in the ability of spermatozoa to penetrate oocytes (Martinez *et al.*, 1993; Wang *et al.*, 1995). Although a Tris-buffered medium (Abeydeera *et al.*, 1996a) as well as bicarbonate-buffered medium (Suzuki *et al.*, 1994) supports sperm penetration of pig oocytes *in vitro*, most IVF systems use a bicarbonate-buffered medium. Bicarbonate causes a loss of sperm surface coating materials (Ashworth *et al.*, 1995) and a major alteration in sperm plasma membrane lipid architecture (Harrison *et al.*, 1996) which may induce an initial permissive event in the capacitation process. However, in a traditional IVF system that uses modified Medium 199, frozen and thawed spermatozoa seem to undergo the capacitation process leisurely, and consequently the population of acrosome-reacted spermatozoa increases very slowly during co-culture of oocytes with spermatozoa compared with the pattern in modified Tris-buffered medium (Fig. 2; H. Funahashi, Y. Kajiwara, K. Sawai and K. Niwa, unpublished). Similar results have been obtained in a capacitation medium that used freshly ejaculated spermatozoa (Mattioli *et al.*, 1996). In most current IVF systems, therefore, pig oocytes may be exposed to a number of freshly capacitated spermatozoa over a long period after insemination. Finding a suitable IVF system that will induce sperm capacitation of nearly all spermatozoa at the same time may reduce the number of newly formed penetrable spermatozoa during IVF. Determining the nature of oviductal secretions around the time of ovulation more completely would help define the conditions needed to induce efficient sperm capacitation. Hyaluronic acid, which is detected in oviductal fluid, seems to induce the sperm capacitation but not the acrosome reaction (H. Rodriguez-Martinez, Y. Han, X. Song, H. Funahashi and K. Niwa, unpublished).

Attempts to reduce the number of spermatozoa binding to oocytes

A short co-preincubation of spermatozoa with pig oviductal epithelial cells (Nagai and Moor, 1990) or preincubation in the presence of a low concentration of pig follicular fluid (Funahashi and Day, 1993) reduces the incidence of polyspermic penetration in pig oocytes, although a long co-culture with oviductal cells or a high concentration of follicular fluid also reduces the incidence of sperm penetration. Similar observations have been demonstrated when oocytes were co-cultured with spermatozoa in the presence of 1% oviductal fluid collected from day 20 and 21 oestrous gilts

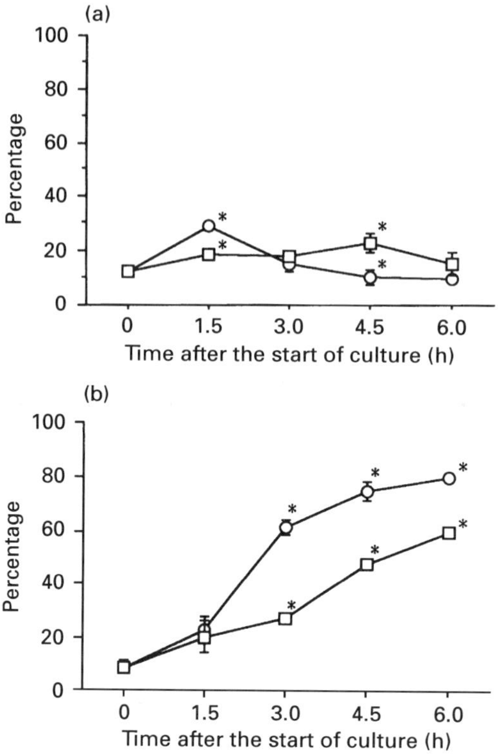

Fig. 2. The incidence of capacitated (a) and acrosome-reacted (b) spermatozoa after the start of culture in modified Tris-buffered medium (○) or modified medium 199 (□). After washing, frozen–thawed spermatozoa were cultured in 500 μl droplets of each medium at 1×10^6 cells ml^{-1}, fixed every 1.5 h and the functional state of the cells was examined using a chlortetracycline fluorescence assay. Values with asterisks are significantly different ($P < 0.05$) within each observation time.

(Kim *et al.*, 1996a) or in a medium conditioned with secretions of pig oviductal epithelial cells (Kano *et al.*, 1994). The pig zona pellucida is composed of three glycoprotein families, pZP1, pZP3-α and pZP3-β. pZP1 is divided into pZP2 and pZP4 by reduction of S-S bonds (Hedrick, 1993; Hasegawa *et al.*, 1994). pZP3-α is a primary receptor for boar spermatozoa (Sacco *et al.*, 1989; Yurewicz *et al.*, 1993) and binds to the acrosome region of acrosome-damaged or partially acrosome-reacted spermatozoa, but not to acrosome-intact or fully acrosome-reacted spermatozoa (Yonezawa *et al.*, 1995). The presence of sulfated polymers such as dextran sulfate, fucoidan, polyvinylsulphate and heparin are known to compete with ZP3 on the zona pellucida to bind the zona-binding glycoproteins of boar spermatozoa and to induce the acrosome reaction (Jones, 1991; Parry *et al.*, 1992). Therefore, oviductal and follicular sulfate compounds in preincubation and(or) fertilization media probably compete with sperm receptors for binding to zona pellucida ligands, stimulate the rate of sperm acrosome reaction, and consequently reduce the number of capacitated spermatozoa attaching to the surface of pig oocytes.

Can the oocyte cortical reaction prevent polyspermic penetration in the current IVF systems?

In pigs, the timing of cortical granule migration in IVM oocytes and the reorganization of microfilaments to the oocyte cortex area (Kim *et al.*, 1996c) are similar to those of oocytes matured *in vivo* (Yoshida *et al.*, 1993b). Even so, the cortical reaction does not seem to be sufficiently effective in pig IVF systems to prevent polyspermy. The amount of pZP1, which is a second receptor for boar spermatozoa, decreases substantially during IVF of pig oocytes (Hatanaka *et al.*, 1992). Sperm penetration is not blocked after electrical stimulation of IVM oocytes (Funahashi *et al.*, 1993; Funahashi *et al.*, 1995) with the incidence of polyspermic oocytes and the number of penetrated spermatozoa increasing gradually with time after insemination (Funahashi *et al.*, 1995), even though cortical granules are released following electrical stimulation (Sun *et al.*, 1992) or microinjection of the G-protein stimulator guanosine-5′-*O*-(3′-thiotriphosphate) (Machaty *et al.*, 1995). Although the cortical granule reaction seems to be slower and incomplete in some oocytes matured and fertilized in historical *in vitro* systems as compared with ovulated oocytes fertilized *in vivo* (Kim *et al.*, 1996b) or *in vitro* (Kim *et al.*, 1996b), the degree of exocytosis in oocytes matured and fertilized in a modified *in vitro* system appears not to be different from that in ovulated oocytes fertilized in the same system (Fig. 3; W. H. Wang and B. N. Day, unpublished). However, since pig oocytes seem to need several hours *in vivo* to complete the release of the cortical granules from the ooplasma (Laurincik *et al.*, 1995; Soede *et al.*, 1995), the preventive mechanism of the cortical reaction may be inadequate to prevent polyspermy in the current IVF system in which pig oocytes are exposed to a number of waves of capacitated spermatozoa. Furthermore, the volume fraction and size of mitochondria and cortical granules of IVM oocytes from prepubertal sheep differ from those of IVM oocytes from adult sheep (O'Brien *et al.*, 1996). Oocytes from prepubertal animals may have an inadequate amount of cortical granule substance to induce effective reaction(s). In summary, the best possible means to reduce the incidence of polyspermic penetration in oocytes from prepubertal gilts may be to reduce the number of capacitated spermatozoa reaching the oocytes.

Developmental Competence of Pig Embryos Matured And Fertilized *In Vitro*

Culture of pig embryos matured and fertilized in vivo

As described by Petters and Wells (1993), recent technological progress in culturing one-cell pig embryos matured and fertilized *in vivo* has been achieved by the use of simple media such as modified Whitten's medium, NCSU-23 or NCSU-37 media and modified Tyrode's medium. Recently, it has been reported that replacement of BSA with FBS in BECM-3 medium by late day 5 (Dobrinsky *et al.*, 1996) or transferring embryos from CZB medium to modified Eagle's minimal essential medium containing 20% FBS (Pollard *et al.*, 1995) improve the incidence of hatched blastocysts.

Attempts to culture IVM–IVF embryos for early embryonic development

Available methods for the successful culture of pig embryos to the blastocyst stage have made it possible to examine the developmental ability of IVM/IVF pig embryos. However, although IVM and IVF pig oocytes develop to the blastocyst stage in simple media, the efficiency is very low, even with improved methods that promote male pronuclear formation and monospermic penetration (Funahashi *et al.*, 1994a; Funahashi *et al.*, 1994c). An increased incidence of the development of IVM/IVF embryos to the blastocyst stage has been shown by culture in the amniotic fluid of developing chick embryos (Ocampo *et al.*, 1994) and in co-culture systems with pig cumulus cells (Nagai and Takahashi, 1992), trophoblastic cells (Nagai and Takahashi, 1992) or oviduct epithelial cell aggregates (Choi *et al.*, 1995). Furthermore, adding IGF-I to the culture medium enhances the cleavage rate of IVM and IVF pig embryos (Xia *et al.*, 1994), and has a stimulating effect on metabolism and cellular proliferation of pig blastocysts (Lewis *et al.*, 1992). The cleavage rate of pig

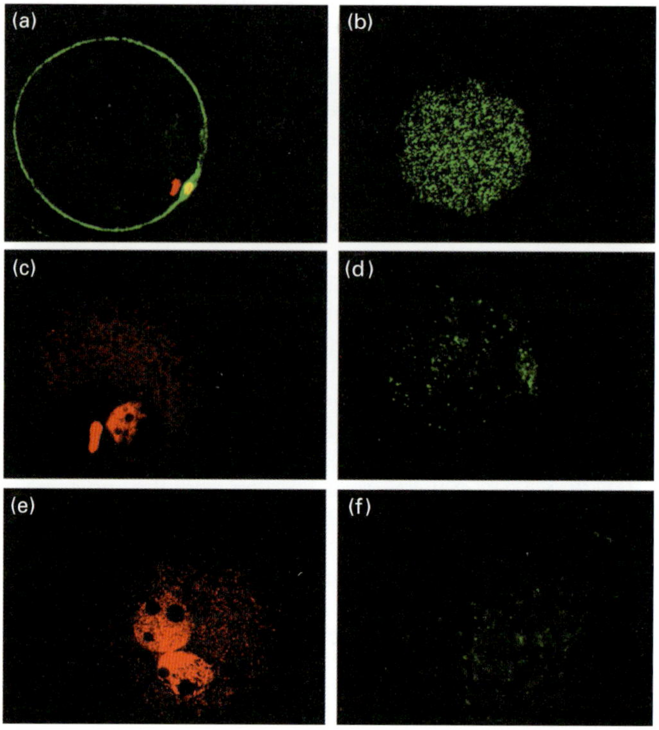

Fig. 3. Confocal micrographs of cortical granules in pig oocytes matured in NCSU23 for 44 h, and then inseminated in modified Tris-buffered medium. Oocytes were examined 6 h (a–d) and 12 h (e and f) after insemination. Green and red images show cortical granules and DNA, respectively. (a,b) Oocyte that was not penetrated by spermatozoa at 6 h after insemination; the oocyte was still at the stage of metaphase II. No cortical granule exocytosis was found. (c,d) An oocyte that was penetrated at 6 h after IVF by a spermatozoon; this oocyte had been activated to form a female pronucleus (c). Some cortical granules were still intact in the cortex (d). (e,f) A penetrated oocyte at 12 h after IVF: both male and female pronuclei had been formed (e) and very few residual cortical granules were present in the cortex (f).

IVM/IVF embryos is also increased if glycine and oviductal cells are present in the medium during early embryonic development (Xia *et al.*, 1995).

The ability of IVM/IVF embryos to develop to the blastocyst stage in simple media has also been improved by modification of the maturation medium to reduce the concentration of sodium chloride (Funahashi *et al.*, 1994a) or supplementation of maturation medium with cysteamine (Grupen *et al.*, 1995) or organic osmolytes (Funahashi *et al.*, 1996a). The presence of glycine in the maturation medium also improves the incidence of cleaved IVM/IVF embryos (Xia *et al.*, 1995). However, there is a significantly smaller number of cells in the trophectoderm of pig IVF embryos compared with those produced *in vivo* (Rath *et al.*, 1995a). Since the presence of BSA in the culture media after day 5 of culture decreases the number of blastomeres of embryos matured and fertilized *in vivo* (Dobrinsky *et al.*, 1996), the smaller number of cells in the trophectoderm of pig IVM/IVF embryos may be due to a detrimental effect of BSA in culture media.

Embryonic development following embryo transfer

The transfer of the IVM and IVF pig embryos at the four-cell stage has often resulted in extended oestrous cycles and failure of pregnancy during later stages of gestation (Funahashi *et al.*, 1996a). Furthermore, piglets from oocytes matured and fertilized *in vitro* have been produced in only a few laboratories (Mattioli *et al.*, 1989; Yoshida *et al.*, 1993c; Funahashi *et al.*, 1996a). In gilts, embryo survival seems to be associated with a close synchrony between the peak concentration of oestradiol and the onset of the LH surge (Blair *et al.*, 1994; Soede *et al.*, 1994) or oestrus (Blair *et al.*, 1994). It has also been suggested that the longer time interval between the onset of oestrus and ovulation is important for the high rate of embryo survival in Meishan pigs (Hunter and Picton, 1995). Furthermore, in polytocous animals such as pigs, it is proposed that asynchronous embryonic development reduces the number of embryos that develop to full term because oestrogen secretion from the more advanced blastocysts adversely affects the development of smaller blastocysts (Pope *et al.*, 1990; Greenwald and Roy, 1994). Therefore, the quality control of oocytes before the start of IVM may be an important factor for synchronized embryonic development and consequently the survival of IVM and IVF pig embryos following embryo transfer.

Oocyte Competence For Early Embryonic Development

Developmental competence of oocytes matured in vivo *and* in vitro

Maternal materials for *in vitro* production of pig embryos have usually been collected from antral follicles of slaughtered prepubertal gilts. Recently, it has been reported that glutamine metabolism of IVM oocytes from prepubertal sheep is lower than that of oocytes from adult sheep and that the mitochondria and cortical granules of IVM oocytes from prepubertal sheep differ from those of IVM oocytes from adult sheep (O'Brien *et al.*, 1996). Furthermore, it has been shown that the low developmental competence of calf oocytes compared with cow oocytes would appear not to be due to a difference in oocyte protein patterns (Khatir *et al.*, 1996), but to a low sensitivity of the inositol 1,4,5-trisphosphate receptor (Damiani *et al.*, 1996). Therefore, the mechanism for signal transduction in oocytes of pigs as well as those of sheep and cattle may not be completed until about the time of puberty or during the follicular phase of the oestrous cycle.

The developmental competence of oocytes matured *in vitro* as determined by *in vitro* development to the blastocyst stage and the number of cells in blastocysts are lower than those matured *in vivo* (Nagashima *et al.*, 1996). However, since culture conditions used for the maturation of pig oocytes *in vitro* affect their subsequent response to oocyte activation (Yamauchi *et al.*, 1996), there may still be a major need for further improvements in the culture conditions during IVM of oocytes. For example, the ability of IVM and IVF pig embryos to cleave and develop to the blastocyst stage *in vitro* has been improved by modification of various conditions between oocyte collection and germinal vesicle breakdown (Table 1; Funahashi *et al.*, 1997a, b).

Variation in the morphology of oocytes at the germinal vesicle stage

Extensive use has been made in IVM and IVF systems of cumulus–oocyte complexes (COCs) with uniform ooplasm and a compact cumulus cell mass collected from antral follicles of slaughtered prepubertal gilts. However, the size of follicles that are selected for *in vitro* production of pig embryos differs among investigators (Day and Funahashi, 1996). There is a large variation in the dictyate stage of the first meiotic prophase among oocytes collected from follicles of slaughtered gilts for *in vitro* production of embryos (Hirao *et al.*, 1995; Funahashi *et al.*, 1997a). Furthermore, this variation in the germinal vesicle morphology seems to cause an increased range in the meiotic stage at the end of the maturation culture (Funahashi *et al.*, 1997b). Since histone H1 kinase activity of aged oocytes is significantly lower with time in culture (Kikuchi *et al.*, 1995), extension of the culture duration to obtain a higher maturation rate in a meiotically asynchronized population of oocytes may reduce the oocyte competence for early embryonic development. In contrast, the germinal vesicle stage of COCs collected from gilts 72 h after injection of equine chorionic gonadotrophins (eCG) is closely synchronized (Funahashi *et al.*, 1996b). For efficient embryo production *in vitro*, therefore, oocytes to be used for IVM may need to be meiotically synchronized and of high quality. Although variation in oocyte quality may also be affected by the method (for example dissection versus aspiration) used to collect COCs (Nagai, 1994) and by the presence of corpora lutea in ovaries (Ocampo *et al.*, 1993), an oocyte population that is harvested from follicles of different ages may be expected to have an increased variation in range of quality of the oocytes. The ability of follicles to secrete steroids and support cytoplasmic maturation of the oocyte is more dependent on age than size of follicles (Ding and Foxcroft, 1994).

Reducing morphological variation appears to enhance the developmental competence of pig oocytes. Preincubation of COCs in maturation medium without gonadotrophins for 12 h before exposing them to gonadotrophins reduces the variation in the morphology of germinal vesicle and enhances the developmental competence of pig oocytes following IVM–IVF (Funahashi *et al.*, 1996a). Furthermore, exposure of COCs to dibutyryl cyclic adenosine 3′,5′-monophosphate (dbcAMP) for the first 20 h of culture for maturation does not affect the maturation of oocytes during a 44 h culture period or sperm penetration but does increase the homogeneity of oocyte nuclear maturation (Funahashi *et al.*, 1997b). This treatment also improves the efficiency of *in vitro* production of pig embryos, and piglets have been produced with a high pregnancy rate and acceptable litter size following surgical embryo transfer of the IVM–IVF embryos at the two–four-cell stages (Funahashi *et al.*, 1997b). Treatment with hypoxanthine may be expected to produce similar effects during IVM because hypoxanthine (Miyano *et al.*, 1995) as well as dbcAMP (Petr *et al.*, 1991; Funahashi *et al.*, 1997b) inhibits meiotic resumption reversibly, and oocytes treated with these inhibitors have a mass of filamentous chromosomes around the nucleolus (Miyano *et al.*, 1995; Funahashi *et al.*, 1997b). Hypoxanthine appears to maintain the oocyte arrest by modulating cAMP concentration through its inhibitory action on cAMP-phosphodiesterase (Downs *et al.*, 1989).

However, since increasing and decreasing cAMP per se also stimulates oocyte phosphorylation via a signal transduction pathway, the stimulatory effect of cAMP may also enhance the competence of oocytes to develop to the blastocyst stage. Inhibiting all tyrosine phosphorylation with the tyrosine-specific protein inhibitor prevents changes in the morphology of the germinal vesicle (Jung *et al.*, 1993). A 42 kDa protein in pig oocytes that increases in amount after 12 h of maturation culture is localized to condensing and condensed chromosomes (Miyano *et al.*, 1996). A further understanding of the mechanisms required to induce developmental competence in oocytes is needed in our efforts to improve the viability of IVM–IVF blastocysts.

Oocyte competence after the germinal vesicle stage

The competence of oocytes to develop to the blastocyst stage also seems to be improved by modification of IVM conditions after the germinal vesicle stage. The presence of tissue inhibitor of matrix metalloproteinase-1 (TIMP-1) from 20 to 44 h of culture for *in vitro* maturation of pig oocytes enhances the oocyte competence to the blastocyst stage without affecting factors associated with fertilization (Table 1; Funahashi *et al.*, 1997c). A combination of techniques using dbcAMP during the first 20 h of IVM and TIMP-1 from 20 to 44 h improves the ability of IVM–IVF embryos to develop to the blastocyst stage to a more acceptable level (34%) for *in vitro* production of embryos (Funahashi *et al.*, 1997c). TIMP-1 is a major secretory protein of pig preovulatory granulosa cells following treatment with hCG (Smith *et al.*, 1994). After the preovulatory gonadotrophin surge, the concentration of mRNA encoding TIMP-1 is known to be increased and localized to the granulosa cells (Smith *et al.*, 1994). Therefore, TIMP-1 may be required for oocytes to obtain the full ability for early embryonic development. However, an understanding of how TIMP-1 promotes the ability of pig oocytes to develop to the blastocyst stage without affecting fertilization is still not known. Cyclin B1 content in pig oocytes is relatively low during the GV stage and increases between 20 and 30 h after the start of IVM (Naito *et al.*, 1995). Mitogen-activated protein kinase is involved in the

Table 1. Recent improvement in blastocyst formation of IVM/IVF porcine embryos

Authors	Percentage of blastocyst formation	Improved technology
Simple culture systems		
Funahashi *et al.* (1994a)	3	Using modified Whitten's medium during IVM and embryo culture
Grupen *et al.* (1995)	12	Supplementation of maturation medium (modified Medium 199) with 500 μmol cysteamine l^{-1}. IVM/IVF embryos were cultured in modified Whitten's medium
Funahashi *et al.* (1997a)	23	Preincubation of cumulus–oocyte complexes in gonadotrophin-free medium (modified NCSU37 medium) before the start of IVM. Culture medium was NCSU37 medium.
Funahashi *et al.* (1997b)	22	No preincubation but adding dibutyryl cAMP and gonadotrophic supplements to maturation medium (modified NCSU37 medium) during the first 20 h period of IVM. Culture medium was modified NCSU37 medium.
Funahashi *et al.* (1997c)	34	A combination of techniques adding dibutyryl cAMP and gonadotrophic supplements to modified NCSU37 medium during the first 20 h of IVM and tissue inhibitor of metalloproteinase 1 from 20 to 44 h of IVM. Culture medium was modified NCSU37 medium.
Co-culture systems and *in vivo* culture system		
Nagai and Takahashi (1992)	10	Co-culture of IVM–IVF embryos with cumulus cells in modified BMOC-II. Maturation medium was modified Waymouth MB752/1 medium.
Ocampo *et al.* (1994)	26	Culture of IVM–IVF embryos in the amniotic fluid in chick embryo. Maturation medium was modified Medium 199.
Abeydeera *et al.* (1996b)	36	Co-culture of cumulus–oocyte complexes with follicular shell piece in modified NCSU23 medium during IVM. IVM–IVF embryos were cultured in NCSU23 medium.

regulation of meiotic maturation of pig oocytes, especially after germinal vesicle breakdown (Inoue *et al.*, 1995). It is of interest that the effective time of adding TIMP-1 is close to the time of initial synthesis of cyclin B1 and mitogen-activated protein kinase in pig oocytes. Further experiments will be required to clarify the mechanism of the stimulating effect by TIMP-1.

A recent study has also demonstrated that co-culture of pig cumulus–oocyte complexes with follicular shell pieces improved early embryonic development to the blastocyst stage after *in vitro* fertilization (Table 1; Abeydeera *et al.*, 1996b). In this IVM system (Abeydeera *et al.*, 1996b), piglets were produced at a relatively high efficiency following surgical embryo transfer of the IVM–IVF embryos. It is also possible that the effect of the follicular shell pieces is mainly due to TIMP-1 secretion from granulosa cells.

The developmental competence of pig oocytes matured *in vivo* is lower following IVF than following *in vivo* fertilization (Rath *et al.*, 1995b). Therefore, there are inadequacies not only in the IVM system but also in the IVF system that impair the development of pig oocytes.

Future Studies

There are large numbers of primordial follicles in a pig ovary (210 000), compared with those of other mammals such as humans (151 000), cattle (105 000) and mice (2 135) (Gosden and Telfer, 1987). Use of primordial oocytes for *in vitro* embryo production would bring about a major reduction in the cost and effort required to produce large numbers of embryos. However, the materials used for *in vitro* embryo production have largely been limited to a few fully grown oocytes from mid-size antral follicles. The development of mature mouse oocytes *in vitro* from the primordial follicle has been reported and it was shown that offspring could be produced from these oocytes (Eppig and O'Brien, 1996). Oocyte maturation and sperm penetration of pig oocytes following culture of preantral follicles was reported by Hirao *et al.* (1994). However, the incidence (40%) of oocytes grown *in vitro* that matured is still low (Hirao *et al.*, 1994) and developmental ability of this type of oocyte following *in vitro* fertilization has not been reported. Isolation and culture of preantral follicles from pig ovaries have been reviewed (Telfer, 1996). Further research with oocytes derived from primordial and growing follicles needs to be undertaken to increase the number of oocytes available from a single ovary for IVM–IVF.

Conclusion

Failure in male pronuclear formation has been overcome by reducing the oxidative stress during IVM. However, there is a considerable need for improved methods to reduce further the rate of polyspermic penetration even though it has been decreased significantly in recent years. The developmental competence of oocytes matured and fertilized *in vitro* has been enhanced through modification of culture conditions of oocytes during *in vitro* maturation. Oocyte competence for early embryonic development appears to be achieved by mimicking active communications between the oocyte and follicular cells. In the most current IVM–IVF system, more than 80% of pig oocytes that were matured and fertilized normally developed to the blastocyst stage. Further research is needed to improve our understanding of the culture conditions required for pig oocytes of both preantral and antral follicles to achieve efficient growth, maturation, fertilization and early development *in vitro*, especially to induce developmental competence in oocytes, as well as to determine how to overcome the problem of polyspermic penetration.

The authors are very grateful to K. Niwa and T. Nagai for critical advice and B. Nichols for secretarial assistance with the preparation of this manuscript.

References

Abeydeera LR, Funahashi H and Day BN (1996a) Penetration of *in vitro* matured pig oocytes by frozen–thawed ejaculated boar sperm in a modified Tris-buffered medium and their subsequent development *Journal of Animal Science* **74** (Supplement 1) 47 (Abstract)

Abeydeera LR, Funahashi H, Cantley TC, Rieke A and Day BN (1996b) Co-culture with follicular shell pieces (FSP) increases the developmental competence of pig oocytes matured *in vitro*. *Biology of Reproduction* **54** (Supplement 1) 81

Ashworth PJC, Harrison RAP, Miller NGA, Plummer JM and Watson PF (1995) Flow cytometric detection of bicarbonate-induced changes in lectin binding in boar and ram sperm populations *Molecular Reproduction and Development* **40** 164–176

Blair RM, Coughlin CM, Minton JE and Davis DL (1994) Peri-oestrous hormone profiles, embryonic survival and variation in embryonic development in gilts and primiparous sows *Journal of Reproduction and Fertility* **101** 167–173

Choi YH, Saito S and Oguri N (1995) *In vitro* development of porcine oocytes fertilized *in vitro* with spermatozoa preincubated in two different media *Theriogenology* **44** 287–294

Cran DG and Esper CR (1990) Cortical granules and the cortical reaction in mammals *Journal of Reproduction and Fertility Supplement* **42** 177–188

Damiani P, Fissore RA, Cibelli JB, Long CR, Balise JJ, Robl JM and Duby RT (1996) Evaluation of developmental competence, nuclear and ooplasmic maturation of calf oocytes *Molecular Reproduction and Development* **45** 521–534

Dandekar P and Talbot P (1992) Perivitelline space of mammalian oocytes: extracellular matrix of unfertilized oocytes and formation of a cortical granule envelope following fertilization *Molecular Reproduction and Development* **31** 135–143

Day BN and Funahashi H (1996) *In vitro* maturation and fertilization of pig oocytes. In *Beltsville Symposia in Agricultural Research XX. Biotechnology's Role In The Genetic Improvement Of Farm Animals* pp 125–144 Eds RH Miller, VG Pursel, and HD Norman. American Society of Animal Science, Savoy, IL

Ding J and Foxcroft GR (1994) Conditioned media produced by follicular shells of different maturity affect maturation of pig oocytes *Biology of Reproduction* **50** 1377–1384

Dobrinsky JR, Johnson LA and Rath D (1996) Development of a culture medium (BECM-3) for porcine embryos: effects of bovine serum albumin and fetal bovine serum on embryo development *Biology of Reproduction* **55** 1069–1074

Downs SM, Daniel SAJ, Bornslaeger EA, Hoppe PC and Eppig JJ (1989) Maintenance of meiotic arrest in mouse oocytes by purines: modulation of cAMP levels and cAMP phosphodiesterase activity *Gamete Research* **23** 323–334

Eppig JJ and O'Brien MJ (1996) Development *in vitro* of mouse oocytes from primordial follicles *Biology of Reproduction* **54** 197–207

Funahashi H and Day BN (1993) Effects of follicular fluid at fertilization *in vitro* on sperm penetration in pig oocytes *Journal of Reproduction and Fertility* **99** 97–103

Funahashi H and Day BN (1995) Effects of cumulus cells on glutathione content of porcine oocytes during *in vitro* maturation *Journal of Animal Science* **73** (Supplement 1) 90 (Abstract)

Funahashi H, Stumpf TT, Terlouw SL and Day BN (1993) Effects of electrical stimulation before or after *in vitro* fertilization on sperm penetration and pronuclear formation in pig oocytes *Molecular Reproduction and Development* **36** 361–367

Funahashi H, Cantley TC, Stumpf TT, Terlouw SL and Day BN (1994a) *In vitro* development of *in vitro* matured porcine oocytes following chemical activation or *in vitro* fertilization *Biology of Reproduction* **50** 1072–1077

Funahashi H, Cantley TC, Stumpf TT, Terlouw SL and Day BN (1994b) Use of low salt culture medium for *in vitro* maturation of porcine oocytes is associated with elevated oocyte glutathione levels and enhanced male pronuclear formation after *in vitro* fertilization *Biology of Reproduction* **51** 633–639

Funahashi H, Stumpf TT, Terlouw SL, Cantley TC, Rieke A and Day BN (1994c) Developmental ability of porcine oocytes matured and fertilized *in vitro*. *Theriogenology* **41** 1425–1433

Funahashi H, Stumpf TT, Cantley TC, Kim N-H and Day BN (1995) Pronuclear formation and intracellular glutathione content of *in vitro*-matured porcine oocytes following *in vitro* fertilization and/or electrical activation *Zygote* **3** 273–281

Funahashi H, Kim N-H, Stumpf TT, Cantley TC and Day BN (1996a) Presence of organic osmolytes in maturation medium enhances cytoplasmic maturation of porcine oocytes *Biology of Reproduction* **54** 1412–1219

Funahashi H, Tatemoto H, Cantley TC and Day BN (1996b) Nuclear morphology of swine oocytes during follicular development following stimulation by eCG injection *Biology of Reproduction* **54** (Supplement 1) 156

Funahashi H, Cantley TC and Day BN (1997a) Preincubation of oocyte–cumulus complexes before exposure to gonadotropins improves the developmental ability of porcine embryos matured and fertilized *in vitro*. *Theriogenology* **47** 679–686

Funahashi H, Cantley TC and Day BN (1997b) Synchronization of meiosis in porcine oocytes by exposure to dibutyryl cyclic AMP improves developmental competence following *in vitro* fertilization *Biology of Reproduction* **47** 679–686

Funahashi H, McIntush EW, Smith MF and Day BN (1997c) Effect of tissue inhibitor of metalloproteinase (TIMP-1) on early development of swine oocytes matured and fertilized *in vitro*. *Theriogenology* **47** 277 (Abstract)

Gosden RG and Telfer E (1987) Number of follicles and oocytes in mammalian ovaries and their allometric relationships *Journal of Zoology* **211** 169–175

Greenwald GS and Roy SK (1994) Follicular development and its control. In *The Physiology of Reproduction* pp 629–724 Ed. E Knobil, JD Neill, GS Greenwald, CL Markert and DW Pfaff. Raven Press, New York.

Grupen CG, Nagashima H and Nottle MB (1995) Cysteamine enhances *in vitro* development of porcine oocytes matured and fertilized *in vitro*. *Biology of Reproduction* **53** 173–178

Harrison RAP, Ashworth PJC and Miller NGA (1996) Bicarbonate/CO_2, an effector of capacitation, induces a rapid and reversible change in the lipid architecture of boar sperm plasma membranes *Molecular Reproduction and Development* **45** 378–391

Hasegawa A, Koyama K, Okazaki Y, Sugimoto M and Isojima S (1994) Amino acid sequence of a porcine zona pellucida glycoprotein ZP4 determined by peptide mapping and cDNA cloning *Journal of Reproduction and Fertility* **100** 245–255

Hatanaka Y, Nagai T, Tobita T and Nakano M (1992) Changes in the properties and composition of zona pellucida of pigs during fertilization *in vitro*. *Journal of Reproduction and Fertility* **95** 431–440

Hedrick JL (1993) The pig zona pellucida: sperm binding ligands, antigens and sequence homologies *Reproductive Immunology* **97** 59–65

Hirao Y, Nagai T, Kubo M, Miyano T, Miyake M and Kato S (1994) *In vitro* growth and maturation of pig oocytes *Journal of Reproduction and Fertility* **100** 333–339

Hirao Y, Tsuji Y, Miyano T, Okano A, Miyake M, Kato S and Moor RM (1995) Association between p34cdc2 levels and meiotic arrest in pig oocytes during early growth *Zygote* **3** 325–332

Hunter MG and Picton HM (1995) Effect of hCG administration at the onset of oestrus on early embryo survival and development in Meishan gilts *Animal Reproduction Science* **38** 231–238

Hunter RHF (1982) Interrelationships between spermatozoa, the female reproductive tract, and the egg investments *Control of Pig Reproduction* pp 49–63 Ed. DJA Cole and GR Foxcroft. Butterworths, London

Hunter RHF (1991) Oviduct function in pigs, with particular reference to the pathological condition of polyspermy *Molecular Reproduction and Development* **29** 385–391

Hunter RHF (1996) Ovarian control of very low sperm/egg ratios at the commencement of mammalian fertilisation to avoid polyspermy *Molecular Reproduction and Development* **44** 417–422

Inoue M, Naito K, Aoki F, Toyoda Y and Sato E (1995) Activation of mitogen-activated protein kinase during meiotic maturation in porcine oocytes *Zygote* **3** 265–271

Jones R (1990) Identification and functions of mammalian sperm–egg recognition molecules during fertilization *Journal of Reproduction and Fertility Supplement* **42** 89–105

Jones R (1991) Interaction of zona pellucida glycoproteins, sulphated carbohydrates and synthetic polymers with proacrosin, the putative egg-binding protein from mammalian spermatozoa *Development* **111** 1155–1163

Jung T, Fulka JJ, Lee C and Moor RM (1993) Effects of the protein phosphorylation inhibitor genistein on maturation of pig oocytes *in vitro. Journal of Reproduction and Fertility* **98** 529–535

Kano K, Miyano T and Kato S (1994) Effect of oviductal epithelial cells on fertilization of pig oocytes *in vitro. Theriogenology* **42** 1061–1068

Khatir H, Lonergan P, Carolan C and Mermillod P (1996) Prepubertal bovine oocytes: a negative model for studying oocyte developmental competence *Molecular Reproduction and Development* **45** 231–239

Kikuchi K, Izaike Y, Noguchi J, Furukawa T, Daen FP, Naito K and Toyoda Y (1995) Decrease of histone H1 kinase activity in relation to parthenogenetic activation of pig follicular oocytes matured and aged *in vitro. Journal of Reproduction and Fertility* **105** 325–330

Kim NH, Simerly C, Funahashi H, Schatten G and Day BN (1996a) Microtubule organization in porcine oocytes during fertilization and parthenogenesis *Biology of Reproduction* **54** 1397–1404

Kim N-H, Funahashi H, Abeydeera LR, Moon SJ, Prather RS and Day BN (1996b) Effects of oviductal fluid on sperm penetration and cortical granule exocytosis during *in vitro* fertilization of porcine oocytes *Journal of Reproduction and Fertility* **107** 79–86

Kim N-H, Funahashi H, Prather RS, Schatten G and Day BN (1996c) Microtubule and microfilament dynamics in porcine oocytes during meiotic maturation *Molecular Reproduction and Development* **43** 248–255

Laurincik J, Hyttel P and Kopecny V (1995) DNA synthesis and pronucleus development in pig zygotes obtained *in vivo*: an autoradiographic and ultrastructural study *Molecular Reproduction and Development* **40** 325–332

Lewis AM, Kaye PL, Lising R and Cameron RDA (1992) Stimulation of protein synthesis and expansion of pig blastocysts by insulin *in vitro. Reproduction, Fertility and Development* **4** 119–123

Machaty Z, Mayes MA and Prather RS (1995) Parthenogenetic activation of porcine oocytes with guanosine-5′-*O*-(3′-thiotriphosphate) *Biology of Reproduction* **52** 753–758

Martinez E, Vazquez JM, Matas C, Roca J, Coy P and Gadea J (1993) Evaluation of boar spermatozoa penetrating capacity using pig oocytes at the germinal vesicle stage *Theriogenology* **40** 547–557

Mattioli M, Bacci ML, Galeati G and Seren E (1989) Developmental competence of pig oocytes matured and fertilized *in vitro. Theriogenology* **31** 1201–1207

Mattioli M, Barboni B, Lucidi P and Seren E (1996) Identification of capacitation in boar spermatozoa by chlortetracycline staining *Theriogenology* **45** 373–381

Mattioli M, Galeati G, Bacci ML and Seren E (1988) Follicular factors influence oocyte fertilizability by modulating the intercellular cooperation between cumulus cells and oocyte *Gamete Research* **21** 223–232

Mburu JN, Einarsson S, Lundeheim N and Rodriguez-Martinez H (1996) Distribution, number and membrane integrity of spermatozoa in the pig oviduct in relation to spontaneous ovulation *Animal Reproduction Science* **45** 109–121

Miyano T, Ebihara M, Goto Y, Hirao Y, Nagai T and Kato S (1995) Inhibitory action of hypoxanthine on meiotic resumption of denuded pig follicular oocytes *in vitro. Journal of Experimental Zoology* **273** 70–75

Miyano T, Moor RM, Wooding FBP and Shiroo M (1996) Localization and function of tyrosine-phosphorylated protein in pig oocytes *Molecular Reproduction and Development* **44** 408–416

Motlik J, Fulka J and Flechon JE (1986) Changes in intercellular coupling between pig oocytes and cumulus cells during maturation *in vivo* and *in vitro. Journal of Reproduction and Fertility* **76** 31–37

Nagai T (1994) Current status and perspectives in IVM–IVF of porcine oocytes *Theriogenology* **41** 73–78

Nagai T (1996) *In vitro* maturation and fertilization of pig oocytes *Animal Reproduction Science* **42** 153–163

Nagai T and Moor RM (1990) Effect of oviduct cells on the incidence of polyspermy in pig eggs fertilized *in vitro. Molecular Reproduction and Development* **26** 377–382

Nagai T and Takahashi M (1992) Culture of *in vitro* matured and fertilized pig oocytes *Proceedings of the 12th International Congress on Animal Reproduction* **3** 1324–1326

Nagai T, Niwa K and Iritani A (1984) Effects of sperm concentration during preincubation in a defined medium on fertilization *in vitro* of pig follicular oocytes *Journal of Reproduction and Fertility* **70** 271–275

Nagashima H, Grupen CG, Ashmen RJ and Nottle MB (1996) Developmental competence of *in vivo* and *in vitro* matured porcine oocytes after subzonal sperm injection *Molecular Reproduction and Development* **45** 359–363

Naito K, Hawkins C, Yamashita M, Nagahama Y, Aoki F, Kohmoto K, Toyoda Y and Moor RM (1995) Association of p34cdc2 and cyclin B1 during meiotic maturation in porcine oocytes *Developmental Biology* **168** 627–634

Navara CS, Simerly C, Zoran S and Schatten G (1995) The sperm centrosome during fertilization in mammals: implications for fertility and reproduction *Reproduction, Fertility and Development* **7** 747–754

Niwa K (1993) Effectiveness of *in vitro* maturation and *in vitro* fertilization techniques in the pig *Journal of Reproduction and Fertility Supplement* **48** 49–59

O'Brien JK, Dwarte D, Ryan JP, Maxwell WMC and Evens G (1996) Developmental capacity, energy metabolism and ultrastructure of mature oocytes from prepubertal and adult sheep *Reproduction, Fertility and Development* **8** 1029–1037

Ocampo MB, Ocampo LC, Ryu IS, Mori T, Ueda J and Kanagawa H (1993) Effects of culture time, ovarian activity, cumulus cells and sera on the nuclear and cytoplasmic maturation of pig oocytes *in vitro. Animal Reproduction Science* **34** 135–146

Ocampo MB, Ocampo LC, Mori T, Ueda J, Shimizu H and Kanagawa H (1994) Blastocyst formation of pig embryos derived from *in vitro* fertilization of *in vitro* matured pig oocytes in the amniotic fluid of a developing chick embryo *Animal Reproduction Science* **37** 65–73

Parry RV, Barker PJ and Jones R (1992) Characterization of low Mr zona pellucida binding proteins from boar spermatozoa and seminal plasma *Molecular Reproduction and Development* **33** 108–115

Perreault SD (1990) Regulation of sperm nuclear reactivation during fertilization. In *Fertilization in Mammals* pp 285–296 Eds BD Bavister, J Cummins and ERS Roldan. Serono Symposia, Norwell, MA

Petr J, Zetova L and Fulka JJ (1991) Influence of dbcAMP on the inhibitory effect of cumulus cell factor(s) *Reproduction, Nutrition and Development* **31** 135–140

Petters RM and Wells KD (1993) Culture of pig embryos *Journal of Reproduction and Fertility Supplement* **48** 61–73

Pollard JW, Plante C and Leibo SP (1995) Comparison of development of pig zygotes and embryos in simple and complex culture media *Journal of Reproduction and Fertility* **103** 331–337

Pope WF, Xie S, Broermann DM and Nephew KP (1990) Causes and consequences of early embryonic diversity in pigs *Journal of Reproduction and Fertility Supplement* **40** 251–260

Rath D, Niemann H, Tao T and Boerjan M (1995a) Ratio and number of inner cell mass and trophoblast cells of *in vitro* and *in vivo* produced porcine embryos *Theriogenology* **43** 304 (Abstract)

Rath D, Niemann H and Torres CRL (1995b) *In vitro* development to blastocysts of early porcine embryos produced *in vivo* or *in vitro. Theriogenology* **43** 913–926

Sacco AG, Yurewicz EM, Subramanian MG and Matzat PD (1989) Porcine zona pellucida: association of sperm receptor activity with the glycoprotein component of the Mr = 55,000 family *Biology of Reproduction* **41** 523–532

Sawai K, Funahashi H and Niwa K (1997) Stage-specific requirement of cysteine during *in-vitro* maturation of porcine oocytes for glutathione synthesis associated with male pronuclear formation *Biology of Reproduction* **57** 1–6

Schatten G (1994) The centrosome and its mode of inheritance: the reduction of the centrosome during gametogenesis and its restoration during fertilization *Developmental Biology* **165** 299–335

Smith MF, Kemper CN, Smith GW, Goetz TL and Jarrell VL (1994) Production of tissue inhibitor of metalloproteinases-1 by porcine follicular and luteal cells *Journal of Animal Science* **72** 1004–1012

Soede NM, Helmond FA and Kemp B (1994) Periovulatory profiles of oestradiol, LH and progesterone in relation to oestrus and embryo mortality in multiparous sows using transrectal ultrasonography to detect ovulation *Journal of Reproduction and Fertility* **101** 633–641

Soede NM, Wetzels CC, Zondag W, Hazeleger W and Kemp B (1995) Effects of a second insemination after ovulation on fertilization rate and accessory sperm count in sows *Journal of Reproduction and Fertility* **105** 135–140

Sun FZ, Hoyland J, Huang X, Mason W and Moor RM (1992) A comparison of intercellular changes in porcine eggs after fertilization and electroactivation *Development* **115** 947–956

Suzuki K, Ebihara M, Nagai T, Clarke NGE and Harrison RAP (1994) Importance of bicarbonate/CO_2 for fertilization of pig oocytes *in vitro,* and synergism with caffeine *Reproduction, Fertility and Development* **6** 221–227

Telfer EE (1996) The development of methods for isolation and culture of preantral follicles from bovine and porcine ovaries *Theriogenology* **45** 101–110

Wang WH, Abeydeera LR, Fraser LR and Niwa K (1995) Functional analysis using chlortetracycline fluorescence and *in vitro* fertilization of frozen–thawed ejaculated boar spermatozoa incubated in a protein-free chemically defined medium *Journal of Reproduction and Fertility* **104** 305–313

Xia P, Tekpetey FR and Armstrong DT (1994) Effect of IGF-I on pig oocyte maturation, fertilization, and early embryonic development *in vitro*, and on granulosa and cumulus cell biosynthetic activity *Molecular Reproduction and Development* **38** 373–379

Xia P, Rutledge J and Armstrong DT (1995) Expression of glycine cleavage system and effect of glycine on *in vitro* maturation, fertilization and early embryonic development in pigs *Animal Reproduction Science* **38** 155–165

Yamauchi N, Sasada H, Sugawara S and Nagai T (1996) Effect of culture conditions on artificial activation of porcine oocytes matured *in vitro. Reproduction, Fertility and Development* **8** 1153–1156

Yonezawa N, Hatanaka Y, Takeyama H and Nakano M (1995) Binding of pig sperm receptor in the zona pellucida to the boar sperm acrosome *Journal of Reproduction and Fertility* **103** 1–8

Yoshida M, Ishigaki K, Nagai T, Chikyu M and Pursel VG (1993a) Glutathione concentration during maturation and after fertilization in pig oocytes: relevance to the ability of oocytes to form male pronucleus *Biology of Reproduction* **49** 89–94

Yoshida M, Cran DG and Pursel VG (1993b) Confocal and fluorescence microscopic study using lectins of the distribution of cortical granules during the maturation and fertilization of pig oocytes *Molecular Reproduction and Development* **36** 462–468

Yoshida M, Mizoguchi Y, Ishigaki K, Kojima T and Nagai T (1993c) Birth of piglets derived from *in vitro* fertilization of pig oocytes matured *in vitro. Theriogenology* **39** 1303–1311

Yurewicz EC, Pack BA, Armant DR and Sacco AG (1993) Porcine zona pellucida ZP3a glycoprotein mediates binding of the biotin-labeled Mr 55,000 family (ZP3) to boar sperm membrane vesicles *Molecular Reproduction and Development* **36** 382–389

Journal of Reproduction and Fertility Supplement **52**, 285–300

Oviductal regulation of fertilization and early embryonic development

W. C. Buhi[1,2,3], I. M. Alvarez[1] and A. J. Kouba[3]

Departments of [1]Obstetrics and Gynecology, [2]Biochemistry and Molecular Biology and [3]Animal Science, University of Florida, Gainesville, FL 32610–0294, USA

During the period of late follicular development and the first four days of the oestrous cycle, the oviduct occupies a central role in the establishment of pregnancy. Oviductal function is regarded as being either 'passive' or biologically active, providing an environment that sustains and enhances fertilization and early cleavage-stage embryonic development. Recent reports have focused on this microenvironment and shown that ovarian steroids induce marked morphological, physiological and biochemical changes. Alterations include changes in the biosynthetic activity and release of macromolecules by the oviductal epithelium which become part of the luminal microenvironment. Furthermore, both regional and temporal differences in activity and protein production occur through hormonal changes during the oestrous cycle and early pregnancy. Studies on identification, characterization and regulation of several proteins synthesized *de novo* have indicated oocyte–oviduct and embryo–oviduct interactions. However, the identification of oviduct-derived proteins, their regulation and their potential function *in vivo* needs to be examined. Studies in other species also suggest roles for growth factors in early embryonic development, but little information is available for the pig. We propose that ovarian hormones control changes in synthetic activity, synthesis of some oviduct-derived proteins and the presence of specific factors in the luminal microenvironment which sustain and enhance fertilization and early cleavage-stage embryonic development.

Introduction

The oviduct in many mammalian species, including the pig, is divided into three distinct regions: the infundibulum, ampulla and isthmus. Each segment has certain histological features that contribute to their particular physiological and biochemical functions. Although fertilization and early cleavage-stage embryonic development in the oviduct are not understood in detail, they are known to involve a highly complex series of cellular interactions within these regions. These interactions are clearly dependent upon the oviduct and its secretions for providing an essential *in vivo* physiological and biochemical microenvironment. The oviduct has been studied at times that are critical for the development of a pregnancy: preovulation, ovulation, fertilization, and early preimplantation development. These are significant for sperm transport, sperm binding and release, capacitation, capture and transport of oocytes, fertilization, and early cleavage-stage embryonic development and extend through the embryonic four-cell stage in the pig oviduct. The oviduct, in part, appears to be programmed for these events by ovarian steroids (Hunter, 1988).

There are two prevailing views regarding oviductal function (Gandolfi, 1995). First, the oviduct provides a 'passive' but optimal environment in terms of temperature, pH, osmotic pressure, nutrients, oxygen tension and other factors. Second, this tissue is biologically active, providing a variety of protein secretory molecules which sustain and regulate or enhance events preceding and during fertilization, and in early embryonic development. Studies on the 'passive' physical environment (see, Hunter, 1988; Leese, 1988) have permitted development of techniques for improving *in vitro* fertilization and embryo development. However, recent studies on sperm–oviduct, oocyte–oviduct and embryo–oviduct interactions suggest regulation of activities through paracrine pathways by factors that may be embryotrophic. Oviductal biosynthetic activities and the identification and characterization of secretory and signalling molecules also suggest the presence of other embryotrophic factors. Here, current literature on the biologically active pig oviductal microenvironment is reviewed..

Functional Histology

The oviductal epithelial lining is composed of two major cell types, ciliated and nonciliated (secretory) cells. Scanning electron microscopy has shown marked cellular structural changes in these cells in pigs during oestrus and dioestrus (Wu *et al.*, 1976; Abe and Oikawa, 1992). Dense ciliation was described in the ampulla and isthmus with well-rounded and bulbous secretory cells at oestrus. Ciliation decreased in the ampulla during the luteal phase without a concomitant change in the isthmus. Secretory granules are observed at all stages of the oestrous cycle but are more abundant at oestrus (Nayak and Zimmerman, 1971a). In pigs and other species, oviductal development is promoted by oestrogen while increased progesterone concentrations antagonize the effects of oestrogen (Nayak and Zimmerman, 1971b; Murray, 1992).

Our studies in sexually mature, cyclic Florida crossbred gilts have confirmed the effect of oestrogen. The effects have been further examined in relation to changes in cell height and cell type in the ampulla and isthmus. Ciliated and secretory cells reach maximal height by day 0 and this persists until day 6 of the oestrous cycle in both ampulla and isthmus. Cell height in both ampulla and isthmus decrease with increased concentrations of progesterone at dioestrus. An increase in cell height begins again at about day 18 in nonpregnant pigs in preparation for a new oestrous cycle. The percentage of ciliated cells (80%) in the ampulla was greatest at oestrus and declined at dioestrus with a concomitant increase in nonciliated cells from 20% to 50%. In contrast, the percentage of ciliated (70–80%) and nonciliated (20–30%) cells did not change throughout the oestrous cycle in the isthmus. These changes suggest a differential response of oviductal epithelium to endocrine regulation and this has been confirmed in studies with ovariectomized gilts treated with hormone replacement. Ovariectomized gilts that were treated with oestrogen showed an increased cell height in both ampulla and isthmus. Treatment of ovariectomized gilts with progesterone led to a lower cell height compared with the effect of oestrogen. In normal cyclic gilts, ovariectomy caused a loss of cilia in the ampulla, while oestrogen administration promoted ciliation. In contrast, progesterone treatment after ovariectomy did increase ciliation in the ampulla but not to the same extent as oestrogen. Progesterone also increased numbers of nonciliated cells to values that were previously seen with intact animals at dioestrus. However, regardless of steroid treatment, the number of ciliated or nonciliated cells in the isthmus did not change, in agreement with results in intact cyclic gilts. In nonciliated secretory cells of the ampulla, secretory granules were considerably more abundant at oestrus than on other days of the cycle (Fig. 1). Similarly, secretory granules were more abundant in epithelial cells in ovariectomized gilts treated with oestrogen compared with other steroid treatments. The appearance of these granules, which contained the porcine oviduct-specific protein (POSP) (Buhi *et al.*, 1993), correlated with increased biosynthetic activity (see below). These changes, development of epithelial cell height, ciliation and the appearance of secretory granules, may reflect functional differences in intraluminal events of fertilization and early embryonic development.

Oviductal Biosynthetic Activity

Biochemical studies have shown the oviduct to be a dynamic tissue, synthesizing and releasing macromolecules into the lumen throughout the oestrous cycle and early pregnancy. These studies have focused on the use of radiolabelled amino acid precursors to examine the biosynthetic activity of the oviduct, a measure of its role in the contribution of macromolecules that may affect

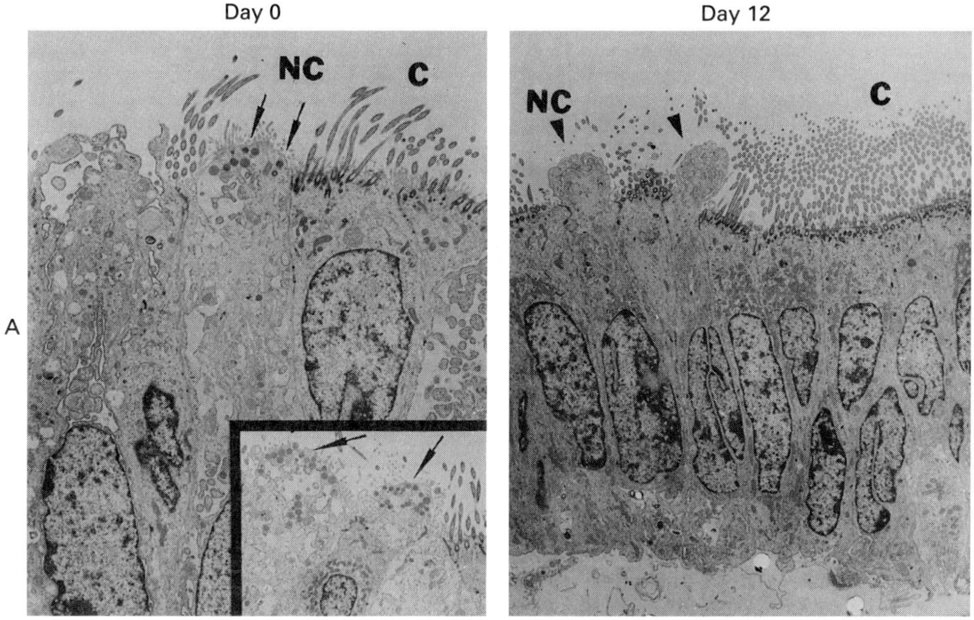

Fig. 1. Electronmicrographs of the oviductal ampullary (A) epithelium from pigs at day 0 (× 3500; insert × 2500) and day 12 (× 2500) of pregnancy. Ciliated (C) and nonciliated secretory (NC) cells are indicated. Secretory granules (arrows) are seen in the supranuclear region of nonciliated secretory cells in the ampulla from gilts at day 0 of pregnancy. Few, if any, secretory granules (arrowheads) are detected in the supranuclear region of nonciliated cells in gilts at day 12 of pregnancy.

fertilization and early cleavage-stage embryonic development. The biosynthetic activity of the oviduct has been measured in the pig during the oestrous cycle (Buhi *et al.*, 1989), early pregnancy (Buhi *et al.*, 1989), or following ovariectomy and steroid replacement (Buhi *et al.*, 1992).

Evidence indicates that the biosynthetic activity of the oviduct is related to the hormonal status of the animal. Oestrogen-dominated animals have significantly greater synthetic activity, measured as incorporation of radiolabelled precursor into secreted nondialysable macromolecules, than progesterone-dominated animals. Greater incorporation was indicated at pro-oestrus, oestrus and metoestrus in pigs (Buhi *et al.*, 1989) than on other days of the cycle. Differences in biosynthetic activity were not detected between cyclic and early pregnant gilts (Buhi *et al.*, 1989), in contrast to observations in rabbits (Roy *et al.*, 1972) in which differences were recognized. Uptake of amino acid precursor and incorporation into protein were reduced in pregnant rabbits, 14–70 h postcoitum, compared with that in nonpregnant rabbits. However, in pigs, the process of mating or the presence of gametes or embryos did not affect oviductal macromolecular protein synthesis. When gilts were unilaterally ovariectomized, biosynthetic differences were not detected in the oviduct ipsilateral or contralateral to the ovary containing the follicle or corpus luteum. These data suggest that circulating concentrations of ovarian hormones produced by one ovary were sufficient to control biosynthetic activities of both oviducts.

Changes in the biosynthetic activity of the whole oviduct and individual functional segments clearly indicated a temporal association of increased circulating concentrations of oestrogen with increased biosynthesis in the infundibulum, ampulla or whole oviduct (Buhi *et al.*, 1989; W. Buhi, unpublished) (Fig. 2). In contrast, the biosynthetic activity did not change in the isthmus during the oestrous cycle or early pregnancy. The infundibulum and ampulla, regardless of hormonal status, had a biosynthetic capacity significantly greater (2–3 times) than that of the isthmus on all days of the oestrous cycle or early pregnancy, suggesting a differential response in biosynthetic activity of the different segments.

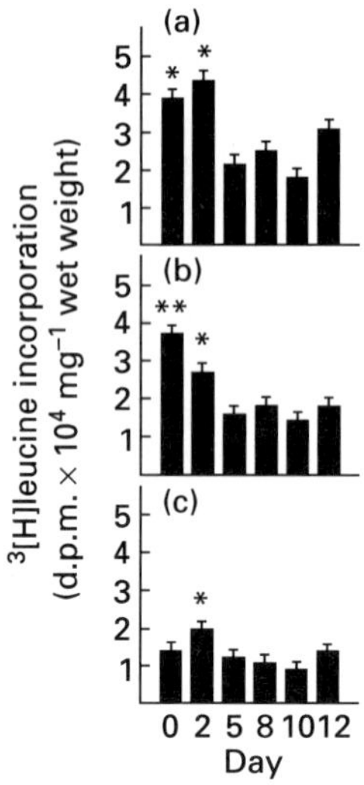

Fig. 2. Least squares means (±SEM) of [^{3}H]leucine incorporation (d.p.m. × 10^4 mg^{-1} wet tissue weight) into nondialysable macromolecules by oviductal tissue in explant culture from early pregnant Large White × Landrace gilts. Tissue collected from the (a) infundibulum, (b) ampulla and (c) isthmus. The incorporation of [^{3}H]leucine is greater by the infundibulum on days 0 and 2 (*$P < 0.05$), by the ampulla on day 0 (**$P > 0.01$) and day 2 (*$P < 0.05$) and by the isthmus on day 2 ($P < 0.05$) than on other days of early pregnancy.

Evidence from these and other studies clearly suggest that steroids, and oestrogen in particular, upregulate the biosynthetic activity of the oviduct. The ovariectomized pig was used as a model to establish that the biosynthetic activity is steroid responsive (Buhi *et al.*, 1992). The whole oviduct, infundibulum and ampulla responded with significantly greater biosynthetic activity after treatment with oestrogen compared with other treatments. The activity of whole oviduct and ampulla were greater than that of the isthmus, but the activity of the ampulla was always greater regardless of treatment.

Proteins Synthesized *De Novo* By The Oviduct

Information on a variety of proteins in oviductal fluids has been discussed in several reviews (Johnson and Foley, 1974; Hunter, 1988; and Leese, 1988). Many of these proteins, derived from serum and other fluids as a transudate, including enzymes and serum proteins such as albumin, transferrin and immunoglobulins, are present at concentrations lower than that found in serum. Total protein content, protein concentration and fluid volume in oviductal luminal fluid varies during the oestrous cycle and early pregnancy (Lippes *et al.*, 1981; Sutton *et al.*, 1984; Hunter, 1988). Increases in protein content and fluid appear to be temporally associated with increased concentrations of oestrogen. A variety of techniques using either oviductal organ explants or cultured epithelial cells have contributed to the identification and characterization of epithelial-derived proteins in the oviductal environment. In the pig oviduct, gametes and fertilized eggs rely on oviductal secretions to provide macromolecules of a nutrient, protective, or stimulatory nature through the four- to eight-cell stage. As described below, some of these proteins are synthesized in a temporal and regional specific manner during the oestrous cycle and early pregnancy.

In pigs, the synthesis *de novo* and release of 14 major radiolabelled proteins (Buhi *et al.*, 1990) have been identified and characterized by isoelectric point and relative molecular mass using two-dimensional gel electrophoresis and fluorography (Fig. 3). Most of these 14 proteins as well as other minor proteins remain to be identified. The most abundant radiolabelled secretory product is a family of oviductal-specific proteins, porcine oviductal secretory proteins (POSP) 1–3 (M_r 75 000–85 000, pI < 4; M_r 100 000, pI 4–5.5; M_r 100 000, pI > 8) the function of which is at present unknown. These proteins in the pig and similar oviduct-specific proteins identified in other species (cow, sheep, human, hamster, mouse and baboon) have been reviewed by Gandolfi (1995) and Buhi (1996). Several studies, including N-terminal amino acid sequence analysis (Buhi *et al.*, 1993; Buhi *et al.*, 1996a; W. Buhi and I. Alvarez, unpublished), carbohydrate analysis and lectin binding (W. Buhi and I. Alvarez, unpublished), western blot analysis (Buhi *et al.*, 1993), *in vitro* translation and northern blot analysis (Buhi *et al.*, 1996a) suggest that the three pig glycoproteins are related and most likely derived from a common precursor molecule. Expression of POSP is temporally associated with increased circulating concentrations of oestrogen during pro-oestrus, oestrus and early pregnancy (Buhi *et al.*, 1989; Buhi *et al.*, 1990). Studies in ovariectomized, steroid-treated gilts indicate that expression of the POSP family is oestrogen dependent and that this dependency is antagonized by progesterone (Buhi *et al.*, 1992).

The complete cDNA sequence for POSP has been determined (Buhi *et al.*, 1996a) and a full-length amino acid sequence deduced. This protein sequence exhibits a significant identity (65–78%) and similarity (78–87%) to oviductal-specific proteins from the cow (Sendai *et al.*, 1994), sheep (DeSouza and Murray, 1995), human (Arias *et al.*, 1994), mouse (Sendai *et al.*, 1995), and hamster (Suzuki *et al.*, 1995) indicating high conservation across species. In addition, sequence analyses revealed a significant similarity of POSP to the chitinase gene family. Products of this family, chitinases, are a well-studied group of enzymes the function of which is to cleave chitin, polymers of *N*-acetylglucosamine. Hydrophobic cluster analysis (Woodcock *et al.*, 1992), a secondary structure predictive method, also suggests a high similarity between POSP and bacterial chitinases. These analyses further indicate that the N-terminal domain of POSP corresponds to the catalytic domain of chitinases, whereas the C-terminal domain corresponds to a C-terminal extension which frequently represents a binding domain found in many cellulases and chitinases. These data suggest that the POSP C-terminal region, which also appears variable among the oviductal-specific proteins, may constitute a binding domain and function in both recognition and species specificity (Buhi *et al.*, 1996a). Although POSP may structurally belong to this growing mammalian chitinase gene subfamily, the N-terminal domain lacks an essential glutamic acid residue, shown to be an absolute requirement for chitinase activity (Watanabe *et al.*, 1993). Thus, while POSP lacks chitinase activity (Buhi *et al.*, 1996a) it may potentially have other carbohydrate cleavage activity.

Steady state amounts of mRNA of POSP in the functional segments during the oestrous cycle correlate with synthesis of protein and biosynthetic activity. Furthermore, the amount of mRNA encoding POSP was significantly greater in the ampulla than in the infundibulum or isthmus regardless of day, and was greatest on day 0 regardless of location. Ovariectomy and hormone replacement confirmed that expression of mRNA encoding POSP was oestrogen dependent, as was synthesis of POSP protein (Buhi *et al.*, 1996a). As with biosynthetic activity, synthesis of POSP protein was not different between cyclic and mated/early pregnant gilts confirming that mating, presence of gametes, or presence of embryos did not affect specific oviductal protein production. However, studies of synthesis of POSP protein showed both a regional distribution (ampulla > infundibulum > isthmus) and a temporal difference (days 1–3 > other days) (Buhi *et al.*, 1990).

Although the function of this protein remains unknown, there are clear indications that POSP interacts with oocytes or embryos, potentially through a binding domain. Use of immunogold electromicroscopy showed that POSP was localized throughout the zona pellucida and perivitelline space, and associated with the vitelline membrane or plasma membrane of ovulated oocytes and oviductal early embryos (Fig. 4). POSP was similarly localized in unhatched uterine embryos (four-cell to blastocyst) and with the hatched embryo only to day 7 of pregnancy (Buhi *et al.*, 1993).

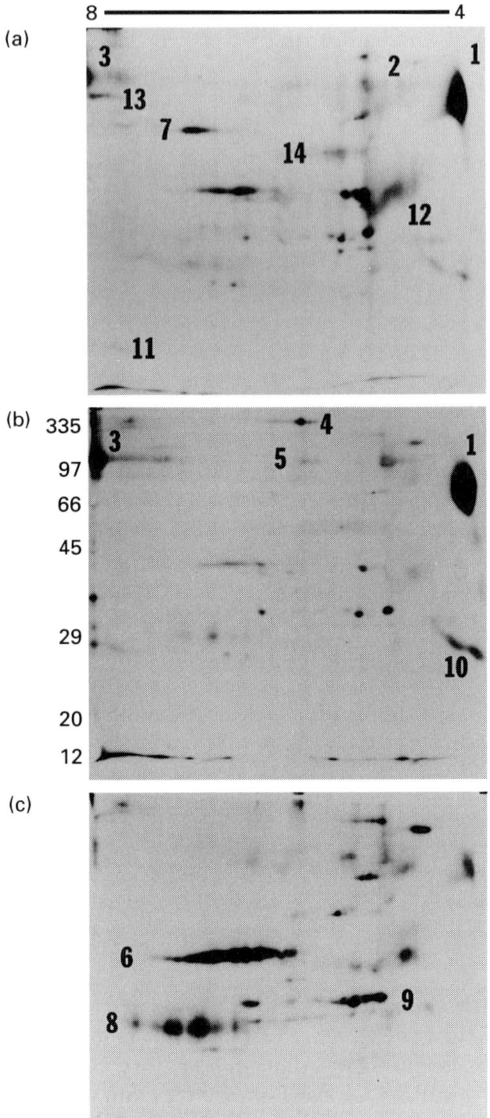

Fig. 3. Representative fluorographs after two-dimensional polyacrylamide gel electrophoretic separation of proteins from explant culture medium conditioned by the infundibulum (a), ampulla (b) or isthmus (c). Tissue was collected from Large White × Landrace gilts on day 0 of pregnancy. Fourteen proteins, characterized by M_r and pI, are indicated. Numbers 1–3, 6 and 8 represent porcine oviductal secretory protein (POSP), tissue inhibitor of metalloproteinase 1 (TIMP-1), and plasminogen activator inhibitor 1 (PAI-1), respectively. Molecular weight standards ($\times 10^3$) are indicated and the pH gradient runs from left (pH 8) to right (pH 4).

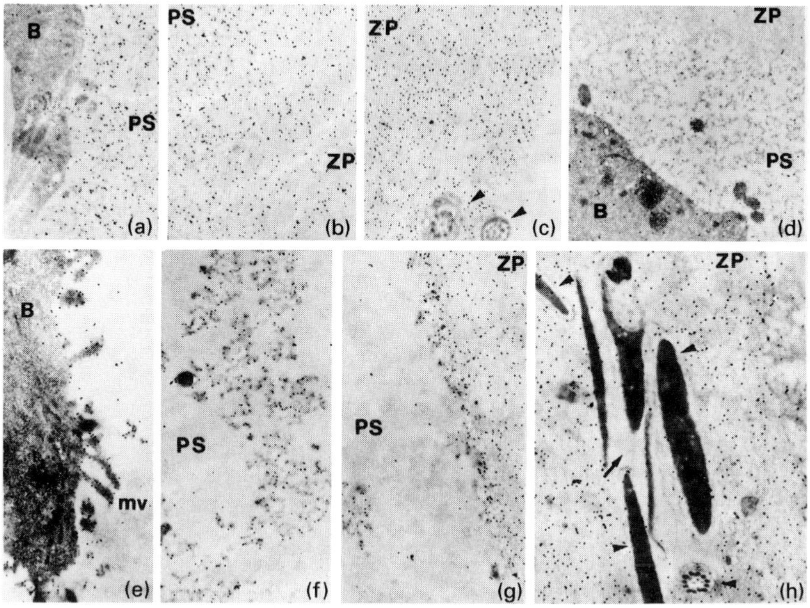

Fig. 4. Colloidal gold immunoelectron micrographs of pig oviductal (a–d) and uterine (e–h) four-cell embryos. Gold particles immunoreactive with porcine oviductal secretory protein (POSP) are associated with flocculent material (a, × 18 000; b, × 18 000) in the perivitelline space (PS) surrounding the blastomere (B) and in the zona pellucida (ZP) (b; c, × 18 000). Arrowheads (c) mark the location of sperm tails in the outer region of the ZP. Oviductal embryo treated with preimmune serum (d, × 18 000). In the four-cell uterine embryos, immunoreactive gold particles were associated with the blastomere membrane and microvilli (mv) (e, × 21 000) and flocculent material in the PS (f, × 21 000) and ZP (g, × 21 000; h, × 20 000). Sperm heads (h, arrowheads) were located in the outer region of the ZP without apparent association of POSP. The clear sperm penetration pathway can be seen (h, arrow). (Reproduced from Buhi *et al.*, 1993, with permission).

An appropriate question is whether the oocyte/embryo-associated POSP, and similar proteins, protect, signal or stimulate embryonic activity. Hill *et al.* (1996), using semi-purified ovine oviduct-specific protein with *in vitro* fertilized and developed embryos, detected improved embryo cleavage rates when these were compared with control embryos. In our laboratory, semi purified POSP and *in vitro* matured and fertilized bovine embryos were used and an increase in numbers of embryos proceeding to the four-cell and 16-cell stage was determined (A. Kouba and W. Buhi, unpublished). Wallenhaupt *et al.* (1996), using *in vivo* fertilized pig embryos and wheat germ agglutinin purified 97 kDa oviductal-specific protein, showed an increase in the rate of incorporation of methionine into protein by four-cell embryos cultured *in vitro* from the one- or two-cell stage.

In contrast to the pronounced association of POSP with oocytes or embryos, data are conflicting concerning an interaction with spermatozoa. In pig (A. Kouba and W. Buhi, unpublished) and human (Reuter *et al.*, 1994) studies, no association of spermatozoa and POSP or human oviduct-specific protein, respectively, has been detected. However, Boatman and Magnoni (1995) and King and Killian (1994) indicated that the hamster and bovine oviductal-specific protein, respectively, bound to sperm membranes.

A recent study (Buhi *et al.*, 1996b; Buhi *et al.*, 1997) reported the identification of a second major protein synthesized *de novo*, tissue inhibitor of metalloproteinase (TIMP)-1, from the oviduct (Fig. 3, Protein 8). TIMPs are specific inhibitors of matrix metalloproteinases such as collagenases, stromelysins, and gelatinases (Werb and Alexander, 1993). The amount of mRNA encodingTIMP-1 showed a significant effect of day and functional segment with greater amounts in the isthmus on day 2 than in other segments or days examined. Although TIMP-1 protein was synthesized *de novo* by all segments in cyclic and pregnant gilts, regional synthesis of TIMP-1 protein correlated with mRNA expression. In addition, TIMP-1 protein in oviductal flushings collected during the oestrous cycle, early pregnancy and after ovariectomy and steroid replacement was measured by radioimmunoassay. Significant differences in luminal fluid TIMP-1 concentrations were found by day, with day 0 being the greatest. Breed differences in TIMP-1 concentrations were also detected between standard Western and the prolific Chinese Meishan breeds. Immunostaining localized TIMP-1 in all segments of the oviduct during early pregnancy (day 0 to day 12). Furthermore, staining intensity correlated with TIMP-1 protein concentrations in oviduct flushing declining toward day 12. Steroid control of expression of TIMP-1 mRNA and protein was examined. Steady state amounts of mRNA encoding TIMP-1 were significantly increased in ovariectomized gilts treated with oestrogen as were TIMP-1 protein synthesis and TIMP-1 protein concentration in oviductal fluid. In contrast, in intact pregnant gilts, the expression of TIMP-1 mRNA and TIMP-1 protein synthesis were greater with basal concentrations of oestrogen. Therefore, the increased concentration of TIMP-1 protein in oviductal fluid at day 0 was proposed to be due to some TIMP-1 protein being derived from serum transudate.

Although the role of TIMP-1 in the pig oviduct and with pig oocytes/embryos is unknown, this protein has the capability to function as a growth promoter or to participate in control of extracellular matrix remodelling. Satoh *et al.* (1994) suggested that TIMP-1 functions as an embryogenic stimulator of *in vitro* fertilized bovine embryos. Other studies have proposed regulatory roles in remodelling and as a growth promoter in the ovary (Smith *et al.*, 1994). Addition of TIMP-1 to pig oocytes during *in vitro* maturation has been shown to improve cleavage rates after fertilization and during culture (Funahashi *et al.*, 1997).

A third major oviductal protein (Fig. 3, protein 6) synthesized and released during the oestrous cycle and early pregnancy has been identified by N-terminal amino acid microsequence analysis as pig plasminogen activator inhibitor (PAI)-1, a member of the serpin family of serine protease inhibitors (Kouba *et al.*, 1997). Although PAI-1 is produced by all oviductal segments, it appears to be differentially expressed, with greater expression of protein by the isthmus than the ampulla or infundibulum. Comparison of PAI-1 synthesis by gilts in early pregnancy by two-dimensional polyacrylamide gel electrophoresis and fluorography confirmed significant differences by segment and between breeds (Kouba *et al.*, 1997).

Both inhibitors, TIMP-1 and PAI-1, and potentially other unidentified protease inhibitors most likely serve to regulate the oviductal microenvironment and protect oocytes and embryos from proteolytic damage by enzymes in the lumen. Mechanistically, they act by preventing activation of enzymes or by binding to active proteinases to form a proteinase–inhibitor complex. These proteins may also be involved in maintaining a non-invasive environment, control extracellular matrix production and remodelling, development of blastomeres, or prevent premature hatching of embryos. As noted by Broermann *et al.* (1989), pig follicular oocytes and embryos exposed to the oviductal microenvironment are more resistant to *in vitro* proteolytic degradation than those not exposed to the oviduct or those recovered from the uterine environment.

Retinol-binding protein, which functions in the transport of retinol, can be found as a minor protein after two-dimensional gel electrophoretic analysis of oviduct-conditioned culture media (W. Buhi and I. Alvarez, unpublished). Harney *et al.* (1994) demonstrated by northern blot analysis that mRNA encoding RBP is present in the oviduct at day 0.

In addition to these known proteins, at least 11 others are yet to be identified. Some of these 11 proteins appear to be differentially expressed, with synthesis and release being greater in the ampulla than in other segments. Others appear to be regulated by oestrogen or progesterone.

Growth Factors and the Oviduct

The precise role of growth factors or cytokines in oviductal function, fertilization and early embryonic development is unclear and remains intriguing. However, it is clear that gametes and

Table 1. Growth factor, cytokine and receptor expression by the oviduct during the oestrous or menstrual cycle or early pregnancy

Factor	Gene expression	Protein expression
IGF-I	b, o, h, m, r	b, h, m, p
IGF-II	b, o, h, m, r, p	b
IGF-IR	o	p, h
IGF-BP (1–6)	h, b (2–5)	h, p (1–4), b (2–5)
EGF	h, m, p	h, m, p, bab
TGFα	b, o, h, m, p	h, m, p, bab
EGF-R	h, p	h, p, bab
TGFβ1	o, h, m	h, m, p
β2	h, m	h, m, p
β3	h, m	h
TGFβ (type I-III)R	h	h, p (Type II)
HB-EGF	h, m	h, m
bFGF	b, o	–
PDGF-AA, BB	h	h
PDGF α, β R	h	h
PDGF-B	b	b
Activin B_A	b, m (also β_B)	b, m (also β_B and α)
TNFα	m	—
CSF-1	m, p	m, p
GM-CSF	h	h
GM-CSF α & β R	h	h

b, bovine; o, ovine; h, human; m, mouse; r, rat ; p, porcine; bab, baboon.

preimplantation embryos of many species reside in an oviductal microenvironment in which they are exposed to growth factors and cytokines. Potentially, these factors operate in an autocrine and paracrine manner to regulate or modulate oviductal function (Gandolfi, 1995; Chegini, 1996) and/or embryonic and fetal growth and development (Schultz and Heyner, 1993). There is a growing body of evidence that a variety of growth factors markedly improve embryo development in culture (Gandolfi, 1995). Development is effected by increased protein and RNA synthetic rates, increased cell division, increased number of cells and increased blastocyst expansion and zona hatching rates (see Watson *et al.*, 1994).

The source of growth factors in the oviduct of the pig and in other species, as with other proteins, is most likely from synthesis by the luminal secretory epithelium as well as from serum transudate. Investigations into the role of growth factors with early embryos stimulated identification of a number of growth factors within the oviductal environment (see Table 1). Expression of mRNA encoding epidermal growth factor (EGF) in the oviduct was measured as early as day 0 and day 1 of oestrus (W. Buhi, unpublished), while other studies have shown the presence of mRNA encoding EGF, EGF receptor (R), amphiregulin, and transforming growth factor (TGF)α on days 4 to 20 of the oestrous cycle in the pig (Kennedy *et al.*, 1994). Swanchara *et al.* (1995) identified and quantitated EGF after extraction of oviductal tissue and collecting flushings in cyclic and pregnant gilts at 48 h after the onset of oestrus. Both EGF and TGFα protein were measured in oviductal flushings from mated and early pregnant (days 0, 2, 5, 8, 10 and 12) gilts (W. Buhi and J. Diehl, unpublished) by radioimmunoassay. Although differences in concentration were not apparent for EGF, TGFα concentrations were significantly greater on day 0 than on other days. In ovariectomized gilts given steroid replacement, TGFα concentrations in oviductal fluids were significantly greater with oestrogen treatment than with any other hormonal treatment. Immunocytochemical studies have localized TGFα, EGF and the EGF-R to epithelial cells of the infundibulum, ampulla and isthmus during the oestrous cycle, early pregnancy (Fig. 5a) and after ovariectomy and hormone replacement (Fig. 5b,c) in gilts. Similarly, Swanchara *et al.* (1995) localized EGF and EGF-R in the ampulla and

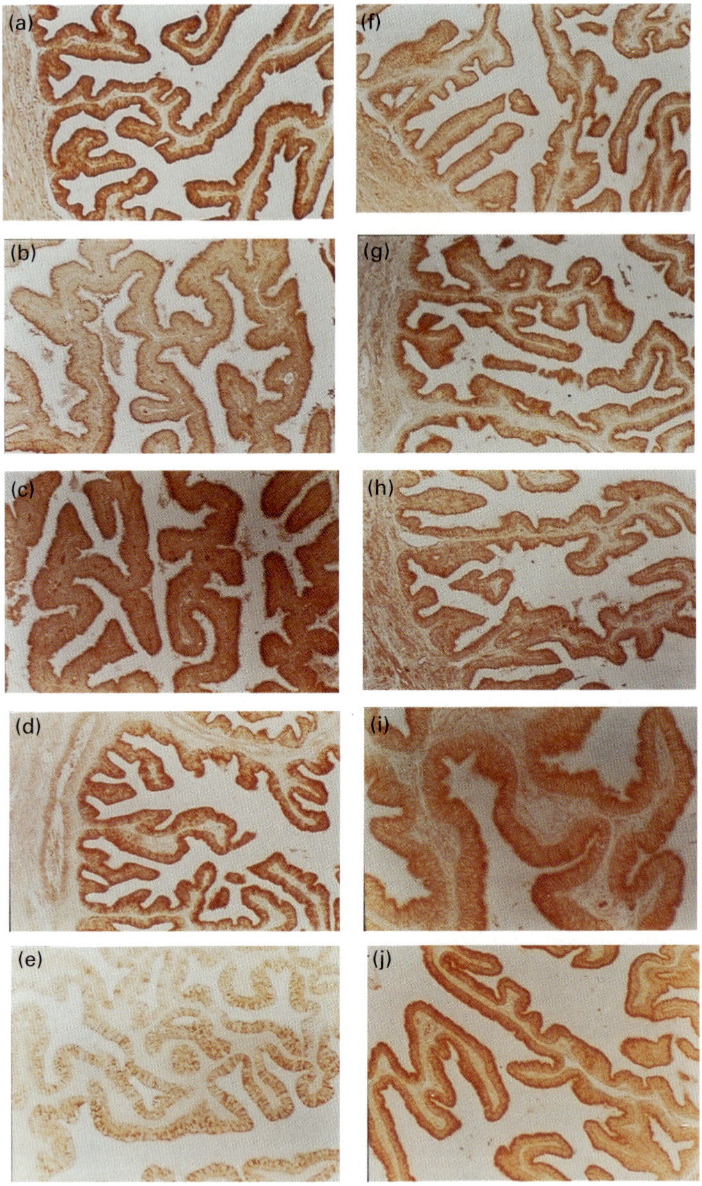

Fig. 5. Immunohistochemical localization of TGFα (a), EGF (b), EGF-R (c), IGF-I (d), IGFBP 1–3 (f–h), TGFβ1 (e), TGFβ2 (i), and TGFβ Type II R (j) in pig oviductal tissue from a variety of reproductive states. Immunoreactive TGFα (a), IGF-I (d), and IGFBP 1–3 (f–h) were localized in the ampullary epithelium on day 0 of pregnancy (IGFBP 4 not shown). TGFβ1 (e), TGFβ2 (i) and TGFβ Type II R (j) were similarly localized in epithelium of ampullary tissue collected on day 1 of the oestrous cycle while EGF(b) and EGF-R(c) were localized in ampullary tissue from gilts ovariectomized and treated with 100 μg oestradiol valerate for 11 consecutive days.

isthmus at day 2 of the oestrous cycle or pregnancy. While the function of EGF and TGFα in the pig remains to be determined, promotion of blastocyst formation and stimulation of embryo development have been shown in co-culture studies with *in vitro* fertilized embryos (mouse and bovine) and oviductal cells, producing EGF and TGFα, or without oviductal cells and added EGF (Morishige *et al.*, 1993; Lonergan *et al.*, 1996).

Insulin-like growth factors (IGF-I and IGF-II) regulate a variety of biological activities in many types of cell (Jones and Clemmons, 1994) and the actions of IGFs have been well characterized in pig uterine secretions (Simmen *et al.*, 1993). Wiseman *et al.* (1992), using radioimmunoassay to measure IGF-I and IGF-II, demonstrated a higher content (concentration × fluid volume) of both factors in oviductal fluid collected at oestrus than at non-oestrous periods. IGF-I and IGF-II have also been measured in primary culture media collected after 48 h from oviductal epithelium at day 2. Oviductal flushes from early pregnant gilts showed a significant increase in IGF-I concentration on days 0 and 2 compared with day 5 while a study with ovariectomized, steroid-treated gilts showed that IGF-I concentration in oviductal flushings was significantly greater with oestrogen treatment when compared with all other treatments (W. Buhi, unpublished). IGF-I has also been immunolocalized in the epithelium of the infundibulum, ampulla and isthmus during early pregnancy (Fig. 5d). IGF transport and function are also modulated by IGF -binding proteins (BP) (Jones and Clemmons, 1994). In the pig, IGFBP 1–4 were immunolocalized on epithelial cells of the infundibulum, ampulla and isthmus from day 0 to day 12 of pregnancy (Fig. 5f–h). Staining intensity of IGFBP-1 and -2 was greater than that of IGFBP-3 and -4. These data indicate that IGFBPs are present and may modulate IGF activity in the oviduct. While a potential role for IGF-I has yet to be determined in the pig oviduct, IGF-I and insulin have each been shown to increase the percentage of bovine embryos developing to morulae by day 5 (Matsui *et al.*, 1995).

Transforming growth factors (TGF)β1, TGFβ2 and TGFβ type II receptor were localized by immunocytochemistry throughout the pig oviduct in all functional segments during the oestrous cycle and early pregnancy (Fig. 5e,i,j). Staining intensity of TGFβ1 and TGFβ type II receptor tended to decrease at dioestrus and increase at pro-oestrus, supporting hormonal control of these (Buhi *et al.*, 1994). However, it is not known whether these factors are available to the developing embryo in oviductal fluid. Transcripts for TGFβ type II receptor were also detected by RT–PCR in oviductal RNA collected on days 0 and 1 of oestrus (W. Buhi, unpublished), while TGFβ1 transcripts have been found in cultured ovine oviductal epithelial cells (Watson *et al.*, 1994). One role for TGFβs may be to modulate or regulate extracellular matrix proteins. A second possible role in the pig oviduct can be postulated from studies in cows which showed that TGFβ, when combined with basic fibroblast growth factor, promoted the formation of blastocysts *in vitro* (Larson *et al.*, 1992a).

Although not yet examined in pigs, expression of colony-stimulating factor (CSF-1), a haematopoietic growth factor, has been found in mouse oviducts during the first five days of pregnancy (Pampfer *et al.*, 1991; Pollard, personal communication). In pigs, immunoreactive CSF-1 protein has been localized on luminal epithelium in early pregnant gilts (not shown). In addition, Tuo *et al.* (1995) detected CSF-1 transcripts in the pig oviduct in late dioestrus by northern blot analysis.

Transcripts and proteins for the above factors have also been determined by northern blot analysis, RT-PCR, *in situ* hybridization and immunoreactive studies in the bovine, human and mouse oviduct (Watson *et al.*, 1992; Chegini, 1996; Dalton *et al.*, 1994) (Table 1). Other growth factors found in the oviduct include platelet-derived growth factor (PDGF)-β and basic fibroblast growth factor (Viuff *et al.*, 1995), activin beta A (Gandolfi *et al.*, 1995) and granulocyte–macrophage (GM)-CSF and GM-CSF receptor (Chegini, 1996). However, it remains to be shown whether these factors are present in the pig oviduct, whether they are available to developing embryos and, if so, what their function may be. One role reported for PDGF was its promotion of bovine embryo development beyond the eight-cell stage (Larson *et al.*, 1992b).

Whether there is a direct effect of growth factors and cytokines on pig oocytes or embryos from fertilization through the 4- to 8-cell stage is not clear. At this developmental stage, the presence of specific receptors on pig oocytes and embryos has not yet been determined. However, one study has reported expression of mRNA encoding EGF-R in pig embryos as early as day 7 of pregnancy

(Vaughan *et al.*, 1992). Similarly, transcripts for insulin and IGF-I receptors were detected in ovine embryos from the one-cell to blastocyst stage (Watson *et al.*, 1994), while transcripts for IGF-I, IGF-II, IGFBP2-4 and PDGF-α receptor were found at all stages of early bovine development from mature oocyte to the blastocyst (Watson *et al.*, 1992; Winger *et al.*, 1997). Thus, these data suggest the potential for autocrine and paracrine growth factor interactions between the oviduct and early cleavage-stage embryos.

Co-culture Studies, Spermatozoa and Zona Pellucida

Although embryos will develop *in vitro*, they will not develop to the same potential when cultured in chemically defined media, or with uterine or other cell types as *in vivo* (Murray *et al.*, 1995). Co-culture experiments with embryos *in vitro* or embryo transfer studies in several species have established the importance of oviductal secretions, oviductal-specific proteins and oviductal epithelial cells for enhancing fertilization and embryonic development (Petters and Wells, 1993). Specifically, pig oviductal secretions (Archibong *et al.*, 1989), oviductal epithelial cells (White *et al.*, 1989), and organ (oviduct) culture (Krisher *et al.*, 1989) were shown to be beneficial for embryo development.

Brown and Cheng (1986) and Hedrick *et al.* (1987) have shown that oviductal glycoproteins were bound to the zona pellucida in pigs after ovulation and fertilization, and during early embryonic development. Recent studies have shown that POSP associates with the zona pellucida (Buhi *et al.*, 1993). Studies by Broermann *et al.* (1989) indicated that pig follicular oocytes or embryos exposed to the oviduct microenvironment were more resistant to proteolysis than oocytes without oviductal exposure or embryos collected from the uterine environment, suggesting the uptake of protective agents from the oviduct. The presence of two identified major protease inhibitors in oviductal fluid and other, as yet unidentified, proteinase inhibitor(s) may account for this increased resistance to proteases by oocytes and embryos (Buhi *et al.*, 1997; Kouba *et al.*, 1997).

Polyspermy is a major complication during *in vitro* fertilization of pig oocytes. Nagai and Moor (1990) reduced the incidence of polyspermy with pig oocytes by incubation of oocytes and spermatozoa with oviductal secretory molecules. Co-culturing spermatozoa with oviductal cells for only 30 min was shown to reduce polyspermy (Dubuc and Sirard, 1995). With natural breeding, sperm interact with oviductal cells, spending up to 36 h at the tubo–uterine junction associated with isthmic epithelium (Hunter, 1988). Incubation of boar spermatozoa with oviductal explants show an intimate association of spermatozoa with epithelium which is independent of oviductal region or day of oestrous cycle (Suarez *et al.*, 1991). A second study observed spermatozoa binding to the isthmus more readily than to the ampulla (Raychoudhury and Suarez, 1991). This binding may be through specific receptors or glycoprotein interactions. Raychoudhury *et al.* (1993) used a variety of lectins to show distribution of glycosylated molecules on the oviductal mucosa at oestrus and dioestrus and after ovariectomy and steroid treatment. The patterns of lectin binding varied with hormonal state and region. In other experiments, De Mott *et al.* (1995) indicated that sialylated oligosaccharides may mediate binding of hamster spermatozoa to oviductal epithelium and suggested that release of spermatozoa may be accomplished by modification of binding components. Furthermore, binding of spermatozoa by oviductal cells was also shown to increase motility and penetration when compared with spermatozoa that had not bound (Dubuc and Sirard, 1995). Two additional studies suggested a sperm–oviductal protein interaction. Boatman and Magnoni (1995) showed that the hamster oviduct-specific glycoprotein enhanced binding of spermatozoa to the zona pellucida and the zona pellucida-induced acrosome reaction. In the second study, O'Day-Bowman *et al.* (1996) showed that incubation of human oviduct-specific glycoprotein with human spermatozoa and hemizona increased binding of spermatozoa to the hemizona. These data clearly indicate a role for the oviduct or its secretions in enhancing sperm–oocyte interactions.

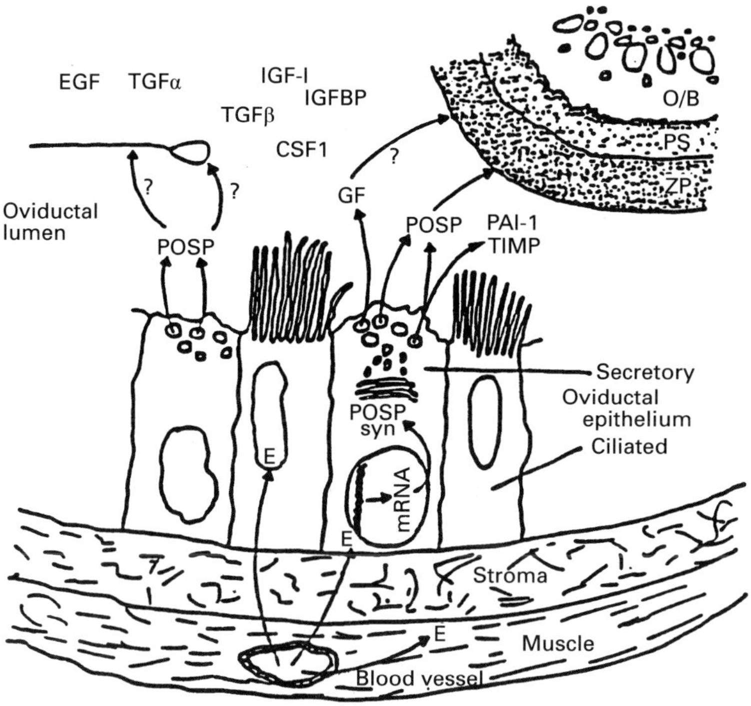

Fig. 6. Ciliated and secretory epithelium achieve maximum development, biosynthetic capacity and secretory activity with oestrogen during the periods of folliculogenesis, ovulation, and early cleavage-stage embryonic development. The secretory products that are contributed to the oviductal luminal microenvironment potentially operate in an autocrine and/or paracrine manner to regulate oviductal and embryonic growth and development.

Conclusions

Studies of the pig oviduct have provided information on its biological (synthetic) activity and a variety of protein secretory molecules contributed to and found within the oviductal microenvironment during periods of ovulation, fertilization and early cleavage-stage embryonic development (Fig. 6). POSP and the equivalent oviduct-specific proteins found in other species with a high conservation of structure suggest important roles in oviductal events. Important questions remain concerning identification of oviductal-derived proteins, purification of individual proteins, their regulation and their biological function. For example, how do those molecules, contributed in a temporal and regional specific manner to oviductal luminal fluid, participate in events of fertilization and early cleavage-stage embryonic development? Although growth factors are multifunctional regulators of cellular activities, a direct demonstration of their involvement with early cleavage-stage (one- to eight-cell stage) embryonic development is lacking in pigs. However, *in vitro* studies in several other species suggest the potential importance of these factors on embryonic development during this time. Complementary studies, which use individually purified or combined oviductal-derived proteins and growth factors in an *in vitro* fertilization and culture system, as well as methods that inhibit expression of these agents *in vivo*, will be necessary to establish function and understand their mechanism of action.

This paper was supported by grants from the USDA (92-37203-7996 to W. C. Buhi and F. A. Simmen and 95-37203-2308 to W. C. Buhi). The authors thank B. Cleaver for statistical analysis and G. Erdos and R. Davis of the Electronmicroscopy Core Laboratory.

References

Abe H and Oikawa T (1992) Examination by scanning electron microscopy of oviductal epithelium of the prolific Chinese Meishan pig at follicular and luteal phases *Anatomical Record* **233** 399–408

Archibong AE, Petters RM and Johnson BH (1989) Development of porcine embryos from one- and two-cell stages to blastocysts in culture medium supplemented with porcine oviductal fluid *Biology of Reproduction* **41** 1076–1083

Arias EB, Verhage HG and Jaffee RC (1994) Complementary deoxyribonucleic acid cloning and molecular characterization of an estrogen-dependent human oviductal glycoprotein *Biology of Reproduction* **51** 685–694

Boatman DE and Magnoni GE (1995) Identification of a sperm penetration factor in the oviduct of the golden hamster *Biology of Reproduction* **52** 199–207

Broermann DM, Xie S, Nephew KP and Pope WF (1989) Effects of the oviduct and wheat germ agglutinin on enzymatic digestion of porcine zona pellucidae *Journal of Animal Science* **67** 1324–1329

Brown CR and Cheng WKT (1986) Changes in composition of the porcine zona pellucida during development of the oocyte to the 2- to 4-cell embryo *Journal of Embryology and Experimental Morphology* **77** 411–417

Buhi WC (1996) Cyclic changes in the oviduct during fertilization and early cleavage-stage embryonic development *Archives of Animal Breeding* **39** 15–25

Buhi WC, Vallet JL and Bazer FW (1989) De novo synthesis and release of polypeptides from cyclic and early pregnant porcine oviductal tissue in explant culture *The Journal of Experimental Zoology* **252** 79–88

Buhi WC, Alvarez IM, Sudhipong V and Dones-Smith MM (1990) Identification and characterization of de novo-synthesized porcine oviductal secretory proteins *Biology of Reproduction* **43** 929–938

Buhi WC, Ashworth CJ, Bazer FW and Alvarez IM (1992) *In vitro* synthesis of oviductal secretory proteins by estrogen-treated ovariectomized gilts *The Journal of Experimental Zoology* **262** 426–435

Buhi WC, O'Brien B, Alvarez IM, Erdos G and Dubois D (1993) Immunogold localization of porcine oviductal secretory proteins within the zona pellucida, perivitelline space, and plasma membrane of oviductal and uterine oocytes and early embryos *Biology of Reproduction* **48** 1274–1283

Buhi WC, Chegini N, Rong H and Alvarez I (1994) Immunolocalization of transforming growth factor (TGF-β) TGF-β1, TGF-β2, and TGF-β type II receptor protein in the porcine oviduct *Society for Gynecological Investigation* Chicago

Buhi WC, Alvarez IM, Choi I, Cleaver BD and Simmen FA (1996a) Molecular cloning and characterization of an estrogen-dependent porcine oviductal secretory glycoprotein *Biology of Reproduction* **55** 1305–1314

Buhi WC, Alvarez IM, Pickard AR, Ashworth CJ, McIntush EW and Smith MF (1996b) Tissue inhibitor of metalloproteinase (TIMP)-1 in the porcine oviduct during early pregnancy (EP) and the estrous cycle (EC) *Biology of Reproduction* **54** (Supplement 1) 103

Buhi WC, Alvarez IM, Pickard AR, McIntush EW, Kouba AJ, Ashworth CJ and Smith MF (1997) Expression of tissue inhibitor of metalloproteinase (TIMP)-1 protein and mRNA by the oviduct of cyclic, early pregnant, and ovariectomized steroid-treated gilts *Biology of Reproduction* **57** 7–15

Chegini N (1996) Oviductal-derived growth factors and cytokines: implication in preimplantation *Seminars in Reproductive Endocrinology* **14** 219–229

Dalton T, Kover K, Dey SK and Andrews GK (1994) Expression of growth factor, interleukin-1, and lactoferrin genes and the distribution of inflammatory leukocytes in the preimplantation mouse oviduct *Biology of Reproduction* **51** 597–606

DeMott RP, Lefebvre R and Suarez SS (1995) Carbohydrates mediate the adherence of hamster sperm to oviductal epithelium *Biology of Reproduction* **52** 1395–1403

DeSouza MM and Murray MK (1995) An estrogen-dependent secretory protein which shares identity with chitinases, is expressed in a temporally and regionally specific manner in the sheep oviduct at the time of fertilization and embryo development *Endocrinology* **136** 2485–2496

Dubuc A and Sirard MA (1995) Effect of coculturing spermatozoa with oviductal cells on the incidence of polyspermy in the pig *in vitro* fertilization *Molecular Reproduction and Development* **41** 360–367

Funahashi H, McIntuch EW, Smith MF and Day BN (1997) Effect of tissue inhibitor of metalloproteinases (TIMP-1) on early development of swine oocytes matured and fertilized *in vitro. Theriogenology* **47** 277

Gandolfi F (1995) Functions of proteins secreted by oviductal cells *Microscopy Research and Technique* **32** 1–12

Gandolfi F, Modina S, Brevini TA, Passoni L, Artini P, Petraglia F and Lauria A (1995) Activin beta A subunit is expressed in bovine oviduct *Molecular Reproduction and Development* **40** 286–291

Harney JP, Smith LC, Simmen RCM, Fliss AE and Bazer FW (1994) Retinol-binding protein: immunolocalization of protein and abundance of messenger ribonucleic acid in conceptus and maternal tissues during pregnancy in pigs *Biology of Reproduction* **50** 1126–1135

Hedrick JL, Wardrip NJ and Berger T (1987) Differences in the macromolecular composition of the zona pellucida isolated from pig oocytes, eggs and zygotes *The Journal of Experimental Zoology* **241** 257–262

Hill JL, Walker SK, Brown GH and Nancarrow CD (1996) The effects of an estrus-associated oviductal glycoprotein on the *in vitro* fertilization and development of ovine oocytes matured *in vitro. Theriogenology* **46** 1379–1388

Hunter RHF (1988) *The Fallopian Tubes: Their Role In Fertility And Infertility* Springer-Verlag, Berlin

Johnson A and Foley (1974) *The Oviduct and its Function* Academic Press, New York

Jones J and Clemmons D (1994) Insulin-like growth factors and their binding proteins: biological actions *Endocrine Reviews* **16** 3–34

Kennedy TG, Brown KD and Vaughan TJ (1994) Expression of genes for the epidermal growth factor receptor and its ligands in porcine oviduct and endometrium *Biology of Reproduction* **50** 751–756

King RS and Killian GJ (1994) Purification of bovine estrus-associated protein and localization of binding on sperm *Biology of Reproduction* **51** 34–42

Kouba AJ, Alvarez IM and Buhi W (1997) Purification and characterization of a 45 kDa protein synthesized *de novo* by oviductal tissue from early pregnant and ovariectomized (OVX) steroid-treated gilts *Biology of Reproduction* **56** (Supplement 1) 187

Krisher RL, Petters RM, Johnson BH, Bavister BD and Archibong AE (1989) Development of porcine embryos from one-cell stage to blastocyst in mouse oviducts maintained in organ culture *The Journal of Experimental Zoology* **249** 235–239

Larson RC, Ignotz GG and Currie WB (1992a) Transforming growth factor β and basic fibroblast growth factor synergistically promote early bovine embryo development during the fourth cell cycle *Molecular Reproduction and Development* **33** 432–435

Larson RC, Ignotz GG and Currie WB (1992b) Platelet derived growth factor (PDGF) stimulates development of bovine embryos during the fourth cell cycle *Development* **115** 821–826

Leese HJ (1988) The formation and function of oviduct fluid *Journal of Reproduction and Fertility* **82** 843–856

Lippes J, Krasner J, Alfonso LA, Dacalos ED and Lucero R (1981) Human oviductal fluid proteins *Fertility and Sterility* **36** 623–629

Lonergan P, Carolan C, Van Langendonckt A, Donnay I, Khatir H and Mermillod P (1996) Role of epidermal growth factor in bovine oocyte maturation and preimplantation embryo development *in vitro. Biology of Reproduction* **54** 1420–1429

Matsui M, Takahashi Y, Hishinuma M and Kanagawa H (1995) Insulin and insulin-like growth factor-I (IGF-I) stimulate the development of bovine embryos fertilized *in vitro. Journal of Veterinary Medical Science* **57** 1109–1111

Morishige K, Kurachi H, Amemiya K, Adachi H, Adachi K, Sakoyama Y, Miyake A and Tanizawa O (1993) Menstrual stage-specific expression of epidermal growth factor and transforming growth factor-alpha in human oviduct epithelium and their role in early embryogenesis *Endocrinology* **133** 199–207

Murray MK (1992) Biosynthesis and immunocytochemical localisation of an estrogen-dependent glycoprotein and associated morphological alterations in the sheep ampulla oviduct *Biology of Reproduction* **47** 889–902

Murray MK, DeSouza MM and Messinger SM (1995) Oviduct during early pregnancy: hormonal regulation and interactions with the fertilized ovum *Microscopy Research and Techniques* **31** 497–506

Nagai T and Moor RM (1990) Effect of oviductal cells on the incidence of polyspermy in pig eggs fertilized *in vitro. Molecular Reproduction and Development* **26** 377–382

Nayak RK and Zimmerman DR (1971a) Ultrastructural changes in porcine oviduct epithelium during the estrous cycle *Journal of Animal Science* **33** 262

Nayak RK and Zimmerman DR (1971b) Effect of estrogen and progesterone on the ultrastructure of porcine oviduct epithelium *Journal of Animal Science* **33** 1161

O'Day-Bowman MB, Mavrogianis PA, Renter LM, Johnson DE, Fazleabas AT and Verhage HG (1996) Association of oviduct-specific glycoproteins with human and baboon (*Papio anubis*) ovarian oocytes and enhancement of human sperm binding to human hemizona following *in vitro* incubation *Biology of Reproduction* **54** 60–69

Pampfer S, Arceci RJ and Pollard JW (1991) Role of colony stimulating factor-1 (CSF-1) and other lympho–hematopoietic growth factors in mouse preimplantation development *BioEssays* **13** 535–540

Petters RM and Wells KD (1993) Culture of pig embryos *Journal of Reproduction and Fertility Supplement* **48** 61–73

Raychoudhury SS and Suarez SS (1991) Porcine sperm binding to oviductal explants in culture *Theriogenology* **36** 1059–1069

Raychoudhury SS, Suarez SS and Buhi WC (1993) Distribution of lectin binding sites in the oviducts of cycling and hormone-treated pigs *The Journal of Experimental Zoology* **265** 659–668

Reuter LM, O'Day-Bowman MB, Mavrogianis PA, Fazleabas AT and Verhage HG (1994) *In vitro* incubation of golden (Syrian) hamster ovarian oocytes and human sperm with a human oviduct specific glycoprotein *Molecular Reproduction and Development* **38** 160–169

Roy SR, Jr, Gupta DN, Roy SK, Karkun JN and Kar AB (1972) Studies in physiology and biochemistry of the fallopian tube. *In vitro* uptake of DL-valine-1-^{14}C by different parts of the rabbit fallopian tube during early pregnancy *Endokrinologie* **59** 221–226

Satoh T, Kobayashi K, Yamashita S, Kikuchi M, Sendai Y and Hoshi H (1994) Tissue inhibitor of metalloproteinases (TIMP-1) produced by granulosa and oviduct cells enhances *in vitro* development of bovine embryos *Biology of Reproduction* **50** 835–844

Schultz GA and Heyner S (1993) Growth factors in preimplantation mammalian embryos *Oxford Reviews of Reproductive Biology* **15** 43–81

Sendai Y, Abe H, Kikuchi M, Satoh T and Hoshi H (1994) Purification and molecular cloning of bovine oviduct specific glycoprotein *Biology of Reproduction* **50** 927–934

Sendai Y, Komiya H, Suzuki K, Onuma T, Kikuchi M, Hoshi H and Araki Y (1995) Molecular cloning and characterization of a mouse oviduct-specific glycoprotein *Biology of Reproduction* **53** 285–294

Simmen RCM, Ko Y and Simmen FA (1993) Insulin-like growth factors and blastocyst development *Theriogenology* **39** 163–175

Smith GW, Goetz TL, Anthony RV and Smith MF (1994) Molecular cloning of an ovarian tissue inhibitor of metalloproteinases: ontogeny of messenger ribonucleic acid expression and in situ localization within preovulatory follicles and luteal tissue *Endocrinology* **134** 344–352

Suarez S, Redfern K, Raynor P, Martin F and Phillips DM (1991) Attachment of boar sperm to mucosal explants of oviduct *in vitro*: possible role in formation of a sperm reservoir *Biology of Reproduction* **44** 998–1004

Sutton R, Nancarrow CD, Wallace ALC and Rigby NW (1984) Identification of an oestrus-associated glycoprotein in oviductal fluid of the sheep *Journal of Reproduction and Fertility* **72** 415–422

Suzuki K, Sendai Y, Onuma T, Hiroyoshi H, Hiroi M and Araki Y (1995) Molecular characterization of a hamster oviduct-specific glycoprotein *Biology of Reproduction* **53** 345–354

Swanchara KW, Henricks DM, Birrenkott GP, Bodine AB and Richardson ME (1995) Expression of epidermal growth factor (EGF) and the EGF receptor in the porcine oviduct *Biology of Reproduction* **53** 911–922

Tuo W, Harney JP and Bazer FW (1995) Colony-stimulating factor-1 in conceptus and uterine tissue in pigs *Biology of Reproduction* **53** 133–142

Vaughan TJ, James PS, Pascall JC and Brown KD (1992) Expression of the genes for TGFα, EGF and the EGF receptor during early pig development *Development* **116** 663–669

Viuff D, Hyttel P, Greve T, Eriksen T and Alexandersen S (1995) Transcription and localization of growth factor mRNA in the bovine oviduct *Molecular Reproduction and Development* **42** 28–34

Wallenhaupt K, Alm H, Tomek W and Brüssow K-P (1996) Studies on the influence of a specific 97 kd porcine oviductal secretory protein on the de novo protein synthesis of preimplantation embryos *Archives of Animal Breeding* **39** 74

Watanabe T, Kobori K, Miyashita K, Fujii T, Saki H, Uchida M and Tanaka H (1993) Identification of glutamic acid 204 and aspartic acid 200 in chitinase A1 of *Bacillus circulans* WL-12 as essential residues for chitinase activity *Journal of Biological Chemistry* **268** 18567–18572

Watson AJ, Hogan A, Hahnel A, Weimer KE and Schultz GA (1992) Expression of growth factor ligand and receptor genes in the preimplantation bovine embryo *Molecular Reproduction and Development* **31** 87–95

Watson AJ, Watson PH, Arcellana-Panlilio M, Warnes D, Walker SK, Schultz GA, Armstrong DT and Seamark RF (1994) A growth factor phenotype map for ovine preimplantation development *Biology of Reproduction* **50** 725–733

Werb Z and Alexander CM (1993) Proteinases and matrix degradation. In *Textbook of Rheumatology* pp 248–268 Eds WN Kelley, ED Harris, Jr, S Ruddy and CB Sledge. WB Saunders, Philadelphia

White KL, Hehnke K, Rickords LF, Southern LL, Thompson DL, Jr and Wood TC (1989) Early embryonic development *in vitro* by coculture with oviductal epithelial cells in pigs *Biology of Reproduction* **41** 425–430

Winger QA, de los Rios P, Han VKM, Armstrong DT, Hill DJ and Watson AJ (1997) Bovine oviductal and embryonic insulin-like growth factor binding proteins: possible regulators of "embryotrophic" insulin-like growth factor circuits *Biology of Reproduction* **56** 1415–1423

Wiseman DL, Henricks DM, Eberhardt DM and Bridges WC (1992) Identification and content of insulin-like growth factors in porcine oviductal fluid *Biology of Reproduction* **47** 126–132

Woodcock S, Mormon J-P and Henrissat B (1992) Detection of secondary structural elements in proteins by hydrophobic cluster analysis *Protein Engineering* **5** 629–635

Wu ASH, Carlson SD and First N (1976) Scanning electron microscopic study of the porcine oviduct and uterus *Journal of Animal Science* **42** 804–809

Journal of Reproduction and Fertility Supplement **52**, 301–312

Cryopreservation of pig embryos

J. R. Dobrinsky

Germplasm & Gamete Physiology Laboratory, Agricultural Research Service US Department of Agriculture, Beltsville, MD 20705, USA

The changing global needs for food and animal products require the development of breeding strategies for maximizing genetic improvement while maintaining genetic diversity. Genetic diversity can be conserved by using separate breeding herds; however, they may be expensive to maintain and inbreeding becomes a major concern. Alternative methods are needed to preserve valuable genetic resources in a reasonable and economic manner. Embryo cryopreservation allows indefinite storage *in vitro* at subambient temperatures where metabolism and other cellular functions are greatly reduced or cease, and upon recovery from storage, normal developmental competence can be resumed. Storage and transportation require little maintenance and there is no expense in animal care and little concern about disease transmission. Although there are methods for routine cryopreservation of germplasm and embryos of most livestock species, development of this technology in the pig industry is far behind and has abated improvements in genetic potential. Pig embryos are very sensitive to hypothermic conditions, and this limits their ability to withstand many conventional methods of preservation. Much research has focused on the high lipid content of pig embryos, and its role in hypothermic sensitivity and cryosurvival. Many studies have reported the conventional freezing of pig embryos, and vitrification has shown promise of eluding the difficulties associated with cooling sensitivity and ice crystallization. Recent research suggests that the embryonic cytoskeleton is susceptible to damage during cryopreservation, and this cellular disruption may be averted by using cytoskeletal stabilizers before preservation. Embryos cryopreserved by conventional freezing and vitrification under the influence of cytoskeletal stabilization have resulted in pregnancies or live offspring from recipient females after surgical transfer. Although cryopreservation technology is less advanced in pigs than in other livestock species, promising research shows evidence that researchers are close to achieving a methodology for preserving pig embryos.

Introduction

Existing genetic diversity has been used to make vast improvements in the quality and production of food and animal products. Advances in animal biotechnology used in conjunction with traditional animal breeding programmes can significantly improve the rate of genetic progress in domestic animals. But while improvements have been made, genetic diversity is being reduced as native animal populations become less isolated and breeding programmes become more globally oriented. Therefore, evolutionary processes for maintaining genetic diversity of the past are not sufficient for the future. Variability is needed to improve populations and to allow populations to adapt to environmental changes. Desirable breeds possessing beneficial production characteristics or valuable disease resistance traits are not in any one country or region, yet need to be globally available to ensure optimal production and disease resistance. Disease transmission and other health concerns limit transport of live animals and subsequent propagation globally. Conservation of genetic resources is essential for availability of desirable genes and germplasm to meet the varying needs of the future. Banking of desirable animals for unique genetic, production and disease resistance traits will facilitate acquisition and characterization of potentially useful germplasm, ensure genetic variation through preservation of selected stocks, and facilitate use of useful germplasm in research and industry in the future.

There are methods that adequately, but not optimally, preserve germplasm and embryos from genetically superior animals of most of livestock species except the pig. While there are methods to preserve boar spermatozoa, there has been little success until recently in preserving pig oocytes and embryos. The use of embryos in addition to spermatozoa represents an increase in the efficiency of transmitting improved genetic potential. Large numbers of pigs are transported by air freight from countries in which nucleus herds are located to countries in which new breeding units are being established (Polge, 1985). Transportation costs are extremely high, and health tests and other requirements are involved. These costs and associated constraints could be considerably reduced by shipping embryos rather than live animals.

Implementation of methodologies for long-term embryo preservation and transfer in pigs would provide a foundation for effective use of the world's best genetic resources on a global basis while modernizing pork production and enhancing genetic improvement programmes. The development of a repeatable method for the long-term preservation of pig embryos would provide numerous practical applications, including transport of maternal germplasm, rapid regeneration or expansion of new and existing lines, the ability to increase selection pressure in nucleus herds, extraction of healthy stock from diseased herds, improve quarantine conditions and provide a method for the international export and import of potential breeding stock (Polge, 1977, 1985; Dobrinsky, 1993; Niemann and Reichelt, 1993). While maintaining genetic resources through embryo banking, successful embryo cryopreservation would enhance the further development of other animal production technologies such as sperm sexing, artificial insemination, *in vitro* fertilization and non-surgical embryo transfer. With the increasing use of swine in human biomedical research, embryos from valuable genetic or transgenic lines of pigs used as organ donors in xenotransplantation could be preserved for future use. Collectively, these technologies could be instrumental in the continuous production of animals of high genetic merit capable of having a significant impact on the improvement of the world pig population and on human medicine. This review will summarize the hypothermic sensitivity of pig embryos and the current progress in pig embryo preservation and emphasize its potential impact on global swine production.

Hypothermic Sensitivity

Pig embryos can be severely altered or permanently damaged by being cooled below physiological temperatures. This chilling sensitivity appears to be dependent on the developmental stage of the embryo and on the conditions under which it develops (Pollard and Leibo, 1994). Wilmut (1972) first demonstrated the apparent sensitivity to hypothermic exposure on pig embryo development. Simply, embryos cooled *in vitro* to 15 or 20°C implanted after surgical transfer to recipients, while embryos cooled to 5 or 10°C did not implant. Polge *et al.* (1974) verified this observation, noting that 15°C was the critical temperature at which damage occurs (Polge, 1977) and that rate of cooling to the critical temperature had no effect on embryo survival (Wilmut, 1986). This sensitivity was further verified by Green *et al.* (1984). Pollard and Leibo (1994) and Leibo *et al.* (1996) observed similar survival of embryos cooled to 15°C, with 90% of pig morulae developing into blastocysts. However, only 5% developed after being cooled to 14°C, while none developed after chilling to 13°C. Pig oocytes are also sensitive to cooling (Didion *et al.*, 1990) as immature oocytes are irreversibly damaged and do not survive following chilling to 15°C or lower or after cryopreservation.

The sensitivity of pig embryos to chilling is not apparent in embryos at the morula to blastocyst stage from other domestic and laboratory species, such as cattle, sheep, rodents and rabbits (for review see Dobrinsky, 1993). However, it is well understood that boar spermatozoa are extremely sensitive to cooling and cryopreservation (Dziuk and Henshaw, 1958; Pursel *et al.*, 1973), suggesting that hypothermic sensitivity may be species specific. The addition of phospholipids to boar semen was beneficial at temperatures below 15°C but was found ineffective with pig embryos (Butler and

Roberts, 1975). The addition of fetal bovine serum (Nagashima *et al.*, 1988a) to embryo storage media helped improve the sustenance and survival of expanded and hatched pig blastocysts cooled below 15°C. Furthermore, Nagashima *et al.* (1988b) observed that pig hatched blastocysts, especially those that had been cultured from expanding blastocysts (blastocysts hatched *in vitro*), had a higher tolerance to cooling than pre-hatching stage embryos. Survival of expanding blastocysts was significantly lower than *in vitro* and *in vivo* hatched blastocysts. Normal fetal development ensued after transfer of cooled (to 11°C) *in vitro* hatched blastocysts, whereas morulae and early blastocysts did not develop. Niemann (1985) attempted to chill pig embryos below 15°C in the presence of cryoprotectants known to be effective in cryopreserving embryos of other domestic species; however, the treatments did not improve hypothermic stability. Collectively, these observations suggest that there are apparent differences in the low temperature tolerance of pig embryos at different developmental stages. This cooling intolerance might be due to the high lipid content found in early developing pig embryos before blastocyst formation (Niemann, 1985; Toner *et al.*, 1986). Perhaps a qualitative change of cells or cell plasma membranes may give rise to tolerance of low temperatures (Nagashima, 1988a,b).

Cryopreservation

Conventional freezing

The sensitivity of pig embryos to chilling appears to be correlated with its sensitivity to cryopreservation (Pollard and Leibo, 1994). Early attempts at pig embryo cryopreservation were few and inconsistent in survival. Nagashima *et al.* (1989a) reported that up to 83% of hatched blastocysts cultured *in vitro* survived freezing to –20°C in 1.5 mol dimethylsulfoxide l^{-1}(DMSO) while morulae and early blastocyst stage pig embryos did not survive cryopreservation. They concluded that development to at least the expanded blastocyst stage was essential for pig embryos to tolerate freezing. Hayashi *et al.* (1989) demonstrated that expanded or hatched blastocyst stage pig embryos survived brief cryopreservation to –35°C. After thawing 11 frozen expanded blastocysts and subsequent transfer, five live piglets were born to one recipient sow. The viability of peri-hatching blastocysts after storage in liquid nitrogen (LN_2) was confirmed. Kameyama *et al.* (1990) cooled expanded blastocysts rapidly from room temperature to seeding temperature, and then slow cooled to –35°C before plunging into LN_2. After transfer of 20 frozen–thawed embryos, two live piglets were born to a recipient sow. Kashiwazaki *et al.* (1991) used conventional slow freezing and reported the birth of four live offspring after 32 frozen–thawed embryos were transferred to a recipient female, while only 11% of the cryopreserved embryos survived *in vitro*. Fujino *et al.* (1993) tested egg-yolk (used in semen cryopreservation) as a cryoprotectant for conventional freezing of pig embryos at various stages of development. After transfer of hatched blastocysts into six recipient gilts, a single recipient delivered one live piglet. They concluded that egg-yolk did not improve the cryosurvivability of pig embryos. Further research using various methods with conventional freezing of pig embryos has shown similar *in vitro* and *in vivo* post-thaw development (Hsu *et al.*, 1990; Jung and Chang, 1990; Jung *et al.*, 1990; Feng *et al.*, 1991; Kuo and Chiang, 1993; Iwasaki *et al.*, 1994). Most recently, Modl *et al.* (1996) reported the birth of two litters of pigs after transfer of conventionally frozen hatched blastocysts.

Although the aforementioned studies demonstrated that embryos could survive cryopreservation, they contained very low numbers of embryos and subsequent embryo transfers, making it difficult to make firm conclusions. This is not surprising, considering the difficulty and expense in producing large numbers of pig embryos for research use and surgical embryo transfer. Most of the reported studies do not share cryopreservation protocols and many have not been repeated. However, an interesting trend emerged from these reports. It was apparent that survival of pig embryos after cryopreservation may be correlated with stage of embryo development, which may be optimized *in vitro* and peaks during peri-hatching blastocyst progression. They hypothesized the role of embryonic lipid phase changes between the morula/early blastocyst and the peri-hatching blastocyst stages and possible effects on embryo survival after cryopreservation. Nagashima *et al.* (1992) made substantial contributions to support these observations. They observed that 150–300 μm blastocysts hatched *in vivo* and *in vitro* developed *in vitro* more frequently than did embryos ≥ 300 μm after cryopreservation. Embryo survival after freezing and thawing was also affected by a period of culture before cryopreservation, the presence of protein in the freezing solution and possible breed differences among embryo donors. This effect of developmental stage on embryonic cryosurvival is generally accepted and has been confirmed in subsequent reports (Cameron *et al.*, 1992; Dobrinsky and Johnson, 1994). In short, while intact morulae and early blastocysts do not survive cryopreservation and harsh osmotic disruptions, peri-hatching blastocyst stage pig embryos possess characteristics amenable to cryopreservation.

Vitrification

Vitrification is the rapid cooling of liquid medium in the absence of ice crystal formation. The solution forms an amorphous glass as a result of rapid cooling by direct submersion of the embryo in a plastic straw into LN_2. The glass retains the normal molecular and ionic distributions of a liquid but remains in an extremely viscous, supercooled form (Rall, 1987). The glass is devoid of all ice crystals, and embryos are not subjected to the physical damage that is associated with ice crystal formation (Rall and Fahy, 1985). Vitrification is simple, inexpensive, and has been developed as a viable alternative to conventional freezing of domestic and laboratory animal embryos. Cattle embryos have been successfully cryopreserved with vitrification (Massip *et al.*, 1986; Dobrinsky *et al.*, 1991). More recently, van Wagtendonk-de Leeuw *et al.* (1994, personal communication) demonstrated that vitrification of bovine embryos can be successfully applied in large numbers under field conditions without a significant reduction in pregnancy rate. Overall pregnancy rates after non-surgical embryo transfer were 44.5% (n = 393) for vitrified embryos and 45.1% (n = 335) for conventionally frozen controls.

Until recently, *Drosophila melanogaster* embryos had not been successfully cryopreserved. Like pig embryos, they contain large amounts of lipid and are extremely sensitive to subzero temperatures when supercooled in the absence of ice (Mazur *et al.*, 1992). Unfortunately, embryos died rapidly if they were cooled slowly enough to prevent intracellular ice formation. Mazur *et al.* (1992) suggested that slow cooling injury in *Drosophila* embryos could result from the loss of synchrony of coupled reactions involved in early embryonic development. They speculated that vitrification reduced the possibility of both intracellular and extracellular ice formation, while rapid cooling of embryos could bypass or 'out-race' detrimental chilling-induced cellular changes that take place during slow cooling. Steponkus *et al.* (1990) used vitrification and reported the first successful cryopreservation of *Drosophila* embryos.

Studies in our laboratory have involved development of vitrification procedures for cryopreserving pig embryos. Some vitrification solutions have been shown to be extremely toxic to pig embryos (Weber *et al.*, 1992; Dobrinsky and Johnson, 1994) possibly limiting their use. Dobrinsky and Johnson (1994) applied established vitrification protocols for rodent or bovine embryos using vitrification media based on glycerol-propylene glycol (Dobrinsky *et al.*, 1992), glycerol (VS3a; Rall, 1992; van Wagtendonk-de Leeuw *et al.*, 1994) or ethylene glycol (Kobayashi *et al.*, 1995a,b) and tested their relative toxicity on non-cryopreserved pig embryos produced *in vivo*. Morulae–early blastocysts survived equally across all treatments, while expanded blastocysts developed best after exposure to ethylene glycol and glycerol based media. Hatched blastocysts developed well after exposure to glycerol-based VS3a only, as exposure to ethylene glycol and glycerol-propylene glycol based solutions was detrimental to further development *in vitro*. Therefore, glycerol/VS3a was used as it was least toxic to embryos regardless of embryonic age. After vitrification in VS3a, expanded (27%, n = 64) and hatched blastocysts (39%, n = 84) modestly survived vitrification, while morulae–early blastocysts did not (0%, n = 59). In a subsequent study, Dobrinsky (1993) showed that a culture period before vitrification had no effect on embryo survival following cryopreservation, which is consistent with earlier reports with conventional freezing of embryos (Nagashima *et al.*, 1989a–c; 1992).

Yoshino *et al.* (1993) tested many different vitrification solutions and their cryophysical properties. After cryopreservation, they reported little or no survival *in vitro* and concluded that their vitrification media were extremely toxic to pig embryos. Weber and Youngs (1994) investigated toxicity of cryoprotectant solutions and typical concentrations of sucrose used in the removal of intracellular permeating cryoprotectants on pooled populations of morula to hatched blastocyst pig embryos. They found that sucrose concentrations over 1.0 m l^{-1} were detrimental to further embryo development *in vitro*. Pig embryos were incubated for 30 s in cryoprotectant solutions containing 10, 20, 30, 40 or 50% glycerol, propylene glycol or ethylene glycol, and then diluted in 1.0 mol sucrose l^{-1}. As the concentration of an individual cryoprotectant increased beyond 30%, embryo development decreased. Embryos exposed to glycerol or propylene glycol exhibited poorer development than did embryos placed into ethylene glycol, especially at concentrations of 40% or higher. It should be noted that in these experiments, embryos were placed directly into high concentrations of cryoprotectants without any prior equilibration.

Kobayashi *et al.* (1995a,b) successfully vitrified pig embryos in ethylene glycol and polyvinylpyrrolidone. After vitrification, the respective survivals at 48 h in culture of embryos vitrified as blastocysts, expanded blastocysts and hatched blastocysts were 26%, 33% and 11%. They concluded that extreme chilling sensitivity can be partially circumvented by cooling the embryos rapidly at advanced blastocyst stages. In a subsequent study, Kobayashi *et al.* (1995) showed improved embryo development when using galactose as a cryoprotectant diluting agent. In this study, the respective survivals at 48 h in culture of embryos vitrified as blastocysts, expanded blastocysts and hatched blastocysts were 35%, 96% and 82%. Since their initial reports of these findings in abstract form, no further report in the literature was found on further development of these vitrification procedures and subsequent embryo development *in vitro* or *in vivo*.

In general, all of the reported vitrification studies investigated *in vitro* development after cryopreservation. As of the preparation of this manuscript, there has been no report of live offspring after conventional vitrification of pig embryos. However, it is interesting to note that the trends in embryonic survival after conventional embryo freezing seem to be evident after vitrification and that development to advanced blastocyst stages is paramount to pig embryo survival after cryopreservation.

Cellular Disruption During Cryopreservation

Cryopreservation can be extremely disruptive to the cellular organization of embryos. Ice crystal formation can lyse plasma membranes, storage in liquid nitrogen can denature critical intracellular functions and organelles, and the central cytoarchitecture of a cell can be destroyed. Documenting cellular damage during or after cryopreservation would provide useful, non-empirical information for understanding cellular sensitivities to cryopreservation and would lead to improved protocols for embryo cryopreservation and a better understanding of domestic animal embryology (Dobrinsky, 1996). Recently membrane lipids, intracellular lipids and the embryonic cytoskeleton have been implicated in the hypothermic sensitivity of pig embryos. In the following section more invasive attempts to analyse these cellular disruptions that occur during preservation and subsequent attempts to avoid such damage will be discussed.

Membrane lipids

Polge (1977) first suggested that chilling-induced damage to pig embryos was associated with lipid phase changes within the plasma membranes of blastomeres. There is evidence of phase changes in plasma membrane lipids as temperature is altered (Edidin and Petit, 1977), and the changes are usually reversible. However, during cryopreservation, proteins may interact during cooling that may render phase changes irreversible. This disruption in membrane lipids was identified as a reason why pig embryos may not withstand cooling or cryopreservation (Polge and

Willadsen, 1978; Wilmut 1986). Many attempts have been made to avert damage to plasma membranes during cooling. As was previously discussed, boar spermatozoa are extremely sensitive to cooling and cryopreservation. The addition of phospholipids to boar semen was found to be beneficial at temperatures below 15°C, but was found ineffective with pig embryos (Butler and Roberts, 1975; Niemann, 1985). The addition of fetal bovine serum (Nagashima *et al.*, 1988a,b) to embryo storage media helped improve the sustenance and survival of chilled, expanded and hatched pig blastocysts. Rubinsky *et al.* (1992) proposed that antifreeze glycoproteins isolated from antarctic fishes preserve the structural integrity of the plasma membrane when added to vitrification media. Their use in pig embryo preservation has been limited. Trehalose, a disaccharide known to stabilize plasma membranes and decrease osmotic shock during removal of cryoprotectants (Valdez *et al.*, 1991), was combined with glycerol for conventional freezing of pig embryos by Cameron *et al.* (1992), who reported viabilities of up to 50% after 24 h of culture following cryopreservation. In all, plasma membrane stability and intracellular lipids in general are linked to hypothermic sensitivity of pig embryos.

Intracellular lipids

Pig embryos contain a large amount of sudanophilic lipid in vesicles that is abundant early in development to the blastocyst stage (Norberg, 1973; Niimura and Ishida, 1980; Nagashima *et al.*, 1994). As development continues to the hatching blastocyst stage, lipid content begins to decline (Niimura and Ishida, 1980). The lipid content continues to decline after hatching, coincident with the stage at which pig embryos are able to be cryopreserved (Nagashima *et al.*, 1989 a–c; 1992; 1994; Dobrinsky and Johnson, 1994). It has been suggested that lipid turnover may be important for hatching and normal development, and could be instrumental in increasing the tolerance of pig embryos to cryopreservation. Cooling intolerance was speculated to be due to the high lipid content found in early developing pig embryos (Niemann, 1985; Toner *et al.*, 1986). While it is evident that pig embryos can withstand the osmotic assault of cryoprotectant permeation and withdrawal, cryoprotectants do not protect embryonic lipids or their integrity.

One way to determine whether intracellular lipid is in any way associated with cooling or cryosensitivity of pig embryos is to remove it from the embryo, or delipation, and then cool or cryopreserve the embryo (Nagashima *et al.*, 1994; 1995). When this was done, over 60% of the delipated eight-cell embryos cleaved after chilling to 4°C, whereas all of the controls lysed within 24 h. Fully delipated embryos survived better than partially delipated embryos, while almost 40% developed to the blastocyst stage. It is clear that pig embryos gain some form of tolerance to chilling when their lipid content is reduced. Nagashima *et al.* (1994) proposed that delipation results in changes in the lipid composition of the plasma membranes, and this in turn may limit the extent of phase separation and resultant damage during cooling discussed above.

Lipid vesicles themselves may have a direct effect on embryo survival during cooling (Nagashima *et al.*, 1994; 1995). When pig embryos were held at 15°C, structural changes to the lipid vesicles were observed where they actually coalesce to form larger vesicles (Edidin and Petit, 1977). Lipid vesicles within the cytoplasm have a close spatial arrangement with the smooth endoplasmic reticulum within the embryo (Hyttel and Niemann, 1990) and are thought to play a role of providing nutrition to the cell as well as modifying the physical properties and functions of the cellular plasma membranes (Stubbs and Smith, 1984). During cryopreservation, as the vesicles are at least partially surrounded by endoplasmic reticulum, the resultant loss of cytoplasmic organization may result in irreversible damage to the embryo (Mohr and Trounson, 1981). Removal of the cytoplasmic lipid could therefore eliminate potential cytoplasmic elements causing disruption of the subcellular localization of organelles (Nagashima *et al.*, 1994) such as mitochondria. Although Nagashima *et al.* (1994) have proposed what might be occurring after delipation in relation to cooling sensitivity, no one has explained how pig embryos, with crucial cytoplasmic components removed via delipation, actually compensate for their cytoplasmic losses and maintain development after cooling.

In 1995, Nagashima *et al.* applied delipation principles to conventional freezing of two- to eight-cell pig embryos. More than half of delipated embryos survived cryopreservation, whether they had been frozen immediately after delipation or after further culture (to determine mitotic competence) prior to freezing, while none of the controls survived. Normal piglets were obtained from at least one recipient after unfrozen delipated and frozen delipated embryos were transferred. This study showed that early cleavage stage embryos survive cryopreservation following delipation, and that somehow, the loss of cytoplasmic lipid is compensated for later in development. These observations of Nagashima and coworkers are landmark observations in understanding the sensitivity of the pig embryo to cooling and cryopreservation, and provide opportunities for studying cryopreservation of oocytes and early stage embryos in other animals that may contain large amounts of cytoplasmic lipid.

Embryonic cytoskeleton

The cytoskeleton is a complex network of protein constituents, actin (microfilaments) and tubulin (microtubules), distributed throughout the cytoplasm of a cell (for review, see Albertini *et al.*, 1987; Dobrinsky, 1996). They give three-dimensionality and mechanical strength to the surface of a cell and provide a system of fibres that impart polarity while regulating cell shape, cell movement, and the plane of cell division. Actin and tubulin filaments bind a variety of accessory proteins that enable them to participate in distinct functions in different regions of the cell, including the plasma membrane. Thus, this highly organized network of filaments forms an internal framework, often referred to as the cytoarchitecture, for the large volume of cytoplasm within the cell.

While the cytoskeleton is important in maintaining structural integrity within the cell, cryoprotectants and cryopreservation disrupt the entire embryonic cytoskeleton. During cryopreservation, cryoprotectants (CPA) are organic solutes used to protect intracellular organelles during long-term preservation in liquid nitrogen. Permeating cryoprotectants, such as glycerol and propylene glycol, among other properties, act to depolymerize microfilaments and microtubules. Depolymerization is beneficial in protecting these cytoskeletal components during osmotic stresses induced by exposure to or removal of cryoprotectants. When coupled with vitrification, however, disruption of the microfilaments or microtubules can be irreversible and lethal to embryos. Total disruption of the plasma membranes of individual blastomeres, as well as lack of blastocoel cavity reformation indicate that many pig embryos cannot tolerate cryopreservation or cryoprotectant treatments and that membrane integrity and metabolic activity are jeopardized.

Some vitrification solutions have been shown to be extremely toxic to pig embryos (Weber *et al.*, 1992; Dobrinsky and Johnson, 1994), and cytoskeletal components react quite differently to CPA exposure (Dobrinsky and Johnson, 1994). Under the influence of high molar concentrations of glycerol needed for vitrification, microtubules in morula to hatched blastocyst stage pig embryos retain polymerization throughout vitrification protocol. Microfilaments are variable in their response to CPA, and the variability is directly related to embryonic stage of development. Prior to cryopreservation of morulae and early blastocyst stage embryos, microfilaments were disrupted by interaction with the CPA. After thawing and subsequent rehydration, cell lysis, membrane disintegration and nuclear damage became evident, microfilaments failed to repolymerize in any organized manner while many nuclei remained crenated and microtubule fluorescence intensity was reduced. Microfilaments of control morulae and early blastocysts equilibrated in vitrification solution repolymerize normally following rehydration while microtubules retain normal staining intensity and nuclei regain normal fluorescent, non-crenated morphology. The irreversible damage of morulae/early blastocyst stage pig embryos is linked to cooling to –196°C but not to the exposure to the CPA used for vitrification. In contrast, some vitrified pig hatched blastocysts retained polymerized microfilaments and microtubules but exhibited regions of intact cortical microfilaments and heterogeneous, cytoplasmic microtubule staining after rehydration. Other vitrified embryos revealed extensive cell lysis and perturbations of normal microfilament and microtubule localization after rehydration. Non-cryopreserved control embryos equilibrated in vitrification solution exhibited normal microfilament repolymerization while microtubules retained normal perinuclear staining patterns.

Table 1. Development of morphologically different stages of swine embryos after cryopreservation by vitrification under the influence of microfilament stabilizer cytochalasin-b

	mVS3a Control			mVS3a + cytochalasin-b		
Stage of development	*n*	dev	% dev	*n*	dev	% dev
MB	17	0	0 [a]	17	1	6 [a]
XB	27	6	22 [ab]	25	15	60 [b;Z]
HB1 (< 400 μm diameter)	36	10	28 [b]	48	43	90 [c;Z]
HB2 (> 400 μm diameter)	14	4	29 [ab]	22	9	41 [b]

Data from Dobrinsky *et al.* (1997).
MB: morulae/blastocysts; XB: expanded blastocysts; HB1: hatched blastocysts < 400 μm in diameter; HB2: hatched blastocysts > 400 μm in diameter.
[a–c]Values with different superscripts within individual columns are significantly different ($P < 0.05$); ANOVA–GLM.
[Z]Value with superscript within a row is significantly different ($P < 0.01$); chi-square analysis.

The differential survival of pig embryos after vitrification seems to be dependent upon developmental stage of the embryo, which is consistent with numerous reports (see above). The high lipid content found in compacted, pre-blastocyst stage pig embryos may undergo irreversible changes rendering them metabolically useless to the embryo, or lysis of cell plasma membranes may release vital lipid out of the embryo upon rehydration creating a deficiency of lipid for utilization by the embryo. Other membrane-bound structures (such as lysozymes) may release their contents into the cytoplasm or be rendered useless to the embryo. Microtubule damage from cryopreservation might also affect the mobility and transport of lipid and other components in the embryo, thus breaking the supply of substrate needed during blastocyst formation in the embryo. Microfilament damage may be so extensive as to disable any structural support to the plasma membrane (for review, see Dobrinsky, 1996).

An intact cytoskeleton is the highway for cytokinesis and karyokinesis, and if disrupted in any way, the mitotic cell cycle will cease. Maintaining the integrity of the cytoarchitecture within an embryo during cryopreservation is of the utmost importance, especially with pig embryos. Because of the cooling sensitivity, and fragile plasma membranes, finding ways to overcome or prevent cytoskeletal disruption will improve the survival of pig embryos during and after cryopreservation.

Cytoskeletal stabilization

Cytoskeletal stabilizers, such as the cytochalasins, have been used extensively in reproductive biotechnology for studies dealing with micromanipulation of embryonic development (McGrath and Solter, 1983; Surani *et al.*, 1984). It is generally accepted that cytochalasins are a microfilament inhibitor (Figure 1) disrupting actin polymerization by blocking monomer addition at the fast-growing end of the polymers (Theodoropoulos *et al.*, 1994), thus preventing cytokinesis without affecting karyokinesis (Modlinski, 1980; Cuthbertson, 1983). Treatment of cells with cytochalasins makes the plasma membrane less rigid and more elastic so microfilaments are not disrupted during extensive micromanipulation (McGrath and Solter, 1983).

Dobrinsky *et al.* (1997; submitted) have documented microfilament damage during swine embryo vitrification and the use of cytochalasin-b before and during cryopreservation (Fig. 1; Table 1) to help deter damage and stabilize the plasma membrane (Prather and First, 1986; Dobrinsky, 1996; Dobrinsky *et al.*, 1997). As reported in previous studies, morulae or early blastocysts do not survive cryopreservation. Treatment with cytochalasin-b did not improve their viability after vitrification. However, cytochalasin-b treatment provided a threefold improvement in survival of

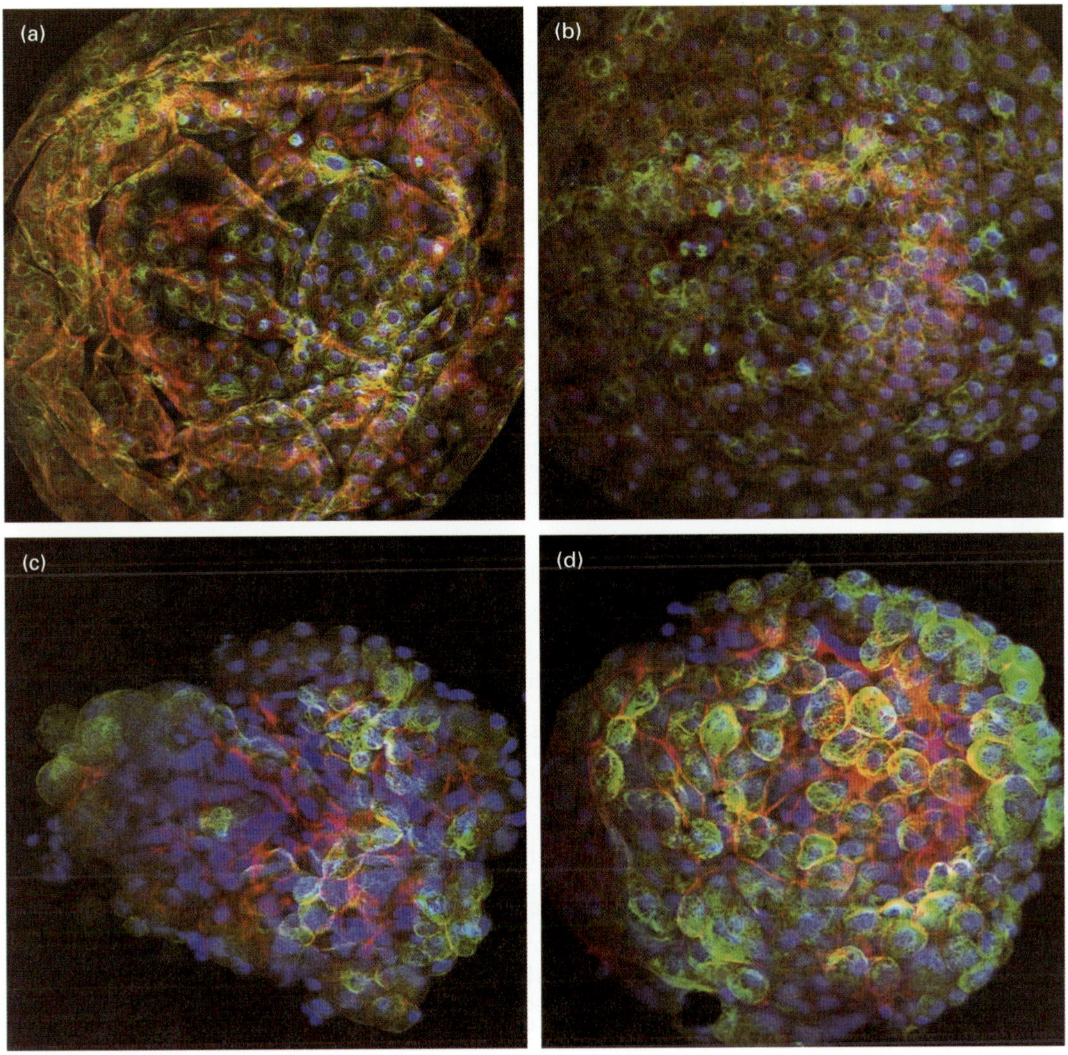

Fig. 1. Laser scanning confocal micrographs of control and cytochalasin-b treated and cryopreserved hatched blastocyst stage pig embryos. In the micrographs, the red label is Texas Red-X phalloidin staining of filamentous actin (microfilaments), the green label is FITC staining of α- and β-tubulin (microtubules), and the blue label is Hoechst 33342 staining of cell nuclei. (a) Non-treated control hatched blastocyst; note the intact actin labelling consistent and prevalent at all cell to cell borders. (b) Cytochalasin-b treated hatched blastocyst: after incubation in cytochalasin, note the reduction of phalloidin staining at the cell to cell borders, indicating depolymerization of microfilaments, while microtubules and the nuclei remain mostly unchanged. (c) Non-treated, vitrified, warmed hatched blastocyst, cultured for 2 h; note the disrupted spherical orientation of the embryo proper, and inconsistency of cellular organization. Part of the embryo remains intact, while the remainder of the embryo has lost all normal cellular organization and staining patterns of the cytoskeleton. (d) Cytochalasin-b treated, vitrified, warmed hatched blastocyst, cultured for 2 h; note the reformation of normal staining patterns of the cytoskeleton, indicating normal repolymerization of microfilaments and microtubules, as well as the reformation of a normal spherical dimension for the embryo.

expanded (60%) and hatched blastocysts less than 400 μm in diameter (90%). Although cytochalasin-b-treated expanded blastocysts had improved viability after vitrification, their development in terms of blastocoel reformation and proliferation in culture following vitrification was still lower than that of hatched blastocysts. These data are the highest rates of *in vitro* development after cryopreservation ever attained in our laboratory, and results are consistent across embryo donors. Cellular analysis with laser scanning confocal microscopy revealed reduced fluorescence intensity of microfilaments in cytochalasin-b treated embryos (Fig. 1). Non-cytochalasin treated vitrified embryos exhibited partial or major cytoskeletal disruptions within 2 h after rehydration, whereas cytochalasin treated vitrified embryos mostly exhibited normal repolymerization of microfilaments and other cytoskeletal components after vitrification. These experiments show that the cytoskeleton is affected during vitrification, and that microfilament depolymerization prior to cryopreservation significantly improves expanded and hatched blastocysts development.

Most recently, we (Dobrinsky, V. Pursel and L. Johnson, unpublished observations) have established four pregnancies after transfer of cytochalasin-b treated vitrified hatched blastocysts into seven recipient gilts. Hatched blastocysts < 400 μm were treated with cytochalasin-b, vitrified and then stored in LN_2 for at least one week. Natural cyclic gilts, with normal oestrous cycles and sound confirmation were selected as recipients and used on day 6 (day 0 = oestrus) of the oestrous cycle. Straws containing embryos were warmed and embryos were diluted from cryoprotectants, rehydrated, cultured in BECM-3+10% FBS for up to 6 h, and then transferred via transoviductal catheterization into a single uterine horn.

Conclusion

Methods for preservation of pig embryos are in early stages of development, as are other technologies in pigs, such as non-surgical embryo transfer (Li *et al.*, 1996), that could make embryo cryopreservation practical and usable by the producer. Technologies available for embryo transfer in cattle, such as non-surgical embryo recovery and *in vitro* embryo production, are not readily available to the pig industry or are in early stages of development, making it impractical for the producer to use these biotechnical advances in swine. Much of the early research in preservation of pig embryos was observational. A few reports mentioned inconsistent rates of embryo survival both *in vitro* and *in vivo*. However, since the late 1980s, techniques for preservation of pig embryos have advanced. The pig embryo is known to be extremely sensitive to hypothermic conditions. Its intracellular lipids play a major role in this sensitivity. Delipation and conventional freezing or vitrification are viable methods for embryo preservation, but the test of these methods is a repeatable production of live pigs. Embryo delipation is somewhat impractical as it relies on invasive manipulation of the embryos before preservation, only to inflate the cost of its application. Vitrification is a simple alternative for cryopreserving pig embryos, but its efficacy is yet to be realised in providing high rates of live offspring after transfer. Other advances in pig reproduction and biotechnology, such as *in vitro* embryo production and non-surgical embryo transfer, will only enhance the use of embryo cryopreservation. Alternative methods are needed to preserve valuable genetic resources in a reasonable and economical manner. More invasive research needs to be conducted to characterize cellular and molecular disruption during and after cryopreservation, so that better methods for preserving pig embryos can be developed.

References

Albertini DF, Overström EW and Ebert KM (1987) Changes in the organization of the actin cytoskeleton during pre-implantation development of the pig embryo *Biology of Reproduction* **37** 441–451

Butler WJ and Roberts TK (1975) Effects of some phosphatidyl compounds on boar spermatozoa following cold shock or slow cooling *Journal of Reproduction and Fertility* **43** 183–187

Cameron RDA, Lising R, Nagashima H and Blackshaw AW (1992) Cryopreservation of cultured and uncultured porcine embryos with glycerol and trehalose *Proceedings of the 12th International Pig Veterinary Society* The Hague, The Netherlands p. 476

Cuthbertson KSR (1983) Parthenogenetic variation of mouse oocytes *in vitro* with ethanol and benzyl alcohol *Journal of Experimental Zoology* **226** 311–314

Didion BA, Pomp D, Martin MJ, Homanics GE and Markert CL (1990) Observations on the cooling and cryopreservation of pig oocytes at the germinal vesicle stage *Journal of Animal Science* **68** 2803–2810

Dobrinsky JR (1993) Cryopreservation of porcine embryos:

current status and perspectives *Embryo Transfer Newsletter* **11** 13–18

Dobrinsky JR (1996) Cellular approach to cryopreservation of embryos *Theriogenology* **45** 17–26

Dobrinsky JR and Johnson LA (1994) Cryopreservation of porcine embryos by vitrification: a study of *in vitro* development *Theriogenology* **42** 25–35

Dobrinsky JR, Hess FF, Duby RT and Robl JM (1991) Cryopreservation of bovine embryos by vitrification *Theriogenology* **35** 194

Dobrinsky JR, Stice SL, Phillips PE, Duby RT and Robl JM (1992) Development of IVM–IVF bovine embryos following vitrification dilution treatments *Theriogenology* **37** 202

Dobrinsky JR, Long CR and Johnson LA (1997) Stability of microfilaments during swine embryos cryopreservation *Theriogenology* **47** 343

Dziuk PJ and Henshaw G (1958) Fertility of boar semen artificially inseminated following *in vitro* storage *Journal of Animal Science* **17** 554

Edidin M and Petit VA (1977) The effect of temperature on the lateral diffusion of plasma membrane proteins. In *Ciba Foundation Symposium* pp 155–174 Ed. ??? Name Excerpta Medica, New York

Feng S, Zhang Y Li S, Ma Z, Wang R and Lu D (1991) Piglets from frozen (–20°C) embryos were born in China *Theriogenology* **35** 199

Fujino Y, Ujisato Y, Endo K, Tomizuka T, Kojima T and Oguri N (1993) Cryoprotective effect of egg yolk in cryopreservation of porcine embryos *Cryobiology* **30** 299–305

Green ST, Boland MP and Gordon I (1984) Studies on the recovery and storage of porcine embryos. In *Research Report 1982–1983* p 100 Faculty of General Agriculture, University College Dublin, Dublin

Hayashi S, Kobayashi K, Mizuno J, Saitoh K and Hirano S (1989) Birth of piglets from frozen embryos *Veterinary Record* **125** 43–44

Hsu T-T, Chen L-R, Luoh Y-S and Sheen Y-W (1990) Comparison of cryopreservation effects between vitrification method and slow cooling method on mammalian whole bisected and fused embryos *5th American Association of Animal Production* **3** 297

Hyttel P and Niemann H (1990) Ultrastructure of porcine embryos following development *in vitro* versus *in vivo* *Molecular Reproduction and Development* **27** 136–144

Iwasaki S, Mizuno J, Kobayashi K, Yoshikane Y and Hayashi T (1994) Changes in morphology and cell number of inner cell mass of porcine blastocysts during freezing *Theriogenology* **42** 841–848

Jung JK and Chang WK (1990) Freezing of porcine embryos *5th American Association of Animal Production* **3** 245

Jung JK, Chang WK and Yoo SH (1990) Studies on the pig embryo freezing *Korean Journal of Animal Science* **32** 445–449

Kameyama K, Takedomi T, Itakura H and Onihara T (1990) Effect of lecithin on cryopreservation of porcine embryos *Proceedings of the 78th Annual Conference of Society for Animal Reproduction* p 22

Kashiwazaki N, Ohtani S, Miyamoto K and Ogawa S (1991) Production of normal piglets from hatched blastocysts frozen at –196°C *Veterinary Record* **128** 256–257

Kobayashi S, Tomita M, Pollard JW and Leibo SP (1995a) Survival of cryopreserved porcine embryos vitrified in ethylene glycol plus polyvinylpyrrolidone *Theriogenology* **41** 228

Kobayashi S, Takei M, Ichikawa AT and Leibo SP (1995b) Effect of dilution of cryoprotectants of 48-h development of vitrified porcine embryos *Cryobiology* **32** 589–590

Kuo Y-H and Chiang J-C (1993) Study of frozen pig embryos to –196°C. In *Pig Research Institute Taiwan Annual Report* pp 45–47

Leibo SP, Martino A, Kobayashi S and Pollard JW (1996) Stage dependent sensitivity of oocytes and embryos to low temperatures *Animal Reproduction Science* **42** 45–53

Li J, Rieke A, Day BN and Prather RS (1996) Technical note: porcine nonsurgical embryo transfer *Journal of Animal Science* **74** 2263–2268

Massip A, van der Zwalmen P, Scheffen B and Ectors, F (1986) Pregnancies following transfer of cattle embryos preserved by vitrification *Cryo-Letters* **7** 270–273

Mazur P, Schneider U and Mahowald AP (1992) Characteristics and kinetics of sub-zero chilling injury in *Drosophila* embryos *Cryobiology* **29** 39–68

McGrath J and Solter D (1983) Nuclear transplantation in the mouse embryo using microsurgery and cell fusion *Science* **220** 1300–1302

Modlinski JA (1980) Preimplantation development of microsurgically obtained haploid and homozygous diploid mouse embryos and effects of pretreatment with cytochalasin-b on enucleated eggs *Journal of Embryology and Experimental Morphology* **60** 153–161

Mödl J, Reichenbach HD, Wolf E and Brem G (1996) Development of frozen–thawed porcine blastocysts *in vitro* and *in vivo* *Veterinary Record* **139** 208–210

Mohr LR and Trounson AO (1981) Structural changes associated with freezing of bovine embryos *Biology of Reproduction* **25** 1009–1025

Nagashima H, Kato Y and Yamakawa H (1988a) Survival of pig hatched blastocysts exposed below 15°C *Japanese Journal of Animal Reproduction* **34** 123–131

Nagashima H, Kato Y, Yamakawa H and Ogawa S (1988b) Survival of pig hatched blastocysts exposed below 15°C *Theriogenology* **29** 280–?

Nagashima H, Kato Y, Yamakawa, Matsumoto T and Ogawa S (1989a) Changes in freezing tolerance of pig blastocysts in peri-hatching stage *Japanese Journal of Animal Reproduction* **35** 130–134

Nagashima H, Kato Y, Yamakawa H and Ogawa S (1989b) Low temperature sensitivity of blastocysts and blastocyst-derived cells in pigs *Theriogenology* **31** 232

Nagashima H, Kato Y, Yamakawa H, Yokozawa K, Kato M, Sawasaki T, Kobayashi K and Ogawa S (1989c) Studies on freezing preservation and sexing of bisected embryos in pigs, cows, goats and rabbits *Japanese Journal of Animal Reproduction* **35** 69P–75P

Nagashima H, Yamakawa H and Niemann H (1992) Freezability of porcine blastocysts at different peri-hatching stages *Theriogenology* **37** 839–850

Nagashima H, Kashiwazaki N, Ashman RJ, Grupen CG, Seamark RF and Nottle MB (1994) Removal of cytoplasmic lipid enhances the tolerance of porcine embryos to chilling *Biology of Reproduction* **51** 618–622

Nagashima H, Kashiwazaki N, Ashman RJ, Grupen CG and Nottle MB (1995) Cryopreservation of porcine embryos *Nature* **374** 416

Niemann H (1985) Sensitivity of pig morulae to DMSO/PVP or glycerol treatment and cooling to 10°C *Theriogenology* **23** 213

Niemann H and Reichelt B (1993) Manipulating early pig

embryos *Journal of Reproduction and Fertility Supplement* **48** 75–94

Niimura S and Ishida K (1980) Histochemical observation of lipid droplets in mammalian eggs during the early development *Japanese Journal of Animal Reproduction* **26** 46–49

Norberg HS (1973) Ultrastructural aspects of the preattached pig embryo: cleavage and early blastocyst stages *Z Anat Entwicklungsgesch* **143** 95–114

Polge C (1977) The freezing of mammalian embryos: perspectives and possibilities. In *Ciba Foundation Symposium* pp 3–18 Ed. Name (??? AUTHOR TO PROVIDE) Excerpta Medica, New York

Polge C (1985) How does embryo manipulation fit into present and future pig reproduction? *Journal of Reproduction and Fertility Supplement* **33** 93–100

Polge C and Willadsen SM (1978) Freezing eggs and embryos of farm animals *Cryobiology* **15** 370–373

Polge C, Wilmut I and Rowson LEA (1974) The low temperature preservation of cow, sheep and pig embryos *Cryobiology* **11** 560

Pollard JW and Leibo SP (1994) Chilling sensitivity of mammalian embryos *Theriogenology* **41** 101–106

Prather RS and First NL (1986) Effect of cytochalasin-b and demecolcine on freeze–thaw survival of murine embryos *in vitro. Journal of Experimental Zoology* **239** 37–40

Pursel VG, Johnson LA and Schulman LL (1973) Fertilizing capacity of boar semen stored at 15°C *Journal of Animal Science* **37** 532–535

Rall WF (1987) Factors affecting the survival of mouse embryos cryopreserved by vitrification *Cryobiology* **24** 387–402

Rall WF (1992) Cryopreservation of oocytes and embryos: methods and applications *Animal Reproduction Science* **28** 237–245

Rall WF and Fahy GM (1985) Ice-free cryopreservation of mouse embryos at –196°C by vitrification *Nature* **313** 573–575

Rubinsky B, Arav A and DeVries AL (1992) The cryoprotective effect of antifreeze glycopeptides from antarctic fishes *Cryobiology* **29** 69–79

Steponkus PL, Myers SP, Lynch DV, Gardner L, Bronshteyn V, Leibo SP, Rall WF, Pitt RE, Lin TT and MacIntyre RJ (1990) Cryopreservation of *Drosophila melanogaster* embryos *Nature* **345** 170–172

Stubbs CD and Smith AD (1984) The modification of mammalian membrane polyunsaturated fatty acid composition in relation to membrane fluidity and function *Biochimica et Biophyica Acta* **779** 89–137

Surani MAH, Barton DS and Norris ML (1984) Development of reconstituted mouse eggs suggests imprinting of the genome during gametogenesis *Nature* **308** 548–550

Theodoropoulos PA, Gravanis A, Tsapara A, Margioris N, Papadogiorgaki E, Galanopoulos V and Stournaras C (1994) Cytochalsin-b may shorten actin filaments by a mechanism independent of barbed end capping *Biochemal Pharmacology* **47** 1875–1881

Toner M, Cravalho EG, Ebert KM and Overström EW (1986) Cryobiophysical properties of porcine embryos *Biology of Reproduction* **34** (Supplement) 98

Valdez CA, Hishinuma M, Takahashi Y and Kanagawa H (1991) Effect of trehalose dilution on the survival of vitrified–thawed mouse morulae *Japanese Journal of Veterinary Research* **39** 23–26

van Wagtendonk-de Leeuw AM, den Daas JHG and Rall WF (1994) Pregnancy rates in a comparative field trial of vitrification and one-step dilution or conventional slow freezing and three-step dilution of bovine embryos are similar *Theriogenology* **41** 326

Weber PK and Youngs CR (1994) Investigation of cryoprotectant toxicity to porcine embryos *Theriogenology* **41** 1291–1298

Weber PK, McGinnis LK and Youngs CR (1992) An evaluation of potential vitrification solutions for cryopreservation of porcine embryos *Theriogenology* **37** 321

Wilmut I (1972) The low temperature preservation of mammalian embryos *Journal of Reproduction and Fertility* **31** 513–514

Wilmut I (1986) Cryopreservation of mammalian eggs and embryos. In *Developmental Biology: Volume 4 Manipulation of Mammalian Development* pp 217–247 Eds LW Browder and RBL Gwatkin. Plenum Press, New York and London

Yoshino J, Kojima T, Shimizu M and Tomizuka T (1993) Cryopreservation of porcine blastocysts by vitrification *Cryobiology* **30** 4113–422

Poster titles

Meinecke B, Wehrend A and Meinecke-Tillmann, S
Initiation of porcine oocyte nuclear maturation *in vitro* occurs independently of H1 and map kinase activation

Rodriguez-Martinez H, Han Y, Song X, Funahashi H, Niwa K and H Pertoft
Production of hyaluronic acid by porcine oocyte–cumulus cell complexes during *in vitro* maturation

Hunter MG and V Brankin
Porcine oocyte maturation in follicular fluid from Meishan or Large-White hybrid pigs

Wang WH, Abeydeera LR, Cantley TC, Rieke A, Prather RS and BN Day
Morphological comparison of porcine oocytes matured *in vivo* and *in vitro*

Rodriguez-Martinez H, Han Y, Song X, Funahashi H and K Niwa
Additive effects of sodium hyaluronate as inducer of capacitation of boar spermatozoa *in vitro*

Kim J-H, Macháty Z, Cabot R and RS Prather
Pig oocyte activation by stimulation of an exogenous G protein coupled receptor

Kikuchi K, Kashiwazaki N, Noguchi J, Shimada A and H Kaneko
Culture conditions under low oxygen or with ß-mercapto-ethanol enhance embryonic development of porcine zygotes produced *in vitro*

Macháty Z, Day BN and RS Prather
Cell numbers of pig embryos developed *in vitro* versus *in vivo*

AUTHOR INDEX

	Page
Alvarez, I. M.	285
Ashman, R. J.	237
Barb, C. R	3
Bazer, F. W.	151
Bowen, J. A.	151
Buhi, W. C.	285
Burghardt, R. C.	151
Cox, N. M.	31
Croy, B. A.	115
Day, B. N.	271
Dobrinsky, J. R.	301
Du, Z. T.	237
Engelhardt, H.	115
Estienne, M. J.	3
Flowers, W. L.	67
Ford, J. J.	181
Ford, S. P.	165
Foxcroft, G. R.	47
Funahashi, H.	271
Geisert, R. D.	133
Grupen, C. G.	237
Harrison, R. A. P.	195
Harter-Dennis, J. M.	3
Holt, W. V.	213
Jarry, H.	19
Johnson, L. A.	255
Kemp, B.	79, 91
King, G. J.	115
Klindt, J.	181

	Page
Knoke, I.	19
Kouba, A. J.	285
Lee, C.	245
Lunstra, D. D.	181
MacIlfatrick, S.	237
Medrano, A.	213
Messer, L. A.	227
Moore, K.	245
Nagashima, H.	237
Newton, G. R.	151
Nottle, M. B.	237
Oetama, B.	245
Piedrahita, J. A.	245
Pitzel, L.	19
Ramsoondar, J.	245
Rothschild, M. F.	227
Soede, N. M.	79, 91
Theiling, K.	19
Thomson, J.	245
Vasquez, J.	245
Verma, P. J.	237
Vincent, A.	227
Waberski, D.	105
Weaks, R.	245
Wise, T. H.	181
Wuttke, W.	19
Yelich, J. V.	133

SUBJECT INDEX

Amino acids, and secretion of pituitary hormones 3

Breed, effect on litter size 165

Conceptus, regulation of development and attachment 133
Corpus luteum, immune–endocrine interactions in 19
Cryopreservation, of embryos 301

Embryo
- cryopreservation of 301
- *in vitro* production of 271

Embryo development, oviductal regulation of 285
Embryo survival, effect of nutrition on 47
Extracellular matrix, uterine, and implantation 151

Fertility
- genetic approach to improving 227
- and sperm membrane characteristics 195

Fertilization
- effects of semen components on 105
- oviductal regulation of 285

Follicular development, control of 31

Gender preselection 255
Genetics, use in improving fertility 227

Immune response, of uterus to pregnancy 115
Implantation, role of uterine epithelium in 151
***In vitro* production,** of embryos 271
Insemination, timing of, relative to ovulation 79

Litter size, in different genotypes 165
Luteolysis 19

Management programme, and semen production 67
Meishan pig, pubertal development in male 181

Neuropeptides, and secretion of pituitary hormones 3
Nutrition, effect on embryo survival 47

Oestrus, relative to ovulation 91
Oviduct, role in fertilization and embryo development 285
Ovulation
- effects of semen components on 105
- relative to oestrus 91
- timing of insemination relative to 79

Ovulation rate, control of 31

Pituitary hormones, neuropeptide and amino acid control of 3
Placenta, efficiency in different genotypes 165
Pregnancy, early, role of uterine immune cells in 115
Puberty, physiology of, in Meishan boar 181

Semen, effects on ovulation and fertilization 105
Semen production, management programmes and 67
Spermatozoa
- assessment of fertilizing ability 213
- membrane characteristics and fertility 195

Transgenic techniques 237, 245